CONVERSION FACTORS FROM ENGLISH TO SI UNITS

Length:	1 ft	= 0.3048 m
	1 ft	= 30.48 cm
	1 ft	= 304.8 mm
	1 in.	= 0.0254 m
	1 in.	= 2.54 cm
	1 in.	= 25.4 mm
Area:	1 ft^2	= 929.03×10^{-4} m^2
	1 ft^2	= 929.03 cm^2
	1 ft^2	= 929.03×10^{2} mm^2
	1 in^2	= 6.452×10^{-4} m^2
	1 in^2	= 6.452 cm^2
	1 in^2	= 645.16 mm^2
Volume:	1 ft^3	= 28.317×10^{-3} m^3
	1 ft^3	= 28.317×10^{3} cm^3
	1 in^3	= 16.387×10^{-6} m^3
	1 in^3	= 16.387 cm^3
Force:	1 lb	= 4.448 N
	1 lb	= 4.448×10^{-3} kN
	1 lb	= 0.4536 kgf
	1 kip	= 4.448 kN
	1 U.S. ton	= 8.896 kN
	1 lb	= 0.4536×10^{-3} metric ton
	1 lb/ft	= 14.593 N/m

Stress:	1 lb/ft^2	= 47.88 N/m^2
	1 lb/ft^2	= 0.04788 kN/m^2
	1 U.S. ton/ft^2	= 95.76 kN/m^2
	1 kip/ft^2	= 47.88 kN/m^2
	1 lb/in^2	= 6.895 kN/m^2
Unit weight:	1 lb/ft^3	= 0.1572 kN/m^3
	1 lb/in^3	= 271.43 kN/m^3
Moment:	1 lb-ft	= 1.3558 N·m
	1 lb-in.	= 0.11298 N·m
Energy:	1 ft-lb	= 1.3558 J
Moment of inertia:	1 in^4	= 0.4162×10^{6} mm^4
	1 in^4	= 0.4162×10^{-6} m^4
Section modulus:	1 in^3	= 0.16387×10^{5} mm^3
	1 in^3	= 0.16387×10^{-4} m^3
Hydraulic conductivity:	1 ft/min	= 0.3048 m/min
	1 ft/min	= 30.48 cm/min
	1 ft/min	= 304.8 mm/min
	1 ft/sec	= 0.3048 m/
	1 ft/sec	= 304.8 mm
	1 in./min	= 0.0254 m/
	1 in./sec	= 2.54 cm/s
	1 in./sec	= 25.4 mm/
Coefficient of consolidation:	1 in^2/sec	= 6.452 cm
	1 in^2/sec	= 20.346 ×
	1 ft^2/sec	= 929.03 cr

Introduction to Geotechnical Engineering

Introduction to Geotechnical Engineering

Second Edition

Braja M. Das

Dean Emeritus,
California State University,
Sacramento, USA

Nagaratnam Sivakugan

Associate Professor and Head,
Discipline of Civil Engineering,
James Cook University,
Australia

Australia • Brazil • Canada • Mexico • Singapore • United Kingdom • United States

Introduction to Geotechnical Engineering, Second Edition

Braja M. Das and Nagaratnam Sivakugan

Publisher, Global Engineering: Timothy L. Anderson

Senior Development Editor: Hilda Gowans

Development Editor: Eavan Cully

Media Assistant: Ashley Kaupert

Team Assistant: Sam Roth

Marketing Manager: Kristin Stine

Director, Content and Media Production: Sharon L. Smith

Senior Content Project Manager: Kim Kusnerak

Production Service: RPK Editorial Services, Inc.

Copyeditor: Harlan James

Proofreader: Martha McMaster

Indexer: Braja M. Das

Compositor: MPS Limited

Senior Art Director: Michelle Kunkler

Internal Designer: RPK Editorial Services, Inc.

Cover Designer: Rokusek Design

Cover Image: Courtesy of N. Sivakugan, James Cook University, Australia

Intellectual Property

Analyst: Christine Myaskovsky

Project Manager: Sarah Shainwald

Text and Image Permissions Researcher: Kristiina Paul

Manufacturing Planner: Doug Wilke

WCN: 01-100-101

Library of Congress Control Number: 2014947846

ISBN: 978-1-305-25732-0

Cengage
200 Pier 4 Boulevard
Boston, MA 02210
USA

Cengage is a leading provider of customized learning solutions with office locations around the globe, including Singapore, the United Kingdom, Australia, Mexico, Brazil, and Japan. Locate your local office at: **www.cengage.com/global**.

To learn more about Cengage platforms and services, register or access your online learning solution, or purchase materials for your course, visit **www.cengage.com**.

Printed in Mexico
Print Number: 02 Print Year: 2020

To Janice, Rohini, Joe,
Valerie, and Elizabeth

Contents

13 Deep Foundations 306

14 Retaining Walls 368

Preface

During the past half century, geotechnical engineering—soil mechanics and foundation engineering—have developed rapidly. Intensive research and observation in the field and laboratory have refined and improved the science of foundation design. *Introduction to Geotechnical Engineering, Second Edition,* uses those materials, and they are presented in a simple and concise form. This book is designed primarily for classroom instruction in Civil Engineering Technology and Construction Management programs. It is non-calculus based. It will be a useful reference book for civil engineering practitioners. It also can be used as a text for students in Civil Engineering programs where soil mechanics and foundations are combined into one course and covered in one semester. However, some supplemental material may be necessary.

The first edition of this text was published in 2008. The majority of the materials in the first edition of the text were drawn from *Principles of Foundation Engineering* and *Principles of Geotechnical Engineering*, which were originally published with 1984 and 1985 copyrights, respectively. These books are now in their eighth editions.

The present edition of the text has added a co-author—Dr. Nagaratnam Sivakugan, Associate Professor and Head of Civil Engineering at James Cook University, Townsville, Australia. In this edition, the following changes and additions have been incorporated.

- Since several users of the first edition preferred SI units, dual units (English and SI) have been used in the text.
- Several additional examples and homework problems have been added. These problems are approximately 50/50 in English and SI units. There are 113 example problems and 246 homework problems.
- Several new photographs have been added to help students visualize the material under discussion.
- Chapter 1, entitled "Geotechnical Engineering," is new and explains what geotechnical engineering is in general terms.
- Chapter 2 presents a general description of soil grain-size and grain-size analysis.
- Soil compaction has now been moved to Chapter 5.
- Based on the review comments, several additions and clarifications have been incorporated into the text.

Instructors must emphasize the difference between soil mechanics and foundation engineering in the classroom. Soil mechanics is the branch of engineering that involves the student of the properties of soils and their behavior under stress and strain under idealized conditions. Foundation engineering is the application of the principles of soil mechanics and geology in the planning, design, and construction of foundations for buildings, highways, dams, and so forth. Approximations and deviations from idealized conditions of soil mechanics become necessary for proper foundation design, because natural soil deposits are not homogeneous in most cases. However, if a structure is to function properly, these approximations only can be made by an engineer with a good background in soil mechanics. This book provides that background.

Instructor Resource Materials

A detailed *Instructor's Solutions Manual* is available for instructors.

MindTap Online Course and Reader

In addition to the print version, this textbook also will be available online through MindTap, which is a personalized learning program. Students who purchase the MindTap version will have access to the book's MindTap Reader and will be able to complete homework and assessment material online, through their desktop, laptop, or iPad. If your class is using a Learning Management System (such as Blackboard, Moodle, or Angel) for tracking course content, assignments, and grading, you can seamlessly access the MindTap suite of content and assessments for this course.

In MindTap, instructors can

- Personalize the Learning Path to match the course syllabus by rearranging content, hiding sections, or appending original material to the textbook content
- Connect a Learning Management System portal to the online course and Reader
- Customize online assessments and assignments
- Track student progress and comprehension with the Progress app
- Promote student engagement through interactivity and exercises

Additionally, students can listen to the text through ReadSpeaker, take notes and highlight content for easy reference, and check their understanding of the material.

Acknowledgements

Thanks are given to the following for their contributions.

- The three reviewers, including Professor Saeed Daniali of the University of Washington, for their comments and helpful suggestions.
- Janice F. Das, in helping to complete the Instructor's Solution Manual and providing other suggestions and advice in the preparation of the revision.

- Several individuals at Cengage Learning, for their assistance and advice in the final development of the book—namely:

  Tim Anderson, Publisher
  Hilda Gowans, formerly Senior Development Editor, Cengage
  Eavan Cully, Development Editor, Cengage

It is also fitting to thank Rose P. Kernan of RPK Editorial Services. She has been instrumental in shaping the style and overseeing the production of this edition of *Introduction to Geotechnical Engineering* as well as the first edition.

Braja M. Das

Nagaratnam Sivakugan

Introduction to Geotechnical Engineering

1 Geotechnical Engineering

Geotechnical engineering is defined as the sub-discipline of civil engineering that involves natural materials found close to the surface of the earth. This text is designed for use as an introduction to the fundamental principles of geotechnical engineering for students enrolled in civil engineering technology and construction management programs. This chapter provides a broad overview of the following topics.

- What is geotechnical engineering?
- An appreciation of different types of geotechnical problems.
- The role of soil testing as a means of determining the design parameters.

1.1 Geotechnical Engineering

Soils, rocks, and aggregates are natural materials of geologic origin derived from the earth. Sometimes, they are known as *geomaterials.* They are the most abundant materials on Earth. Rocks can break down into aggregates and soils. Soils and aggregates can turn into rocks through geologic processes over thousands of years. The mineralogy of the three is similar.

The term "geo" means earth. Geotechnical engineering deals with the engineering aspects of soils, rocks, and aggregates. It is a relatively young branch of civil engineering. With the growth of science and technology, the need for better and more economical structural design and construction became critical. During the early part of the twentieth century, this need led to a detailed study of the nature and properties of soil as they related to engineering. The publication of *Erdbaumechanik* by Karl Terzaghi in 1925 gave birth to modern soil mechanics as we know it today. Being a young discipline, there has been substantial new developments in the past two decades, and there will be more in the next few decades. *Environmental geotechnology* is one of the growing areas that is related to geotechnical engineering.

All buildings, structures, facilities, and infrastructure have to be placed on the ground. To ensure that they perform satisfactorily during their design lives, it is necessary to ensure that there are no excessive settlements or deformations and, worse, any failures in the underlying soils. This requires a thorough understanding of soil behaviour and its response to the structures or facilities built on them.

Geotechnical engineering includes *soil mechanics* and *foundation engineering.* Some exposure to *rock mechanics* and *geology* will strengthen your geotechnical engineering skills. *Mechanics* is the physical science that deals with forces, stresses, moments, displacements, strains, and equilibrium. Soil mechanics is a subject where the principles of mechanics are applied to soils, treating them as *continuous media* or *continuum* for simplicity. Geotechnical engineering is sometimes called *geomechanics* or *geo-engineering* in different parts of the world.

1.2 Geotechnical Engineering Applications

There is a wide range of geotechnical applications. The list keeps growing with new challenges such as hazardous waste disposal. The major applications are

a. *Foundations:* Footings, piles, and piers supporting buildings
b. *Retaining walls:* Concrete or masonry walls preventing lateral spread of soils
c. *Sheet piles:* Water-tight steel walls to prevent lateral spread of soils and retain soils and/or water
d. *Dams:* Large structures made of concrete or geomaterials for impounding water in reservoirs
e. *Earthworks:* Site preparation work prior to building a structure
f. *Slope stability:* Stability of the slope of an embankment or excavation
g. *Geosynthetics:* Natural or synthetic materials used to improve the soil behavior

As an example of geotechnical engineering work, Figure 1.1 shows the land reclamation at a port, where the dredged spoils are pumped (see inset) into the containment paddocks in the form of slurry.

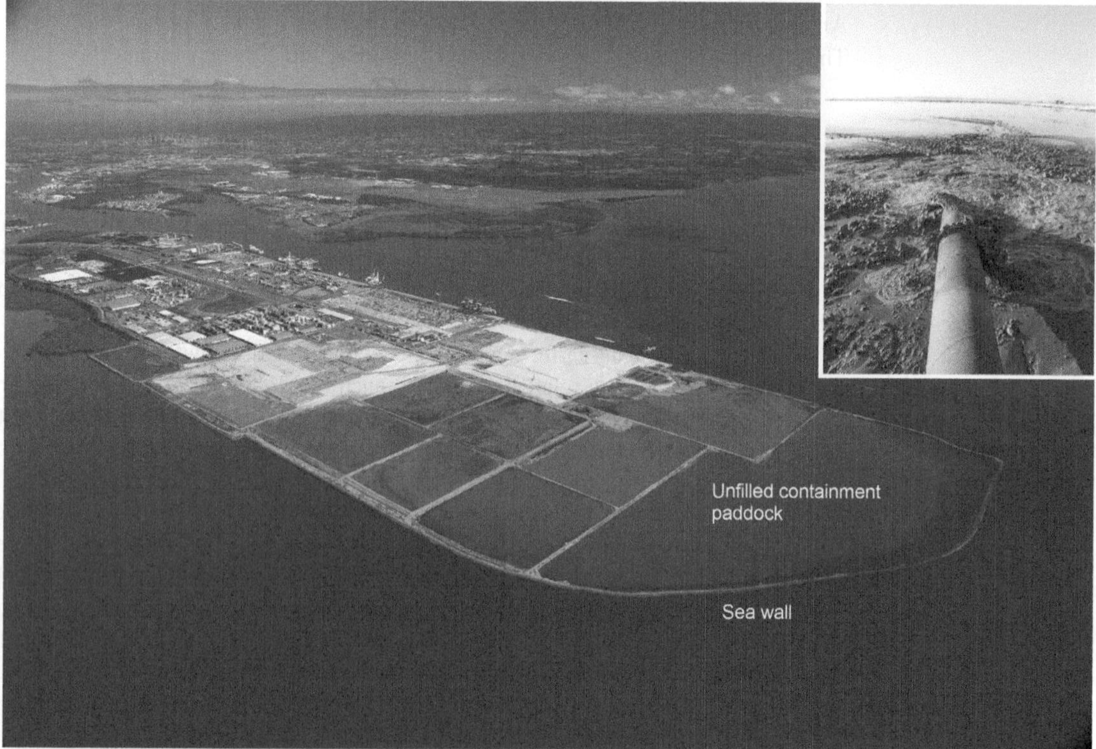

Figure 1.1 Land reclamation containment paddocks (Inset: Dredge spoil pumped into the containment paddock) (*Courtesy of N. Sivakugan, James Cook University, Australia*)

(a)

(b)

Figure 1.2 Earthwork: (a) during a site preparation; (b) after completion of the site preparation (*Courtesy of N. Sivakugan, James Cook University, Australia*)

The slurry settles and forms the reclaimed land for future developments. Similarly, Figure 1.2 shows the earthworks during a site preparation using a range of earthmoving equipment such as excavators, spreaders, and different types of rollers to carry out the compaction. Figure 1.3 shows the Leaning Tower of Pisa in Italy. The serviceability limit state is well and truly violated here, and the tower has become a world wonder and a major tourist attraction. In March 1990, the tower was closed to the

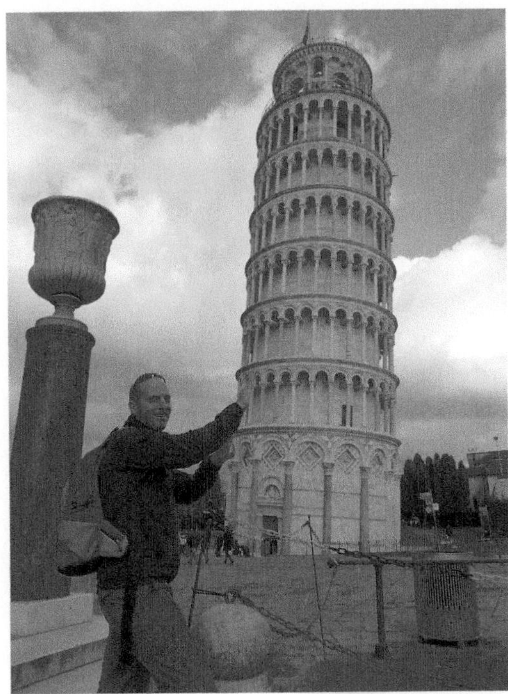

Figure 1.3 Leaning tower of Pisa (*Courtesy of Briony Rankine, Golder Associates, Australia*)

public for safety reasons, and after some years of remedial measures, it was reopened in 2001. After the recent foundation restoration work, the 56 m high tower is still leaning 4° to the vertical. The top of the tower is displaced horizontally by about 3.9 m.

1.3 Soil Parameters

All of the geotechnical applications discussed in Section 1.2 involve *design and analysis* that have to be carried out prior to the *construction.* To carry out proper design and analysis, it is necessary to have the *soil properties*, which are known as *design parameters.* The soil properties are generally determined from field (in situ) tests and laboratory tests on samples collected from the field.

References

TERZAGHI, K. (1926). *Erdbaumechanik auf Bodenphysikalisher Grundlage*, Deuticke, Vienna.

2 Grain-Size Analysis

2.1 Introduction

For engineering purposes, *soil* is defined as an uncemented aggregate of mineral grains and decayed organic matter (solid particles) with liquid and gas in the empty spaces between the solid particles. Soil is used as a construction material in various civil engineering projects, and it supports structural foundations. Thus, civil engineers must study the properties of soil, such as its origin, grain-size distribution, ability to drain water, compressibility, shear strength, and load-bearing capacity. This chapter relates to a general overview of grain sizes and their distribution in soils.

The study of grain sizes present in a given soil is an important and integral part of geotechnical engineering. The grain-size distribution in soil influences its physical properties, such as compressibility and shear strength. This chapter discusses the different types of soils on the basis of grain size and the means of quantifying the grain sizes. At the end of this chapter, you will have learned the following.

- What are the size ranges for gravels, sands and fines?
- How are soils formed?
- Some special names for specific soil types.
- Developing the grain-size distribution curve.

2.2 Soil-Grain Size

The sizes of grains that make up soil may vary over a wide range. Soils are generally called *gravel, sand, silt*, or *clay*, depending on the predominant size of grains within the soil. To describe soils by their grain size, several organizations have developed *soil-separate-size limits*. Table 2.1 shows the soil-separate-size limits developed by the Massachusetts Institute of Technology, the U.S. Department of Agriculture, the American Association of State Highway and Transportation Officials, and the U.S. Army Corps of Engineers and U.S. Bureau of Reclamation. In this table, the MIT system has been presented for illustration purposes only. This system is important in the history of the development of soil-separate-size limits. However, presently, the Unified System is almost universally accepted. The

Table 2.1 Soil-Separate-Size Limits

Name of organization	Grain size (mm)			
	Gravel	**Sand**	**Silt**	**Clay**
Massachusetts Institute of Technology (MIT)	>2	2 to 0.06	0.06 to 0.002	<0.002
U.S. Department of Agriculture (USDA)	>2	2 to 0.05	0.05 to 0.002	<0.002
American Association of State Highway and Transportation Officials (AASHTO)	76.2 to 2	2 to 0.075	0.075 to 0.002	<0.002
Unified Soil Classification System (U.S. Army Corps of Engineers, U.S. Bureau of Reclamation)	76.2 to 4.75	4.75 to 0.075	Fines (i.e., silts and clays) <0.075	

Unified Soil Classification System has now been adopted by the American Society for Testing and Materials. Figure 2.1 shows size limits in a graphical form.

Gravels are pieces of rocks with occasional grains of quartz, feldspar, and other minerals.

Sand grains are mostly made of quartz and feldspar. Other mineral grains may also be present at times.

Silts are the microscopic soil fractions that consist of very fine quartz grains and some flake-shaped grains that are fragments of micaceous minerals.

Clays are mostly flake-shaped microscopic and submicroscopic grains of mica, clay minerals, and other minerals. As shown in Table 2.1, clays are generally defined as grains less than 0.002 mm in size. However, in some cases, grains between 0.002 mm and 0.005 mm in size are also referred to as clay. Grains are classified as *clay* on the basis of their size; they do not necessarily contain clay minerals. Clays have been defined as those particles "which develop plasticity when mixed with a limited amount of water" (Grim, 1953). (Plasticity is the putty-like property of clays when they contain

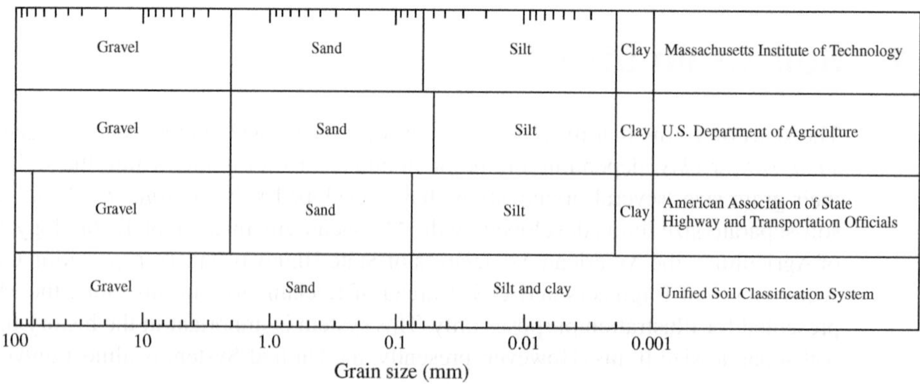

Figure 2.1 Soil-separate-size limits by various systems

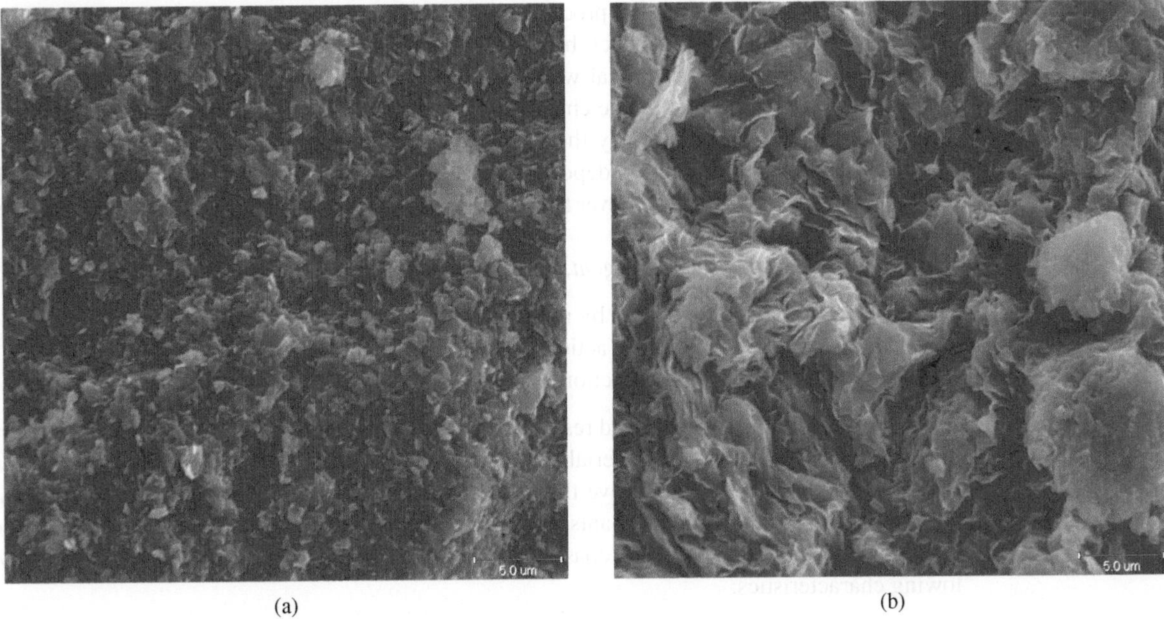

(a) (b)

Figure 2.2 Scanning electron micrographs of clay fabric: (a) kaolinite; (b) montmorillonite (*Courtesy: David White, Iowa State University, Ames, Iowa*)

a certain amount of water.) Non-clay soils can contain grains of quartz, feldspar, or mica that are small enough to be within the clay classification. Hence, it is appropriate for soil grains smaller than 0.002 mm, or 0.005 mm as defined under different systems, to be called "clay-sized" grains rather than "clay." Clay grains are mostly of colloidal size range (<0.001 mm), and 0.002 mm appears to be the upper limit. The clay minerals, which are a product of chemical weathering of feldspars, ferromagnesians, and micas, are the minerals whose presence gives the plastic property to soils. There are three major types of clay mineral: (1) *kaolinite,* (2) *illite,* and (3) *montmorillonite.*

Individual grains of clay cannot be seen by the naked eye. It requires a microscope to see them. Figure 2.2 shows the scanning electron micrographs (same scale) of kaolinite and montmorillonite clays. Unlike the gravels, sands and silts, grains of clay are flaky with very large surface areas, as seen in the micrographs. It can be seen that montmorillonite grains have a much larger surface area than kaolinite. Their mineralogy, flakiness, and the large surface areas make the clays plastic and cohesive. Montmorillonite clays can swell in the presence of water, which enters between the layers. Such clays are known as expansive clays and cause billions of dollars worth of damage annually to roads and low-rise buildings.

2.3 General Soil Deposits

Most of the soils that cover the earth are formed by the weathering of various rocks. There are two general types of weathering: (1) *mechanical weathering*, and (2) *chemical weathering*.

Mechanical weathering is the process by which rocks are broken down into smaller and smaller pieces by physical forces. These physical forces may be running water, wind, ocean waves, glacier ice, frost action, and expansion and contraction caused by gain and loss of heat.

Chemical weathering is the process of chemical decomposition of the original rock. In the case of mechanical weathering, the rock breaks down into smaller pieces without a change of chemical composition. However, in chemical weathering, the original material may be changed to something entirely different. For example, the chemical weathering of feldspar can produce clay minerals.

The soil that is produced by the weathering process of rocks can be transported by physical agents to other places. These soil deposits are called *transported soils*. In contrast, some soils stay in the place of their formation and cover the rock surface from which they derive. These soils are referred to as *residual soils*.

Based on the *transporting agent*, transported soils can be subdivided into three major categories:

1. *Alluvial*, or *fluvial*: deposited by running water
2. *Glacial*: deposited by glacier action
3. *Aeolian*: deposited by wind action

In addition to transported and residual soils, there are *peats* and *organic soils*, which derive from the decomposition of organic materials. Organic soils are usually found in low-lying areas where the ground water table is near or above the ground surface. The presence of a high ground water table helps in the growth of aquatic plants that, when decomposed, form organic soil. This type of soil deposit is usually encountered in coastal areas and in glaciated regions. Organic soils show the following characteristics:

1. The natural moisture content may range from 200% to 300%.
2. They are highly compressible.
3. Laboratory tests have shown that, under loads, a large amount of settlement is derived from secondary consolidation.

2.4 Some Local Terms for Soils

Soils are sometimes referred to by local terms. Following are a few of these terms with a brief description of each.

1. *Caliche*: a Spanish word derived from the Latin word *calix*, meaning *lime*. It is mostly found in the desert southwest of the United States. It is a mixture of sand, silt, and gravel bonded together by *calcareous deposits*. The calcareous deposits are brought to the surface by a net upward migration of water. The water evaporates in the high local temperature. Because of the sparse rainfall, the carbonates are not washed out of the top layer of soil.
2. *Gumbo*: a highly plastic, clayey soil.
3. *Adobe*: a highly plastic, clayey soil found in the southwestern United States.
4. *Terra Rossa*: residual soil deposits that are red in color and derive from limestone and dolomite.
5. *Muck*: organic soil with a very high moisture content.
6. *Muskeg*: organic soil deposit.
7. *Saprolite*: residual soil deposit derived from mostly insoluble rock.
8. *Loam*: a mixture of soil grains of various sizes, such as sand, silt, and clay.
9. *Laterite*: characterized by the accumulation of iron oxide (Fe_2O_3) and aluminum oxide (Al_2O_3) near the surface, and the leaching of silica. Lateritic soils in Central America contain about 80–90% of clay and silt-size grains. In the United States, lateritic soils can be found in the southeastern states, such as Alabama, Georgia, and the Carolinas.

2.5 Grain-Size Analysis

In a natural soil deposit, the size of grains may vary over a wide range. Determining the nature of distribution of the grain size in a given soil mass is important for all design and construction purposes. At the same time, the degree of plasticity of soils due to the presence of clay minerals dictates the physical behavior and properties of soil as they relate to the design of various civil engineering structures. If the grain-size distribution and the plasticity of soil are known, they can be used for engineering classification (Chapter 4).

Grain-size analysis for grain sizes larger than 0.075 mm is done by sieve analysis. Sieve analysis consists of shaking the soil sample through a set of sieves that have progressively smaller openings. U.S. standard sieve numbers and the sizes of openings are given in Table 2.2.

The sieves used for soil analysis are generally 8 in. (203 mm) in diameter. To conduct a sieve analysis, one must first oven-dry the soil and then break all lumps into small grains. The soil then is shaken through a stack of sieves with openings of decreasing size from top to bottom (a pan is placed below the stack). Figure 2.3 shows a set of sieves in a shaker used for conducting the test in the laboratory. The smallest-sized sieve that should be used for this type of test is the U.S. No. 200 sieve. After the soil is shaken, the mass of soil retained on each sieve is determined. When cohesive soils are analyzed, breaking the lumps into individual grains may be difficult. In this case, the soil may be mixed with water to make a slurry and then washed through the sieves. Portions retained on each sieve are collected separately and oven-dried before the mass retained on each sieve is measured.

Referring to Figure 2.4, we can step through the calculation procedure for a sieve analysis:

1. Determine the mass of soil retained on each sieve (i.e., $M_1, M_2, \ldots M_n$) and in the pan (i.e., M_p) (Figure 2.4a and 2.4b).
2. Determine the total mass of the soil: $M_1 + M_2 + \cdots + M_i + \cdots + M_n + M_p = \Sigma M$.

Table 2.2 U.S. Standard Sieve Sizes

Sieve no.	Opening (mm)	Sieve no.	Opening (mm)
4	4.75	35	0.500
5	4.00	40	0.425
6	3.35	50	0.355
7	2.80	60	0.250
8	2.36	70	0.212
10	2.00	80	0.180
12	1.70	100	0.150
14	1.40	120	0.125
16	1.18	140	0.106
18	1.00	170	0.090
20	0.850	200	0.075
25	0.710	270	0.053
30	0.600		

Figure 2.3 A set of sieves for a test in the laboratory (*Photo courtesy of Braja Das*)

3. Determine the cumulative mass of soil retained above each sieve. For the *i*th sieve, it is $M_1 + M_2 + \cdots + M_i$ (Figure 2.4c).
4. The mass of soil passing the *i*th sieve is $\Sigma M - (M_1 + M_2 + \cdots + M_i)$.
5. The percent of soil passing the *i*th sieve (or *percent finer*) (Figure 2.4) is

$$F = \frac{\Sigma M - (M_1 + M_2 + \cdots + M_i)}{\Sigma M} \times 100$$

Once the percent finer for each sieve is calculated (step 5), the calculations are plotted on semilogarithmic graph paper (Figure 2.5) with percent finer as the ordinate (arithmetic scale) and sieve opening size as the abscissa (logarithmic scale). This plot is referred to as the *grain-size distribution curve.*

Analysis for grains smaller than 0.075 mm usually is done by hydrometer analysis, which is based on the principle of sedimentation of soil grains in water. A detailed procedure of hydrometer analysis is given by ASTM (2014) under Test Designation D-422.

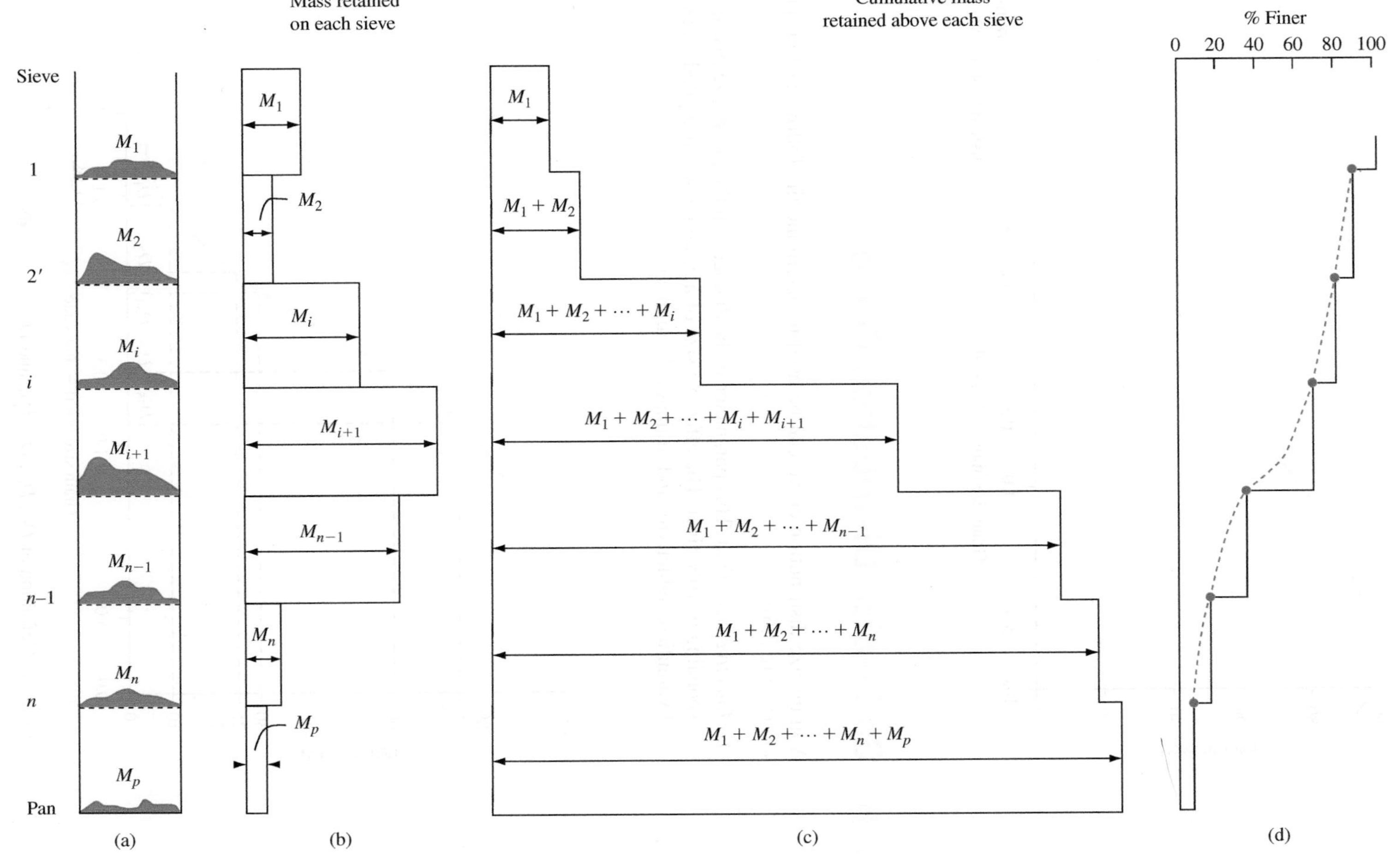

Figure 2.4 Sieve analysis

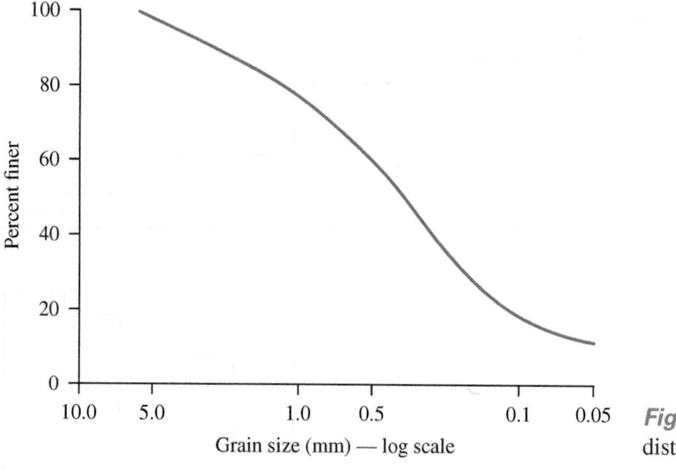

Figure 2.5 Grain-size distribution curve

2.6 Grain-Size Distribution Curve

A grain-size distribution curve can be used to determine the following four parameters for a given soil (Figure 2.6):

1. *Effective size* (D_{10}): This parameter is the diameter in the grain-size distribution curve corresponding to 10% finer. The effective size of a granular soil is a good measure to estimate the hydraulic conductivity and drainage through soil.

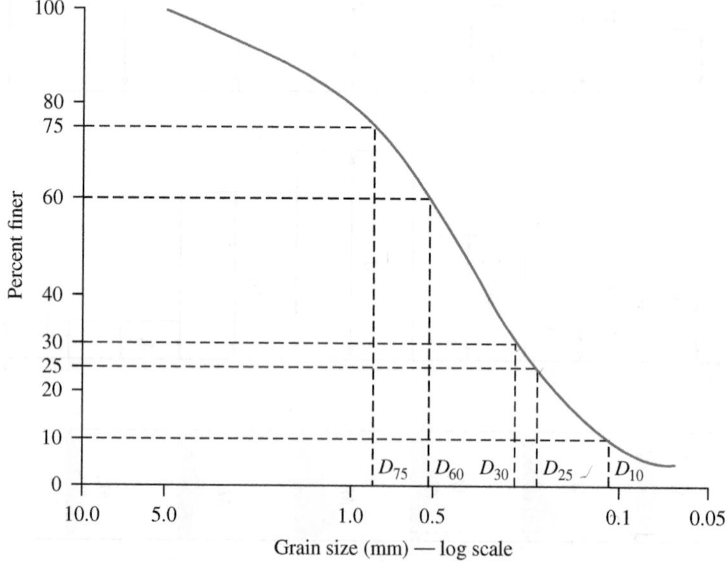

Figure 2.6 Definition of D_{75}, D_{60}, D_{30}, D_{25}, and D_{10}

2. *Uniformity coefficient* (C_u): This parameter is defined as

$$C_u = \frac{D_{60}}{D_{10}} \quad (2.1)$$

where D_{60} = diameter corresponding to 60% finer.

3. *Coefficient of gradation* (C_c): This parameter is defined as

$$C_c = \frac{D_{30}^2}{D_{60} \times D_{10}} \quad (2.2)$$

4. *Sorting coefficient* (S_0): This parameter is another measure of uniformity and is generally encountered in geologic works and expressed as

$$S_0 = \sqrt{\frac{D_{75}}{D_{25}}} \quad (2.3)$$

The sorting coefficient is not frequently used as a parameter by geotechnical engineers.

The grain-size distribution curve shows not only the range of grain sizes present in a soil, but also the type of distribution of various-size particles. Such types of distributions are demonstrated in Figure 2.7. Curve I represents a type of soil in which most of the soil grains are the same size. This is called *poorly graded* soil. Curve II represents a soil in which the grain sizes are distributed over a

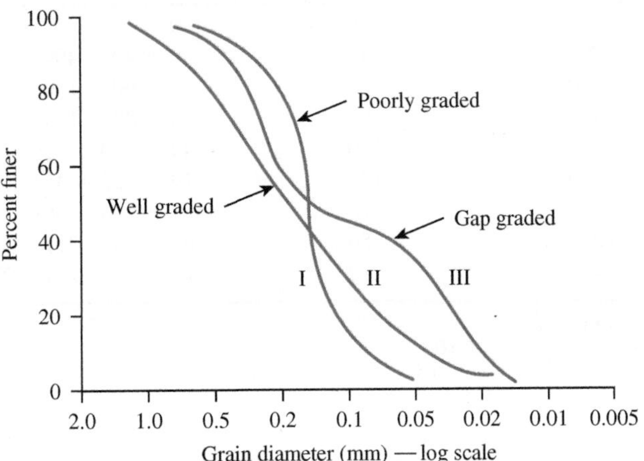

Figure 2.7 Different types of grain-size distribution curves

wide range, termed *well graded.* A well-graded soil has a uniformity coefficient greater than about 4 for gravel and 6 for sands, and a coefficient of gradation between 1 and 3 (for gravels and sands). A soil might have a combination of two or more uniformly graded fractions. Curve III represents such a soil. This type of soil is termed *gap graded.*

Example 2.1

Following are the results of a sieve analysis. Make the necessary calculations and draw a grain-size distribution curve.

U.S. sieve size	Mass of soil retained on each sieve (g)
4	0
10	40
20	60
40	89
60	140
80	122
100	210
200	56
Pan	12

Solution

The following table can now be prepared. Figure 2.8 shows the grain-size distribution.

U.S. sieve (1)	Opening (mm) (2)	Mass retained on each sieve (g) (3)	Cumulative mass retained above each sieve (g) (4)	Percent finer[a] (5)
4	4.75	0	0	100
10	2.00	40	0 + 40 = 40	94.5
20	0.850	60	40 + 60 = 100	86.3
40	0.425	89	100 + 89 = 189	74.1
60	0.250	140	189 + 140 = 329	54.9
80	0.180	122	329 + 122 = 451	38.1
100	0.150	210	451 + 210 = 661	9.3
200	0.075	56	661 + 56 = 717	1.7
Pan	—	12	717 + 12 = 729 = ΣM	0

[a] $\dfrac{\Sigma M - \text{col. 4}}{\Sigma M} \times 100 = \dfrac{729 - \text{col. 4}}{729} \times 100$

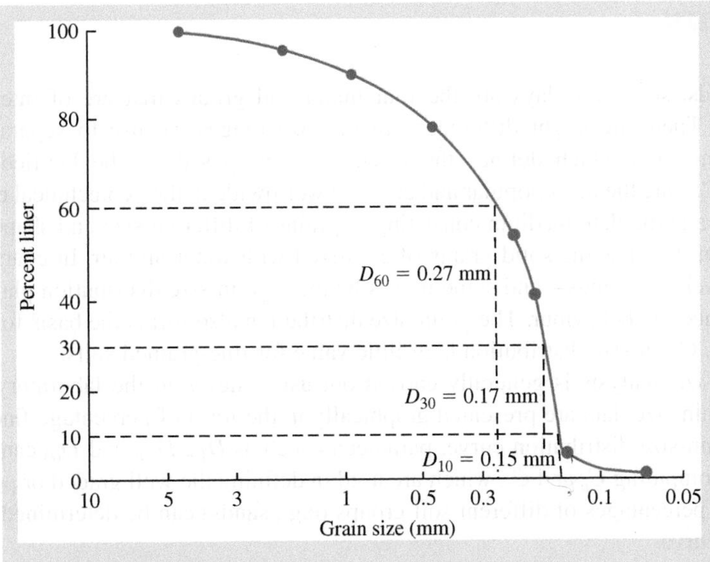

Figure 2.8 Grain-size distribution curve

Example 2.2

For the grain-size distribution curve shown in Figure 2.8, determine

a. D_{10}, D_{30}, and D_{60}
b. Uniformity coefficient, C_u
c. Coefficient of gradation, C_c

Solution

Part a

From Figure 2.8,

$$D_{10} = \mathbf{0.15\ mm}$$
$$D_{30} = \mathbf{0.17\ mm}$$
$$D_{60} = \mathbf{0.27\ mm}$$

Part b

$$C_u = \frac{D_{60}}{D_{10}} = \frac{0.27}{0.15} = \mathbf{1.8}$$

Part c

$$C_c = \frac{D_{30}^2}{D_{60} \times D_{10}} = \frac{(0.17)^2}{(0.27)(0.15)} = \mathbf{0.71}$$ ■

2.7 Summary

Gravels, sands, silts, and clays are the four major soil groups that are of interest in geotechnical engineering. There are slight differences in the size ranges we use to separate them, depending on the organization which defines them. The ones proposed by the Unified Soil Classification System (USCS) are the most popular and are used worldwide in the geotechnical engineering practice.

Soil is a particulate medium consisting of grains of different size and shapes. It is also a three-phase medium that has the solid grains often mixed with water and air. In coarse grained soils that includes gravels and sands—and sometimes silts too—grain-size distribution plays an important role in their engineering behaviour. The grain-size distribution also forms the basis for classifying coarse-grained soils. Grain-size distribution is of little value for fine-grained soils.

Grain-size analysis is generally carried out using sieves in the laboratory for coarse-grained soils. The grain-size data are presented graphically in the form of percentage finer versus grain size. From the grain-size distribution curve, parameters such as D_{10}, D_{30}, and D_{60} can be derived that can be used for computing C_u and C_c, which are used in defining the well graded or poorly graded soil. In addition, the percentages of different soil groups (e.g., sands) can be determined from the grain-size distribution curve.

Problems

2.1 State whether the following are true or false.

a. D_{30} is greater than D_{60}.
b. Coefficient of uniformity C_u is always greater than 1.
c. Coefficient of gradation C_c cannot be less than 1.
d. In sieve analysis, the sieve sizes increase from top to bottom.
e. No. 10 sieve has larger openings than No. 20 sieve.

2.2 Following are the results of a sieve analysis:

U.S. sieve no.	Mass of soil retained on each sieve (g)
4	0
10	21.6
20	49.5
40	102.6
60	89.1
100	95.6
200	60.4
pan	31.2

a. Determine the percent finer than each sieve size and plot a grain-size distribution curve.
b. Determine D_{10}, D_{30}, D_{60} from the grain-size distribution curve.
c. Calculate the uniformity coefficient, C_u.
d. Calculate the coefficient of gradation, C_c.

2.3 Repeat Problem 2.2 with the following results of a sieve analysis:

U.S. sieve no.	Mass of soil retained on each sieve (g)
4	0
6	30
10	48.7
20	127.3
40	96.8
60	76.6
100	55.2
200	43.4
pan	22

2.4 Repeat Problem 2.2 with the following results of a sieve analysis:

U.S. sieve no.	Mass of soil retained on each sieve (g)
4	0
6	0
10	0
20	9.1
40	249.4
60	179.8
100	22.7
200	15.5
pan	23.5

2.5 The grain-size characteristics of a soil are given in the table. Draw the grain-size distribution curve and find the percentages of gravel, sand, silt, and clay according to the MIT system (Table 2.1).

Size (mm)	Percent finer by weight
0.850	100.0
0.425	92.1
0.250	85.8
0.150	77.3
0.075	62.0
0.040	50.8
0.020	41.0
0.010	34.3
0.006	29.0
0.002	23.0

2.6 Redo Problem 2.5 according to the USDA system (Table 2.1).

2.7 Redo Problem 2.5 according to the AASHTO system (Table 2.1).

2.8 The grain-size characteristics of a soil are given in the table. Find the percentages of gravel, sand, silt, and clay according to the MIT system (Table 2.1).

Size (mm)	Percent finer by weight
0.850	100.0
0.425	100.0
0.250	94.1
0.150	79.3
0.075	34.1
0.040	28.0
0.020	25.2
0.010	21.8
0.006	18.9
0.002	14.0

2.9 Redo Problem 2.8 according to the USDA system (Table 2.1).

2.10 Redo Problem 2.8 according to the AASHTO system (Table 2.1).

2.11 The grain-size distributions of two soils, *A* and *B*, are shown in Figure 2.9.

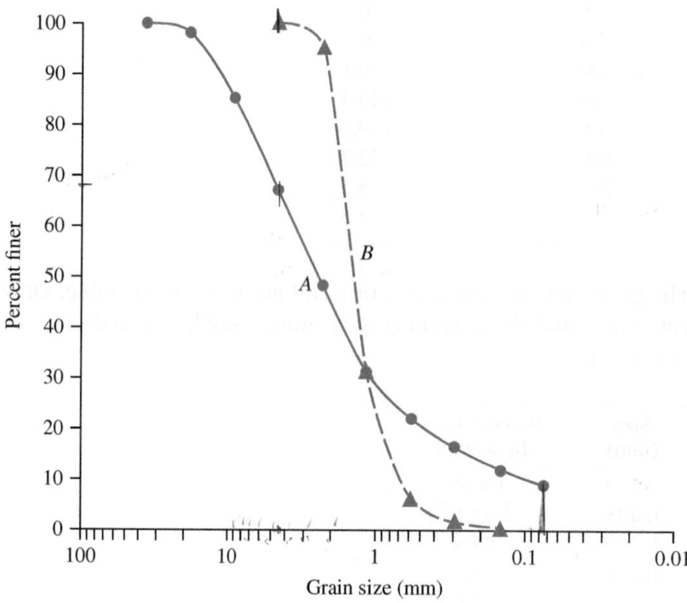

Figure 2.9

Find the percentages of gravel, sand and fine in the two soils. Use the Unified Soil Classification System (Table 2.1).

2.12 The sieve analysis data for a granular soil are given below.
% passing 19 mm sieve = 100
% passing 4.75 mm sieve (No. 4) = 63
% passing 0.075 mm sieve (No. 200) = 16

What are the percentages of gravel, sand, and fines within this soil? Use the Unified Soil Classification System (Table 2.1).

2.13 The grain size distributions of four soils A, B, C and D are shown in Figure 2.10.

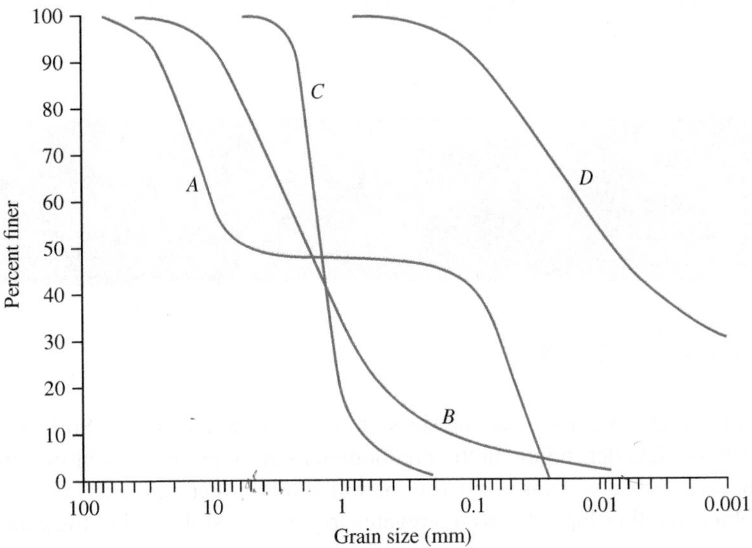

Figure 2.10

Without carrying out any computations, answer the following.

a. Which of these four soils has the largest percentage of gravel?

b. What is the percentage of sand in soil C?

c. Is there clay in any of these four soils?

d. In soil A, do you see grains in a specific size range missing? What is this size range? Use the Unified Soil Classification System (Table 2.1).

References

AMERICAN SOCIETY FOR TESTING AND MATERIALS (2014). *ASTM Book of Standards*, Vol. 04.08, West Conshohocken, PA.

GRIM, R. E. (1953). *Clay Mineralogy*, McGraw-Hill, New York.

3 Weight–Volume Relationships

3.1 Introduction

In Chapter 2, we discussed the sizes of grains that make up soil. Soil generally is called gravel, sand, silt, or clay, depending on the predominant size of grains present in it. In natural occurrences, soils are three-phase systems consisting of soil solids, water, and air. This chapter discusses the weight–volume relationships of soils aggregates in general. At the end of the chapter you would have learned the following.

- The terms used in the weight-volume relationships.
- The interrelationships among these terms.
- Relative density as a measure of grain packing density in granular soils.
- Computations of weights and volumes of the three phases in a soil.

3.2 Volume Relationships

Figure 3.1a shows an element of soil of volume V and weight W as it would exist in a natural state. To develop the weight–volume relationships, we must separate the three phases (that is, solid, water, and air) as shown in Figure 3.1b. Thus, the total volume of a given soil sample can be expressed as

$$V = V_s + V_v = V_s + V_w + V_a \tag{3.1}$$

where

V_s = volume of soil solids
V_v = volume of voids
V_w = volume of water in the voids
V_a = volume of air in the voids

Assuming that the weight of the air is negligible, we can give the total weight of the sample as

$$W = W_s + W_w \tag{3.2}$$

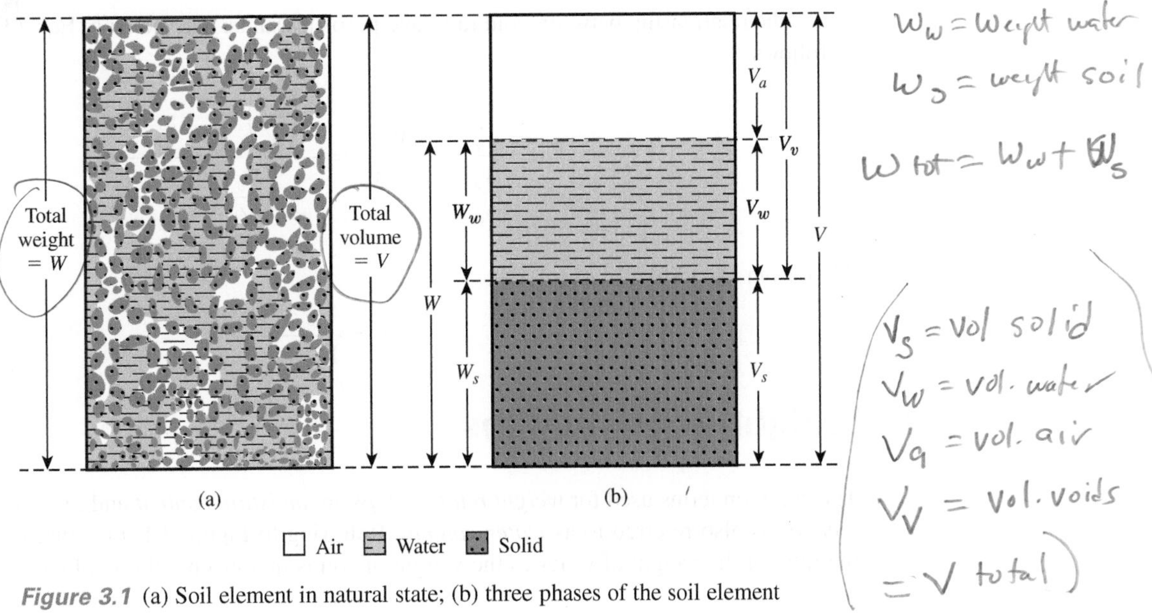

Figure 3.1 (a) Soil element in natural state; (b) three phases of the soil element

where

W_s = weight of soil solids
W_w = weight of water

The common terms used in *volume relationships* for the three phases in a soil element are *void ratio, porosity,* and *degree of saturation. Void ratio* (*e*) is defined as the ratio of the volume of voids to the volume of solids. Thus,

$$e = \frac{V_v}{V_s} \tag{3.3}$$

Porosity (*n*) is defined as the ratio of the volume of voids to the total volume, or

$$n = \frac{V_v}{V} \tag{3.4}$$

The *degree of saturation* (*S*) is defined as the ratio of the volume of water to the volume of voids, or

$$S = \frac{V_w}{V_v} \tag{3.5}$$

It is commonly expressed as a percentage.

The relationship between void ratio and porosity can be derived from Eqs. (3.1), (3.3), and (3.4) as follows:

$$e = \frac{V_v}{V_s} = \frac{V_v}{V - V_v} = \frac{\left(\frac{V_v}{V}\right)}{1 - \left(\frac{V_v}{V}\right)} = \frac{n}{1 - n} \tag{3.6}$$

Also, from Eq. (3.6),

$$n = \frac{e}{1 + e} \tag{3.7}$$

3.3 Weight Relationships

The common terms used for *weight relationships* are *moisture content* and *unit weight. Moisture content (w)* is also referred to as *water content*. Referring to Figure 3.1, moisture content is defined as the ratio of the weight of water to the weight of solids in a given volume of soil:

$$w = \frac{W_w}{W_s} \tag{3.8}$$

Unit weight (γ) is the weight of soil per unit volume. Thus,

$$\gamma = \frac{W}{V} \tag{3.9}$$

The unit weight can also be expressed in terms of the weight of soil solids, the moisture content, and the total volume. From Eqs. (3.2), (3.8), and (3.9),

$$\gamma = \frac{W}{V} = \frac{W_s + W_w}{V} = \frac{W_s\left[1 + \left(\frac{W_w}{W_s}\right)\right]}{V} = \frac{W_s(1 + w)}{V} \tag{3.10}$$

Soils engineers sometimes refer to the unit weight defined by Eq. (3.9) as the *moist unit weight*.

Often, to solve earthwork problems, one must know the weight per unit volume of soil, excluding water. This weight is referred to as the *dry unit weight*, γ_d. Thus,

$$\gamma_d = \frac{W_s}{V} \tag{3.11}$$

From Eqs. (3.10) and (3.11), the relationship of unit weight, dry unit weight, and moisture content can be given as

$$\gamma_d = \frac{\gamma}{1 + w} \tag{3.12}$$

Table 3.1 Void Ratio, Moisture Content, and Dry Unit Weight for Some Typical Soils in a Natural State

Type of soil	Void ratio, e	Natural moisture content in a saturated state (%)	Dry unit weight, γ_d	
			lb/ft^3	kN/m^3
Loose uniform sand	0.8	30	92	14.5
Dense uniform sand	0.45	16	115	18.1
Loose angular-grained silty sand	0.65	25	102	16.0
Dense angular-grained silty sand	0.4	15	121	19.0
Stiff clay	0.6	21	108	17.0
Soft clay	0.9–1.4	30–50	73–93	11.5–14.6
Loess	0.9	25	86	13.5
Soft organic clay	2.5–3.2	90–120	38–51	6.0–8.0
Glacial till	0.3	10	134	21.1

Table 3.1 gives the void ratio, moisture content, and dry unit weight of some typical soils in a natural state.

Unit weight is expressed in English units (a gravitational system of measurement) as pounds per cubic foot (lb/ft^3). In SI (Système International), the unit used is kilo Newtons per cubic meter (kN/m^3). Because the Newton is a derived unit, working with mass densities (ρ) of soil may sometimes be convenient. The SI unit of mass density is kilograms per cubic meter (kg/m^3). We can write the density equations [similar to Eqs. (3.9) and (3.11)] as

$$\rho = \frac{M}{V} \tag{3.13}$$

and

$$\rho_d = \frac{M_s}{V} \tag{3.14}$$

where

ρ = density of soil (kg/m^3)
ρ_d = dry density of soil (kg/m^3)
M = total mass of the soil sample (kg)
M_s = mass of soil solids in the sample (kg)

The unit of total volume, V, is m^3.

The unit weight in kN/m^3 can be obtained from densities in kg/m^3 as

$$\gamma(\text{kN/m}^3) = \frac{g\rho(\text{kg/m}^3)}{1000}$$

and

$$\gamma_d(\text{kN/m}^3) = \frac{g\rho_d(\text{kg/m}^3)}{1000}$$

where g = acceleration due to gravity = 9.81 m/sec^2.

Note that unit weight of water (γ_w) is equal to 9.81 kN/m^3 or 62.4 lb/ft^3 or 1000 kgf/m^3.

3.4 Specific Gravity of Soil Solids

Specific gravity of soils solids, G_s, is used in calculating weight–volume relationship. It is defined as

$$G_s = \frac{\text{weight of soil solids}}{\text{weight of equal volume of water}} \tag{3.15}$$

The procedure for laboratory determination of G_s is given in ASTM Test Designation. D-854 (ASTM, 2014). Table 3.2 shows the value of G_s for some common minerals found in soils. Most of the minerals fall within a general range of 2.6 to 2.9. The specific gravity of solids of light colored sand, which is mostly made of quartz, may be estimated to be about 2.65; for clayey and silty soils, it may vary from 2.6 to 2.9.

In referring to Figure 3.1b, in terms of G_s,

$$W_s = G_s \gamma_w V_s \tag{3.16}$$

$$W_w = \gamma_w V_w \tag{3.17}$$

where γ_w = unit weight of water (= 62.4 lb/ft^3, or 1000 kgf/m^3, or 9.81 kN/m^3).

Table 3.2 Specific Gravity of Important Minerals

Mineral	Specific gravity G_s
Quartz	2.65
Kaolinite	2.6
Illite	2.8
Montmorillonite	2.65–2.80
Halloysite	2.0–2.55
Potassium feldspar	2.57
Sodium and calcium feldspar	2.62–2.76
Chlorite	2.6–2.9
Biotite	2.8–3.2
Muscovite	2.76–3.1
Hornblende	3.0–3.47
Limonite	3.6–4.0
Olivine	3.27–3.37

Example 3.1

A soil from a borrow area is excavated and used to make a compacted fill beneath an embankment. The void ratio of the soil at the borrow area and the compacted fill are 0.86 and 0.54, respectively. If the volume of excavation is 100,000 m^3, what would be the volume of compacted fill that can be made from this soil?

Solution

While the void ratio is changing between the borrow area and the compacted fill, the volume (or mass) of the soil grains remain the some—they are incompressible!

Using Eq. (3.3) for the borrow area,

$$e = \frac{V_v}{V_s}$$

$$\therefore \quad 0.86 = \frac{100{,}000 - V_s}{V_s}$$

$$V_s = 53{,}763.4 \text{ m}^3$$

Applying Eq. (3.3) to the compacted fill,

$$0.54 = \frac{V - 53{,}763.4}{53{,}763.4}$$

$$V = 82{,}795.6 \text{ m}^3$$

∴ Volume of the compacted fill that can be made from the excavated soil is **82,795.6 m³**. ■

Example 3.2

In natural state, a moist soil has a volume of 0.33 ft³ and weighs 39.93 lb. The oven dry weight of the soil is 34.54 lb. If $G_s = 2.71$, calculate the moisture content, moist unit weight, dry unit weight, void ratio, porosity, and degree of saturation.

Solution

Refer to Figure 3.2.

Moisture Content [Eq. (3.8)]

$$w = \frac{W_w}{W_s} = \frac{W - W_s}{W_s} = \frac{39.93 - 34.54}{34.54} = \frac{5.39}{34.54} \times 100$$

$$= \mathbf{15.6\%}$$

Moist Unit Weight [Eq. (3.9)]

$$\gamma = \frac{W}{V} = \frac{39.93}{0.33} = \mathbf{121.0\ lb/ft^3}$$

Dry Unit Weight [Eq. (3.11)]

$$\gamma_d = \frac{W_s}{V} = \frac{34.54}{0.33} = \mathbf{104.67\ lb/ft^3} \simeq \mathbf{104.7\ lb/ft^3}$$

Void Ratio [Eq. (3.3)]

$$e = \frac{V_v}{V_s}$$

From Eq. (3.16),

$$V_s = \frac{W_s}{G_s \gamma_w} = \frac{34.54}{2.71 \times 62.4} = 0.204 \text{ ft}^3$$

$$V_v = V - V_s = 0.33 - 0.204 = 0.126 \text{ ft}^3$$

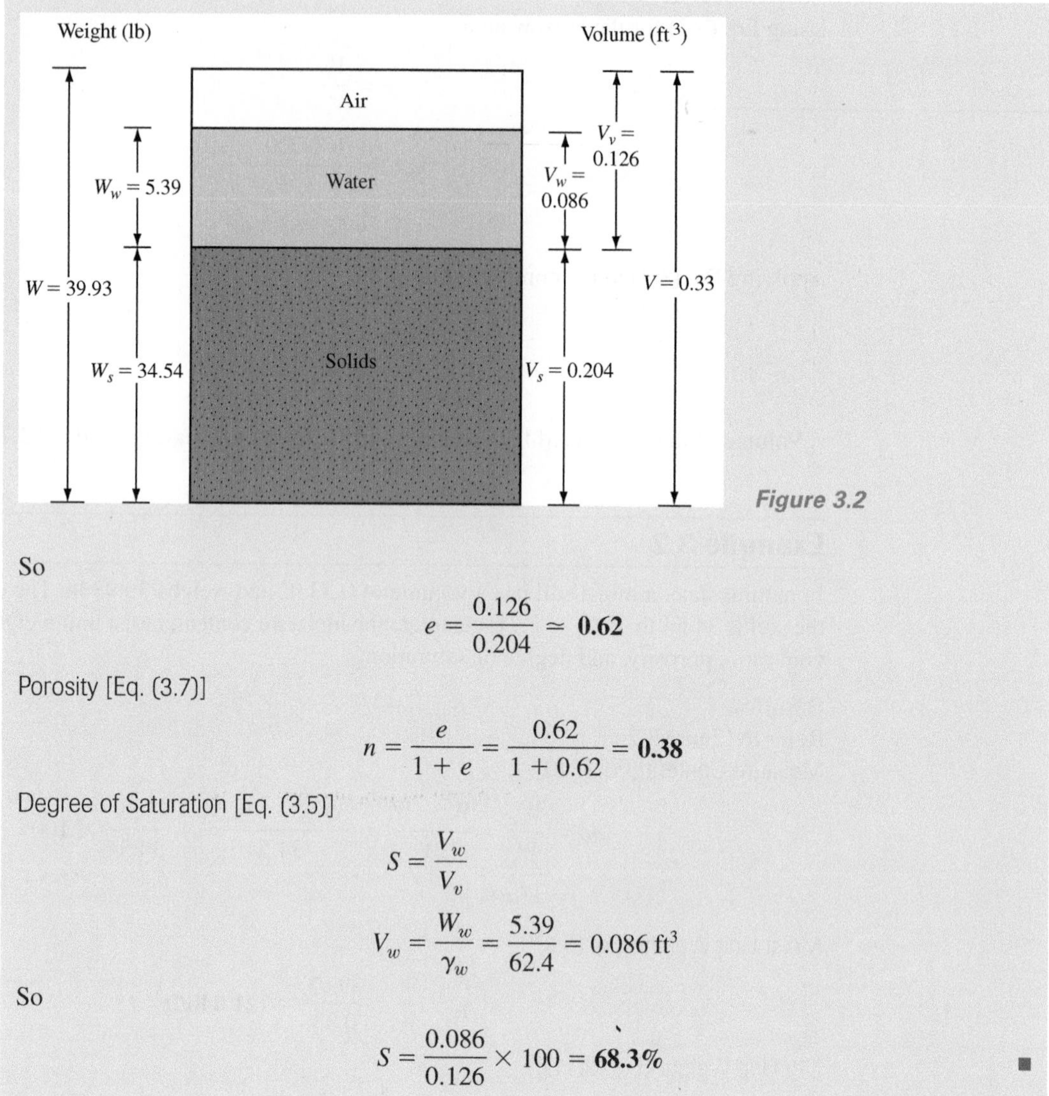

Figure 3.2

So

$$e = \frac{0.126}{0.204} \simeq \mathbf{0.62}$$

Porosity [Eq. (3.7)]

$$n = \frac{e}{1+e} = \frac{0.62}{1+0.62} = \mathbf{0.38}$$

Degree of Saturation [Eq. (3.5)]

$$S = \frac{V_w}{V_v}$$

$$V_w = \frac{W_w}{\gamma_w} = \frac{5.39}{62.4} = 0.086 \text{ ft}^3$$

So

$$S = \frac{0.086}{0.126} \times 100 = \mathbf{68.3\%}$$

■

3.5 Relationships among Unit Weight, Void Ratio, Moisture Content, and Specific Gravity

To obtain a relationship among unit weight (or density), void ratio, and moisture content, let us consider a volume of soil in which the volume of the soil solids is one, as shown in Figure 3.3. If the volume of the soil solids is one, then the volume of voids is numerically equal to the void ratio, e [from Eq. (3.3)]. The weights of soil solids and water can be given as

$$W_s = G_s\gamma_w$$
$$W_w = wW_s = wG_s\gamma_w$$

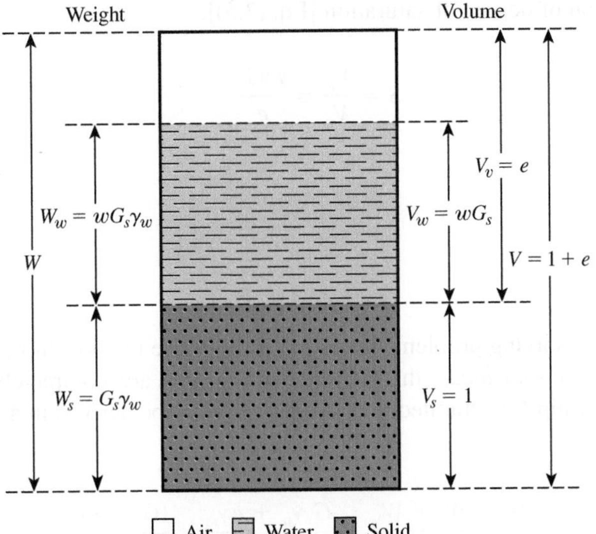

Figure 3.3 Three separate phases of a soil element with volume of soil solids equal to one

where

G_s = specific gravity of soil solids
w = moisture content
γ_w = unit weight of water

Now, using the definitions of unit weight and dry unit weight [Eqs. (3.9) and (3.11)], we can write

$$\gamma = \frac{W}{V} = \frac{W_s + W_w}{V} = \frac{G_s\gamma_w + wG_s\gamma_w}{1+e} = \frac{(1+w)G_s\gamma_w}{1+e} \tag{3.18}$$

and

$$\gamma_d = \frac{W_s}{V} = \frac{G_s\gamma_w}{1+e} \tag{3.19}$$

or

$$e = \frac{G_s\gamma_w}{\gamma_d} - 1 \tag{3.20}$$

Because the weight of water for the soil element under consideration is $wG_s\gamma_w$, the volume occupied by water is

$$V_w = \frac{W_w}{\gamma_w} = \frac{wG_s\gamma_w}{\gamma_w} = wG_s$$

Hence, from the definition of degree of saturation [Eq. (3.5)],

$$S = \frac{V_w}{V_v} = \frac{wG_s}{e}$$

or

$$Se = wG_s \tag{3.21}$$

This equation is useful for solving problems involving three-phase relationships.

If the soil sample is *saturated*—that is, the void spaces are completely filled with water (Figure 3.4)—the relationship for saturated unit weight (γ_{sat}) can be derived in a similar manner:

$$\gamma_{sat} = \frac{W}{V} = \frac{W_s + W_w}{V} = \frac{G_s\gamma_w + e\gamma_w}{1 + e} = \frac{(G_s + e)\gamma_w}{1 + e} \tag{3.22}$$

Also, from Eq. (3.21) with $S = 1$,

$$e = wG_s \tag{3.23}$$

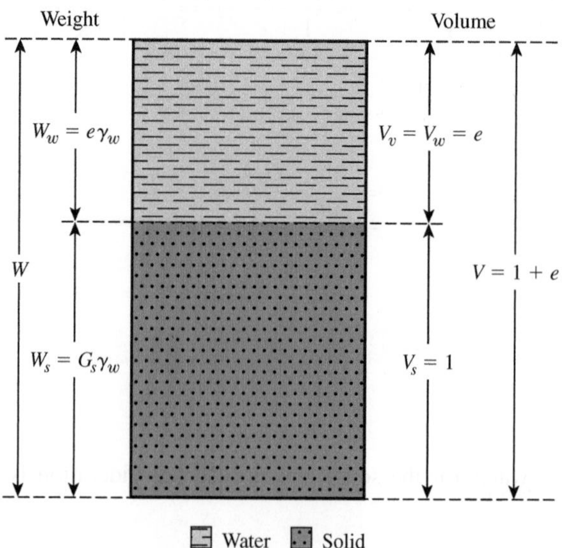

Figure 3.4 Saturated soil element with volume of soil solids equal to one

Example 3.3

The dry unit weight of a compacted earthwork is measured as 17.5 kN/m^3. The water content is 16.5%, and the specific gravity of the soil grains is 2.68. What is the degree of saturation of this soil?

Solution

From Eq. (3.20),

$$e = \frac{G_s \gamma_w}{\gamma_d} - 1 = \frac{2.68 \times 9.81}{17.5} - 1 = 0.502$$

From Eq. (3.21),

$$S = \frac{wG_s}{e} = \frac{0.165 \times 2.68}{0.502} = \mathbf{0.881 \text{ or } 88.1\%}$$

■

Example 3.4

A thin-walled sampler with a 50-mm inner diameter is pushed into the wall of an excavation, and a 150-mm-long intact sample with a mass of 490 g was obtained. When dried in the oven for 24 hours, the sample weighed 415 g. The specific gravity of the soil grains was 2.70. Determine the moist unit weight, dry unit weight, void ratio, moisture content, and the degree of saturation of the in situ soil.

If the ground gets flooded, assuming the soil is saturated, what would be the new moisture content?

Solution

$$\text{Volume of the intact sample} = \frac{\pi}{4} \times 0.05^2 \times 0.15 = 0.000295 \text{ m}^3$$

$$\text{Weight of the specimen} = 0.490 \times 9.81 = 4.807 \text{ N}$$

$$\text{The moist unit weight} = (4.807/0.000295) \div 1000 = \mathbf{16.29 \text{ kN/m}^3}$$

$$\text{Dry weight of the sample} = 0.415 \times 9.81 = 4.071 \text{ N}$$

$$\text{The dry unit weight} = (4.071/0.000295) \div 1000 = \mathbf{13.80 \text{ kN/m}^3}$$

From Eq. (3.20),

$$e = \frac{G_s \gamma_w}{\gamma_d} - 1 = \frac{2.70 \times 9.81}{13.8} - 1 = \mathbf{0.919}$$

From Eq. (3.8),

$$\text{Moisture content } w = \frac{W_w}{W_s} = \frac{W - W_s}{W_s} = \frac{490 - 415}{415} = \mathbf{0.181 \text{ or } 18.1\%}$$

From Eq. (3.21),

$$Se = wG_s$$

Hence, the degree of saturation is,

$$S = \frac{wG_s}{e} = \frac{0.181 \times 2.70}{0.919} = \mathbf{0.532 \text{ or } 53.2\%}$$

If the ground is saturated (i.e., $S = 1$) and $e = 0.919$, from Eq. (3.23), the saturated moisture content is

$$w = \frac{e}{G_s} = \frac{0.919}{2.70} = \mathbf{0.340\ or\ 34.0\%}$$

■

3.6 Relationships among Unit Weight, Porosity, and Moisture Content

The relationship among *unit weight, porosity,* and *moisture content* can be developed in a manner similar to that presented in the preceding section. Consider a soil that has a total volume equal to one, as shown in Figure 3.5. From Eq. (3.4),

$$n = \frac{V_v}{V}$$

If V is equal to 1, then V_v is equal to n, so $V_s = 1 - n$. The weight of soil solids (W_s) and the weight of water (W_w) can then be expressed as follows:

$$W_s = G_s\gamma_w(1 - n) \tag{3.24}$$

$$W_w = wW_s = wG_s\gamma_w(1 - n) \tag{3.25}$$

So, the dry unit weight equals

$$\gamma_d = \frac{W_s}{V} = \frac{G_s\gamma_w(1 - n)}{1} = G_s\gamma_w(1 - n) \tag{3.26}$$

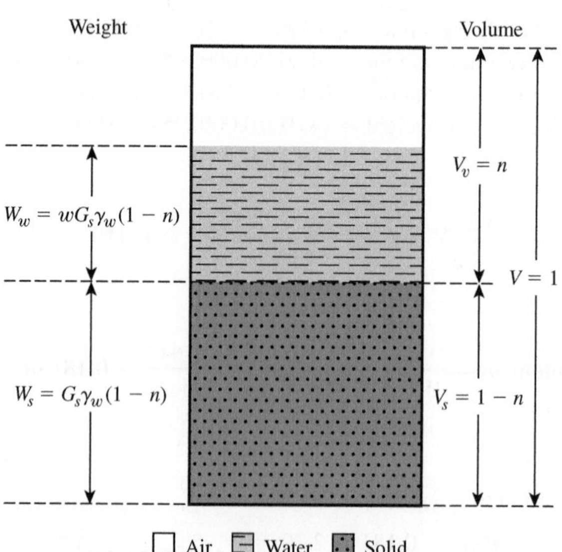

Figure 3.5 Soil element with total volume equal to one

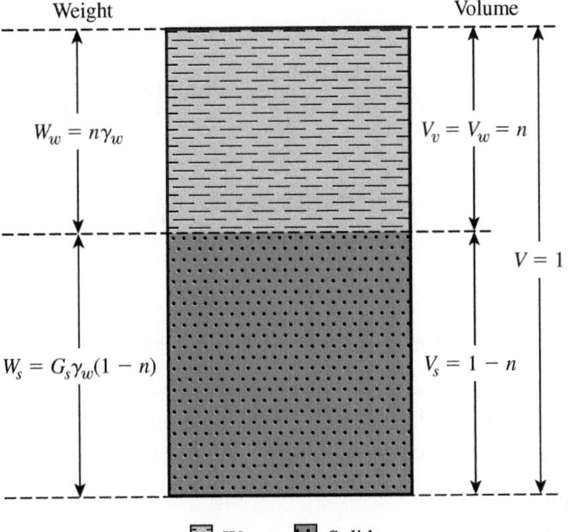

Figure 3.6 Saturated soil element with total volume equal to one

The moist unit weight equals

$$\gamma = \frac{W_s + W_w}{V} = G_s\gamma_w(1 - n)(1 + w) \tag{3.27}$$

Figure 3.6 shows a soil sample that is saturated and has $V = 1$. According to this figure,

$$\gamma_{sat} = \frac{W_s + W_w}{V} = \frac{(1 - n)G_s\gamma_w + n\gamma_w}{1} = [(1 - n)G_s + n]\gamma_w \tag{3.28}$$

The moisture content of a saturated soil sample can be expressed as

$$w = \frac{W_w}{W_s} = \frac{n\gamma_w}{(1 - n)\gamma_w G_s} = \frac{n}{(1 - n)G_s} \tag{3.29}$$

Example 3.5

For a soil, given: $e = 0.75$; $w = 22\%$; $G_s = 2.66$. Calculate the porosity, moist unit weight, dry unit weight, and degree of saturation.

Solution

Porosity [Eq. (3.7)]

$$n = \frac{e}{1 + e} = \frac{0.75}{1 + 0.75} = \mathbf{0.43}$$

Moist Unit Weight [Eq. (3.18)]

$$\gamma = \frac{(1 + w)G_s\gamma_w}{1 + e} = \frac{(1 + 0.22)2.66 \times 62.4}{1 + 0.75}$$
$$= \mathbf{115.7\ lb/ft^3}$$

Dry Unit Weight [Eq. (3.19)]

$$\gamma_d = \frac{G_s\gamma_w}{1+e} = \frac{2.66 \times 62.4}{1+0.75} = \mathbf{94.9\ lb/ft^3}$$

Degree of Saturation [Eq. (3.21)]

$$S(\%) = \frac{wG_s}{e} \times 100 = \frac{0.22 \times 2.66}{0.75} \times 100 = \mathbf{78\%}$$

■

Example 3.6

The moist weight of 0.2 ft^3 of a soil is 23 lb. The moisture content and the specific gravity of soil solids are determined in the laboratory to be 11% and 2.7, respectively. Calculate the following:

a. Moist unit weight (lb/ft^3)
b. Dry unit weight (lb/ft^3)
c. Void ratio
d. Porosity
e. Degree of saturation (%)

Solution

Part a

From Eq. (3.9),

$$\gamma = \frac{W}{V} = \frac{23}{0.2} = \mathbf{115\ lb/ft^3}$$

Part b

From Eq. (3.12),

$$\gamma_d = \frac{\gamma}{1+w} = \frac{115}{1+\left(\frac{11}{100}\right)} = \mathbf{103.6\ lb/ft^3}$$

Part c

From Eq. (3.20),

$$e = \frac{G_s\gamma_w}{\gamma_d} - 1 = \frac{(2.7)(62.4)}{103.6} - 1 = \mathbf{0.626}$$

Part d

From Eq. (3.7),

$$n = \frac{e}{1+e} = \frac{0.626}{1+0.626} = \mathbf{0.385}$$

Part e

From Eq. (3.21),

$$S(\%) = \left(\frac{wG_s}{e}\right)(100) = \frac{(0.11)(2.7)}{0.626}(100) = \mathbf{47.4\%}$$

■

Example 3.7

The dry unit weight of a sand with a porosity of 0.387 is 99.84 lb/ft^3. Find the void ratio of the soil and the specific gravity of the soil solids.

Solution

Void ratio

From $n = 0.387$ and Eq. (3.6),

$$e = \frac{n}{1-n} = \frac{0.387}{1-0.387} = \mathbf{0.631}$$

Specific gravity of soil solids

From Eq. (3.19),

$$\gamma_d = \frac{G_s\gamma_w}{1+e}$$

Thus,

$$99.84 = \frac{G_s(62.4)}{1+0.631}$$

$$G_s = \mathbf{2.61}$$

■

Example 3.8

The moist unit weight of a soil is 105 lb/ft^3. Given that $w = 15\%$ and $G_s = 2.7$, determine

a. Dry unit weight
b. Porosity
c. Degree of saturation
d. Weight of water, in lb/ft^3, to be added to reach full saturation

Solution

Part a

From Eq. (3.12),

$$\gamma_d = \frac{\gamma}{1+w} = \frac{105}{1+\left(\frac{15}{100}\right)} = \mathbf{91.3\ lb/ft^3}$$

Part b

From Eqs. (3.20) and (3.7), respectively,

$$e = \frac{G_s\gamma_w}{\gamma_d} - 1 = \frac{(2.7)(62.4)}{91.3} - 1 = 0.845$$

$$n = \frac{e}{1+e} = \frac{0.845}{1+0.845} = \mathbf{0.458}$$

Part c

From Eq. (3.21),

$$S = \frac{wG_s}{e} = \frac{\left(\frac{15}{100}\right)(2.7)}{0.845}(100) = \mathbf{47.9\%}$$

Part d

From Eq. (3.22),

$$\gamma_{sat} = \frac{(G_s + e)\gamma_w}{1 + e} = \frac{(2.7 + 0.845)(62.4)}{1 + 0.845} = \mathbf{119.9\ lb/ft^3}$$

Thus, the water to be added is

$$\gamma_{sat} - \gamma = 119.9 - 105 = \mathbf{14.9\ lb/ft^3}$$

■

3.7 Relative Density

The term *relative density* is commonly used to indicate the *in situ* denseness or looseness of granular soil. It is defined as

$$D_r = \frac{e_{max} - e}{e_{max} - e_{min}} \tag{3.30}$$

where

D_r = relative density, usually given as a percentage
e = *in situ* void ratio of the soil
e_{max} = void ratio of the soil in the loosest state
e_{min} = void ratio of the soil in the densest state

The values of D_r may vary from a minimum of 0% for very loose soil to a maximum of 100% for very dense soils. Soils engineers qualitatively describe the granular soil deposits according to their relative densities, as shown in Table 3.3. In-place soils seldom have relative densities less than 20 to 30%. Compacting a granular soil to a relative density greater than about 85% is difficult.

Table 3.3 Qualitative Description of Granular Soil Deposits

Relative density (%)	Description of soil deposit
0–15	Very loose
15–50	Loose
50–70	Medium
70–85	Dense
85–100	Very dense

By using the definition of dry unit weight given in Eq. (3.19), we can express relative density in terms of maximum and minimum possible dry unit weights. Thus,

$$D_r = \frac{\left[\frac{1}{\gamma_{d(\min)}}\right] - \left[\frac{1}{\gamma_d}\right]}{\left[\frac{1}{\gamma_{d(\min)}}\right] - \left[\frac{1}{\gamma_{d(\max)}}\right]} = \left[\frac{\gamma_d - \gamma_{d(\min)}}{\gamma_{d(\max)} - \gamma_{d(\min)}}\right]\left[\frac{\gamma_{d(\max)}}{\gamma_d}\right] \tag{3.31}$$

where

$\gamma_{d(\min)}$ = dry unit weight in the loosest condition (at a void ratio of $e_{\max}$)
γ_d = *in situ* dry unit weight (at a void ratio of e)
$\gamma_{d(\max)}$ = dry unit weight in the densest condition (at a void ratio of $e_{\min}$)

ASTM Test Designations D-4253 and 4254 provide procedures for the determination of the maximum and minimum dry unit weights of granular soils so that they can be used in Eq. (3.31) to measure the relative density of compaction in the field.

3.8 Summary

In laboratory tests and earthworks, it is often required to compute the weights and volumes of the three phases. In this chapter, terms such as void ratio, porosity, degree of saturation, moisture content, and different forms of unit weights (e.g., dry unit weight, moist unit weight) were introduced. Some simple equations relating these terms were also developed. They can be used for computing the weights and volumes of the soil grains, water, and air. Relative density is a measure of how densely the grains are packed in a granular soil. It varies between 0 for very loose soil and 100% for very dense soil.

Problems

3.1 State whether the following are true or false.
a. Moisture content must be less than 100%.
b. Saturated unit weight is always greater than the dry unit weight.
c. Porosity of a soil can exceed 100%.
d. A dry soil must have zero degree of saturation.
e. The higher the relative density, the higher is the void ratio.

3.2 The moist weight of 0.1 ft^3 of soil is 12.2 lb. If the moisture content is 12% and the specific gravity of soil solids is 2.72, find the following:
a. Moist unit weight (lb/ft^3).
b. Dry unit weight (lb/ft^3).
c. Void ratio.
d. Porosity.
e. Degree of saturation (%).
f. Weight of water in the void (lb).

3.3 A cylindrical clay specimen extruded from a compaction mould has a volume of 943.9 cm^3 and a mass of 1741.2 g. The moisture content of the specimen was 18.5%, G_s = 2.72. Find the void ratio, degree of saturation, and porosity of the compacted soil specimen.

3.4 The moist unit weight of a soil is 122.1 lb/ft³. Given $G_s = 2.69$ and moisture content, $w = 9.8\%$, determine:
a. Dry unit weight (lb/ft³).
b. Void ratio.
c. Porosity.
d. Degree of saturation (%).

3.5 A saturated soil below the water table a has unit weight of 20.5 kN/m³ and a water content of 26.5%. Assuming a G_s of 2.70, find the void ratio of the soil.

3.6 For a saturated soil, given $w = 40\%$ and $G_s = 2.71$, determine the saturated and dry unit weights in lb/ft³.

3.7 The moisture content and void ratio of a soil are 25.5% and 0.73, respectively. If $G_s = 2.68$, determine the degree of saturation and the moist unit weight.

3.8 The mass of a moist soil sample collected from the field is 465 grams, and its oven dry mass is 405.76 grams. The specific gravity of the soil solids was determined in the laboratory to be 2.68. If the void ratio of the soil in the natural state is 0.83, find the following:
a. The moist unit weight of the soil in the field (lb/ft³).
b. The dry unit weight of the soil in the field (lb/ft³).
c. The weight of water, in pounds, to be added per cubic foot of soil in the field for saturation.

3.9 The soil at a site has a void ratio of 0.73 and a water content of 15.5%. If the specific gravity of the soil grains is 2.72, find the moist unit weight of the in situ soil.

3.10 A soil has a unit weight of 126.8 lb/ft³. Given $G_s = 2.67$ and $w = 12.6\%$, determine:
a. Dry unit weight (lb/ft³).
b. Void ratio.
c. Porosity.
d. The weight of water in lb/ft³ of soil needed for full saturation.

3.11 A large piece of dry rock has a volume of 1.26 m³. The rock mineral grains have a specific gravity of 2.85. The mass of the rock is 3451 kg. What is the porosity of the rock?

3.12 The saturated unit weight of a soil is 128 lb/ft³. Given $G_s = 2.74$, determine:
a. γ_{dry} (lb/ft³)
b. e
c. n
d. w (%)

3.13 The maximum and minimum dry unit weights of a granular soil are 18.4 kN/m³ and 14.3 kN/m³, respectively. $G_s = 2.65$. What would be the relative density of the same sand at void ratio of 0.62?

3.14 For a soil, given $e = 0.86$, $w = 28\%$, and $G_s = 2.72$, determine:
a. Moist unit weight (lb/ft³).
b. Degree of saturation (%).

3.15 A 500-mm-thick soil layer with a void ratio of 0.84 is compacted into a 450-mm-thick layer. What is the new void ratio?

3.16 For a saturated soil, given $\gamma_d = 97.3$ lb/ft³ and $w = 21\%$, determine:
a. γ_{sat} (lb/ft³)
b. e
c. G_s
d. γ_{moist} when the degree of saturation is 50%

3.17 The maximum and minimum void ratios of a sand are 0.8 and 0.41, respectively. What is the void ratio of the soil corresponding to a relative density of 48%?

3.18 For a sand, the maximum and minimum possible void ratios were determined in the laboratory to be 0.94 and 0.33, respectively. Find the moist unit weight of sand in lb/ft^3 compacted in the field at a relative density of 60% and moisture content of 10%. Given: $G_s = 2.65$. Also calculate the maximum and minimum possible dry unit weights that the sand can have.

References

AMERICAN SOCIETY FOR TESTING AND MATERIALS (2014). *Annual Book of ASTM Standards*, Vol. 04.08, Conshohocken, PA.

4 Plasticity and Soil Classification

4.1 Introduction

In Chapter 2, we discussed the different size limits for the gravel, sand, silt, and clay. In addition, we also discussed the grain-size analysis, which is mostly carried out for coarse grained soils for which the engineering properties are significantly influenced by the grain-size distribution. In clay soils, their flaky grains with large surface areas and the mineralogy make them plastic (putty-like property) when they are wet. As a result, plasticity is more important than the grain-size distribution for clays.

Grain-size distribution and plasticity play a key role in the classification of coarse and fine grained soils, respectively. A soil classification system is the arrangement of different soils with similar properties into groups and sub-groups based on their application. Classification systems provide a common language to briefly express the general characteristics of soil, which are infinitely varied, without detailed descriptions. In this chapter, you will learn the following.

- The consistency limits, known as Atterberg limits, which define the different states at which a fine grained soil can exist.
- The laboratory tests for the Atterberg limits.
- The properties used in soil classification of coarse and fine grained soils.
- The two major soil classification systems, namely AASHTO and USCS.

4.2 Consistency of Soil—Atterberg Limits

When clay minerals are present in fine-grained soil, the soil can be remolded in the presence of some moisture without crumbling. This cohesive nature is caused by the adsorbed water surrounding the clay particles. In the early 1900s, a Swedish scientist named Atterberg developed a method to describe the consistency of fine-grained soils with varying moisture contents. At a very low moisture content, soil behaves more like a solid. When the moisture content is very high, the soil and water may flow like a liquid. Hence, on an arbitrary basis, depending on the moisture content, the behavior of soil can be divided into four basic states—*solid, semisolid, plastic*, and *liquid*—as shown in Figure 4.1.

The moisture content, in percent, at which the transition from solid to semisolid state takes place is defined as the *shrinkage limit*. The moisture content at the point of transition from semisolid to

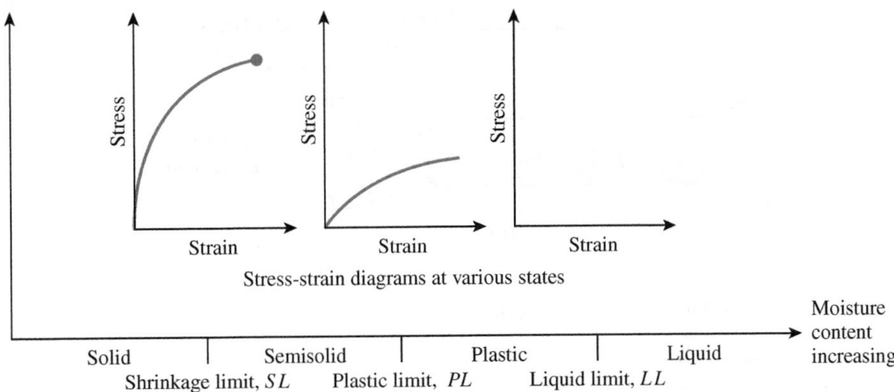

Figure 4.1 Atterberg limits

plastic state is the *plastic limit*, and from plastic to liquid state is the *liquid limit*. These parameters are also known as *Atterberg limits*. In the following sections, we describe the procedures for laboratory determination of Atterberg limits.

4.3 Liquid Limit (*LL*)

A schematic diagram (side view) of a liquid limit device is shown in Figure 4.2a. This device consists of a brass cup and a hard rubber base. The brass cup can be dropped onto the base by a cam operated by a crank. To perform the liquid limit test, one must place a soil paste in the cup. A groove is then cut at the center of the soil pat with the standard grooving tool (Figure 4.2b). By the use of the crank-operated cam, the cup is lifted and dropped from a height of 0.394 in. (10 mm). The moisture content, in percent, required to close a distance of 0.5 in. (12.7 mm) along the bottom of the groove (see Figures 4.2c and 4.2d) after 25 blows is defined as the *liquid limit*.

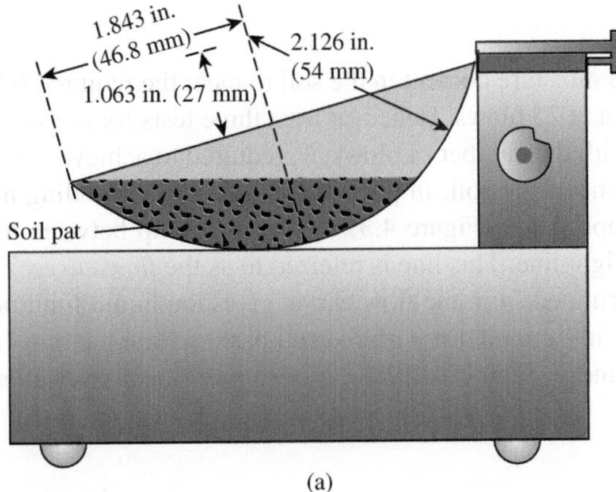

Figure 4.2 Liquid limit test: (a) liquid limit device

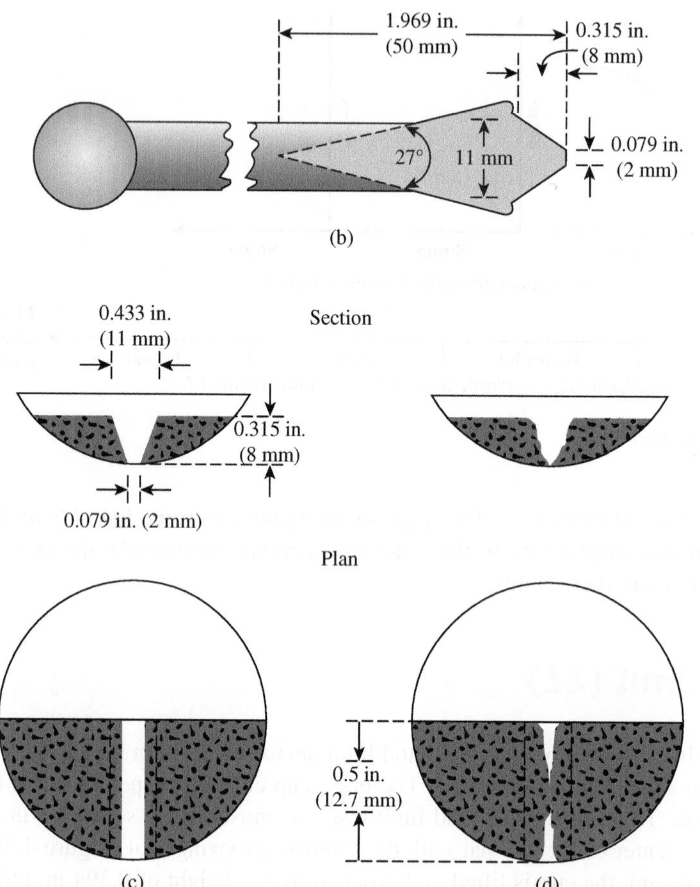

Figure 4.2 (*continued*) (b) grooving tool; (c) soil pat before test; (d) soil pat after test

It is difficult to adjust the moisture content in the soil to meet the required 0.5-in. (12.7 mm) closure of the groove in the soil pat at 25 blows. Hence, at least three tests for the same soil are conducted at varying moisture contents, with the number of blows, N, required to achieve closure varying between 15 and 35. The moisture content of the soil, in percent, and the corresponding number of blows are plotted on semilogarithmic graph paper (Figure 4.3). The relationship between moisture content and log N is approximated as a straight line. This line is referred to as the *flow curve*. The moisture content corresponding to $N = 25$, determined from the flow curve, gives the liquid limit of the soil.

Figure 4.4 shows the liquid limit device with the soil paste within the brass cup. In Figure 4.4a, the groove has been just cut, and in Figure 4.4b, the groove has closed over a length of 12.7 mm at the bottom after several blows.

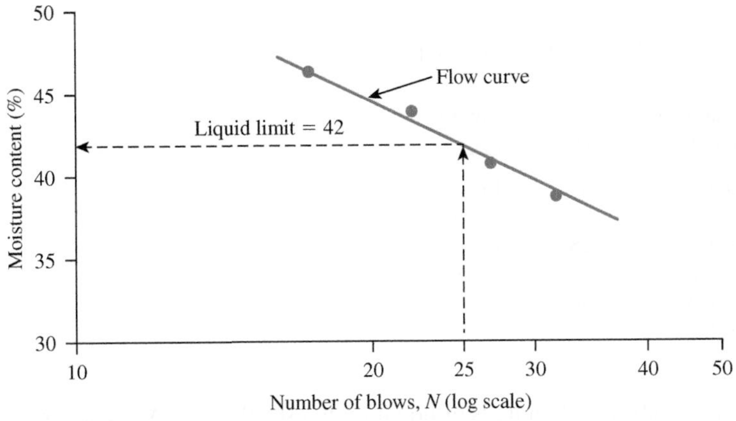

Figure 4.3 Flow curve for liquid limit determination of a clayey silt

(a)

(b)

Figure 4.4 Atterberg limit test using Casagrande cup: (a) groove, (b) groove closing over 12.7 mm (*Courtesy of N. Sivakugan, James Cook University, Australia*)

Table 4.1 Values of $\left(\frac{N}{25}\right)^{0.121}$ [see Eq. (4.1)]

N	$\left(\frac{N}{25}\right)^{0.121}$	N	$\left(\frac{N}{25}\right)^{0.121}$
20	0.973	26	1.005
21	0.979	27	1.009
22	0.985	28	1.014
23	0.990	29	1.018
24	0.995	30	1.022
25	1.000		

From the analysis of hundreds of liquid limit tests, the U.S. Army Corps of Engineers (1949) at the Waterways Experiment Station in Vicksburg, Mississippi, proposed an empirical equation of the form (also see Table 4.1)

$$LL = w_N \left(\frac{N}{25}\right)^{\tan\beta} \tag{4.1}$$

where

N = number of blows in the liquid limit device for a 0.5-in. (12.7 mm) groove closure
w_N = corresponding moisture content
$\tan\beta$ = 0.121 (but note that $\tan\beta$ is not equal to 0.121 for all soils)

Equation (4.1) generally yields good results for the number of blows between 20 and 30. For routine laboratory tests, it may be used to determine the liquid limit when only one test is run for a soil. This procedure is generally referred to as the *one-point method* and was also adopted by ASTM under designation D-4318.

4.4 Plastic Limit (*PL*)

The *plastic limit* is defined as the moisture content in percent, at which the soil crumbles, when rolled into threads of $\frac{1}{8}$ in. (3.2 mm) in diameter. The plastic limit is the lower limit of the plastic stage of soil. The plastic limit test is simple and is performed by repeated rollings of an ellipsoidal-sized soil mass by hand on a ground glass plate. Soil threads at plastic limit, crumbling at $\frac{1}{8}$ in. (3.2 mm), is shown in Figure 4.5. The procedure for the plastic limit test is given by ASTM in Test Designation D-4318.

The *plasticity index (PI)* is the difference between the liquid limit and the plastic limit of a soil, or

$$PI = LL - PL \tag{4.2}$$

It is the range of moisture content over which the soil remains plastic.

Table 4.2 gives the range of liquid limit, plastic limit, of some clay minerals (Mitchell, 1976; Skempton, 1953).

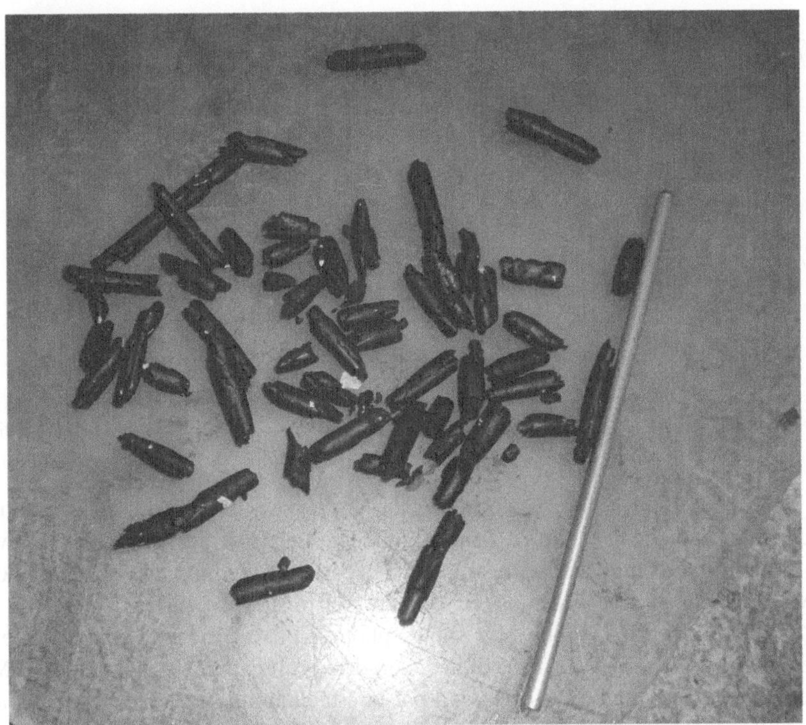

Figure 4.5 Crumbling of soil threads at plastic limit with 1/8 in. (3.2 mm) diameter. The aluminum rod nearby is for comparison. (*Courtesy of N. Sivakugan, James Cook University, Australia*)

4.5 Shrinkage Limit (*SL*)

Soil shrinks as moisture is gradually lost from it. With continuing loss of moisture, a stage of equilibrium is reached at which more loss of moisture will result in no further volume change (Figure 4.6). The moisture content, in percent, at which the volume of the soil mass ceases to change is defined as the *shrinkage limit*.

Table 4.2 Typical Values of Liquid Limit and Plastic Limit of Some Clay Minerals

Mineral	Liquid limit, *LL*	Plastic limit, *PL*
Kaolinite	35–100	20–40
Illite	60–120	35–60
Montmorillonite	100–900	50–100
Halloysite (hydrated)	50–70	40–60
Halloysite (dehydrated)	40–55	30–45
Attapulgite	150–250	100–125
Allophane	200–250	120–150

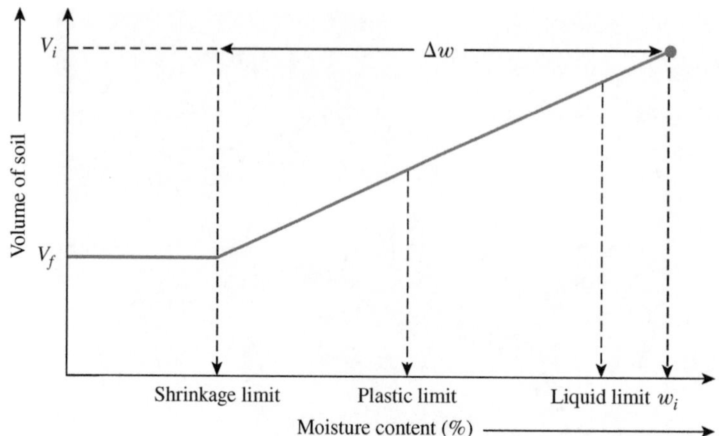

Figure 4.6 Definition of shrinkage limit

Shrinkage limit tests are performed in the laboratory with a porcelain dish about 1.75 in. (44.5 mm) in diameter and about $\frac{1}{2}$ in. (12.7 mm) high. The inside of the dish is coated with petroleum jelly and is then filled completely with wet soil. Excess soil standing above the edge of the dish is struck off with a straightedge. The mass of the wet soil inside the dish is recorded. The soil pat in the dish is then oven-dried. Figure 4.7 shows photographs of the wet soil and the oven-dried soil in the dish. The volume and the mass of the oven-dried soil pat are determined. ASTM D-4943 describes a method of dipping the oven-dried soil pat in a melted pot of wax. The wax-coated soil pat is then cooled. Its volume is determined by submerging it in water.

By reference to Figure 4.6, the shrinkage limit can be determined as

$$SL = w_i(\%) - \Delta w(\%) \tag{4.3}$$

where

w_i = initial moisture content when the soil is placed in the shrinkage limit dish
Δw = change in moisture content (that is, between the initial moisture content and the moisture content at the shrinkage limit)

However,

$$w_i\,(\%) = \frac{M_1 - M_2}{M_2} \times 100 \tag{4.4}$$

(a)

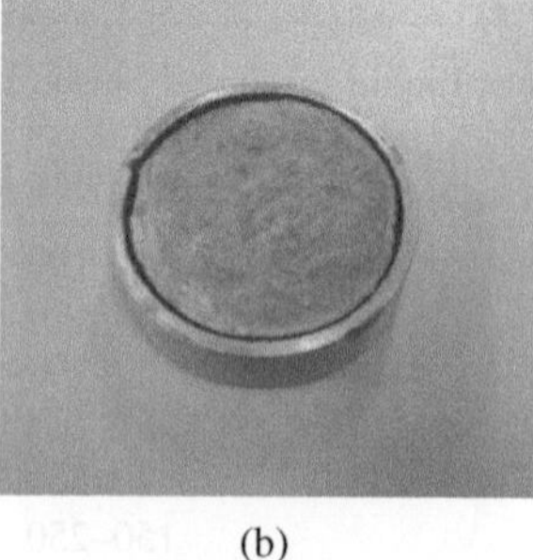

(b)

Figure 4.7 Shrinkage limit test: (a) wet specimen in the dish, (b) dried specimen (*Courtesy of Braja M. Das*)

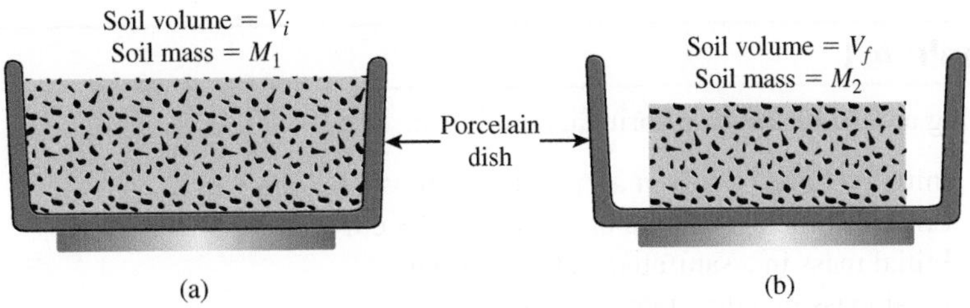

Figure 4.8 Shrinkage limit test: (a) soil pat before drying; (b) soil pat after drying

where

M_1 = mass of the wet soil pat in the dish at the beginning of the test (g)
M_2 = mass of the dry soil pat (g) (see Figure 4.8)

Also,

$$\Delta w\ (\%) = \frac{(V_i - V_f)\rho_w}{M_2} \times 100 \tag{4.5}$$

where

V_i = initial volume of the wet soil pat (that is, inside volume of the dish, cm^3)
V_f = volume of the oven-dried soil pat (cm^3)
ρ_w = density of water (g/cm^3)

Finally, combining Eqs. (4.3), (4.4), and (4.5) gives

$$SL = \left(\frac{M_1 - M_2}{M_2}\right)(100) - \left(\frac{V_i - V_f}{M_2}\right)(\rho_w)(100) \tag{4.6}$$

Typical values of shrinkage limit for some clay minerals are as follows in Table 4.3 (Mitchell, 1976).

Table 4.3 Typical Values of Shrinkage Limit

Mineral	Shrinkage limit
Montmorillonite	8.5–15
Illite	15–17
Kaolinite	25–29

Example 4.1

Following are the results of a shrinkage limit test:

- Initial volume of soil in a saturated state = 16.2 cm^3
- Final volume of soil in a dry state = 10.8 cm^3
- Initial mass in a saturated state = 44.6 g
- Final mass in a dry state = 32.8 g

Determine the shrinkage limit of the soil.

Solution

From Eq. (4.6),

$$SL = \left(\frac{M_1 - M_2}{M_2}\right)(100) - \left(\frac{V_i - V_f}{M_2}\right)(\rho_w)(100)$$

$$M_1 = 44.6\text{g} \qquad V_i = 16.2\ \text{cm}^3 \qquad \rho_w = 1\ \text{g/cm}^3$$

$$M_2 = 32.8\text{g} \qquad V_f = 10.8\ \text{cm}^3$$

$$SL = \left(\frac{44.6 - 32.8}{32.8}\right)(100) - \left(\frac{16.2 - 10.8}{32.8}\right)(1)(100)$$

$$= 35.97 - 16.46 = \mathbf{19.51\%}$$

■

4.6 Engineering Classification of Soil

Currently, two elaborate classification systems are commonly used by soils engineers. Both systems take into consideration the grain-size distribution and Atterberg limits. They are the American Association of State Highway and Transportation Officials (AASHTO) classification system and the Unified Soil classification system. The AASHTO classification system is used mostly by state and county highway departments. Geotechnical engineers generally prefer the Unified system. Classification of soils by the AASHTO and Unified systems are given in the following sections.

4.7 AASHTO Classification System

The AASHTO system of soil classification was developed in 1929 as the Public Road Administration classification system. It has undergone several revisions, with the present version proposed by the Committee on Classification of Materials for Subgrades and Granular Type Roads of the Highway Research Board in 1945 (ASTM designation D-3282; AASHTO method M145).

The AASHTO classification in present use is given in Table 4.4. According to this system, soil is classified into seven major groups: A-1 through A-7. Soils classified under groups A-1, A-2, and A-3 are granular materials of which 35% or less of the particles pass through the No. 200 sieve. Soils of which more than 35% pass through the No. 200 U.S. sieve are classified under groups A-4, A-5,

Table 4.4 Classification of Highway Subgrade Materials

General classification	Granular materials (35% or less of total sample passing No. 200)						
	A-1			A-2			
Group classification	A-1-a	A-1-b	A-3	A-2-4	A-2-5	A-2-6	A-2-7
Sieve analysis (percentage passing)							
No. 10	50 max.						
No. 40	30 max.	50 max.	51 min.				
No. 200	15 max.	25 max.	10 max.	35 max.	35 max.	35 max.	35 max.
Characteristics of fraction passing No. 40							
Liquid limit				40 max.	41 min.	40 max.	41 min.
Plasticity index	6 max.		NP	10 max.	10 max.	11 min.	11 min.
Usual types of significant constituent materials	Stone fragments, gravel and sand		Fine sand	Silty or clayey gravel, and sand			
General subgrade rating	Excellent to good						

General classification	Silt-clay materials (more than 35% of total sample passing No. 200)			
Group classification	A-4	A-5	A-6	A-7 A-7-5[a] A-7-6[b]
Sieve analysis (percentage passing)				
No. 10				
No. 40				
No. 200	36 min.	36 min.	36 min.	36 min.
Characteristics of fraction passing No. 40				
Liquid limit	40 max.	41 min.	40 max.	41 min.
Plasticity index	10 max.	10 max.	11 min.	11 min.
Usual types of significant constituent materials	Silty soils		Clayey soils	
General subgrade rating	Fair to poor			

[a]For A-7-5, $PI \leq LL - 30$

[b]For A-7-6, $PI > LL - 30$

A-6, and A-7. These soils are mostly silt and clay-type materials. This classification system is based on the following criteria:

1. *Grain size*
   a. *Gravel*: fraction passing the 75-mm (3-in.) sieve and retained on the No. 10 (2-mm) U.S. sieve
   b. *Sand*: fraction passing the No. 10 (2-mm) U.S. sieve and retained on the No. 200 (0.075-mm) U.S. sieve
   c. *Silt and clay*: fraction passing the No. 200 U.S. sieve

2. *Plasticity*: The term *silty* is applied when the fine fractions of the soil have a plasticity index of 10 or less. The term *clayey* is applied when the fine fractions have a plasticity index of 11 or more.
3. If cobbles and *boulders* (size larger than 75 mm) are encountered, they are excluded from the portion of the soil sample from which classification is made. However, the percentage of such material is recorded.

To classify a soil according to Table 4.4, one must apply the test data from left to right. By process of elimination, the first group from the left into which the test data fit is the correct classification.

To evaluate the quality of a soil as a highway subgrade material, one must also incorporate a number called the *group index (GI)* with the groups and subgroups of the soil. This index is written in parentheses after the group or subgroup designation. The group index is given by the equation

$$GI = (F_{200} - 35)[0.2 + 0.005(LL - 40)] + 0.01(F_{200} - 15)(PI - 10) \tag{4.7}$$

where

F_{200} = percentage passing through the No. 200 U.S. sieve
LL = liquid limit
PI = plasticity index

The first term of Eq. (4.7)—that is, $(F_{200} - 35)[0.2 + 0.005(LL - 40)]$—is the partial group index determined from the liquid limit. The second term—that is, $0.01\,(F_{200} - 15)(PI - 10)$—is the partial group index determined from the plasticity index. Following are some rules for determining the group index:

1. If Eq. (4.7) yields a negative value for GI, it is taken as 0.
2. The group index calculated from Eq. (4.7) is rounded off to the nearest whole number (for example, $GI = 3.4$ is rounded off to 3; $GI = 3.5$ is rounded off to 4).
3. There is no upper limit for the group index.
4. The group index of soils belonging to groups A-1-a, A-1-b, A-2-4, A-2-5, and A-3 is always 0.
5. When calculating the group index for soils that belong to groups A-2-6 and A-2-7 (granular soils), use the partial group index for PI, or

$$GI = 0.01(F_{200} - 15)(PI - 10) \tag{4.8}$$

In general, the quality of performance of a soil as a subgrade material is inversely proportional to the group index.

Example 4.2

Classify the following soils by the AASHTO classification system.

	Soil				
Description	**A**	**B**	**C**	**D**	**E**
Percent finer than No. 10 sieve	83	100	48	90	100
Percent finer than No. 40 sieve	48	92	28	76	82
Percent finer than No. 200 sieve	20	86	6	34	38
Liquid limit[a]	20	70	—	37	42
Plasticity index[a]	5	32	Nonplastic	12	23

[a]Plasticity for the minus 40 fraction

Solution

Soil A

According to Table 4.4, because 20% of the soil is passing through the No. 200 sieve, it falls under granular material classification—that is, A-1, A-3, or A-2. Proceeding from left to right, we see that it falls under category A-1-b. The group index for A-1-b is 0. So, the classification is **A-1-b(0)**.

Soil B

The percentage passing through the No. 200 sieve is 86, so the soil is a silty-clay material (that is, A-4, A-5, A-6, or A-7), as shown in Table 4.4. Proceeding from left to right, we see that it falls under category A-7. In this case, $PI = 32 < LL - 30$. Thus, the category is A-7-5. From Eq. (4.7),

$$GI = (F_{200} - 35)[0.2 + 0.005(LL - 40)] + 0.01(F_{200} - 15)(PI - 10)$$

For this soil, $F_{200} = 86$, $LL = 70$, and $PI = 32$, so

$$GI = (86 - 35)[0.2 + 0.005(70 - 40)] + 0.01(86 - 15)(32 - 10)$$
$$= 33.47 \simeq 33$$

Thus, the soil is type **A-7-5(33)**.

Soil C

The percentage passing through the No. 200 sieve is less than 35, so the soil is a granular material. Proceeding from left to right in Table 4.4, we find that it is type A-1-a. The group index is 0, so the soil is type **A-1-a(0)**.

Soil D

The percentage passing through the No. 200 sieve is less than 35, so the soil is a granular material. From Table 4.4, we see that it is type A-2-6. From Eq. (4.8),

$$GI = 0.01(F_{200} - 15)(PI - 10)$$

For this soil, $F_{200} = 34$ and $PI = 12$, so

$$GI = 0.01(34 - 15)(12 - 10) = 0.38 \simeq 0$$

Thus, the soil is type **A-2-6(0)**.

Soil E

The percentage passing through the No. 200 sieve is 38, which is greater than 35%, so the soil is a silty clay material. Proceeding from left to right in Table 4.4, we see that it falls under category A-7. In this case, $PI > LL - 30$, so the soil is type A-7-6. From Eq. (4.7),

$$GI = (F_{200} - 35)[0.2 + 0.005(LL - 40)] + 0.01(F_{200} - 15)(PI - 10)$$

For this soil, $F_{200} = 38$, $LL = 42$, and $PI = 23$, so

$$GI = (38 - 35)[0.2 + 0.005(42 - 40)] + 0.01(38 - 15)(23 - 10)$$
$$= 3.62 \simeq 4$$

Hence, the soil is type **A-7-6(4)**. ■

4.8 Unified Soil Classification System

The original form of this system was proposed by Casagrande in 1942 for use in the airfield construction works undertaken by the Army Corps of Engineers during World War II. In cooperation with the U.S. Bureau of Reclamation, this system was revised in 1952. At present, it is used widely by engineers (ASTM Test Designation D-2487). The Unified classification system (USCS) is presented in Table 4.5.

This system classifies soils into two broad categories:

1. Coarse-grained soils that are gravelly and sandy in nature with less than 50% passing through the No. 200 sieve. The group symbols start with a prefix of G or S. G stands for gravel or gravelly soil, and S for sand or sandy soil.
2. Fine-grained soils are with 50% or more passing through the No. 200 sieve. The group symbols start with prefixes of M, which stands for inorganic silt, C for inorganic clay, or O for organic silts and clays. The symbol Pt is used for peat, muck, and other highly organic soils.

Other symbols used for the classification are

- W—well graded
- P—poorly graded
- L—low plasticity (liquid limit less than 50)
- H—high plasticity (liquid limit more than 50)

For proper classification according to this system, some or all of the following information must be known.

1. Percent of gravel—that is, the fraction passing the 76.2-mm sieve and retained on the No. 4 sieve (4.75-mm opening)
2. Percent of sand—that is, the fraction passing the No. 4 sieve (4.75-mm opening) and retained on the No. 200 sieve (0.075-mm opening)
3. Percent of silt and clay—that is, the fraction finer than the No. 200 sieve (0.075-mm opening)
4. Uniformity coefficient (C_u) and the coefficient of gradation (C_c)
5. Liquid limit and plasticity index of the portion of soil passing the No. 40 sieve (0.425-mm opening)

The group symbols for coarse-grained gravelly soils are GW, GP, GM, GC, GC-GM, GW-GM, GW-GC, GP-GM, and GP-GC. Similarly, the group symbols for fine-grained soils are CL, ML, OL, CH, MH, OH, CL-ML, and Pt.

In using Table 4.5, one needs to always remember that, in a given soil,

- Fine fraction = percent passing No. 200 sieve
- Coarse fraction = percent retained on No. 200 sieve
- Gravel fraction = percent retained on No. 4 sieve
- Sand fraction = (percent retained on No. 200 sieve) − (percent retained on No. 4 sieve)

Table 4.5 Unified Soil Classification System (Based on Material Passing 76.2-mm Sieve)

Criteria for assigning group symbols				Group symbol
Coarse-grained soils More than 50% of retained on No. 200 sieve	**Gravels** More than 50% of coarse fraction retained on No. 4 sieve	Clean Gravels Less than 5% fines[a]	$C_u \geq 4$ and $1 \leq C_c \leq 3$[c]	GW
			$C_u < 4$ and/or $1 > C_c > 3$[c]	GP
		Gravels with Fines More than 12% fines[a,d]	$PI < 4$ or plots below "A" line (Figure 4.9)	GM
			$PI > 7$ and plots on or above "A" line (Figure 4.9)	GC
	Sands 50% or more of coarse fraction passes No. 4 sieve	Clean Sands Less than 5% fines[b]	$C_u \geq 6$ and $1 \leq C_c \leq 3$[c]	SW
			$C_u < 6$ and/or $1 > C_c > 3$[c]	SP
		Sands with Fines More than 12% fines[b,d]	$PI < 4$ or plots below "A" line (Figure 4.9)	SM
			$PI > 7$ and plots on or above "A" line (Figure 4.9)	SC
Fine-grained soils 50% or more passes No. 200 sieve	**Silts and clays** Liquid limit less than 50	Inorganic	$PI > 7$ and plots on or above "A" line (Figure 4.9)[e]	CL
			$PI < 4$ or plots below "A" line (Figure 4.9)[e]	ML
		Organic	$\frac{\text{Liquid limit} - \text{oven dried}}{\text{Liquid limit} - \text{not dried}} < 0.75$; see Figure 4.9; OL zone	OL
	Silts and clays Liquid limit 50 or more	Inorganic	PI plots on or above "A" line (Figure 4.9)	CH
			PI plots below "A" line (Figure 4.9)	MH
		Organic	$\frac{\text{Liquid limit} - \text{oven dried}}{\text{Liquid limit} - \text{not dried}} < 0.75$; see Figure 4.9; OL zone	OH
Highly Organic Soils	Primarily organic matter, dark in color, and organic odor			Pt

[a]Gravels with 5 to 12% fine require dual symbols: GW-GM, GW-GC, GP-GM, GP-GC.

[b]Sands with 5 to 12% fines require dual symbols: SW-SM, SW-SC, SP-SM, SP-SC.

[c]$C_u = \frac{D_{60}}{D_{10}}$; $C_c = \frac{(D_{30})^2}{D_{60} \times D_{10}}$

[d]If $4 \leq PI \leq 7$ and plots in the hatched area in Figure 4.9, use dual symbol GC-GM or SC-SM.

[e]If $4 \leq PI \leq 7$ and plots in the hatched area in Figure 4.9, use dual symbol CL-ML.

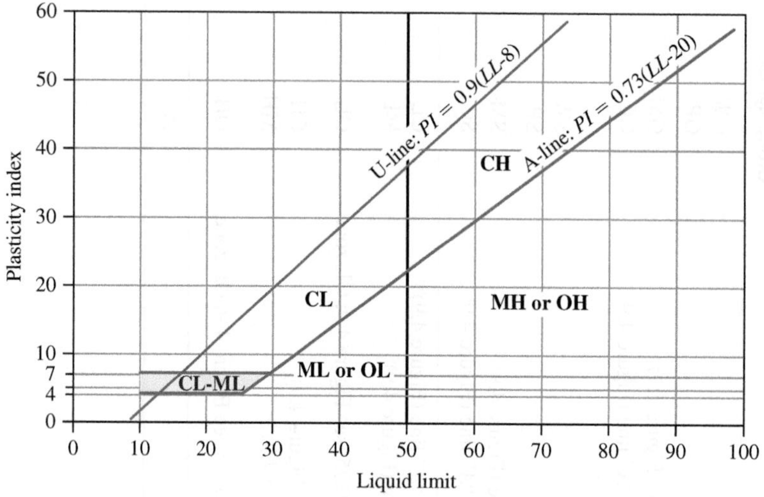

Figure 4.9 Plasticity chart

Example 4.3

Figure 4.10 gives the grain-size distribution of two soils. The liquid and plastic limits of minus No. 40 sieve fraction of the soil are as follows:

	Soil A	Soil B
Liquid limit	30	26
Plastic limit	22	20

Determine the group symbols according to the Unified Soil Classification System and describe the soils.

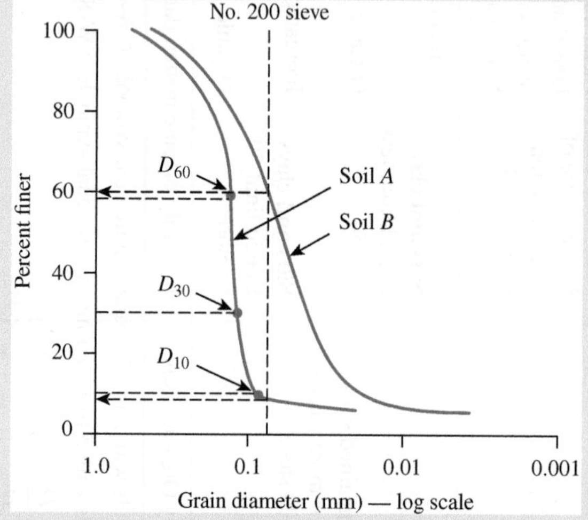

Figure 4.10 Grain-size distribution of two soils

Solution

Soil A

The grain-size distribution curve (Figure 4.10) indicates that percent passing No. 200 sieve is 8. According to Table 4.5 it is a coarse-grained soil. Also, from Figure 4.10 the percent retained on No. 4 sieve is zero. Hence, it is a sandy soil.

From Figure 4.10, $D_{10} = 0.085$ mm, $D_{30} = 0.12$ m, and $D_{60} = 0.135$ mm. Thus,

$$C_u = \frac{D_{60}}{D_{10}} = \frac{0.135}{0.085} = 1.59 < 6$$

$$C_c = \frac{D_{30}^2}{D_{60} \times D_{10}} = \frac{(0.12)^2}{(0.135)(0.085)} = 1.25 > 1$$

With $LL = 30$ and $PI = 30 - 22 = 8$ (which is greater than 7), it plots above the *A*-line in Figure 4.9. Hence, the group symbol is **SP-SC**. It can be described as poorly graded clayey sand.

Soil B

The grain-size distribution curve in Figure 4.10 shows that percent passing No. 200 sieve is 61 (>50%); hence, it is a fine-grained soil. Given: $LL = 26$ and $PI = 26 - 20 = 6$. In Figure 4.9, the *PI* plots in the hatched area. So, from Table 4.5, the group symbol is **CL-ML**. It can be described as sandy clayey silt or sandy silty clay of low plasticity. ■

Example 4.4

For a given soil:

- Percent passing No. 4 sieve = 100
- Percent passing No. 200 sieve = 86
- Liquid limit = 55
- Plasticity index = 28

Classify the soil using the Unified classification system giving the group symbol and give its description.

Solution

We are given that the percent passing the No. 200 sieve is 86 (i.e., >50%), so it is a fine-grained soil. Using Table 4.5 and Figure 4.9, we find the group symbol to be **CH**. The soil has 14% sand. Thus, it can be described as sandy clay of high plasticity. ■

4.9 Summary

Atterberg limits are applicable to fine-grained soils, especially clays. They are the borderline moisture contents that separate the different states (e.g., semisolid, plastic) at which the clay can exist. Determination of the liquid limit and plastic limit in the laboratory is fairly straightforward.

Plasticity index, which is the difference between liquid limit and plastic limit, is a measure of the soil plasticity.

Coarse-grained soils are often classified based on their grain-size distribution, and fine-grained soils are classified by their Atterberg limits.

Geotechnical engineers generally use the Unified Soil Classification System (USCS). However, for roadwork, the American Association of State Highway and Transportation Officials (AASHTO) classification system is used mostly. The two systems were discussed in detail in this chapter.

Problems

4.1 State whether the following are true or false.
   a. Liquid limit is always greater than plastic limit.
   b. Plasticity index can be less than plastic limit.
   c. Atterberg limits cannot exceed 100.
   d. In AASHTO, A-7-5(30) is a better subgrade material than A-7-5(15).
   e. In USCS, the symbol for clayey gravel is CG.

4.2 The following data were obtained from the liquid and plastic limit tests for a soil:
Liquid limit test:

Number of blows (*N*)	Moisture content (%)
15	42.0
20	40.8
28	39.1

Plastic limit test: Moisture content = 18.7%
   a. Draw the flow curve from the liquid limit test data and find the liquid limit.
   b. What is the plasticity index of the soil?

4.3 A saturated soil with a volume of 19.65 cm^3 has a mass of 36 grams. When the soil was dried, its volume and mass were 13.5 cm^3 and 25 grams, respectively. Determine the shrinkage limit for the soil.

4.4 Repeat Problem 4.2 for the following:
Liquid limit test:

Number of blows (*N*)	Moisture content (%)
17	42.1
22	38.2
27	36.2
32	34.1

Plastic limit test: Moisture content = 21.3%

4.5 The sieve analysis of ten soils and the liquid and plastic limits of the fraction passing through the No. 40 sieve are given below. Classify the soils by the AASHTO classification system and give the group indexes.

	Sieve Analysis, % Finer				
Soil no.	**No. 10**	**No. 40**	**No. 200**	**Liquid limit**	**Plastic limit**
1	98	80	50	38	29
2	100	92	80	56	23
3	100	88	65	37	22
4	85	55	45	28	20
5	92	75	62	43	28
6	97	60	30	25	16
7	100	55	8	—	NP
8	94	80	63	40	21
9	83	48	20	20	15
10	100	92	86	70	38

4.6 Classify soils 1 through 6 given in Problem 4.5 by the Unified classification system.

4.7 Classify the following soils using the AASHTO classification system. Give group indexes also.

	Sieve Analysis, % Finer				
Soil	**No. 10**	**No. 40**	**No. 200**	**Liquid limit**	**Plasticity index**
A	48	28	6	—	NP
B	87	62	30	32	8
C	90	76	34	37	12
D	100	78	8	—	NP
E	92	74	32	44	9

4.8 Classify the following soils using the Unified classification system.

	Percent Passing				
Size	**A**	**B**	**C**	**D**	**E**
Sieve No. 4	94	98	100	100	100
No. 10	63	86	100	100	100
No. 20	21	50	98	100	100
No. 40	10	28	93	99	94
No. 60	7	18	88	95	82
No. 100	5	14	83	90	66
No. 200	3	10	77	86	45
0.01 mm	—	—	65	42	26
0.002 mm	—	—	60	47	21
Liquid limit	—	—	63	55	36
Plasticity index	NP	NP	25	28	22

4.9 Classify the soils given in Problem 4.8 by the AASHTO classification system. Give group indexes.

4.10 Classify the following soils by the Unified classification system.

Soil	Sieve Analysis, % Finer		Liquid limit	Plasticity index
	No. 4	No. 200		
A	92	48	30	8
B	60	40	26	4
C	99	76	60	32
D	90	60	41	12
E	80	35	24	2

4.11 The grain-size distributions of three soils *A*, *B*, and *C* are shown in Figure 4.11. The liquid limit and the plastic limit of soil *C* are 74 and 42, respectively. Classify the three soils using the Unified soil classification system, giving their group symbols and descriptions.

4.12 Classify the following soils using Unified soil classification system, giving their group symbols and descriptions.

a. Percent passing 19 mm = 100; percent passing No. 4 sieve = 53; percent passing No. 200 sieve = 25
LL = 56 and PL = 26

b. Percent passing 9.5 mm = 100; percent passing No. 4 sieve = 86; percent passing No. 200 sieve = 60
LL = 62 and PL = 37

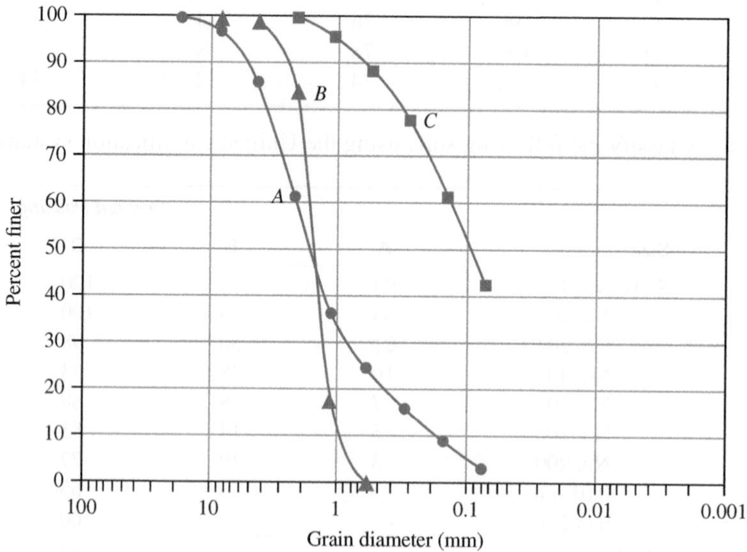

Figure 4.11 Grain-size distribution of soils *A*, *B*, and *C*

c. Percent passing 19 mm = 100; percent passing No. 4 sieve = 65; percent passing No. 200 sieve = 10
$C_u = 34.5$ and $C_c = 2.1$
$LL = 34$ and $PL = 18$

References

American Association of State Highway and Transportation Officials (1982). *AASHTO Materials, Part I, Specifications*, Washington, D.C.

American Society for Testing and Materials (2014). *Annual Book of ASTM Standards*, Sec. 4, Vol. 04.08, West Conshohoken, Pa.

Casagrande, A. (1948). "Classification and Identification of Soils," *Transactions*, ASCE, Vol. 113, 901–930.

Mitchell, J. K. (1976). *Fundamentals of Soil Behavior*, Wiley, New York.

Skempton, A. W. (1953). "The Colloidal Activity of Clays," *Proceedings*, 3rd International Conference on Soil Mechanics and Foundation Engineering, London, Vol. 1, 57–61.

5 Soil Compaction

5.1 Introduction

In the construction of highway embankments, earth dams, and many other engineering structures, loose soils must be compacted to increase their unit weights. Compaction increases the strength characteristics of soils, which increase the bearing capacity of foundations constructed over them. Compaction also decreases the amount of undesirable settlement of structures and increases the stability of slopes of embankments. Smooth-wheel rollers, sheepsfoot rollers, rubber-tired rollers, and vibratory rollers are generally used in the field for soil compaction. Vibratory rollers are used mostly for the densification of granular soils; and sheepsfoot rollers are effective in clay. This chapter discusses in some detail the principles of soil compaction in the laboratory and in the field.

In this chapter you will learn the following.

- The effects of moisture content and the compaction effort on the behaviour of the compacted soil.
- The moisture content versus dry unit weight plots for compaction, optimum moisture content, and maximum dry unit weight.
- Standard and modified Proctor compaction tests.
- Different types of compaction equipment for the field.
- Compaction specification and control.
- Measurements of moisture content and dry unit weight in the field.

One of the early stages of any construction activity involves earthworks where the site is cleared, cut, and filled as necessary in preparation for the proposed structure. Earthworks (Figure 1.2) in general require different types of heavy machinery. Some of the most common earth-moving equipment are shown in Figure 5.1.

5.2 Compaction—General Principles

Compaction, in general, is the densification of soil by removal of air, which requires mechanical energy. The degree of compaction of a soil is measured in terms of its dry unit weight. When water is added to the soil during compaction, it acts as a softening agent on the soil grains. The soil grains

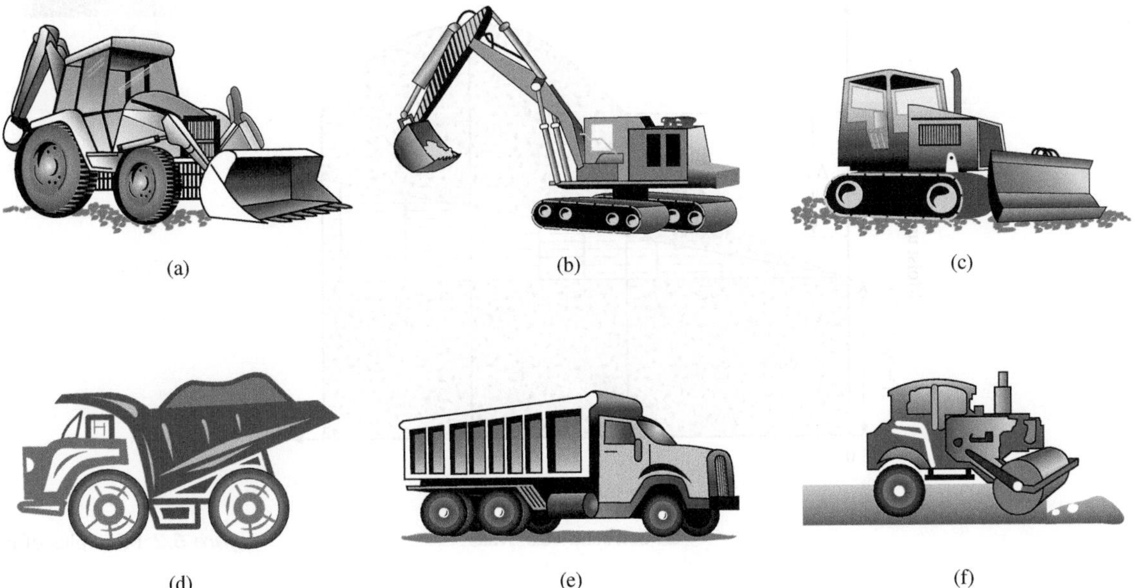

Figure 5.1 Some earthmoving machinery: (a) backhoe, (b) excavator, (c) spreader or dozer, (d) truck, (e) dump truck, and (f) roller

slip over each other and move into a densely packed position. The dry unit weight after compaction first increases as the moisture content increases. (See Figure 5.2.) Note that at a moisture content $w = 0$, the moist unit weight (γ) is equal to the dry unit weight (γ_d), or

$$\gamma = \gamma_{d(w=0)} = \gamma_1$$

When the moisture content is gradually increased and the same compactive effort is used for compaction, the weight of the soil solids in a unit volume gradually increases. For example, at $w = w_1$,

$$\gamma = \gamma_2$$

However, the dry unit weight at this moisture content is given by

$$\gamma_{d(w=w_1)} = \gamma_{d(w=0)} + \Delta\gamma_d$$

Beyond a certain moisture content $w = w_2$ (Figure 5.2), any increase in the moisture content tends to reduce the dry unit weight. This phenomenon occurs because the water takes up the spaces that would have been occupied by the solid grains. The moisture content at which the maximum dry unit weight is attained is generally referred to as the *optimum moisture content,* which is at w_2 in Figure 5.2.

The laboratory test generally used to obtain the maximum dry unit weight of compaction and the optimum moisture content is called the *Proctor compaction test* (Proctor, 1933). The procedure for conducting this type of test is described in the following section.

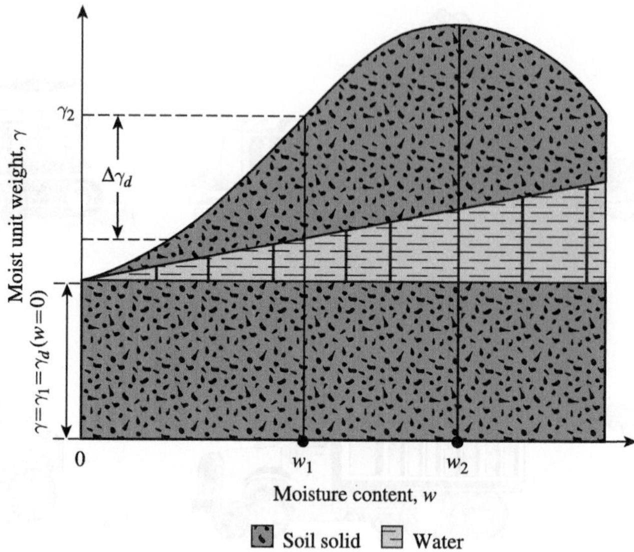

Figure 5.2 Principles of compaction

5.3 Standard Proctor Test

In the Proctor test, the soil is compacted in a mold that has a volume of $\frac{1}{30}$ ft^3 (0.00094 m^3). The diameter of the mold is 4 in. (101.6 mm). During the laboratory test, the mold is attached to a baseplate at the bottom and to an extension at the top (Figure 5.3a). The soil is mixed with varying amounts of water and then compacted in three equal layers by a hammer (Figures 5.3b and 5.3d) that delivers 25 blows to each layer. The hammer has a weight of 5.5 lb (mass = 2.50 kg) and has a drop of 12 in. (304.8 mm). Figure 5.3c is a photograph of a laboratory compaction test in progress.

For each test, the moist unit weight of compaction, γ, can be calculated as

$$\gamma = \frac{W}{V_{(m)}} \tag{5.1}$$

where

W = weight of the compacted soil in the mold
$V_{(m)}$ = volume of the mold [1/30 ft^3 (0.00094 m^3)]

For each test, the moisture content of the compacted soil is determined in the laboratory: With the known moisture content, the dry unit weight can be calculated as [See Eq. (3.12).]

$$\gamma_d = \frac{\gamma}{1 + \dfrac{w(\%)}{100}} \tag{5.2}$$

where w (%) = percent of moisture content.

The values of γ_d determined from Eq. (5.2) can be plotted against the corresponding moisture contents to obtain the maximum dry unit weight and the optimum moisture content for the soil. Figure 5.4 shows such a plot for a silty-clay soil.

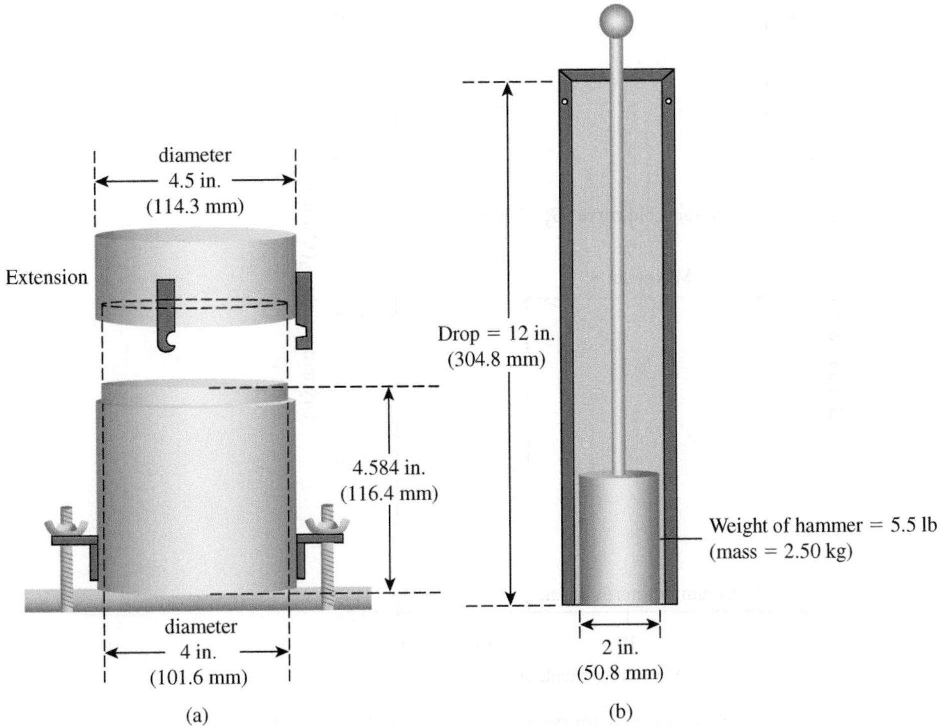

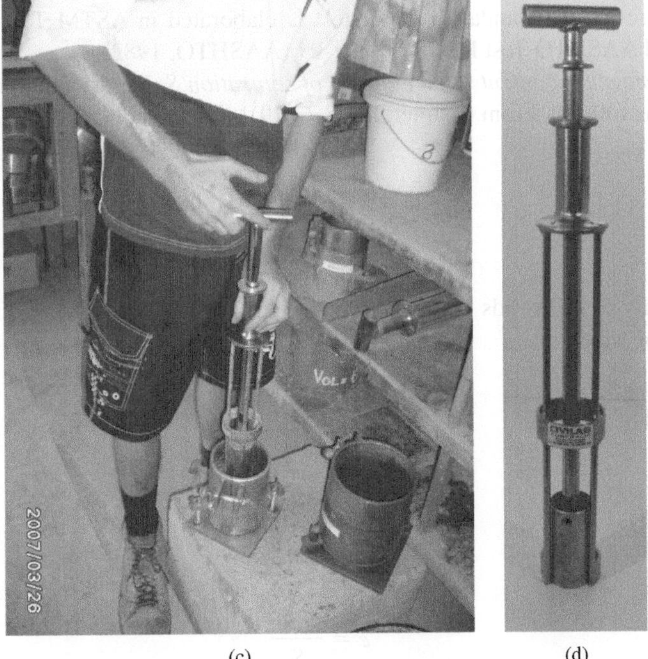

Figure 5.3 Standard Proctor compaction test: (a) mold; (b) hammer; (c) photograph of a compaction test in progress; (d) photograph of a hammer (*parts c and d – courtesy of N. Sivakugan, James Cook University, Australia*)

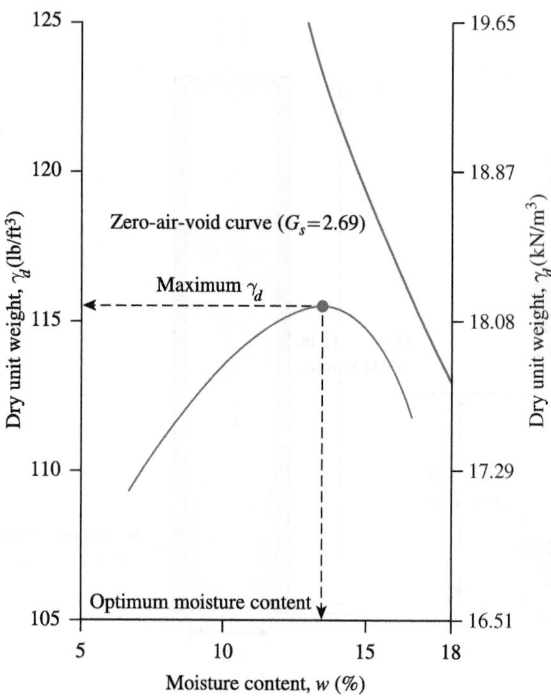

Figure 5.4 Standard Proctor compaction test results for a silty clay

The procedure for the standard Proctor test is elaborated in ASTM Test Designation D-698 (ASTM, 2014) and AASHTO Test Designation T-99 (AASHTO, 1982).

For a given *moisture content* w and *degree of saturation* S, the dry unit weight of compaction can be calculated as follows: From Chapter 3 [Eq. (3.19)], for any soil,

$$\gamma_d = \frac{G_s \gamma_w}{1 + e}$$

where

G_s = specific gravity of soil solids
γ_w = unit weight of water
e = void ratio

and, from Eq. (3.21),

$$Se = G_s w$$

or

$$e = \frac{G_s w}{S}$$

Thus,

$$\gamma_d = \frac{G_s \gamma_w}{1 + \frac{G_s w}{S}} \tag{5.3}$$

For a given moisture content, the theoretical maximum dry unit weight is obtained when no air is in the void spaces—that is, when the degree of saturation equals 100%. Hence, the maximum dry unit weight at a given moisture content with zero air voids can be obtained by substituting $S = 1$ into Eq. (5.3), or

$$\gamma_{zav} = \frac{G_s \gamma_w}{1 + wG_s} = \frac{\gamma_w}{w + \frac{1}{G_s}} \tag{5.4}$$

where γ_{zav} = zero-air-void unit weight.

To obtain the variation of γ_{zav} with moisture content, use the following procedure:

1. Determine the specific gravity of soil solids.
2. Know the unit weight of water (γ_w).
3. Assume several values of w, such as 5%, 10%, 15%, and so on.
4. Use Eq. (5.4) to calculate γ_{zav} for these values of w.

Figure 5.4 also shows the variation of γ_{zav} with moisture content and its relative location with respect to the compaction curve. Under no circumstances should any part of the compaction curve lie to the right of the zero-air-void curve.

5.4 Factors Affecting Compaction

The preceding section showed that moisture content has a strong influence on the degree of compaction achieved by a given soil. Besides moisture content, other important factors that affect compaction are soil type and compaction effort (energy per unit volume). The importance of each of these two factors is described in more detail in the following two sections.

Effect of Soil Type

The soil type—that is, grain-size distribution, shape of the soil grains, specific gravity of soil solids, and amount and type of clay minerals present—has a great influence on the maximum dry unit weight and optimum moisture content. Figure 5.5 shows typical compaction curves obtained from four soils. The laboratory tests were conducted in accordance with ASTM Test Designation D-698.

Note also that the bell-shaped compaction curve shown in Figure 5.4 is typical of most clayey soils. Figure 5.5 shows that for sands, the dry unit weight has a general tendency first to decrease as moisture content increases and then to increase to a maximum value with further increase of moisture. The initial decrease of dry unit weight with increase of moisture content can be attributed to the capillary tension effect. At lower moisture contents, the capillary tension in the pore water inhibits the tendency of the soil grains to move around and be compacted densely.

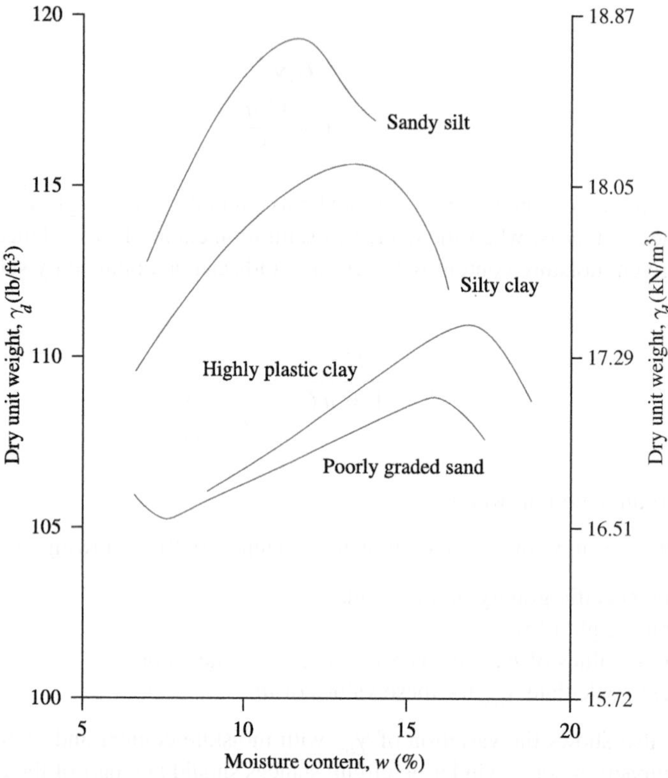

Figure 5.5 Typical compaction curves for four soils (ASTM D-698)

Lee and Suedkamp (1972) studied compaction curves for 35 soil samples. They observed that four types of compaction curves can be found. These curves are shown in Figure 5.6. Type A compaction curves are those that have a single peak. This type of curve is generally found for soils that have a liquid limit between 30 and 70. Curve type B is a one-and-one-half-peak curve, and curve type C is a double-peak curve. Compaction curves of types B and C can be found for soils that have a liquid limit less than about 30. Compaction curves of type D do not have a definite peak. They are termed *odd shaped.* Soils with a liquid limit greater than about 70 may exhibit compaction curves of type C or D. Such soils are uncommon.

Effect of Compaction Effort

The compaction energy per unit volume used for the standard Proctor test described in Section 5.3 can be given as

$$E = \frac{\begin{pmatrix}\text{Number}\\\text{of blows}\\\text{per layer}\end{pmatrix} \times \begin{pmatrix}\text{Number}\\\text{of}\\\text{layers}\end{pmatrix} \times \begin{pmatrix}\text{Weight}\\\text{of}\\\text{hammer}\end{pmatrix} \times \begin{pmatrix}\text{Height of}\\\text{drop of}\\\text{hammer}\end{pmatrix}}{\text{Volume of mold}} \tag{5.5}$$

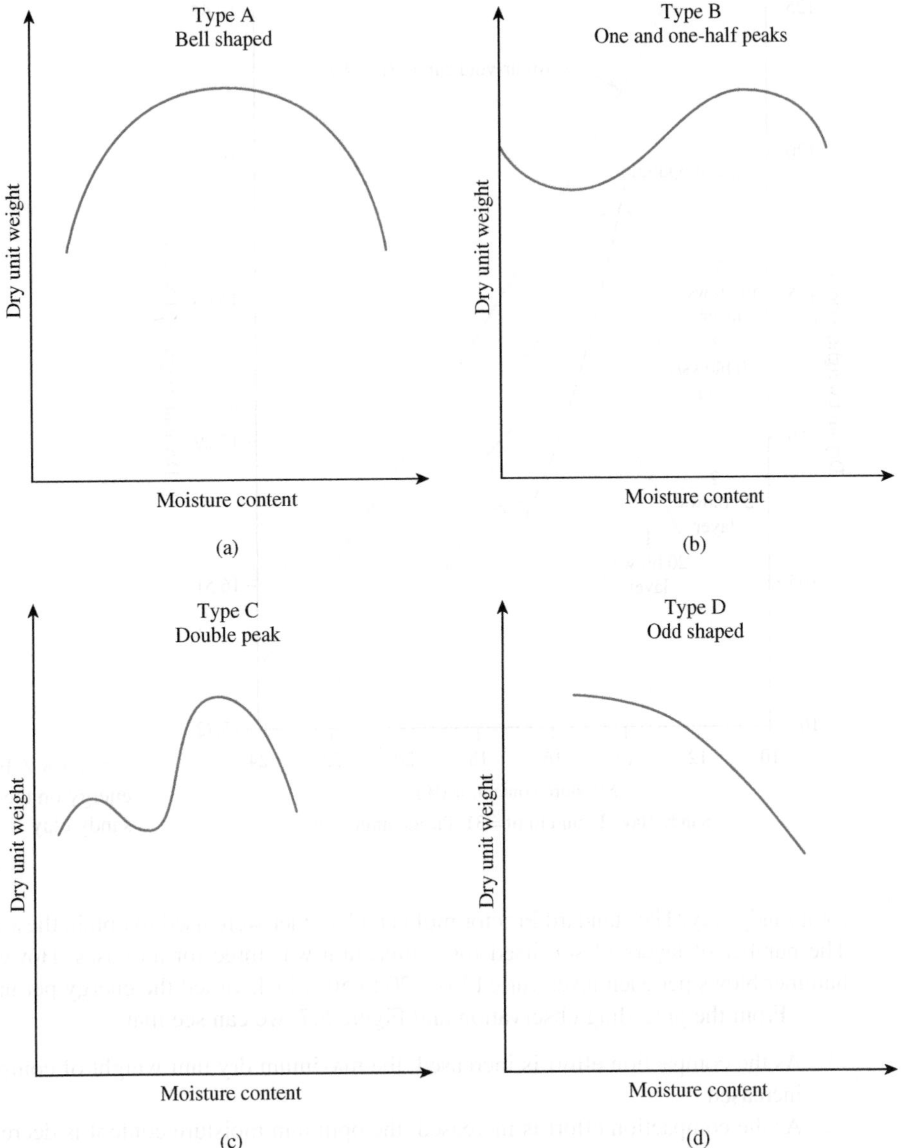

Figure 5.6 Various types of compaction curves encountered in soils

or,

$$E = \frac{(25)(3)(5.5)(1)}{\left(\frac{1}{30}\right)} = 12{,}375 \text{ ft-lb/ft}^3 = 12{,}400 \text{ ft-lb/ft}^3 \text{ (593.9 kN-m/m}^3\text{)}$$

If the compaction effort per unit volume of soil is changed, the moisture-unit weight curve also changes. This fact can be demonstrated with the aid of Figure 5.7, which shows four compaction curves

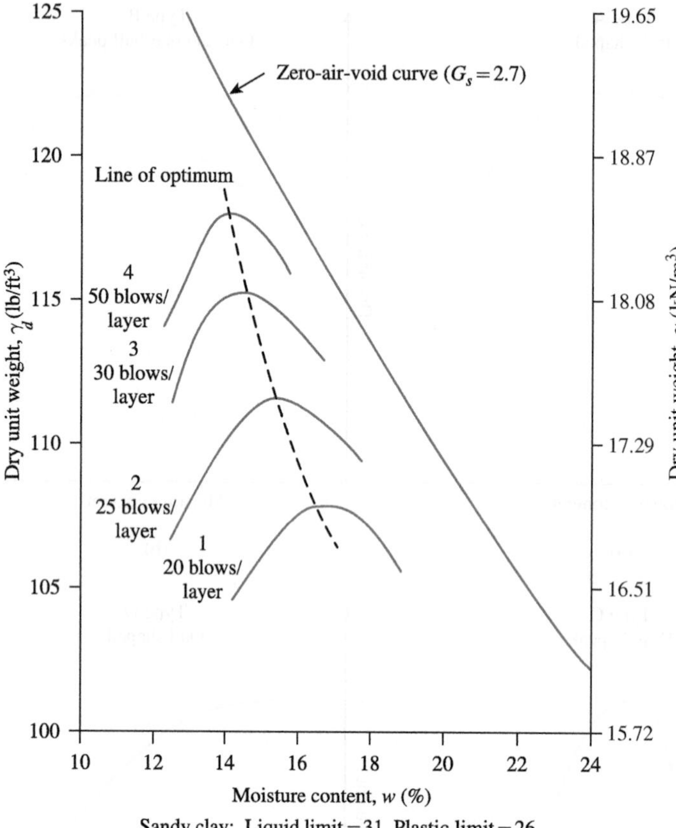

Figure 5.7 Effect of compaction energy on the compaction of a sandy clay

for a sandy clay. The standard Proctor mold and hammer were used to obtain these compaction curves. The number of layers of soil used for compaction was three for all cases. However, the number of hammer blows per each layer varied from 20 to 50, which varied the energy per unit volume.

From the preceding observation and Figure 5.7, we can see that

1. As the compaction effort is increased, the maximum dry unit weight of compaction is also increased.
2. As the compaction effort is increased, the optimum moisture content is decreased to some extent.

The preceding statements are true for all soils. Note, however, that the degree of compaction is not directly proportional to the compaction effort.

5.5 Modified Proctor Test

With the development of heavy rollers and their use in field compaction, the standard Proctor test was modified to better represent field conditions. This revised version sometimes is referred to as the *modified Proctor test* (ASTM Test Designation D-1557 and AASHTO Test Designation T-180). For conducting the modified Proctor test, the same mold is used with a volume of 1/30 ft^3 (0.00094 m^3), as in the case of the standard Proctor test. However, the soil is compacted in five layers by a hammer that has a weight of 10 lb (4.54 kg mass). The drop of the hammer is 18 in. (457.2 mm). The number of hammer blows for each layer is kept at 25 as in the case of the standard Proctor test.

The compaction energy for this type of compaction test can be calculated as 56,000 ft-lb/lb^3 (2687.5 kN-m/m^3).

Because it increases the compactive effort, the modified Proctor test results in an increase in the maximum dry unit weight of the soil. The increase in the maximum dry unit weight is accompanied by a decrease in the optimum moisture content.

In the preceding discussions, the specifications given for Proctor tests adopted by ASTM and AASHTO regarding the volume of the mold and the number of blows are generally those adopted for fine-grained soils that pass through the U.S. No. 4 sieve. However, under each test designation, there are three suggested methods that reflect the mold size, the number of blows per layer, and the maximum particle size in a soil aggregate used for testing. A summary of the test methods is given in Table 5.1.

Table 5.1 Summary of Standard and Modified Proctor Compaction Test Specifications (ASTM D-698 and D-1557)

	Description	Method A	Method B	Method C
Physical data for the tests	Material	Passing No. 4 sieve	Passing $\frac{3}{8}$ in. (9.5 mm) sieve	Passing $\frac{3}{4}$ in. (19.1 mm) sieve
	Use	Used if 20% or less by weight of material is retained on No. 4 (4.75 mm) sieve	Used if more than 20% by weight of material is retained on No. 4 (4.75 mm) sieve and 20% or less by weight of material is retained on $\frac{3}{8}$ in. (9.5 mm) sieve	Used if more than 20% by weight of material is retained on $\frac{3}{8}$ in. (9.5 mm) sieve and less than 30% by weight of material is retained on $\frac{3}{4}$ in. (19.1 mm) sieve
	Mold volume	$\frac{1}{30}$ ft^3 (0.00094 m^3)	$\frac{1}{30}$ ft^3 (0.00094 m^3)	$\frac{1}{30}$ ft^3 (0.00094 m^3)
	Mold diameter	4 in. (101.6 mm)	4 in. (101.6 mm)	4 in. (101.6 mm)
	Mold height	4.584 in. (116.4 mm)	4.584 in. (116.4 mm)	4.584 in. (116.4 mm)
Standard Proctor test	Weight of hammer	5.5 lb (2.5 kg mass)	5.5 lb (2.5 kg mass)	5.5 lb (2.5 kg mass)
	Height of drop	12 in. (304.8 mm)	12 in. (304.8 mm)	12 in. (304.8 mm)
	Number of soil layers	3	3	3
	Number of blows/layer	25	25	56

(continued)

Table 5.1 (*continued*)

	Description	Method A	Method B	Method C
Modified Proctor test	Weight of hammer	10 lb (4.54 kg mass)	10 lb (4.54 kg mass)	10 lb (4.54 kg mass)
	Height of drop	18 in. (457.2 mm)	18 in. (457.2 mm)	18 in. (457.2 mm)
	Number of soil layers	5	5	5
	Number of blows/layer	25	25	56

Example 5.1

The laboratory test data for a standard Proctor test are given below. Find the maximum dry unit weight and optimum moisture content.

Volume of Proctor mold (ft^3)	Weight of wet soil in mold (lb)	Moisture content (%)
$\frac{1}{30}$	3.88	12
$\frac{1}{30}$	4.09	14
$\frac{1}{30}$	4.23	16
$\frac{1}{30}$	4.28	18
$\frac{1}{30}$	4.24	20
$\frac{1}{30}$	4.19	22

Solution

The following table can now be prepared:

Volume, V (ft^3)	Weight of wet soil, W (lb)	Moist unit weight, γ[a] (lb/ft^3)	Moisture content, w (%)	Dry unit weight, γ_d[b] (lb/ft^3)
$\frac{1}{30}$	3.88	116.4	12	103.9
$\frac{1}{30}$	4.09	122.7	14	107.6
$\frac{1}{30}$	4.23	126.9	16	109.4
$\frac{1}{30}$	4.28	128.4	18	108.8
$\frac{1}{30}$	4.24	127.2	20	106.0
$\frac{1}{30}$	4.19	125.7	22	103.0

[a] $\gamma = \dfrac{W}{V}$

[b] $\gamma_d = \dfrac{\gamma}{1 + \dfrac{w\%}{100}}$

The plot of γ_d against w is shown in Figure 5.8. From the graph

Maximum dry unit weight = **109.5 lb/ft³**

Optimum moisture content = **16.3%**.

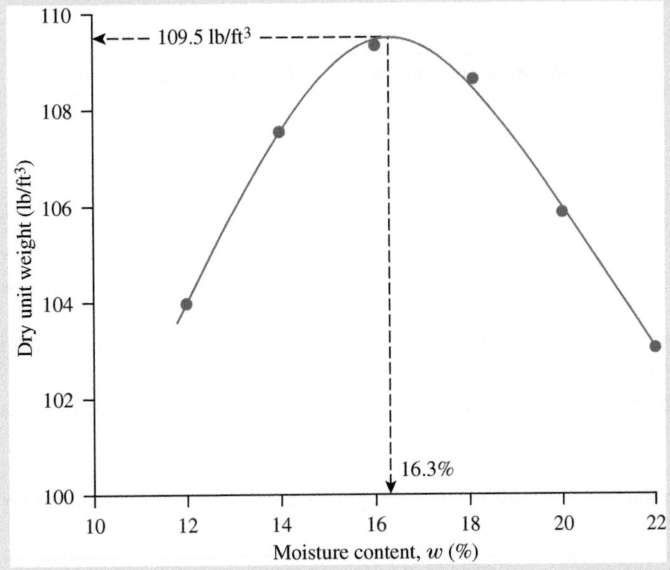

Figure 5.8

Example 5.2

Calculate the zero-air-void unit weights (in lb/ft³) for a soil with $G_s = 2.68$ at moisture contents of 5, 10, 15, 20, and 25%.

Solution

From Eq. (5.4),

$$\gamma_{zav} = \frac{\gamma_w}{w + \dfrac{1}{G_s}}$$

$\gamma_w = 62.4$ lb/ft³; $G_s = 2.68$.

Refer to the following table:

w (%)	γ_{zav} (lb/ft³)
5	**147.47**
10	**131.89**
15	**119.28**
20	**108.88**
25	**100.14**

Example 5.3

A modified Proctor compaction test was carried out using a cylindrical mold with a 1000 ml volume. The moisture content and the mass of the moist specimen extruded from the mold are summarized here.

Moisture content (%)	Mass of wet specimen (g)
10.5	2033.2
11.8	2146.6
13.0	2203.5
14.9	2183.1
16.4	2147.6
17.6	2116.8

Plot the dry unit weight versus moisture content, and find the maximum dry unit weight and the optimum moisture content.

If $G_s = 2.75$, plot the zero air void curve along with the above compaction curve.

Solution

From the given data, moist and dry unit weights are calculated and are summarized as follows.

Moisture content (%)	Weight of wet specimen (g)	Moist unit weight γ (kN/m^3)	Dry unit weight γ_d (kN/m^3)
10.5	2033.2	19.95	18.05
11.8	2146.6	21.06	18.84
13.0	2203.5	21.62	19.13
14.9	2183.1	21.42	18.64
16.4	2147.6	21.07	18.10
17.6	2116.8	20.77	17.66

For the zero air void curve, γ_{zav} values are calculated at four different moisture contents as follows.

Moisture content (%)	γ_{zav} (kN/m^3)
13	19.87
14	19.48
16	18.73
18	18.05

■

The plot of dry unit weight versus moisture content is shown in Figure 5.9. It can be seen from the figure that

$$\text{Maximum dry unit weight} = \mathbf{19.15\ kN/m^3}$$

$$\text{Optimum moisture content} = \mathbf{12.9\%}$$

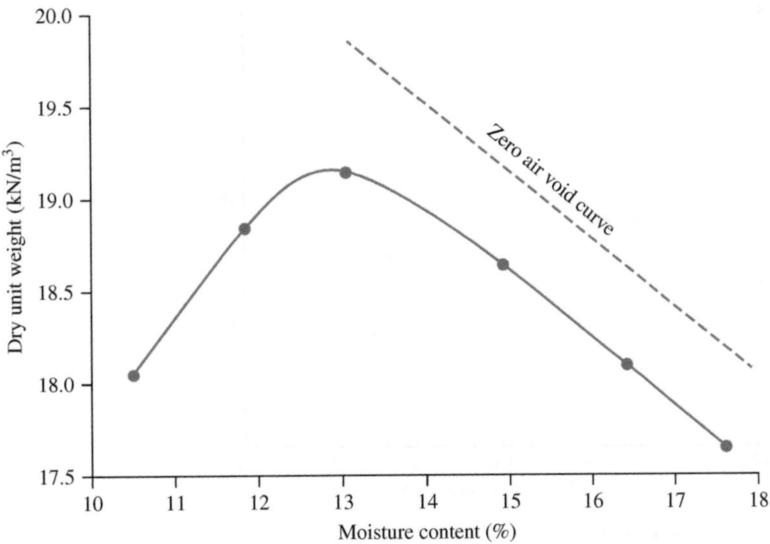

Figure 5.9 Dry unit weight versus moisture content for Example 5.3

5.6 Effect of Compaction on Cohesive Soil Properties

Compaction induces variations in the structure of cohesive soils. Clays compacted at a moisture content less than the optimum moisture content will have a flocculated fabric (i.e., grains randomly oriented in all directions), and when the clay is compacted at a moisture contents greater that the optimum moisture content, the clay fabric becomes more oriented (i.e., grains are oriented in parallel, generally perpendicular to the direction of compaction). Results of these structural variations include changes in hydraulic conductivity, compressibility, and strength. Figure 5.10 shows the results of permeability tests (discussed in Chapter 6) on a sandy clay. The samples used for the tests were compacted at various moisture contents by the same compactive effort. The hydraulic

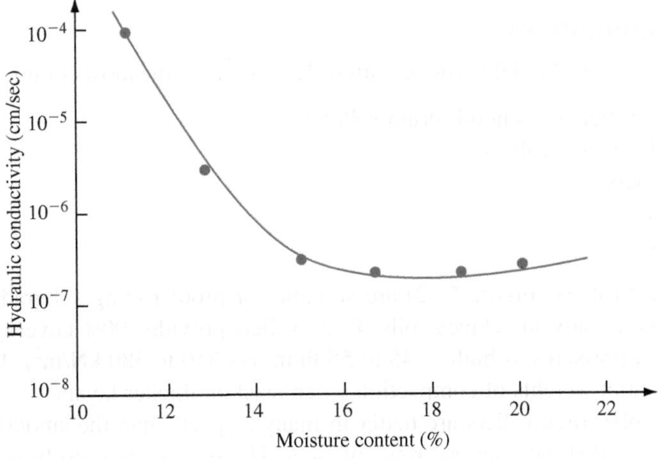

Figure 5.10 Permeability test results on sandy clay

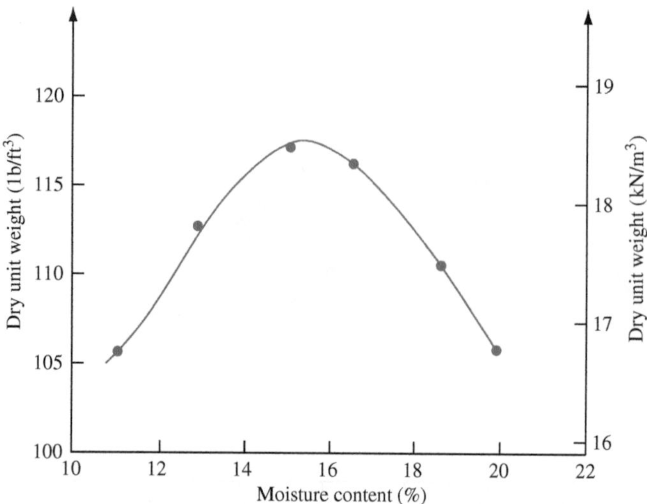

Figure 5.10 *(continued)*

conductivity, which is a measure of how easily water flows through soil, decreases with the increase of moisture content. It reaches a minimum value at approximately the optimum moisture content. Beyond the optimum moisture content, the hydraulic conductivity increases slightly.

The strength of compacted clayey soils (discussed in Chapter 9) generally decreases with the molding moisture content. This is shown in Figure 5.11. Note that at approximately optimum moisture content there is a great loss of strength. This means that, if two specimens are compacted to the same dry unit weight, one of them on the dry side of the optimum and the other on the wet side of the optimum, the specimen compacted on the dry side of the optimum (that is, with flocculent structure) will exhibit greater strength.

5.7 Field Compaction

Compaction Equipment

Most of the compaction in the field is done with rollers. Some of the most common types of rollers are

1. Smooth-wheel rollers (or smooth-drum rollers)
2. Pneumatic rubber-tired rollers
3. Sheepsfoot rollers
4. Vibratory rollers
5. Impact rollers

Smooth-wheel rollers (Figure 5.12) are suitable for proof rolling subgrades and for finishing operation of fills with sandy and clayey soils. These rollers provide 100% coverage under the wheels, with ground contact pressures as high as 45 to 55 lb/in^2 ($\approx$ 310 to 380 kN/m^2). They are not suitable for producing high unit weights of compaction when used on thicker layers.

Pneumatic rubber-tired rollers are better in many respects than the smooth-wheel rollers. The former are heavily loaded with several rows of tires. These tires are closely spaced—four to six in a row. The contact pressure under the tires can range from 85 to 100 lb/in^2 (585 to 690 kN/m^2), and

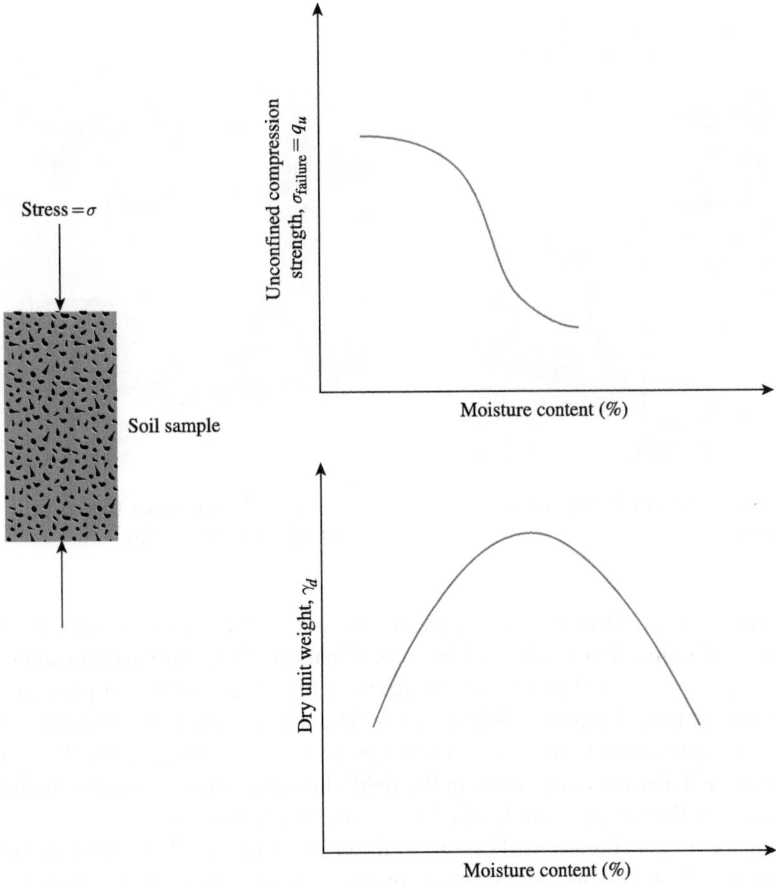

Figure 5.11 Effect of compaction on the strength of clayey soils

Figure 5.12 Smooth wheeled roller (*Courtesy of N. Sivakugan, James Cook University, Australia*)

Figure 5.13 Sheepsfoot roller (*Courtesy of N. Sivakugan, James Cook University, Australia*)

Figure 5.14 Impact roller (*Courtesy of N. Sivakugan, James Cook University, Australia*)

they produce about 70 to 80% coverage. Pneumatic rollers can be used for sandy and clayey soil compaction. Compaction is achieved by a combination of pressure and kneading action.

Sheepsfoot rollers (Figure 5.13) are drums with a large number of projections. The area of each projection may range from 4 to 13 in^2 ($\approx$ 26 to 84 cm^2). These rollers are most effective in compacting clayey soils. The contact pressure under the projections can range from 200 to 1000 lb/in^2 ($\approx$ 1380 to 6900 kN/m^2). During compaction in the field, the initial passes compact the lower portion of a lift. Compaction at the top and middle of a lift is done at a later stage.

Impact rollers (Figure 5.14) come in different shapes as three, four, or five sided rollers. The one in Figure 5.14 has three sides, which produce an impact every time they are in contact with the soil, and impart very high compaction energy to the underlying ground.

Vibratory rollers are extremely efficient in compacting granular soils. Vibrators can be attached to smooth-wheel, pneumatic rubber-tired, or sheepsfoot rollers to provide vibratory effects to the soil. Figure 5.15 demonstrates the principles of vibratory rollers. The vibration is produced by rotating off-center weights.

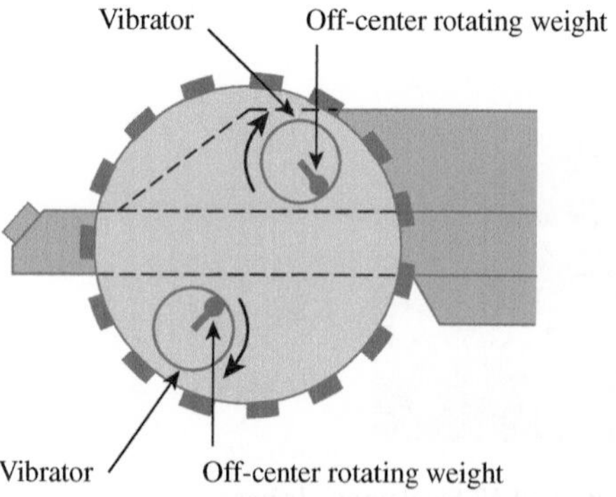

Figure 5.15 Principles of vibratory rollers

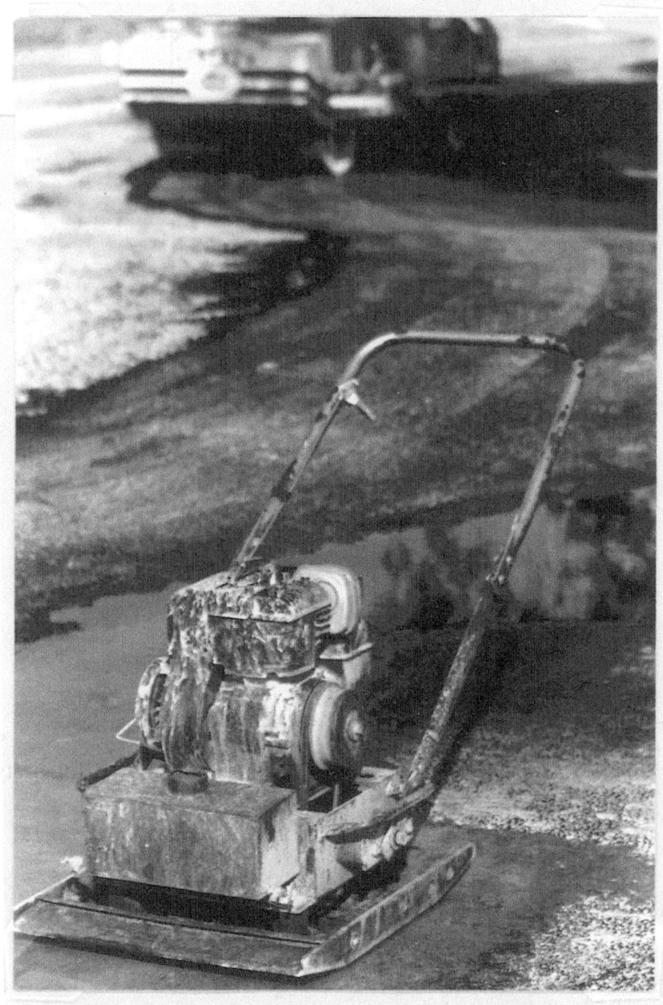

Figure 5.16 Hand-held vibratory plate (*Photo courtesy of Braja Das*)

Hand-held vibrating plates (Figure 5.16) can be used for effective compaction of granular soils over a limited area. Vibrating plates are also gang-mounted on machines. These plates can be used in less restricted areas.

In addition to soil type and moisture content, other factors must be considered to achieve the desired unit weight of compaction in the field. These factors include the thickness of lift, the intensity of pressure applied by the compacting equipment, and the area over which the pressure is applied. These factors are important because the pressure applied at the surface decreases with depth, which results in a decrease in the degree of soil compaction. During compaction, the dry unit weight of soil also is affected by the number of roller passes. Figure 5.17 shows the growth curves for a silty clay soil. The dry unit weight of a soil at a given moisture content increases to a certain point with the number of roller passes. Beyond this point, it remains approximately constant. In most cases, about 10 to 15 roller passes yield the maximum dry unit weight economically attainable.

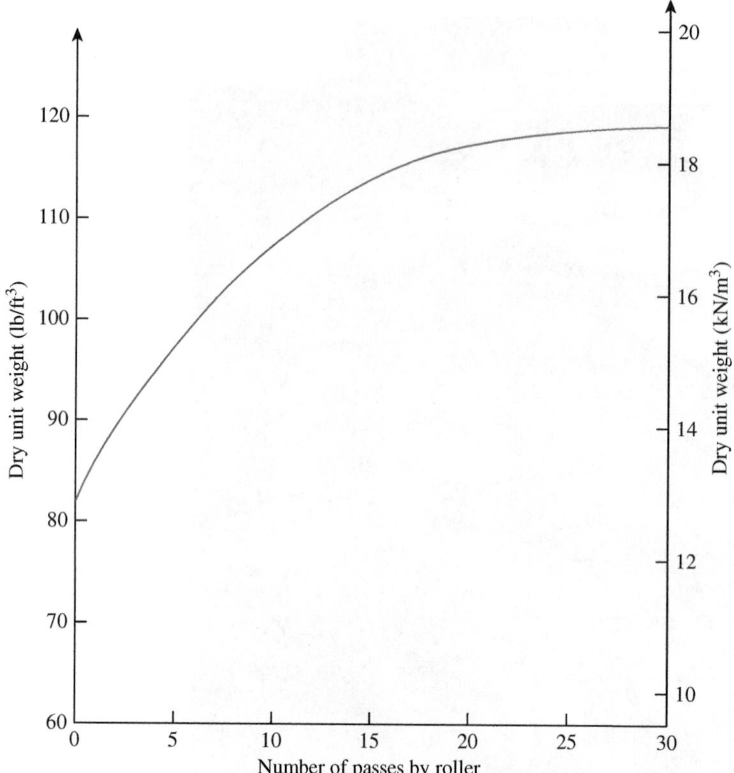

Figure 5.17 Growth curve for a silty clay soil

5.8 Specifications for Field Compaction

In most specifications for earthwork, the contractor is instructed to achieve a compacted field dry unit weight of 90 to 95% of the maximum dry unit weight determined in the laboratory by either the standard or modified Proctor test. This is a specification for relative compaction, which can be expressed as

$$R(\%) = \frac{\gamma_{d(\text{field})}}{\gamma_{d(\text{max}-\text{lab})}} \times 100 \tag{5.6}$$

where R = relative compaction.

For the compaction of granular soils, specifications sometimes are written in terms of the required relative density D_r or the required relative compaction. Relative density should not be confused with relative compaction. From Chapter 3, we can write

$$D_r = \left[\frac{\gamma_{d(\text{field})} - \gamma_{d(\text{min})}}{\gamma_{d(\text{max})} - \gamma_{d(\text{min})}}\right]\left[\frac{\gamma_{d(\text{max})}}{\gamma_{d(\text{field})}}\right] \tag{5.7}$$

Comparing Eqs. (5.6) and (5.7), we see that

$$R = \frac{R_0}{1 - D_r(1 - R_0)} \tag{5.8}$$

where

$$R_0 = \frac{\gamma_{d(\min)}}{\gamma_{d(\max)}} \tag{5.9}$$

On the basis of observation of 47 soil samples, Lee and Singh (1971) devised a correlation between R and D_r for granular soils:

$$R = 80 + 0.2D_r \tag{5.10}$$

The specification for field compaction based on relative compaction or on relative density is an end-product specification. The contractor is expected to achieve a minimum dry unit weight regardless of the field procedure adopted. The most economical compaction condition can be explained with the aid of Figure 5.18. The compaction curves A, B, and C are for the same soil with varying compactive effort. Let curve A represent the conditions of maximum compactive effort that can be obtained from the existing equipment. Let the contractor be required to achieve a minimum dry unit weight of $\gamma_{d(\text{field})} = R\gamma_{d(\max)}$. To achieve this, the contractor must ensure that the moisture content w falls between w_1 and w_2. As can be seen from compaction curve C, the required $\gamma_{d(\text{field})}$ can be achieved with a lower compactive effort at a moisture content $w = w_3$. However, for most practical conditions, a compacted field unit weight of $\gamma_{d(\text{field})} = R\gamma_{d(\max)}$ cannot be achieved by the minimum compactive effort. Hence, equipment with slightly more than the minimum compactive effort should be used. The compaction curve B represents this condition. Now we can see from Figure 5.18 that the

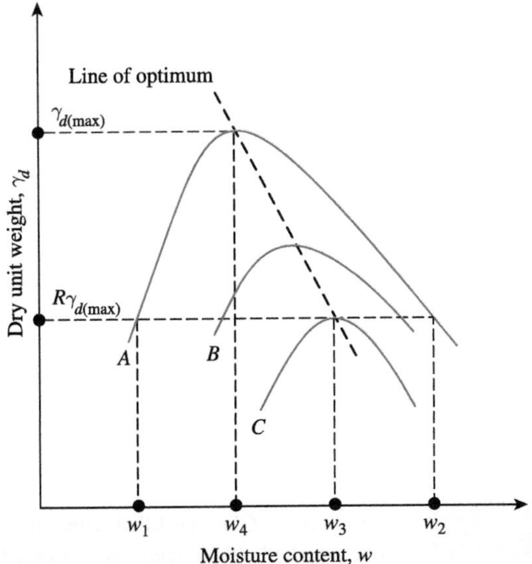

Figure 5.18 Most economical compaction condition

most economical moisture content is between w_3 and w_4. Note that $w = w_4$ is the optimum moisture content for curve A, which is for the maximum compactive effort.

The concept described in the preceding paragraph, along with Figure 5.18, is attributed historically to Seed (1964), who was a legend in modern geotechnical engineering. This concept is elaborated on in more detail in Holtz and Kovacs (1981).

5.9 Determination of Field Unit Weight of Compaction

When the compaction work is progressing in the field, knowing whether the specified unit weight has been achieved is useful. The standard procedures for determining the field unit weight of compaction include

1. Sand cone method
2. Rubber balloon method
3. Nuclear method

Following is a brief description of each of these methods.

Sand Cone Method (ASTM Designation D-1556)

The sand cone device consists of a glass or plastic jar with a metal cone attached at its top (Figure 5.19). The jar is filled with uniform dry Ottawa sand. The combined weight of the jar, the cone, and the sand filling the jar is determined (W_1). In the field, a small hole is excavated in the area where the soil has been compacted. If the weight of the moist soil excavated from the hole (W_2) is

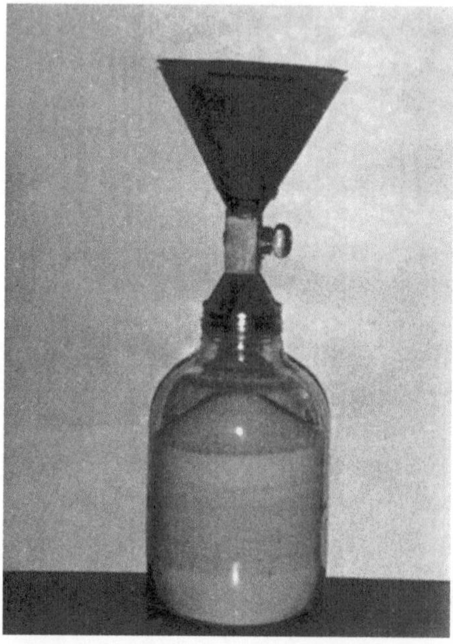

Figure 5.19 Glass jar filled with Ottawa sand with sand cone attached (*Photo courtesy of Braja Das*)

determined and the moisture content of the excavated soil is known, the dry weight of the soil can be obtained as

$$W_3 = \frac{W_2}{1 + \frac{w(\%)}{100}} \tag{5.11}$$

where w = moisture content.

After excavation of the hole, the cone with the sand-filled jar attached to it is inverted and placed over the hole (Figure 5.20). Sand is allowed to flow out of the jar to fill the hole and the cone. After that, the combined weight of the jar, the cone, and the remaining sand in the jar is determined (W_4), so

$$W_5 = W_1 - W_4 \tag{5.12}$$

where W_5 = weight of sand to fill the hole and cone.

The volume of the excavated hole can then be determined as

$$V = \frac{W_5 - W_c}{\gamma_{d(\text{sand})}} \tag{5.13}$$

where

W_c = weight of sand to fill the cone only
$\gamma_{d(\text{sand})}$ = dry unit weight of Ottawa sand used

Figure 5.20 Field unit weight determined by sand cone method: (a) schematic diagram; (b) a test in progress in the field (*Photo courtesy of Braja Das*)

The values of W_c and γ_d(sand) are determined from the calibration done in the laboratory. The dry unit weight of compaction made in the field can then be determined as follows:

$$\gamma_d = \frac{\text{Dry weight of the soil excavated from the hole}}{\text{Volume of the hole}} = \frac{W_3}{V} \tag{5.14}$$

Rubber Balloon Method (ASTM Designation D-2167)

The procedure for the rubber balloon method is similar to that for the sand cone method; a test hole is made and the moist weight of soil removed from the hole and its moisture content are determined. However, the volume of the hole is determined by introducing into it a rubber balloon filled with water from a calibrated vessel, from which the volume can be read directly. The dry unit weight of the compacted soil can be determined by using Eq. (5.14). Figure 5.21 shows a calibrated vessel that would be used with a rubber balloon.

Nuclear Method

Nuclear density meters are often used for determining the compacted dry unit weight of soil. The density meters operate either in drilled holes or from the ground surface. The instrument measures the weight of wet soil per unit volume and the weight of water present in a unit volume of soil. The dry unit weight of compacted soil can be determined by subtracting the weight of water from the moist unit weight of soil. Figure 5.22 shows a photograph of a nuclear density meter.

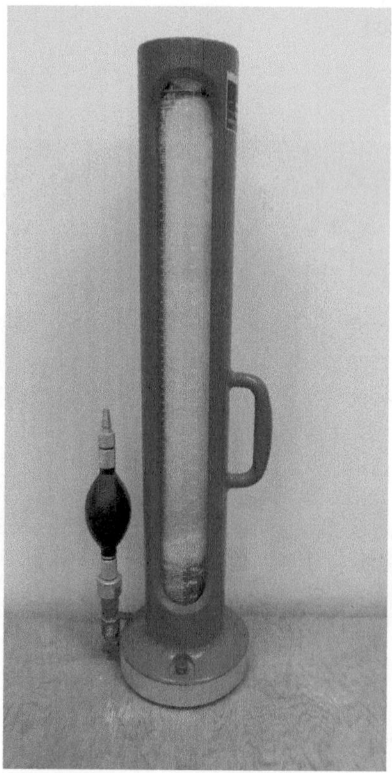

Figure 5.21 Calibrated vessel for the rubber-balloon method for determination of field unit weight (*Courtesy of Braja M. Das*)

Figure 5.22 Nuclear density meter (*Courtesy of N. Sivakugan, James Cook University, Australia*)

Example 5.4

Following are the results of a field unit weight determination test using the sand cone method:

- Calibrated dry unit weight of Ottawa sand = 104 lb/ft^3
- Weight of Ottawa sand to fill the cone = 0.258 lb
- Weight of jar + cone + sand (before use) = 13.21 lb
- Weight of jar + cone + sand (after use) = 6.2 lb
- Weight of moist soil from hole = 7.3 lb
- Moisture content of moist soil = 11.6%

Determine the dry unit weight of compaction in the field.

Solution

The weight of the sand needed to fill the hole and cone is

$$13.21 - 6.2 = 7.01 \text{ lb}$$

The weight of the sand used to fill the hole is

$$7.01 - 0.258 = 6.752 \text{ lb}$$

So the volume of the hole is

$$V = \frac{6.752}{\text{dry unit weight of Ottawa sand}} = \frac{6.752}{104} = 0.0649 \text{ ft}^3$$

From Eq. (5.11), the dry weight of soil from the field is

$$W_3 = \frac{W_2}{1 + \frac{w(\%)}{100}} = \frac{7.3}{1 + \frac{11.6}{100}} = 6.54 \text{ lb}$$

Hence, the dry unit weight of compaction is

$$\gamma_d = \frac{W_3}{V} = \frac{6.54}{0.0649} = \mathbf{100.77\ lb/ft^3}$$

■

5.10 Summary

The early stage of most civil engineering projects involve earthworks, which includes site preparation. Often, the current soil conditions are not favourable in their present form and would require some form of ground improvement. Compaction is one of the simplest forms of ground improvement.

Water is added to a soil when it is compacted. It acts as a lubricant, making the compaction more effective. Under a certain compaction energy, the densest possible packing is obtained at a specific moisture content, known as the optimum moisture content, which gives the maximum dry unit weight. Developing the moisture–density relationship using standard Proctor and modified Proctor laboratory compaction tests were discussed.

There are different types of rollers (e.g., sheepsfoot roller) to compact the different soil types. The engineering behaviour of the compacted soil is sensitive to the moisture content; therefore, it is necessary to ensure that the compaction is carried out correctly in the field. This is ensured by specifying a minimum allowable dry unit weight and a tight range for the moisture content, which are checked randomly on the compacted soil.

Problems

5.1 State whether the following are true or false.
   a. The optimum moisture content increases with increasing compaction effort.
   b. Sheepsfoot rollers are suitable for compacting sandy soils.
   c. Relative compaction cannot exceed 100%.
   d. Modified Proctor compaction effort is greater the standard Proctor one.
   e. Strength of a compacted clay is greater when compacted wet of optimum than dry of optimum.

5.2 It is proposed to carry out a laboratory compaction test on a 2.5 kg soil sample, currently with a moisture content of 10.0%. How much water should be added to raise the moisture content to 16.0%?

5.3 The laboratory test data for a standard Proctor test are given below. Find the maximum dry unit weight and optimum moisture content.

Volume of Proctor mold (ft^3)	Weight of wet soil in the mold (lb)	Moisture content (%)
1/30	3.88	12
1/30	4.09	14
1/30	4.23	16
1/30	4.28	18
1/30	4.24	20
1/30	4.19	22

5.4 Calculate the zero-air-void unit weights for a soil with $G_s = 2.6$ at moisture contents of 5, 10, 15, 20, and 25%. Plot a graph of γ_{zav} vs. moisture content.

5.5 Repeat Problem 5.4 for $G_s = 2.76$.

5.6 Repeat Problem 5.3 with the following laboratory test results.

Volume of Proctor mold (ft^3)	Weight of wet soil in the mold (lb)	Moisture content (%)
1/30	3.69	12
1/30	3.82	14
1/30	3.88	16
1/30	3.87	18
1/30	3.81	20
1/30	3.77	21

5.7 A modified Proctor compaction test was carried out on a clayey sand in a cylindrical mold that has a volume of 1000 ml. The specific gravity of the soil grains is 2.68. The moisture content and the mass of the six compacted specimens are given here.

Moisture content (%)	5.0	7.0	9.5	11.8	14.1	17.0
Mass of specimen (g)	1850	1970	2090	2110	2090	2060

a. Plot the compaction test data and determine the optimum moisture content and the maximum dry unit weight.
b. Plot the void ratio and the degree of saturation against the moisture content.
c. What are the void ratio and degree of saturation at the optimum moisture content?

5.8 The maximum and minimum dry unit weights of a sand were determined in the laboratory to be 116.5 lb/ft^3 and 97 lb/ft^3. What would be the relative compaction in the field if the relative density is 64%?

5.9 Following are the results of a field unit weight determination test by using the sand cone method:

a. Calibrated dry unit weight of Ottawa sand = 97.97 lb/ft^3
b. Calibrated weight of Ottawa sand to fill the cone = 1.202 lb
c. Weight of jar + cone + sand (before use) = 16.733 lb
d. Weight of jar + cone + sand (after use) = 10.538 lb
e. Weight of moist soil from hole = 6.629 lb
f. Moisture content of moist soil = 10.2%

Determine the dry unit weight of compaction of the soil in the field.

5.10 Figure 5.23 shows the moisture content–dry unit weight relationship established from a modified Proctor compaction test for a compacted clay embankment section. The specific gravity of the soil grains is 2.72. The compaction specification requires that the clay be compacted in the field to relative compaction of 95% with respect to modified Proctor compaction effort and that the moisture content lies in the range of 16.0% to 19.0%.

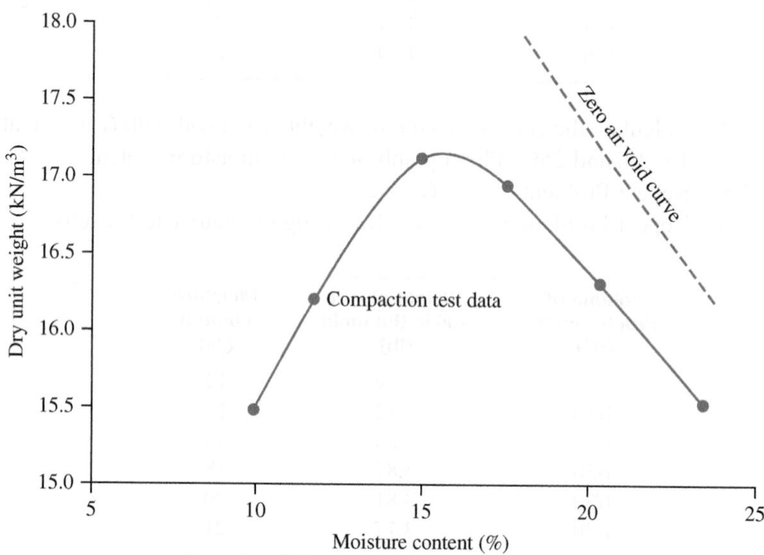

Figure 5.23

Once the clay is compacted in the field, control tests were carried out at four different locations using nuclear density meter, and the moisture content and dry unit weight were measured as follows.

Control test number	1	2	3	4
Moisture content %	13.1	17.2	19.5	18.1
Dry unit weight (kN/m^3)	17.2	16.5	17.9	17.1

Check whether the field compaction meets the specifications at the four locations. Discuss the control test data.

5.11 The modified Proctor compaction test curve is shown in Figure 5.24. The specific gravity of the soil grains is 2.75. A compacted subgrade for roadwork is made from this soil and the specification require that:

a. The moisture content be within ± 1.5% of the optimum moisture content, and
b. The relative compaction, with respect to modified Proctor compaction effort, is greater than 95%.

A field density test was carried out to check whether the compaction meets the specifications. A hole was dug in the compacted ground, and the volume was determined to be 1015 cm^3. The soil removed from the hole has a mass of 2083 g. The entire soil was dried in the oven, weighing 1845 g. Check whether the compaction meets the specifications.

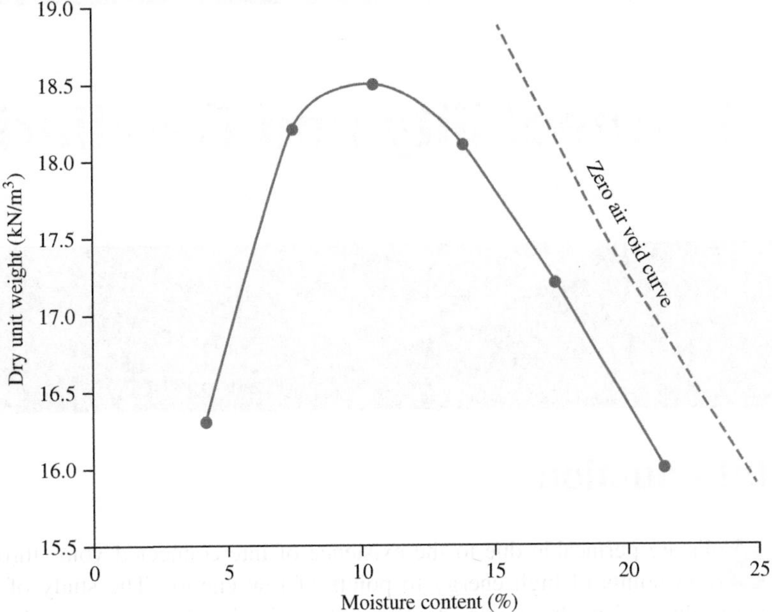

Figure 5.24

References

American Association of State Highway and Transportation Officials (1982). *AASHTO Materials, Part II*, Washington, D.C.

American Society for Testing and Materials (2014). *ASTM Standards*, Vol 04.08, West Conshohocken, Pa.

Holtz, R. D. and Kovacs, W. D. (1981). *An Introduction to Geotechnical Engineering*, Prentice-Hall, Englewood Cliffs, N.J.

Lee, K. W. and Singh, A. (1971). "Relative Density and Relative Compaction," *Journal of the Soil Mechanics and Foundations Division*, ASCE, Vol. 97, No. SM7, 1049–1052.

Lee, P. Y. and Suedkamp, R. J. (1972). "Characteristics of Irregularly Shaped Compaction Curves of Soils," *Highway Research Record No. 381*, National Academy of Sciences, Washington, D.C., 1–9.

Proctor, R. R. (1933). "Design and Construction of Rolled Earth Dams," *Engineering News Record*, Vol. 3, 245–248, 286–289, 348–351, 372–376.

Seed, H. B. (1964). Lecture Notes, CE 271, Seepage and Earth Dam Design, University of California, Berkeley.

6 Permeability and Capillarity

6.1 Introduction

Soils are permeable due to the existence of interconnected voids through which water can flow from points of high energy to points of low energy. The study of the flow of water through permeable soil media is important in soil mechanics. It is necessary for estimating the quantity of underground seepage under various hydraulic conditions, for investigating problems involving the pumping of water for underground construction (e.g., dewatering), and for making stability analyses of earth dams and earth-retaining structures that are subject to seepage forces. In this chapter, you will learn the following.

- Hydraulic-conductivity (or permeability) of a soil as a measure of how easily water can flow through soils
- Relevance of hydraulic conductivity to geotechnical engineering applications
- Typical values and empirical correlations for estimating the hydraulic conductivity
- Laboratory and field tests for measuring the hydraulic conductivity of soils
- Capillary rise in soils

6.2 Darcy's Law

Figure 6.1 shows the flow of water through a soil specimen. Open standpipes called piezometers are installed at points *A* and *B*. The pressure heads at *A* and *B* (which are the heights of the vertical columns of water in the piezometers) are p_A and p_B, respectively. According to Bernoulli's equation in fluid mechanics, the energy of a water molecule can be expressed in terms of head (with unit of length) as

$$\text{Total head} = \text{Elevation head} + \text{Pressure head} + \text{Velocity head}$$

For flow through soils, velocity of flow is very small, hence, the velocity head is negligible. Therefore, the total head at a point can be written as

$$\text{Total head } (h) = \text{Elevation head } (z) + \text{Pressure head } (p) \tag{6.1}$$

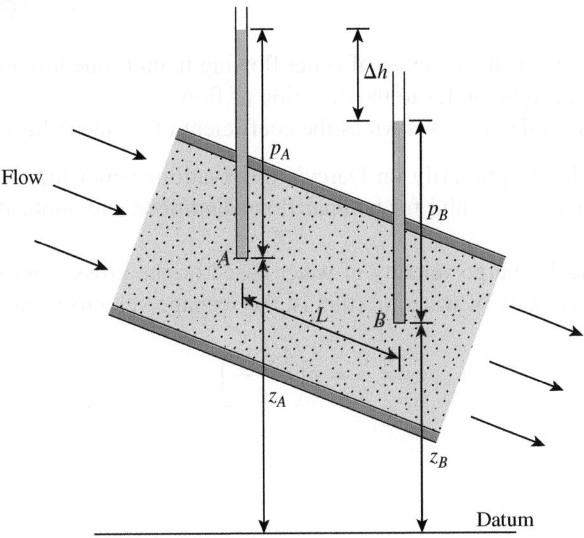

Figure 6.1 Flow of water through a soil specimen

Here, elevation head z is the height of the point above the selected datum. Depending on the datum selected, the elevation head and hence the total head at a point will vary. Pressure head p, as measured from a piezometer, is independent of the selected datum.

The flow of water takes place due to the difference in total head between two points.

From Eq. (6.1), at point A total head h_A = elevation head z_A + pressure head p_A.

At point B, total head h_A = elevation head z_B + pressure head p_B.

The total head loss from point A to point B is the difference in the water levels at the piezometers, given by Δh.

$$\Delta h = h_A - h_B \tag{6.2}$$

The head loss, Δh, can be expressed in a nondimensional form as

$$i = \frac{\Delta h}{L} \tag{6.3}$$

where

i = hydraulic gradient (dimensionless)
L = distance between points A and B—that is, the length of flow over which the loss of head occurred

According to Darcy (1856), the discharge velocity of water through saturated soils may be expressed as

$$v = ki \tag{6.4}$$

where

v = discharge velocity, which is the quantity of water flowing in unit time through a unit gross cross-sectional area of soil at right angles to the direction of flow
k = hydraulic conductivity (otherwise known as the coefficient of permeability)

This equation was based primarily on Darcy's observations about the flow of water through clean sands. Note that Eq. (6.4) is valid for laminar flow conditions and applicable for a wide range of soils.

In Eq. (6.4), v is the discharge velocity of water based on the gross cross-sectional area of the soil. However, the actual velocity of water, v_s (that is, the seepage velocity), can be given as

$$v_s = \frac{v}{n} = v\left(\frac{1+e}{e}\right) \tag{6.5}$$

where

n = porosity
e = void ratio

6.3 Hydraulic Conductivity

Hydraulic conductivity is generally expressed in cm/s, m/s, ft/min, or ft/day. The hydraulic conductivity of soils depends on several factors: fluid viscosity, pore-size distribution, grain-size distribution, void ratio, roughness of mineral particles, and degree of soil saturation. In clayey soils, structure plays an important role in hydraulic conductivity. Other major factors that affect the permeability of clays are the ionic concentration in the pore fluid and the thickness of layers of water held to the clay particles.

The value of hydraulic conductivity (k) varies widely for different soils. Some typical values for saturated soils are given in Table 6.1. The hydraulic conductivity of unsaturated soils is lower and increases rapidly with the degree of saturation.

Figure 6.2 shows the typical values of saturated hydraulic conductivity for the different soil groups. A simple classification based on permeability as suggested by Terzaghi and peck (1967) is also shown.

Table 6.1 Typical Values of Hydraulic Conductivity of Saturated Soils

	k	
Soil type	**cm/s**	**ft/min**
Clean gravel	100 – 1.0	200 – 2.0
Coarse sand	1.0 – 0.01	2.0 – 0.02
Fine sand	0.01 – 0.001	0.02 – 0.002
Silty clay	0.001 – 0.00001	0.002 – 0.00002
Clay	<0.000001	<0.000002

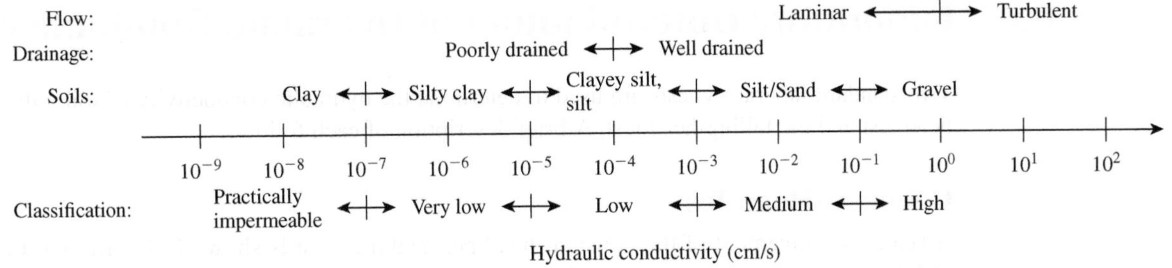

Figure 6.2 Typical values of saturated hydraulic conductivity for different types of soils.

Hydraulic conductivity is a function of the unit weight and the viscosity of water, which is in turn a function of the temperature at which the test is conducted or

$$\frac{k_{T_1}}{k_{T_2}} = \left(\frac{\eta_{T_2}}{\eta_{T_1}}\right)\left(\frac{\gamma_{w(T_1)}}{\gamma_{w(T_2)}}\right) \tag{6.6}$$

where

k_{T_1}, k_{T_2} = hydraulic conductivity at temperatures T_1 and T_2, respectively
η_{T_1}, η_{T_2} = viscosity of water at temperatures T_1 and T_2, respectively
$\gamma_{w(T_1)}, \gamma_{w(T_2)}$ = unit weight of water at temperatures T_1 and T_2, respectively

It is conventional to express the value of k at a temperature of 20°C. Within the range of test temperatures, we can assume that $\gamma_w(T_1) \simeq \gamma_w(T_2)$. So, from Eq. (6.6)

$$k_{20^\circ\mathrm{C}} = \left(\frac{\eta_{T^\circ\mathrm{C}}}{\eta_{20^\circ\mathrm{C}}}\right) k_{T^\circ\mathrm{C}} \tag{6.7}$$

The variation of $\eta_{T^\circ\mathrm{C}}/\eta_{20^\circ\mathrm{C}}$ with the test temperature T varying from 15 to 30°C is given in Table 6.2.

Table 6.2 Variation of $\eta_{T^\circ\mathrm{C}}/\eta_{20^\circ\mathrm{C}}$

Temperature, *T* (°C)	$\eta_{T^\circ\mathrm{C}}/\eta_{20^\circ\mathrm{C}}$	**Temperature, *T* (°C)**	$\eta_{T^\circ\mathrm{C}}/\eta_{20^\circ\mathrm{C}}$
15	1.135	23	0.931
16	1.106	24	0.910
17	1.077	25	0.889
18	1.051	26	0.869
19	1.025	27	0.850
20	1.000	28	0.832
21	0.976	29	0.814
22	0.953	30	0.797

6.4 Laboratory Determination of Hydraulic Conductivity

Two standard laboratory tests are used to determine the hydraulic conductivity of soil—the constant-head test and the falling-head test. A brief description of each follows.

Constant-Head Test

A typical arrangement of the constant-head permeability test is shown in Figure 6.3. In this type of laboratory setup, the water supply at the inlet is adjusted in such a way that the difference of head (h) between the inlet and the outlet remains constant during the test period. After a constant flow rate is established, water is collected in a graduated flask for a known duration.

The total volume of water collected may be expressed as

$$Q = Avt = A(ki)t \tag{6.8}$$

where

Q = volume of water collected
A = area of cross section of the soil specimen
t = duration of water collection

Since

$$i = \frac{h}{L} \tag{6.9}$$

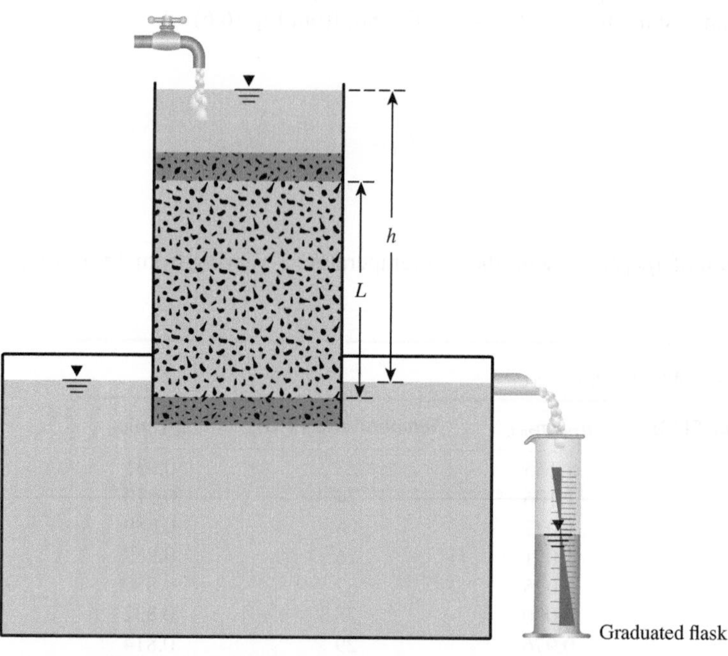

Figure 6.3 Constant-head permeability test

where L = length of the specimen and h = head loss across the specimen, Eq. (6.9) can be substituted into Eq. (6.8) to yield

$$Q = A\left(k\frac{h}{L}\right)t \tag{6.10}$$

or

$$k = \frac{QL}{Aht} \tag{6.11}$$

Falling-Head Test

A typical arrangement of the falling-head permeability test is shown in Figure 6.4. Water from a standpipe flows through the soil. The initial head difference h_1 at time $t = 0$ is recorded, and water is allowed to flow through the soil specimen such that the final head difference at time t is h_2. The hydraulic conductivity can now be calculated as

$$k = 2.303\frac{aL}{At}\log_{10}\frac{h_1}{h_2} \tag{6.12}$$

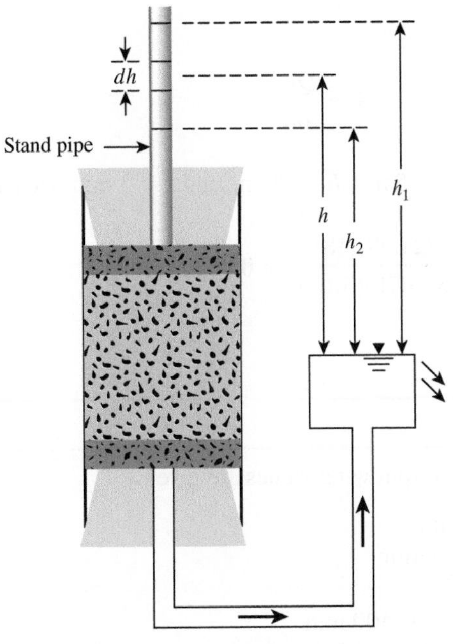

Figure 6.4 Falling-head permeability test

where

a = cross-sectional area of the standpipe
A = cross-sectional area of the soil specimen
h = head loss across the specimen at time t

Constant-head tests are more suited for coarse-grained soils, where a significant volume of water can be collected in few minutes. This is not the case in clay soils, where it can take a very long time to collect a measurable volume of water. Falling-head tests are commonly carried out on clay soils. They are not suitable for coarse-grain soil, where the water level would change very quickly from h_1 to h_2, making it difficult to measure the time correctly.

Example 6.1

Refer to the constant-head permeability test arrangement shown in Figure 6.3. A test gives these values:

- L = 18 in.
- A = area of the specimen = 3.5 in.^{2}
- Constant-head difference, h = 28 in.
- Water collected in a period of 3 min = 21.58 in.^{3}

Calculate the hydraulic conductivity in in./s.

Solution

From Eq. (6.11)

$$k = \frac{QL}{Aht}$$

Given Q = 21.58 in.^{3}, L = 18 in., A = 3.5 in.^{2}, h = 28 in., and t = 3 min, we have

$$k = \frac{(21.58)(18)}{(3.5)(28)(3)(60)} = \mathbf{0.022\ in./s}$$

■

Example 6.2

For a falling-head permeability test, the following values are given:

- Length of specimen = 200 mm
- Area of soil specimen = 1000 mm^2
- Area of standpipe = 40 mm^2
- Head difference at time t = 0 s is 500 mm
- Head difference at time t = 180 s is 300 mm

Determine the hydraulic conductivity of the soil in cm/s.

Solution

From Eq. (6.12)

$$k = 2.303 \frac{aL}{At} \log_{10}\left(\frac{h_1}{h_2}\right)$$

We are given $a = 40\ \text{mm}^2$, $L = 200$ mm, $A = 1000\ \text{mm}^2$, $t = 180$ s, $h_1 = 500$ mm, and $h_2 = 300$ mm,

$$k = 2.303 \frac{(40)(200)}{(1000)(180)} \log_{10}\left(\frac{500}{300}\right) = 0.0227\ \text{mm/s}$$

$$= \mathbf{2.27 \times 10^{-2}\ cm/s}$$

■

Example 6.3

A permeable soil layer is underlain by an impervious layer, as shown in Figure 6.5a. With $k = 4.8 \times 10^{-3}$ cm/s for the permeable layer, calculate the rate of seepage through it in ft³/hr/ft width if $H = 10$ ft and $\alpha = 5°$.

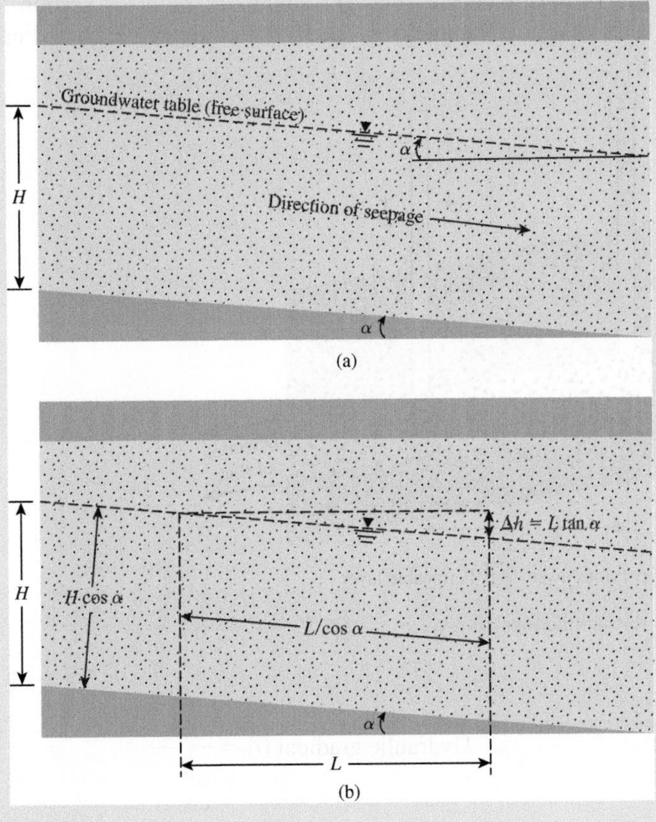

Figure 6.5

Solution

From Figure 6.5b,

$$i = \frac{\text{head loss}}{\text{length}} = \frac{L \tan \alpha}{\left(\dfrac{L}{\cos \alpha}\right)} = \sin \alpha$$

$$q = kiA = (k)(\sin \alpha)(10 \cos \alpha)(1)$$

$$k = 4.8 \times 10^{-3} \text{ cm/s} = 0.000158 \text{ ft/s}$$

$$q = (0.000158)(\sin 5°)(10 \cos 5°)(\underset{\uparrow}{3600}) = \mathbf{0.493\ ft^3/hr/ft}$$

To change to ft/hr

Example 6.4

Find the flow rate in $m^3/s/m$ length (at right angles to the cross section shown) through the permeable soil layer shown in Figure 6.6 given $H = 8$ m, $H_1 = 3$ m, $h = 4$ m, $L = 50$ m, $\alpha = 8°$, and $k = 0.08$ cm/s.

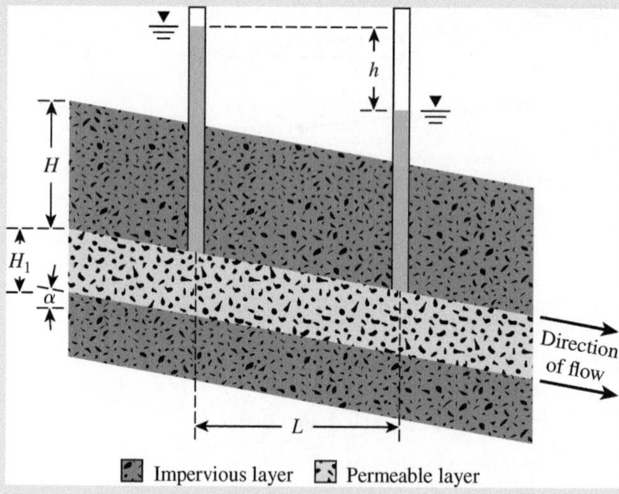

Figure 6.6 Flow through permeable layer

Solution

$$\text{Hydraulic gradient } (i) = \frac{h}{\dfrac{L}{\cos \alpha}}$$

From Eqs. (6.9) and (6.10),

$$q = kiA = k\left(\frac{h \cos \alpha}{L}\right)(H_1 \cos \alpha \times 1)$$

$$= (0.08 \times 10^{-2} \text{ m/s})\left(\frac{4 \cos 8°}{50}\right)(3 \cos 8° \times 1)$$

$$= 0.19 \times 10^{-3} \text{ m}^3\text{/s/m}$$

■

6.5 Relationships for Hydraulic Conductivity–Granular Soil

For fairly uniform sand (that is, sand with a small uniformity coefficient), Hazen (1930) proposed an empirical relationship for hydraulic conductivity in the form

$$k(\text{cm/s}) = cD_{10}^2 \tag{6.13}$$

where

c = a constant that varies from 1.0 to 1.5
D_{10} = the effective size, in mm

Equation (6.13) is based primarily on Hazen's observations of loose, clean, filter sands. A small quantity of silts and clays, when present in a sandy soil, may change the hydraulic conductivity substantially.

Over the last several years, experimental observations have shown that the magnitude of c for various types of granular soils may vary by three orders of magnitude (Carrier, 2003) and, hence, Eq. (6.13) is not very reliable.

Another form of equation that gives fairly good results in estimating the hydraulic conductivity of sandy soils is based on the Kozeny-Carman equation (Kozeny, 1927; Carman, 1938, 1956). The derivation of this equation is not presented here. Interested readers are referred to any advanced soil mechanics book. According to the Kozeny-Carman equation

$$k = \frac{1}{C_s S_s^2 T^2} \frac{\gamma_w}{\eta} \frac{e^3}{1+e} \tag{6.14}$$

where

C_s = shape factor, which is a function of the shape of flow channels
S_s = specific surface area per unit volume of particles
T = tortuosity of flow channels
γ_w = unit weight of water
η = viscosity of permeant
e = void ratio

Equation (6.14) suggests that

$$k \propto \frac{e^3}{1+e} \tag{6.15}$$

More recently, Chapuis (2004) proposed an empirical relationship for k in conjunction with Eq. (6.15) as

$$k(\text{cm/s}) = 2.4622\left[D_{10}^2 \frac{e^3}{(1+e)}\right]^{0.7825} \tag{6.16}$$

where D_{10} = effective size (mm).

The preceding equation is valid for natural, uniform sand and gravel to predict k that is in the range of 10^{-1} to 10^{-3} cm/s. This can be extended to natural, silty sands without plasticity. It is not valid for crushed materials or silty soils with some plasticity.

On the basis of laboratory experiments, the U.S. Department of Navy (1971) provided an empirical correlation between k and D_{10} (mm) for granular soils with the uniformity coefficient varying between 2 and 12 and $D_{10}/D_5 < 1.4$. This correlation is shown in Figure 6.7.

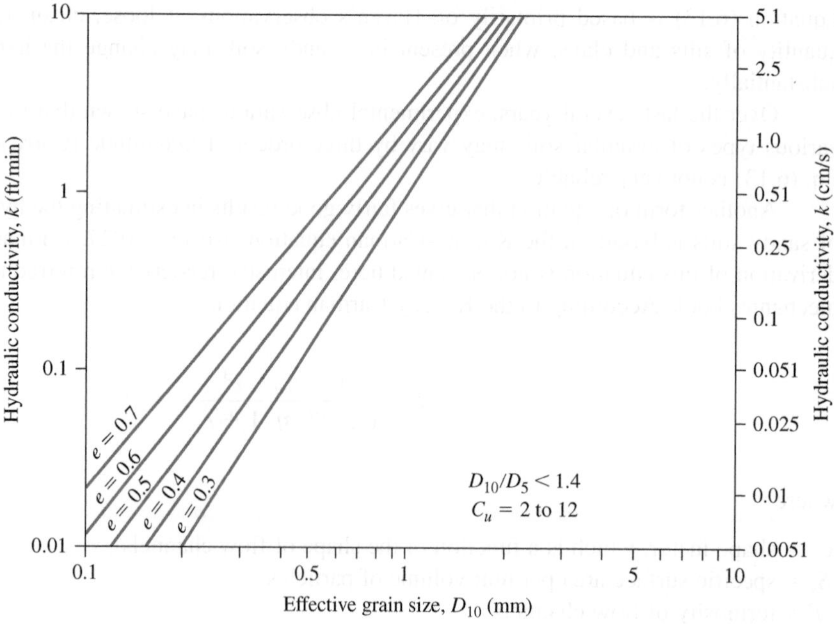

Figure 6.7 Hydraulic conductivity of granular soils (Adapted from U.S. Department of Navy, 1971)

Example 6.5

The hydraulic conductivity of a sand at a void ratio of 0.62 is 0.03 cm/s. Estimate its hydraulic conductivity at a void ratio of 0.48.

Solution

From Eq. (6.15),

$$\frac{k_1}{k_2} = \frac{\dfrac{e_1^3}{1 + e_1}}{\dfrac{e_2^3}{1 + e_2}}$$

$$\frac{0.03}{k_2} = \frac{\dfrac{(0.62)^3}{1 + 0.62}}{\dfrac{(0.48)^3}{1 + 0.48}}$$

$$k_2 = \mathbf{0.015\ cm/s}$$

■

6.6 Relationships for Hydraulic Conductivity–Cohesive Soils

According to their experimental observations, Samarasinghe, et al. (1982) suggested that the hydraulic conductivity of normally consolidated clays (see Chapter 8 for definition) can be given by

$$k = C\left(\frac{e^n}{1 + e}\right) \tag{6.17}$$

where C and n are constants to be determined experimentally. C has the unit of hydraulic conductivity. This equation can be rewritten as

$$\log[k(1 + e)] = \log C + n \log e \tag{6.18}$$

Hence, for any given clayey soil, if the variation of k with the void ratio is known, a log-log graph can be plotted with $k(1 + e)$ against e to determine the values of C and n.

Some other empirical relationships for estimating the hydraulic conductivity in clayey soils are given in Table 6.3. One should keep in mind, however, that any empirical relationship of this type is for estimation only, because the magnitude of k is a highly variable parameter and depends on several factors.

Table 6.3 Empirical Relationships for Estimating Hydraulic Conductivity in Clayey Soil

Type of Soil	Source	Relationship[a]
Clay	Mesri and Olson (1971)	$\log k = A' \log e + B'$
	Taylor (1948)	$\log k = \log k_0 - \dfrac{e_0 - e}{C_k}$
		$C_k \approx 0.5e_0$

[a] k_0 = in situ hydraulic conductivity at void ratio e_0
k = hydraulic conductivity at void ratio e
C_k = hydraulic conductivity change index

Example 6.6

For a normally consolidated clay soil, the following values are given:

Void ratio	*k* (cm/s)
1.1	0.302×10^{-7}
0.9	0.12×10^{-7}

Estimate the hydraulic conductivity of the clay at a void ratio of 0.75. Use Eq. (6.17).

Solution

From Eq. (6.17)

$$k = C\left(\frac{e^n}{1+e}\right)$$

$$\frac{k_1}{k_2} = \frac{\left(\dfrac{e_1^n}{1+e_1}\right)}{\left(\dfrac{e_2^n}{1+e_2}\right)}$$

$$\frac{0.302 \times 10^{-7}}{0.12 \times 10^{-7}} = \frac{\dfrac{(1.1)^n}{1+1.1}}{\dfrac{(0.9)^n}{1+0.9}}$$

$$2.517 = \left(\frac{1.9}{2.1}\right)\left(\frac{1.1}{0.9}\right)^n$$

$$2.782 = (1.222\,)^n$$

$$n = \frac{\log(2.782)}{\log(1.222)} = \frac{0.444}{0.087} = 5.1$$

so

$$k = C\left(\frac{e^{5.1}}{1+e}\right)$$

To find C,

$$0.302 \times 10^{-7} = C\left[\frac{(1.1)^{5.1}}{1+1.1}\right] = \left(\frac{1.626}{2.1}\right)C$$

$$C = \frac{(0.302 \times 10^{-7})(2.1)}{1.626} = 0.39 \times 10^{-7} \text{ cm/s}$$

Hence,

$$k = (0.39 \times 10^{-7} \text{ cm/s})\left(\frac{e^{n}}{1+e}\right)$$

At a void ratio of 0.75,

$$k = (0.39 \times 10^{-7})\left(\frac{0.75^{5.1}}{1+0.75}\right) = \mathbf{0.514 \times 10^{-8} \text{ cm/s}}$$ ■

6.7 Permeability Test in the Field by Pumping from Wells

In the field, the average hydraulic conductivity of a soil deposit in the direction of flow can be determined by performing pumping tests from wells. Figure 6.8 shows a case where the top permeable layer, whose hydraulic conductivity has to be determined, is unconfined and underlain by an

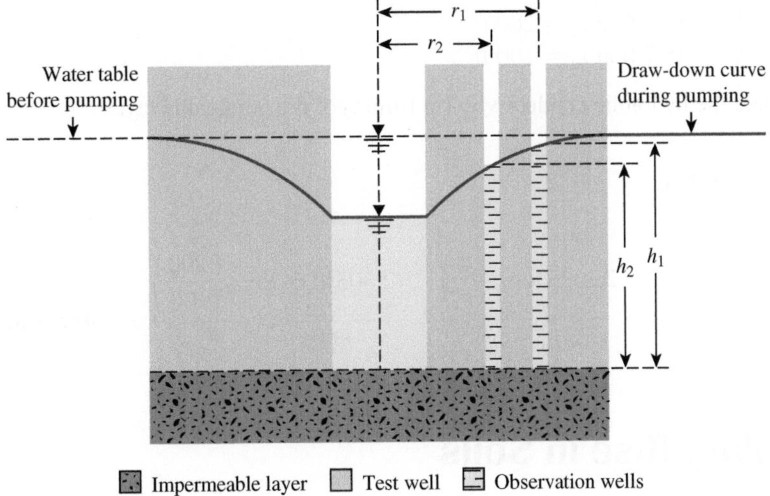

Figure 6.8 Pumping test from a well in an unconfined permeable layer underlain by an impermeable stratum.

impermeable layer. During the test, water is pumped out at a constant rate from a test well that has a perforated casing. Several observation wells at various radial distances are made around the test well.

Continuous observations of the water level in the test well and in the observation wells are made after the start of pumping until a steady state is reached while the pumping continues. The steady state is established when the water level in the test and observation wells becomes constant.

The hydraulic conductivity now can be estimated as

$$k = \frac{2.303q \log_{10}\left(\frac{r_1}{r_2}\right)}{\pi(h_1^2 - h_2^2)} \tag{6.19}$$

From field measurements, if q, r_1, r_2, h_1, and h_2 are known, the hydraulic conductivity can be calculated. This equation can also be written as

$$k\ (\text{cm/s}) = \frac{2.303q \log_{10}\left(\frac{r_1}{r_2}\right)}{14.7\pi(h_1^2 - h_2^2)} \tag{6.20}$$

where q is in gpm and h_1 and h_2 are in ft.

Example 6.7

Consider the case of pumping from a well in an unconfined permeable layer underlain by an impermeable stratum (see Figure 6.8). Given:

- $q = 26\ \text{ft}^3/\text{min}$
- $h_1 = 18.0$ ft at $r_1 = 200$ ft
- $h_2 = 15.7$ ft at $r_2 = 100$ ft

Calculate the hydraulic conductivity (in ft/min) of the permeable layer.

Solution

From Eq. (6.19),

$$k = \frac{2.303q \log_{10}\left(\frac{r_1}{r_2}\right)}{\pi(h_1^2 - h_2^2)} = \frac{(2.303)(26) \log_{10}\left(\frac{200}{100}\right)}{\pi(18^2 - 15.7^2)} = \mathbf{0.074\ ft/min}$$ ■

6.8 Capillary Rise in Soils

The continuous void spaces in soil can behave as bundles of capillary tubes of variable cross section. Because of surface tension force, water may rise above the phreatic surface.

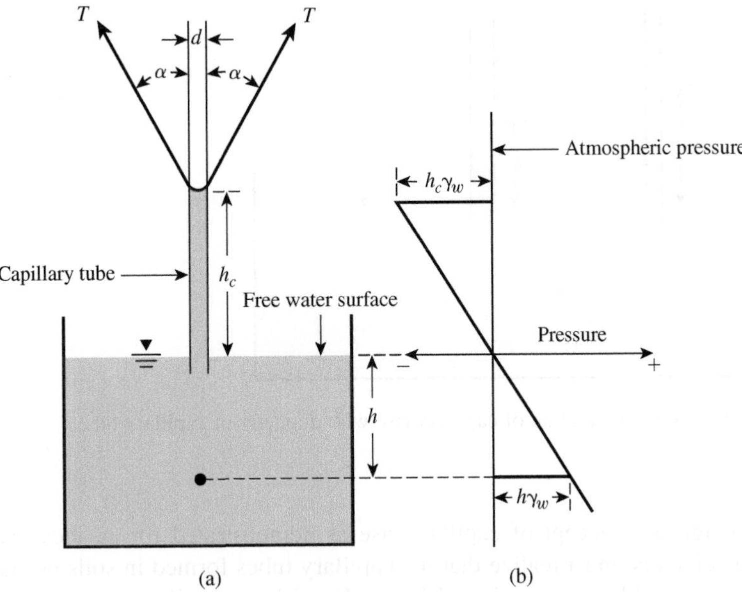

Figure 6.9 (a) Rise of water in the capillary tube; (b) pressure within the height of rise in the capillary tube (atmospheric pressure taken as datum)

Figure 6.9 shows the fundamental concept of the height of rise in a capillary tube. The height of rise of water in the capillary tube can be given by summing the forces in the vertical direction, or

$$\left(\frac{\pi}{4}d^2\right)h_c\gamma_w = \pi dT\cos\alpha$$

$$h_c = \frac{4T\cos\alpha}{d\gamma_w} \tag{6.21}$$

where

T = surface tension (force/length)
α = angle of contact
d = diameter of capillary tube
γ_w = unit weight of water

For pure water and clean glass, $\alpha = 0$. Thus, Eq. (6.21) becomes

$$h_c = \frac{4T}{d\gamma_w} \tag{6.22}$$

From Eq. (6.22), we see that the height of capillary rise

$$h_c \propto \frac{1}{d} \tag{6.23}$$

Thus, the smaller the capillary tube diameter, the larger the capillary rise. This fact is shown in Figure 6.10.

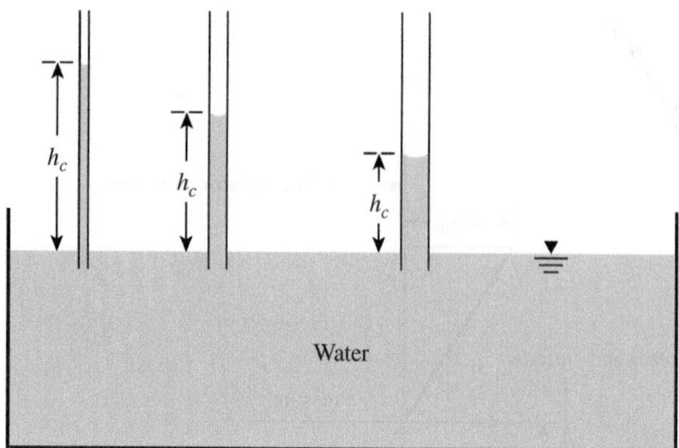

Figure 6.10 Nature of variation of capillary rise with diameter of capillary tube

Although the concept of capillary rise as demonstrated for an ideal capillary tube can be applied to soils, one must realize that the capillary tubes formed in soils because of the continuity of voids have variable cross sections. The results of the nonuniformity on capillary rise can be seen when a dry column of sandy soil is placed in contact with water (Figure 6.11a). After the lapse of a given amount of time, the variation of the degree of saturation with the height of the soil column caused by capillary rise is approximately as shown in Figure 6.11b. The degree of saturation is about 100% up to a height of h_2, and this corresponds to the largest voids. Beyond the height h_2, water can occupy only the smaller voids; hence, the degree of saturation is less than 100%. The

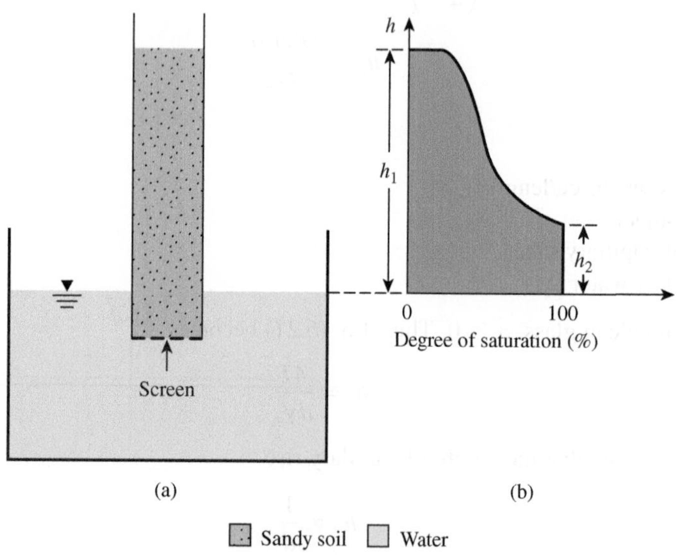

Figure 6.11 Capillary effect in sandy soil: (a) a soil column in contact with water; (b) variation of degree of saturation in the soil column

Table 6.4 Approximate Range of Capillary Rise in Soils

	Range of capillary rise	
Soil type	**ft**	**m**
Coarse sand	0.3–0.6	0.09–0.18
Fine sand	1–4	0.31–1.22
Silt	2.5–25	0.76–7.6
Clay	25–75	7.6–22.9

maximum height of capillary rise corresponds to the smallest voids. Hazen (1930) gave a formula for the approximation of the height of capillary rise in the form.

$$h_1(\text{mm}) = \frac{C}{eD_{10}} \tag{6.24}$$

where

D_{10} = effective size (mm)
e = void ratio
C = a constant that varies from 10 to 50 mm^2

Equation (6.24) has an approach similar to that of Eq. (6.23). With the decrease of D_{10}, the pore size in soil decreases, which causes higher capillary rise. Table 6.4 shows the approximate range of capillary rise that is encountered in various types of soils.

Capillary rise is important in the formation of some types of soils such as *caliche*, which can be found in the desert Southwest of the United States. Caliche is a mixture of sand, silt, and gravel bonded by calcareous deposits. These deposits are brought to the surface by a net upward migration of water by capillary action. The water evaporates in the high local temperature. Because of sparse rainfall, the carbonates are not washed out of the top soil layer.

6.9 Summary

Hydraulic conductivity or permeability, measured in cm/s or ft/min, is a measure of how easily water can flow through a soil. It varies in a wide range from very low values for clays to large values for gravels. It is measured in the laboratory through a constant-head test for coarse-grained soils and a falling-head test for fine-grained soils. In the field, it can be determined from the flow rate when pumping the water out at steady state.

When water flows through soils, hydraulic gradient is defined as the head loss per unit length. It is a dimensionless quantity. According to Darcy's law for flow through soils, the discharge velocity is proportional to the hydraulic gradient. This is the basis of most computations carried out in seepage problems.

Capillary effects are more predominant in clays than in sands. Here, the interconnected voids act as capillary tubes by allowing the water to rise above the free water surface.

Problems

6.1 State whether the following are true or false.
   a. Hydraulic conductivity is greater for sands than clays.
   b. Larger the void ratio, larger is the hydraulic conductivity.
   c. The magnitude of the total head depends on the selected datum.
   d. Capillary rise is greater in sands than in clays.
   e. Capillary rise is proportional to D_{10} of a soil.

6.2 Refer to the constant-head arrangement shown in Figure 6.3. For a test, the following are given:
   - $L = 20$ in.
   - A = area of the specimen = 4.6 in.^{2}
   - Constant-head difference = h = 35 in.
   - Water collected in 3 min = 20 in.^{3}

   Calculate the hydraulic conductivity (in./s).

6.3 Refer to Figure 6.3. For a constant-head permeability test in a sand, the following are given:
   - L = 350 mm
   - A = 125 cm^2
   - h = 420 mm
   - Water collected in 3 min = 580 cm^3
   - Void ratio of sand = 0.61

   Determine
   a. Hydraulic conductivity, k (cm/s)
   b. Seepage velocity

6.4 In a constant-head permeability test in the laboratory, the following are given: L = 250 mm and A = 105 cm^2. If the value of k = 0.014 cm/s and a flow rate of 120 cm^3/min must be maintained through the soil, what is the head difference, h, across the specimen? Also, determine the discharge velocity under the test conditions.

6.5 Two cylindrical soils specimens are subjected to a modified constant head test as shown in Figure 6.12, where the two specimens are tested simultaneously. The diameter of the specimens is 75 mm. In 10 minutes, 650 g of water is collected in a bucket. Find the hydraulic conductivity of the two soil specimens.

6.6 For a falling-head permeability test, the following are given:
   - Length of the soil specimen = 15 in.
   - Area of the soil specimen = 3 in.^{2}
   - Area of the standpipe = 0.15 in.^{2}
   - Head difference at time t = 0 is 25 in.
   - Head difference at time t = 8 min is 12 in.

   a. Determine the hydraulic conductivity of the soil (in./min).
   b. What was the head difference at time t = 4 min?

6.7 For a falling-head permeability test, the following are given: length of specimen = 380 mm, area of specimen = 6.5 cm^2, and k = 0.175 cm/min. What should be the area of the standpipe for the head to drop from 650 cm to 300 cm in 8 min?

6.8 A site consists of a homogeneous dense silty sand layer of 10 m thickness, which is underlain by an impervious stiff clay stratum. The initial water table was at 3.0 m depth below the ground level. A pumping test was carried out by pumping out water at the rate of 0.5 m^3/min. Observation wells were dug into the ground at 20 m and 40 m distances from the centerline

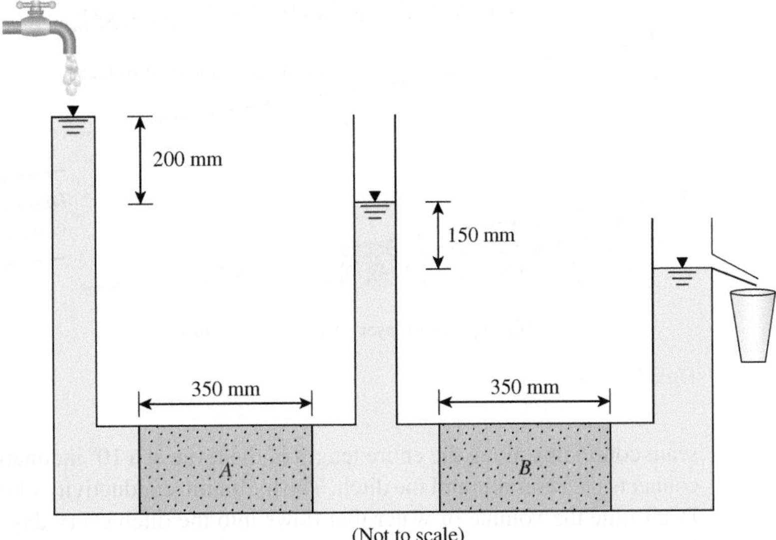

Figure 6.12

of the well. At steady state (i.e., when there was no change in water levels in the observation wells), the water levels in the two wells dropped by 500 mm and 150 mm, respectively. Determine the hydraulic conductivity of the clayey sand layer.

6.9 A sand layer of the cross-sectional area shown in Fig. 6.13 has been determined to exist for a 500-m length of the levee. The hydraulic conductivity of the sand layer is 3 m/day. Determine the quantity of water which flows into the ditch in m^3/min.

6.10 A permeable soil layer is underlain by an impervious layer, as shown in Figure 6.14. With $k = 6.8 \times 10^{-4}$ cm/s for the permeable layer, calculate the rate of seepage through it in m^3/hr/m length if $H = 4.2$ m and $\alpha = 10°$.

6.11 A 500-m long levee made of compacted clay impounds water in a reservoir as shown in Figure 6.15. There is a 1-m thick (measured in the direction perpendicular to the seam) sand

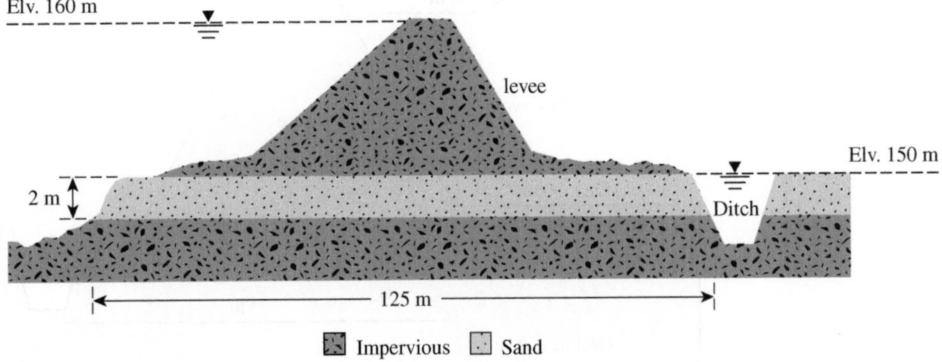

Figure 6.13

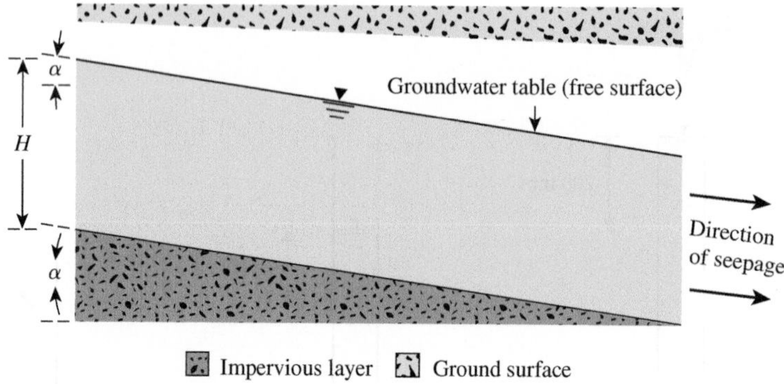

Figure 6.14

seam continuing along the entire length of the levee at a 10° inclination to the horizontal, which connects the reservoir and the ditch. The hydraulic conductivity of the sand is 2.6×10^{-3} cm/s. Determine the volume of water that flows into the ditch every day.

6.12 Refer to Figure 6.16. Find the flow rate in $m^3/s/m$ length (at right angles to the cross section shown) through the permeable soil layer. Given: $H = 5.5$ m, $H_1 = 3$ m, $h = 2.8$ m, $L = 52$ m, $\alpha = 5°$, and $k = 0.05$ cm/s.

6.13 The hydraulic conductivity of a sand at a void ratio of 0.48 is 0.022 cm/s. Estimate its hydraulic conductivity at a void ratio of 0.7. Use Eq. (6.15).

6.14 For a sand, the following are given: porosity, $n = 0.31$ and $k = 0.13$ ft/min. Determine k when $n = 0.4$. Use Eq. (6.15).

6.15 For a normally consolidated clay, the following are given:

Void ratio, e	k (cm/s)
0.8	1.2×10^{-6}
1.4	3.6×10^{-6}

Estimate the hydraulic conductivity of the clay at a void ratio, $e = 0.62$. Use Eq. (6.17).

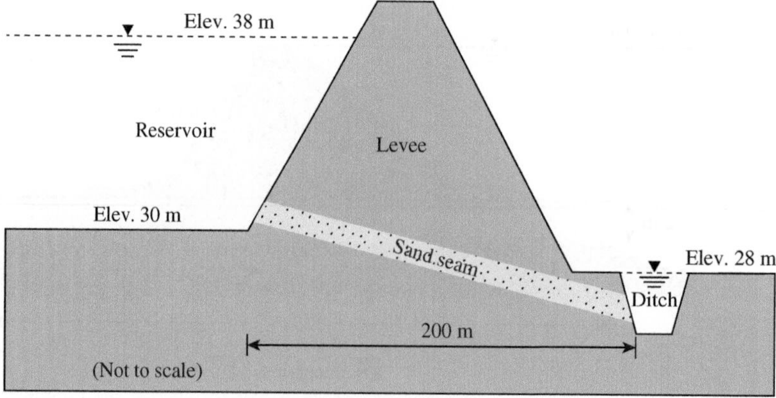

Figure 6.15

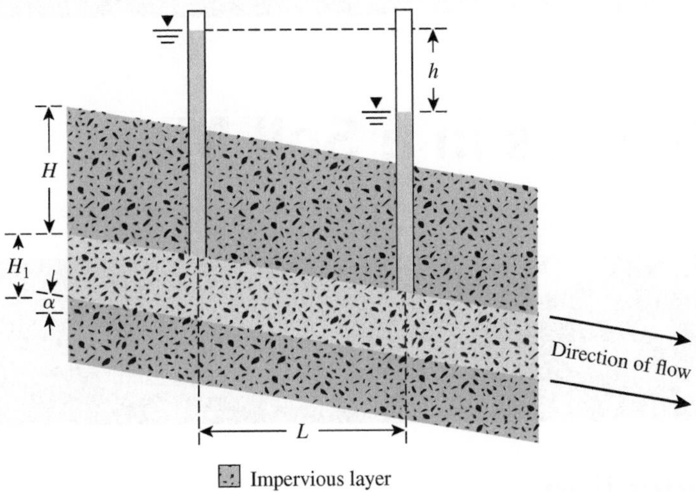

Figure 6.16

6.16 For a normally consolidated clay, the following are given:

Void ratio, e	k (cm/s)
1.2	0.2×10^{-6}
1.9	0.9×10^{-6}

Estimate the magnitude of k of the clay at a void ratio, $e = 0.9$. Use Eq. (6.17).

References

CARMAN, P. C. (1938). "The Determination of the Specific Surface of Powders." *J. Soc. Chem. Ind. Trans.*, Vol. 57, 225.

CARMAN, P. C. (1956). *Flow of Gases through Porous Media*, Butterworths Scientific Publications, London.

DARCY, H. (1856). *Les Fontaines Publiques de la Ville de Dijon*, Dalmont, Paris.

HAZEN, A. (1930). "Water Supply," in *American Civil Engineers Handbook*, Wiley, New York.

KOZENY, J. (1927). "Ueber kapillare Leitung des Wassers im Boden." *Wien, Akad. Wiss.*, Vol. 136, No. 2a, 271.

MESRI, G. and Olson, R. E. (1971). "Mechanism Controlling the Permeability of Clays," *Clay and Clay Minerals*, Vol. 19, 151–158.

SAMARASINGHE, A. M., HUANG, Y. H., AND DRNEVICH, V. P. (1982). "Permeability and Consolidation of Normally Consolidated Soils," *Journal of the Geotechnical Engineering Division*, ASCE, Vol. 108, No. GT6, 835–850.

TAYLOR, D. W. (1948). *Fundamentals of Soil Mechanics*, Wiley, New York.

TERZAGHI, K. and PECK, R. B. (1967). *Soil Mechanics in Engineering Practice*, 2nd ed., Wiley, New York.

U.S. DEPARTMENT OF NAVY (1971). "Design Manual—Soil Mechanics, Foundations, and Earth Structures," *NAVFAC DM-7*, U.S. Government Printing Office, Washington, D.C.

7 Stresses in a Soil Mass

7.1 Introduction

As described in Chapter 3, soils are multiphase systems. In a given volume of soil, the solid grains are distributed randomly with void spaces in between. The void spaces are continuous and are occupied by water, air, or both. To analyze problems such as compressibility of soils, bearing capacity of foundations, stability of embankments, and lateral pressure on earth-retaining structures, engineers need to know the nature of the distribution of stress along a given cross section of the soil profile—that is, what fraction of the normal stress at a given depth in a soil mass is carried by water in the void spaces and what fraction is carried by the soil skeleton at the points of contact of the soil grains. This issue is referred to as the *effective stress concept*, and it is discussed in the first part of this chapter.

When a foundation is constructed, changes take place in the soil under the foundation. The net stress usually increases. This net stress increase in the soil depends on the load per unit area to which the foundation is subjected, the depth below the foundation at which the stress estimation is made, and other factors. It is necessary to estimate the net increase of vertical stress in soil that occurs at a certain depth as a result of the construction of a foundation so that settlement can be calculated. The second part of this chapter discusses the principles for estimating the vertical stress increase in soil caused by various types of loading, based on the theory of elasticity. Although natural soil deposits are not fully elastic, isotropic, or homogeneous materials, calculations for estimating increases in vertical stress yield fairly good results for practical work. In this chapter, you will learn the following,

- Terzaghi's effective stress principle: total stress, effective stress, and pore water pressure.
- Variation of in situ stresses with depth within a soil mass.
- Stresses in soils due to upward flow and quick condition.
- Stress increase within a soil mass due to surface loads.

Effective Stress Concept

7.2 Stresses in Saturated Soil without Seepage

Figure 7.1a shows a column of saturated soil mass with no seepage of water in any direction. The total stress (σ) at the elevation of point A can be obtained from the saturated unit weight of the soil and the unit weight of water above it. Thus,

$$\sigma = H\gamma_w + (H_A - H)\gamma_{sat} \tag{7.1}$$

where

σ = total stress at the elevation of point A
γ_w = unit weight of water
γ_{sat} = saturated unit weight of the soil
H = height of water table from the top of the soil column
H_A = depth of point A below the water table

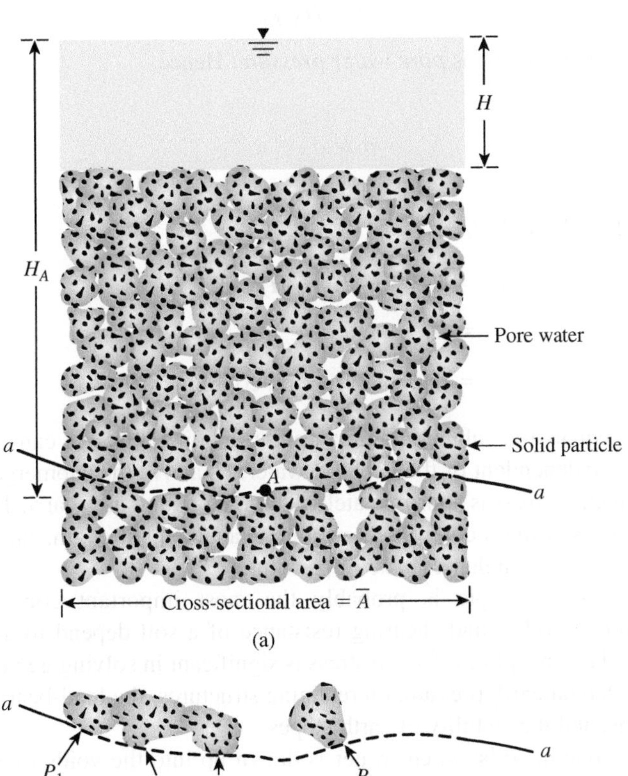

Figure 7.1 (a) Effective stress consideration for a saturated soil column without seepage; (b) forces acting at the points of contact of soil particles at the level of point A

The total stress, σ, given by Eq. (7.1) can be divided into two parts:

1. A portion is carried by water in the continuous void spaces. This portion acts with equal intensity in all directions.
2. The rest of the total stress is carried by the soil grains at their points of contact. The sum of the vertical components of the forces developed at the points of contact of the solid particles per unit cross-sectional area of the soil mass is called the *effective stress*.

This can be seen by drawing a wavy line, *a–a*, through point *A* that passes only through the points of contact of the solid grains. Let $P_1, P_2, P_3, \ldots, P_n$ be the forces that act at the points of contact of the soil particles (Figure 7.1b). The sum of the vertical components of all such forces over the unit cross-sectional area is equal to the effective stress σ', or

$$\sigma' = \frac{P_{1(v)} + P_{2(v)} + P_{3(v)} + \cdots + P_{n(v)}}{\overline{A}} \tag{7.2}$$

where $P_{1(v)}, P_{2(v)}, P_{3(v)}, \ldots, P_{n(v)}$ are the vertical components of $P_1, P_2, P_3, \ldots, P_n$, respectively, and $\overline{A}$ is the cross-sectional area of the soil mass under consideration.

According to Terzaghi (1925, 1936), the portion of the total stress carried by water (u) is approximately equal to the hydrostatic pressure at A. Or

$$u = H_A \gamma_w \tag{7.3}$$

The term u is generally referred to as *pore water pressure*. Hence

$$\sigma = \sigma' + u \tag{7.4}$$

Combining Eqs. (7.1), (7.3), and (7.4)

$$\begin{aligned}\sigma' &= [H\gamma_w + (H_A - H)\gamma_{sat}] - H_A\gamma_w \\ &= (H_A - H)(\gamma_{sat} - \gamma_w) \\ &= (\text{height of the soil column}) \times \gamma' \end{aligned} \tag{7.5}$$

where $\gamma' = \gamma_{sat} - \gamma_w$ equals the submerged unit weight of soil. Thus, we can see that the effective stress at any point A is independent of the depth of water, H, above the submerged soil.

In summary, effective stress is approximately the force per unit area carried by the soil skeleton. The effective stress in a soil mass controls its volume change and strength. Increasing the effective stress induces soil to move into a denser state of packing.

The effective-stress principle is probably the most important concept in geotechnical engineering. The compressibility and shearing resistance of a soil depend to a great extent on the effective stress. Thus, the concept of effective stress is significant in solving geotechnical engineering problems, such as the lateral earth pressure on retaining structures, the load-bearing capacity and settlement of foundations, and the stability of earth slopes.

In the capillary zone in soils, where water is drawn up into the voids by capillary action, the pore water pressure is negative. Also, when a dense soil skeleton is subjected to loading, there can be volume increase, which can lead to negative pore water pressure. When the pore water pressure is negative, the effective stress is greater than the total stress [see Eq. (7.4)].

Example 7.1

A soil profile is shown in Figure 7.2. Calculate the total stress, pore water pressure, and effective stress at points *A*, *B*, and *C*.

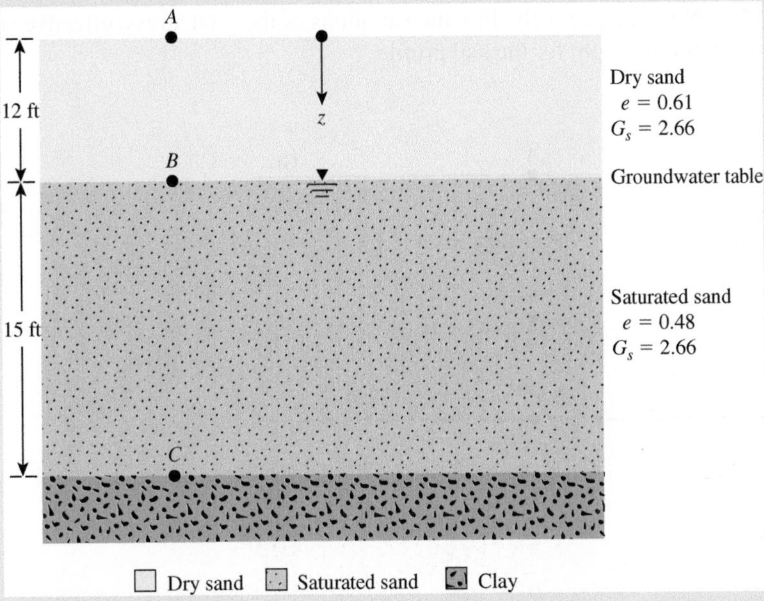

Figure 7.2 Soil profile

Solution

Unit weight calculation: For $z = 0$ to 12 ft,

$$\gamma_d = \frac{G_s \gamma_w}{1 + e} = \frac{(2.66)(62.4)}{1 + 0.61} = 103.1 \text{ lb/ft}^3$$

For $z = 12$ to 27 ft,

$$\gamma_{sat} = \frac{(G_s + e)\gamma_w}{1 + e} = \frac{(2.66 + 0.48)(62.4)}{1 + 0.48} = 132.4 \text{ lb/ft}^3$$

Now, the following table can be prepared:

Point	Total stress, σ (lb/ft^2)	Pore water pressure, u (lb/ft^2)	Effective stress, σ' (lb/ft^2)
A	**0**	**0**	**0**
B	$12\gamma_d = (12)(103.1) =$ **1237.2**	**0**	**1237.2**
C	$12\gamma_d + 15\gamma_{sat} = 1237.2 + (15)(132.4) =$ **3223.2**	$15\gamma_w = (15)(62.4) =$ **936**	**2287.2**

■

Example 7.2

Figure 7.3 shows the soil profile for the top 9 m at a site, where the water table is at a depth of 5 m below the ground level. The top 3 m consists of dry silty gravel with a unit weight of 17.9 kN/m^3. The next 6 m consists of sand where the unit weights above and below the water table are 17.0 kN/m^3 and 19.5 kN/m^3, respectively. Plot the variations of the total stress, effective stress, and pore water pressure with the depth for the soil profile.

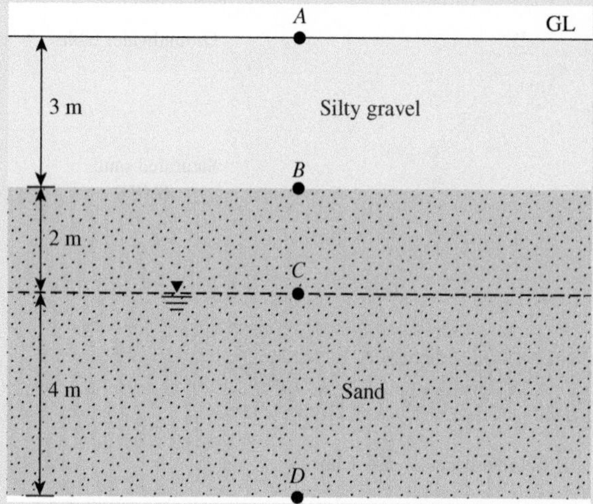

Figure 7.3 Soil profile

Solution

Let's select four points *A* (at the ground level), *B* (silty gravel – sand interface), *C* (water table), and *D* (bottom of sand) and compute the values at these points.

Point	Depth below GL (m)	Total Stress, σ (kN/m^2)	Pore Water Pressure, u (kN/m^2)	Effective Stress, σ' (kN/m^2)
A	0	0	0	0
B	3	$3 \times 17.9 = 53.7$	0	53.7
C	5	$3 \times 17.9 + 2 \times 17.0 = 87.7$	0	87.7
D	9	$3 \times 17.9 + 2 \times 17.0 + 4 \times 19.5 = 165.7$	$4 \times 9.81 = 39.2$	126.5

The values of σ, u, and σ' are plotted against depth in Figure 7.4. While the unit weights remain the same, the stresses increase linearly with depth. The break in the gradient occurs only when the unit weight changes.

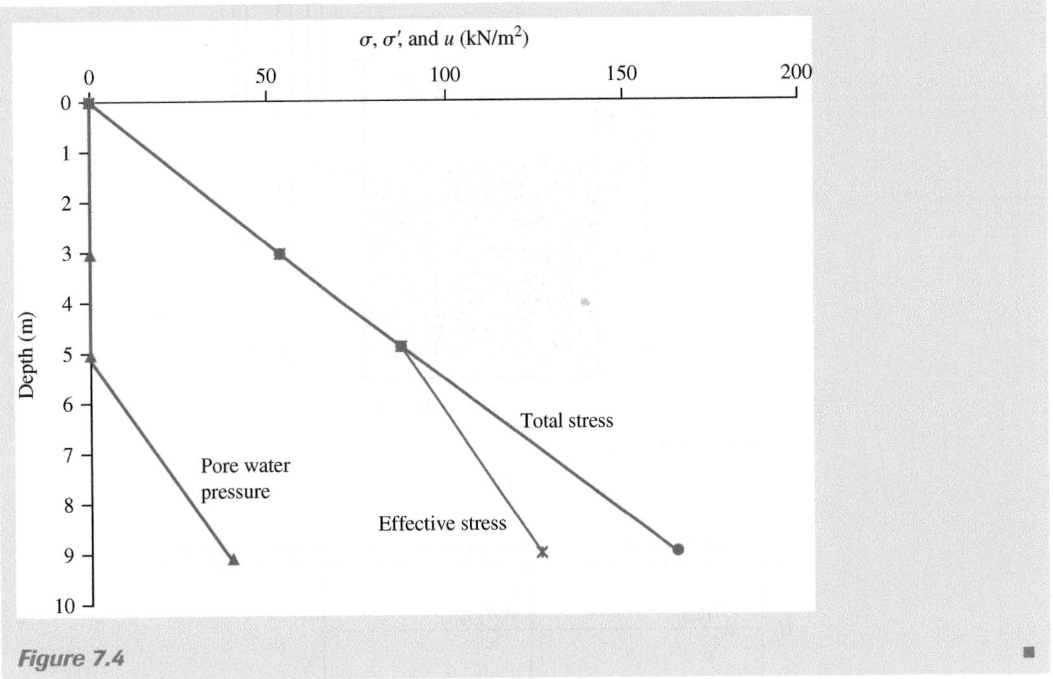

Figure 7.4

7.3 Stresses in Saturated Soil with Upward Seepage

If water is seeping, the effective stress at any point in a soil mass will differ from that in the static case. It will increase or decrease, depending on the direction of seepage.

Figure 7.5a shows a layer of granular soil in a tank where upward seepage is caused by adding water through the valve at the bottom of the tank. The rate of water supply is kept constant. The height of water in the standpipes give the pressure heads, which give the pore water pressures when multiplied by γ_w. The loss of head caused by upward seepage between the levels of A and B is h. Keeping in mind that the total stress at any point in the soil mass is due solely to the weight of soil and water above it, we find that the effective stress calculations at points A and B are as follows:

At *A*,

- Total stress: $\sigma_A = H_1\gamma_w$
- Pore water pressure: $u_A = H_1\gamma_w$
- Effective stress: $\sigma'_A = \sigma_A - u_A = 0$

At *B*,

- Total stress: $\sigma_B = H_1\gamma_w + H_2\gamma_{sat}$
- Pore water pressure: $u_B = (H_1 + H_2 + h)\gamma_w$
- Effective stress: $\sigma'_B = \sigma_B - u_B$

$$= H_2(\gamma_{sat} - \gamma_w) - h\gamma_w$$
$$= H_2\gamma' - h\gamma_w$$

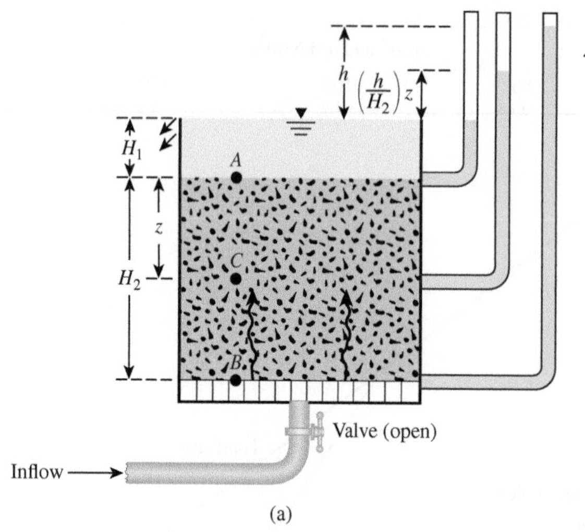

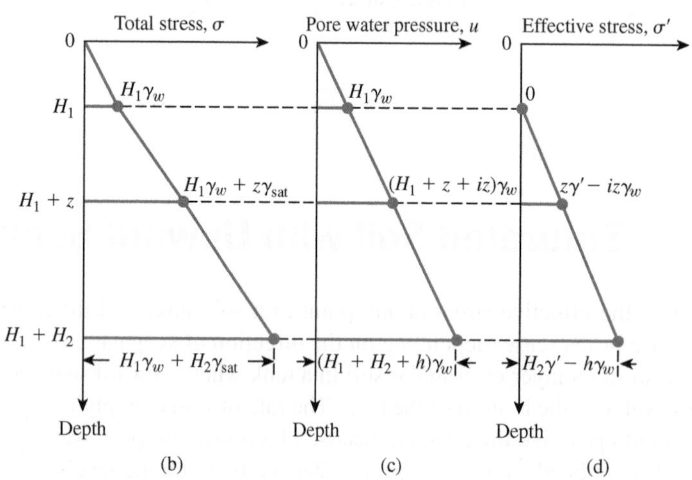

Figure 7.5 (a) Layer of soil in a tank with upward seepage. Variation of (b) total stress; (c) pore water pressure; and (d) effective stress with depth for a soil layer with upward seepage

Similarly, the effective stress at a point C located at a depth z below the top of the soil surface can be calculated as follows:

At *C*,

- Total stress: $\sigma_C = H_1\gamma_w + z\gamma_{sat}$
- Pore water pressure: $u_C = \left(H_1 + z + \frac{h}{H_2}z\right)\gamma_w$
- Effective stress: $\sigma'_C = \sigma_C - u_C$

$$= z(\gamma_{sat} - \gamma_w) - \frac{h}{H_2}z\gamma_w$$

$$= z\gamma' - \frac{h}{H_2}z\gamma_w$$

Note that h/H_2 is the hydraulic gradient i caused by the flow, and therefore,

$$\sigma'_C = z\gamma' - iz\gamma_w \tag{7.6}$$

The variations of total stress, pore water pressure, and effective stress with depth are plotted in Figures 7.5b through 7.5d, respectively.

If the rate of seepage and thereby the hydraulic gradient gradually are increased, a limiting condition will be reached, at which point

$$\sigma'_C = z\gamma' - i_{cr}z\gamma_w = 0 \tag{7.7}$$

where i_{cr} = critical hydraulic gradient (for zero effective stress).

Under such a situation, soil stability is lost. This situation generally is referred to as *boiling* or a *quick condition*.

From Eq. (7.7),

$$i_{cr} = \frac{\gamma'}{\gamma_w} \tag{7.8}$$

For most soils, the value of i_{cr} varies from 0.9 to 1.1, with an average of 1.

Example 7.3

A 27 ft thick layer of stiff saturated clay is underlain by a layer of sand (Figure 7.6). The sand is under artesian pressure. Calculate the maximum depth of cut H that can be made in the clay.

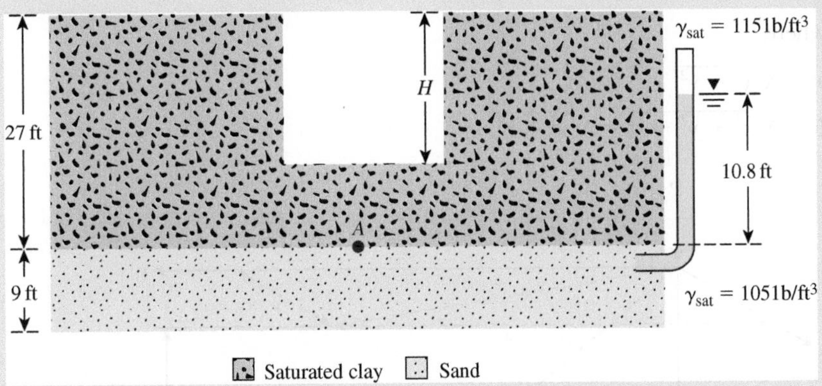

Figure 7.6

Solution

Due to excavation, there will be unloading of the overburden pressure. Let the depth of the cut be H, at which point the bottom will heave. Let us consider the stability of point A at that time:

$$\sigma_A = (27 - H)\gamma_{sat(clay)}$$

$$u_A = 10.8\gamma_w$$

For heave to occur, σ'_A should be 0. So

$$\sigma_A - u_A = (27 - H)\gamma_{sat(clay)} - 10.8\gamma_w$$

or

$$(27 - H)115 - (10.8)62.4 = 0$$

$$H = \frac{(27)115 - (10.8)62.4}{115} = \mathbf{21.13\ ft}$$

■

Vertical Stress Increase Due to Various Types of Loading

7.4 Stress Caused by a Point Load

Soil is a particulate medium consisting of soil grains. For simplicity, we treat it as a continuous medium (or continuum) which behaves as a linear elastic material.

Boussinesq (1883) solved the problem of stresses produced at any point in a homogeneous, elastic, and isotropic medium as the result of a point load applied on the surface of an infinitely large half-space. According to Figure 7.7, Boussinesq's solution for normal vertical stresses at a point caused by the point load P is

$$\Delta\sigma_z = \frac{3P}{2\pi}\frac{z^3}{L^5} = \frac{3P}{2\pi}\frac{z^3}{(r^2 + z^2)^{5/2}} \tag{7.9}$$

where

$r = \sqrt{x^2 + y^2}$

$L = \sqrt{x^2 + y^2 + z^2} = \sqrt{r^2 + z^2}$

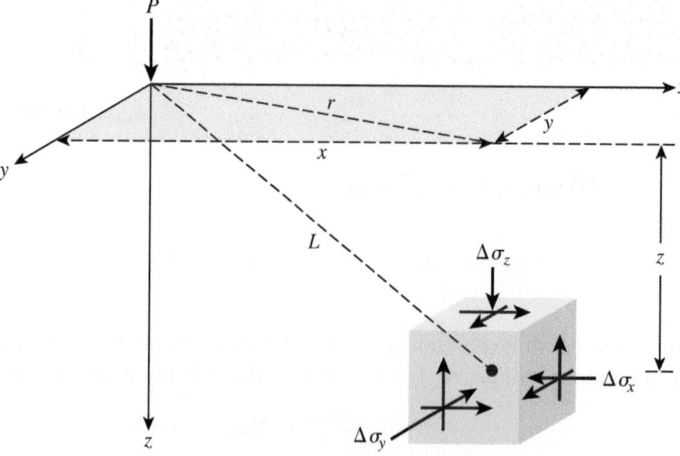

Figure 7.7 Stresses in an elastic medium caused by a point load

Table 7.1 Variation of I_1 for Various Values of r/z [Eq. (7.11)]

r/z	I_1	r/z	I_1	r/z	I_1
0	0.4775	0.36	0.3521	1.80	0.0129
0.02	0.4770	0.38	0.3408	2.00	0.0085
0.04	0.4765	0.40	0.3294	2.20	0.0058
0.06	0.4723	0.45	0.3011	2.40	0.0040
0.08	0.4699	0.50	0.2733	2.60	0.0029
0.10	0.4657	0.55	0.2466	2.80	0.0021
0.12	0.4607	0.60	0.2214	3.00	0.0015
0.14	0.4548	0.65	0.1978	3.20	0.0011
0.16	0.4482	0.70	0.1762	3.40	0.00085
0.18	0.4409	0.75	0.1565	3.60	0.00066
0.20	0.4329	0.80	0.1386	3.80	0.00051
0.22	0.4242	0.85	0.1226	4.00	0.00040
0.24	0.4151	0.90	0.1083	4.20	0.00032
0.26	0.4050	0.95	0.0956	4.40	0.00026
0.28	0.3954	1.00	0.0844	4.60	0.00021
0.30	0.3849	1.20	0.0513	4.80	0.00017
0.32	0.3742	1.40	0.0317	5.00	0.00014
0.34	0.3632	1.60	0.0200		

The relationship for $\Delta\sigma_z$ can be rewritten as

$$\Delta\sigma_z = \frac{P}{z^2}\left\{\frac{3}{2\pi}\frac{1}{[(r/z)^2 + 1]^{5/2}}\right\} = \frac{P}{z^2}I_1 \tag{7.10}$$

where

$$I_1 = \frac{3}{2\pi}\frac{1}{[(r/z)^2 + 1]^{5/2}} \tag{7.11}$$

The variation of I_1 for various values of r/z is given in Table 7.1.

Example 7.4

Consider a point load $P = 1000$ lb (Fig. 7.7). Calculate the variation of vertical stress increase $\Delta\sigma_z$ with depth due to the point load below the ground surface with $x = 3$ ft and $y = 4$ ft.

Solution

$$r = \sqrt{x^2 + y^2} = \sqrt{3^2 + 4^2} = 5 \text{ ft}$$

The following table can now be prepared:

r (ft)	z (ft)	$\frac{r}{z}$	I_1^a	$\Delta\sigma_z = \frac{P}{z^2}I_1^b$ (lb/ft²)
5.0	0	∞	0	0
	2	2.5	0.0034	0.85
	4	1.25	0.0424	2.65
	6	0.83	0.1295	3.60
	10	0.5	0.2733	2.73
	15	0.33	0.3713	1.65
	20	0.25	0.4103	1.03

[a]Eq. (7.11)

[b]Eq. (7.10). *Note:* $P = 1000$ lb

Example 7.5

A point load of 500 kN is applied at ground level. Plot the lateral variation of the vertical stress increase $\Delta\sigma_z$ at depths of 2 m, 3 m, and 4 m below the ground level.

Solution

$\Delta\sigma_z$ values were computed from Eq. (7.10) and plotted for the three depths in Figure 7.8. It can be seen that $\Delta\sigma_z$ decays with depth and lateral distance.

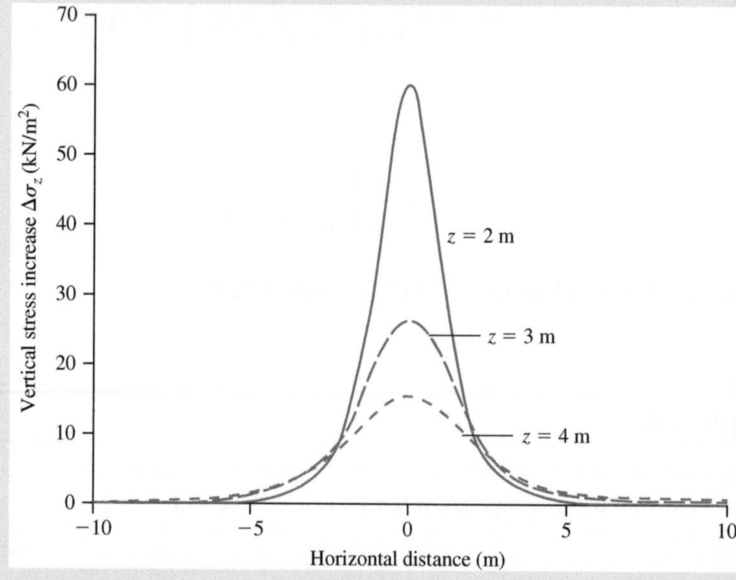

Figure 7.8

7.5 Vertical Stress Below the Center of a Uniformly Loaded Circular Area

Figure 7.9 shows a flexible circular area of radius R. Let the pressure on this area be q. Using Boussinesq's solution for vertical stress $\Delta\sigma_z$ caused by a point load [Eq. (7.10)], one can develop an expression for the vertical stress below the center of the loaded area as,

$$\Delta\sigma_z = q\left\{1 - \frac{1}{[(R/z)^2 + 1]^{3/2}}\right\} \tag{7.12}$$

The variation of $\Delta\sigma_z/q$ with z/R as obtained from Eq. (7.12) is given in Table 7.2. A plot of this also is shown in Figure 7.10. The value of $\Delta\sigma_z$ decreases rapidly with depth, and at $z = 5R$, it is about 6% of q, which is the intensity of pressure at the ground surface.

Table 7.2 Variation of $\Delta\sigma_z/q$ with z/R [Eq. (7.12)]

z/R	$\Delta\sigma_z/q$	z/R	$\Delta\sigma_z/q$
0	1	1.0	0.6465
0.02	0.9999	1.5	0.4240
0.05	0.9998	2.0	0.2845
0.10	0.9990	2.5	0.1996
0.2	0.9925	3.0	0.1436
0.4	0.9488	4.0	0.0869
0.5	0.9106	5.0	0.0571
0.8	0.7562		

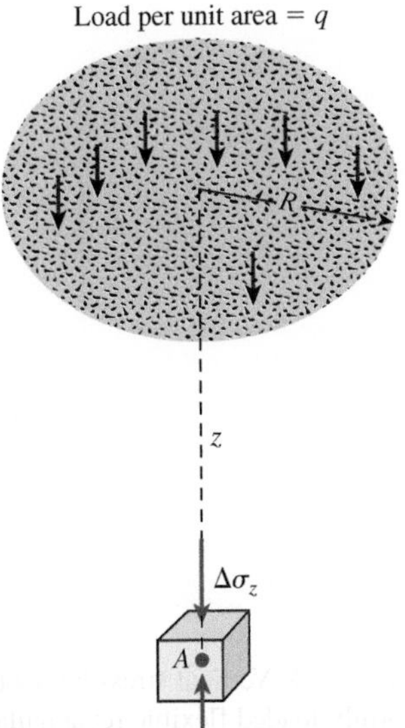

Figure 7.9 Vertical stress below the center of a uniformly loaded flexible circular area

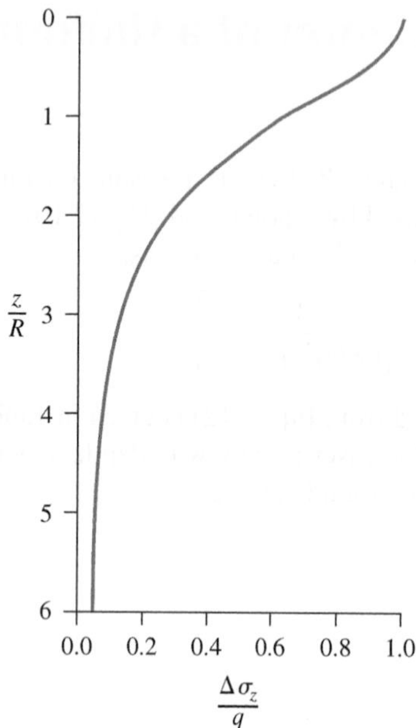

Figure 7.10 Stress under the center of a uniformly loaded flexible circular area

7.6 Vertical Stress Caused by a Uniform Rectangular Load

Boussinesq's solution also can be used to calculate the vertical stress increase below a flexible rectangular loaded area as shown in Figure 7.11. The loaded area is located at the ground surface and has length L and width B. The uniformly distributed load per unit area is equal to q. The increase in the vertical stress ($\Delta\sigma_z$) at point A, which is located at depth z below the corner of the rectangular area can be expressed as

$$\Delta\sigma_z = qI_2 \tag{7.13}$$

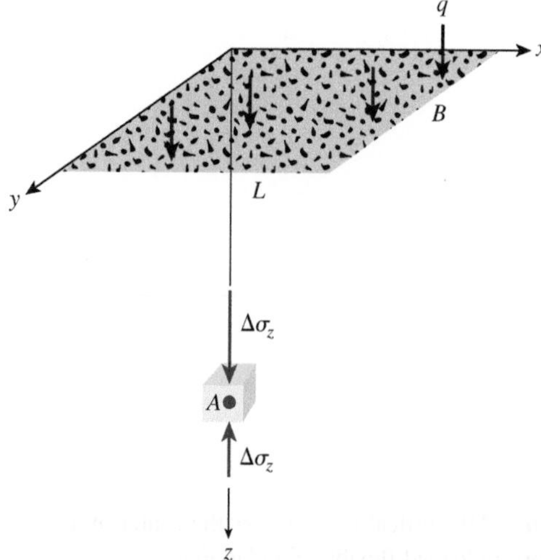

Figure 7.11 Vertical stress below the corner of a uniformly loaded flexible rectangular area

where

$$I_2 = \frac{1}{4\pi}\left[\frac{2mn\sqrt{m^2+n^2+1}}{m^2+n^2+m^2n^2+1}\left(\frac{m^2+n^2+2}{m^2+n^2+1}\right) + \tan^{-1}\left(\frac{2mn\sqrt{m^2+n^2+1}}{m^2+n^2-m^2n^2+1}\right)\right] \tag{7.14}$$

$$m = \frac{B}{z} \tag{7.15}$$

$$n = \frac{L}{z} \tag{7.16}$$

The variation of I_2 with m and n is shown in Table 7.3 and Figure 7.12. When using Table 7.3 or Figure 7.12, m and n are interchangable.

The increase in the stress at any point below a rectangularly loaded area can be found by using Eq. (7.13). This can be explained by reference to Figure 7.13. Let us determine the stress at a point below point A' at depth z. The loaded area can be divided into four rectangles as shown. The point A' is the corner common to all four rectangles. The increase in the stress at depth z below point A' due to each rectangular area can now be calculated by using Eq. (7.13). The total stress increase caused by the entire loaded area can be given by

$$\Delta\sigma_z = q[I_{2(1)} + I_{2(2)} + I_{3(3)} + I_{4(4)}] \tag{7.17}$$

where $I_{2(1)}$, $I_{2(2)}$, $I_{2(3)}$, and $I_{2(4)}$ = values of I_2 for rectangles 1, 2, 3, and 4, respectively.

Table 7.3 Variation of I_2 with m and n [Eq. (7.14)]

	m									
n	**0.1**	**0.2**	**0.3**	**0.4**	**0.5**	**0.6**	**0.7**	**0.8**	**0.9**	**1.0**
0.1	0.0047	0.0092	0.0132	0.0168	0.0198	0.0222	0.0242	0.0258	0.0270	0.0279
0.2	0.0092	0.0179	0.0259	0.0328	0.0387	0.0435	0.0474	0.0504	0.0528	0.0547
0.3	0.0132	0.0259	0.0374	0.0474	0.0559	0.0629	0.0686	0.0731	0.0766	0.0794
0.4	0.0168	0.0328	0.0474	0.0602	0.0711	0.0801	0.0873	0.0931	0.0977	0.1013
0.5	0.0198	0.0387	0.0559	0.0711	0.0840	0.0947	0.1034	0.1104	0.1158	0.1202
0.6	0.0222	0.0435	0.0629	0.0801	0.0947	0.1069	0.1168	0.1247	0.1311	0.1361
0.7	0.0242	0.0474	0.0686	0.0873	0.1034	0.1169	0.1277	0.1365	0.1436	0.1491
0.8	0.0258	0.0504	0.0731	0.0931	0.1104	0.1247	0.1365	0.1461	0.1537	0.1598
0.9	0.0270	0.0528	0.0766	0.0977	0.1158	0.1311	0.1436	0.1537	0.1619	0.1684
1.0	0.0279	0.0547	0.0794	0.1013	0.1202	0.1361	0.1491	0.1598	0.1684	0.1752
1.2	0.0293	0.0573	0.0832	0.1063	0.1263	0.1431	0.1570	0.1684	0.1777	0.1851
1.4	0.0301	0.0589	0.0856	0.1094	0.1300	0.1475	0.1620	0.1739	0.1836	0.1914
1.6	0.0306	0.0599	0.0871	0.1114	0.1324	0.1503	0.1652	0.1774	0.1874	0.1955
1.8	0.0309	0.0606	0.0880	0.1126	0.1340	0.1521	0.1672	0.1797	0.1899	0.1981
2.0	0.0311	0.0610	0.0887	0.1134	0.1350	0.1533	0.1686	0.1812	0.1915	0.1999
2.5	0.0314	0.0616	0.0895	0.1145	0.1363	0.1548	0.1704	0.1832	0.1938	0.2024
3.0	0.0315	0.0618	0.0898	0.1150	0.1368	0.1555	0.1711	0.1841	0.1947	0.2034
4.0	0.0316	0.0619	0.0901	0.1153	0.1372	0.1560	0.1717	0.1847	0.1954	0.2042
5.0	0.0316	0.0620	0.0901	0.1154	0.1374	0.1561	0.1719	0.1849	0.1956	0.2044
6.0	0.0316	0.0620	0.0902	0.1154	0.1374	0.1562	0.1719	0.1850	0.1957	0.2045

(*continued*)

Table 7.3 (continued)

m									
1.2	**1.4**	**1.6**	**1.8**	**2.0**	**2.5**	**3.0**	**4.0**	**5.0**	**6.0**
0.0293	0.0301	0.0306	0.0309	0.0311	0.0314	0.0315	0.0316	0.0316	0.0316
0.0573	0.0589	0.0599	0.0606	0.0610	0.0616	0.0618	0.0619	0.0620	0.0620
0.0832	0.0856	0.0871	0.0880	0.0887	0.0895	0.0898	0.0901	0.0901	0.0902
0.1063	0.1094	0.1114	0.1126	0.1134	0.1145	0.1150	0.1153	0.1154	0.1154
0.1263	0.1300	0.1324	0.1340	0.1350	0.1363	0.1368	0.1372	0.1374	0.1374
0.1431	0.1475	0.1503	0.1521	0.1533	0.1548	0.1555	0.1560	0.1561	0.1562
0.1570	0.1620	0.1652	0.1672	0.1686	0.1704	0.1711	0.1717	0.1719	0.1719
0.1684	0.1739	0.1774	0.1797	0.1812	0.1832	0.1841	0.1847	0.1849	0.1850
0.1777	0.1836	0.1874	0.1899	0.1915	0.1938	0.1947	0.1954	0.1956	0.1957
0.1851	0.1914	0.1955	0.1981	0.1999	0.2024	0.2034	0.2042	0.2044	0.2045
0.1958	0.2028	0.2073	0.2103	0.2124	0.2151	0.2163	0.2172	0.2175	0.2176
0.2028	0.2102	0.2151	0.2184	0.2206	0.2236	0.2250	0.2260	0.2263	0.2264
0.2073	0.2151	0.2203	0.2237	0.2261	0.2294	0.2309	0.2320	0.2323	0.2325
0.2103	0.2183	0.2237	0.2274	0.2299	0.2333	0.2350	0.2362	0.2366	0.2367
0.2124	0.2206	0.2261	0.2299	0.2325	0.2361	0.2378	0.2391	0.2395	0.2397
0.2151	0.2236	0.2294	0.2333	0.2361	0.2401	0.2420	0.2434	0.2439	0.2441
0.2163	0.2250	0.2309	0.2350	0.2378	0.2420	0.2439	0.2455	0.2461	0.2463
0.2172	0.2260	0.2320	0.2362	0.2391	0.2434	0.2455	0.2472	0.2479	0.2481
0.2175	0.2263	0.2324	0.2366	0.2395	0.2439	0.2460	0.2479	0.2486	0.2489
0.2176	0.2264	0.2325	0.2367	0.2397	0.2441	0.2463	0.2482	0.2489	0.2492

Note: m and n are interchangable.

In most cases, the vertical stress increase below the center of a rectangular area is the maximum (Figure 7.14) and hence is important. This stress increase can be given by the relationship

$$\Delta\sigma_z = qI_3 \tag{7.18}$$

where

$$I_3 = \frac{2}{\pi}\left[\frac{m_1 n_1}{\sqrt{1 + m_1^2 + n_1^2}} \frac{1 + m_1^2 + 2n_1^2}{(1 + n_1^2)(m_1^2 + n_1^2)} + \sin^{-1}\frac{m_1}{\sqrt{m_1^2 + n_1^2}\sqrt{1 + n_1^2}}\right] \tag{7.19}$$

$$m_1 = \frac{L}{B} \tag{7.20}$$

$$n_1 = \frac{z}{b} \tag{7.21}$$

$$b = \frac{B}{2} \tag{7.22}$$

The variation of I_3 with m_1 and n_1 is given in Table 7.4.

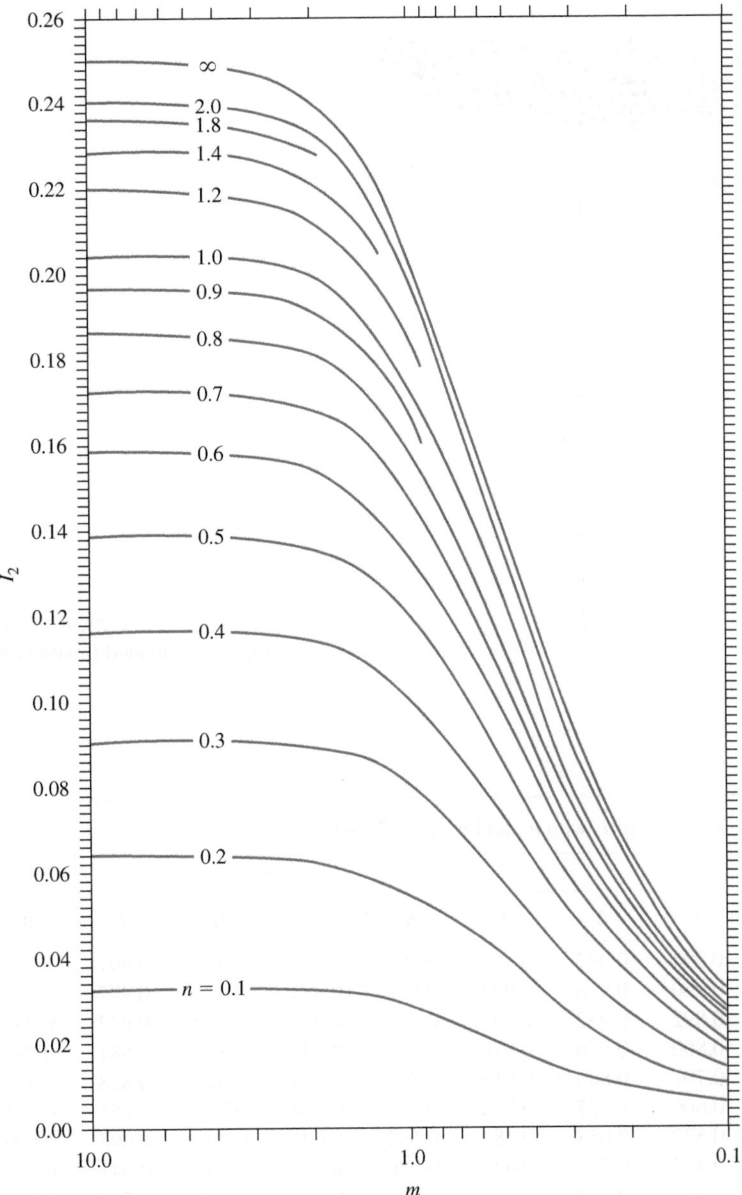

Figure 7.12 Variation of I_2 with m and n

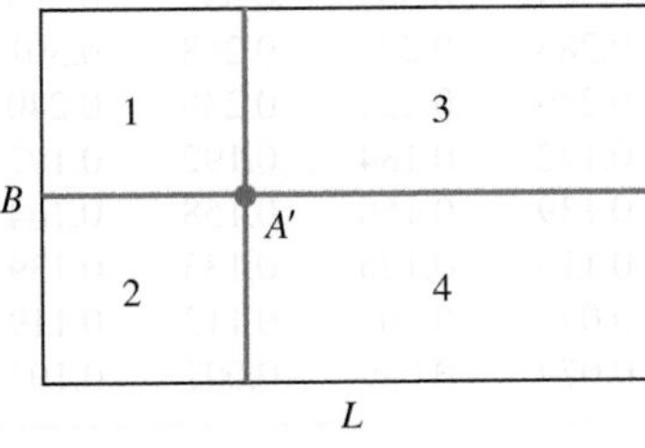

Figure 7.13 Increase of stress at any point below a rectangularly loaded flexible area

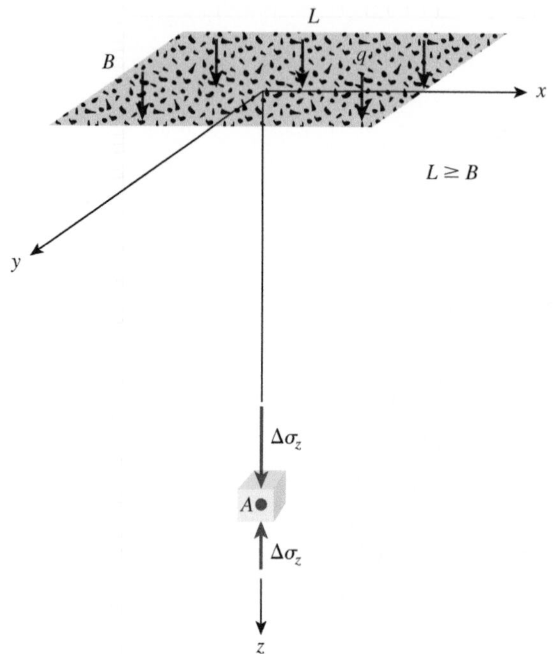

Figure 7.14 Vertical stress below the center of a uniformly loaded flexible rectangular area

Table 7.4 Variation of I_3 with m_1 and n_1 [Eq. (7.19)]

n_1	m_1 = 1	2	3	4	5	6	7	8	9	10
0.20	0.994	0.997	0.997	0.997	0.997	0.997	0.997	0.997	0.997	0.997
0.40	0.960	0.976	0.977	0.977	0.977	0.977	0.977	0.977	0.977	0.977
0.60	0.892	0.932	0.936	0.936	0.937	0.937	0.937	0.937	0.937	0.937
0.80	0.800	0.870	0.878	0.880	0.881	0.881	0.881	0.881	0.881	0.881
1.00	0.701	0.800	0.814	0.817	0.818	0.818	0.818	0.818	0.818	0.818
1.20	0.606	0.727	0.748	0.753	0.754	0.755	0.755	0.755	0.755	0.755
1.40	0.522	0.658	0.685	0.692	0.694	0.695	0.695	0.696	0.696	0.696
1.60	0.449	0.593	0.627	0.636	0.639	0.640	0.641	0.641	0.641	0.642
1.80	0.388	0.534	0.573	0.585	0.590	0.591	0.592	0.592	0.593	0.593
2.00	0.336	0.481	0.525	0.540	0.545	0.547	0.548	0.549	0.549	0.549
3.00	0.179	0.293	0.348	0.373	0.384	0.389	0.392	0.393	0.394	0.395
4.00	0.108	0.190	0.241	0.269	0.285	0.293	0.298	0.301	0.302	0.303
5.00	0.072	0.131	0.174	0.202	0.219	0.229	0.236	0.240	0.242	0.244
6.00	0.051	0.095	0.130	0.155	0.172	0.184	0.192	0.197	0.200	0.202
7.00	0.038	0.072	0.100	0.122	0.139	0.150	0.158	0.164	0.168	0.171
8.00	0.029	0.056	0.079	0.098	0.113	0.125	0.133	0.139	0.144	0.147
9.00	0.023	0.045	0.064	0.081	0.094	0.105	0.113	0.119	0.124	0.128
10.00	0.019	0.037	0.053	0.067	0.079	0.089	0.097	0.103	0.108	0.112

Example 7.6

The cross section and plan of a loaded flexible area is shown in Figure 7.15. Determine the vertical stress increase ($\Delta\sigma_z$) at a depth 10 ft at point A.

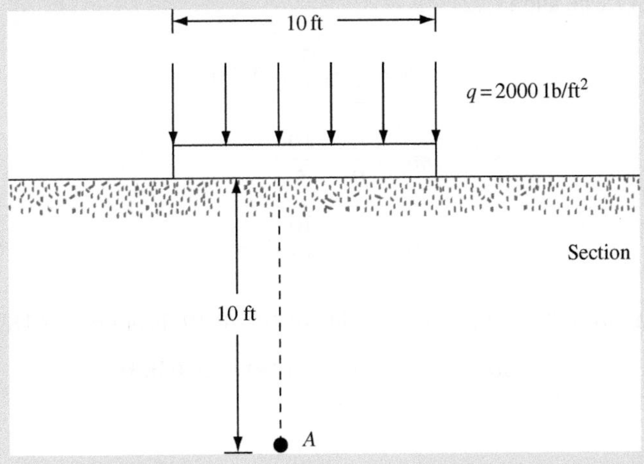

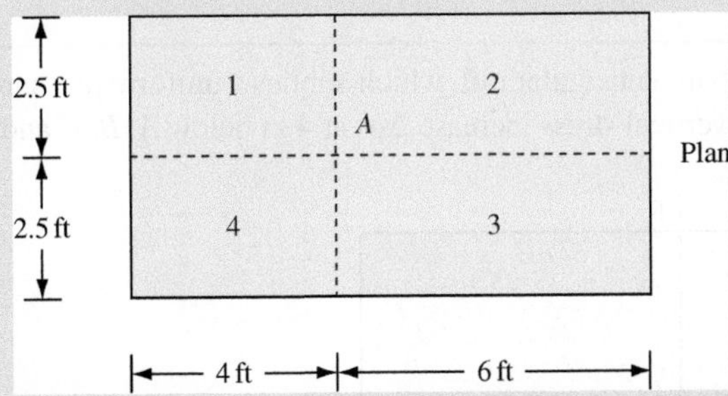

Figure 7.15

Solution

We can prepare the following table.

Area No.	B (ft)	L (ft)	z (ft)	$m = \frac{B}{z}$	$n = \frac{L}{z}$	I_2 (Table 7.3)
1	2.5	4	10	0.25	0.4	≈ 0.04
2	2.5	6	10	0.25	0.6	≈ 0.053
3	2.5	6	10	0.25	0.6	≈ 0.053
4	2.5	4	10	0.25	0.4	≈ 0.04
						$\Sigma I_2 = 0.186$

$$\Delta\sigma_z = (q)(\Sigma I_2) = (2000)(0.186) = \mathbf{372\ lb/ft^2}$$ ■

Example 7.7

Refer to Figure 7.15. Using Eq. (7.18), determine the vertical stress increase below the center of the flexible loaded area at a depth of 10 ft.

Solution

From Eqs. (7.20), (7.21), and (7.22),

$$b = \frac{B}{2} = \frac{5}{2} = 2.5 \text{ ft}$$

$$m_1 = \frac{L}{B} = \frac{10}{5} = 2$$

$$n_1 = \frac{z}{b} = \frac{10}{2.5} = 4$$

From Table 7.4, for $m_1 = 2$ and $n_1 = 4$, the value of $I_3 = 0.19$. From Eq. (7.18),

$$\Delta\sigma_z = qI_3 = (2000)(0.19) = \mathbf{380\ lb/ft^2}$$

■

Example 7.8

Figure 7.16 shows a 8 m × 16 m rectangular raft, which applies a uniform pressure of 80 kN/m² to the underlying ground. Find the vertical stress increase $\Delta\sigma_z$ at 4 m below *A, B, C* and *D* using Eq. (7.13).

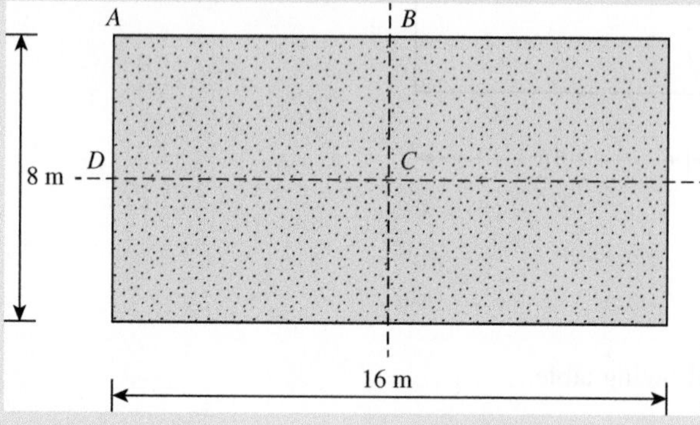

Figure 7.16

Solution

Point *A*:

Considering the whole raft, $B = 8$ m, $L = 16$ m, and $z = 4$ m. Thus,

$$m = B/z = 8/4 = 2.0 \text{ and } n = L/z = 16/4 = 4.0$$

From Table 7.3, $I_2 = 0.2391$

Therefore, $\Delta\sigma_z = 0.2391 \times 80 = \mathbf{19.1\ kN/m^2}$

Point *B*:
Considering the left half of the raft, $B = L = 8$ m and $z = 4$ m. Thus,

$$m = n = 8/4 = 2$$

From Table 7.3, $I_2 = 0.2325$
Therefore, $\Delta\sigma_z = 2 \times 0.2325 \times 80 =$ **37.2 kN/m²**

Point *C*:
Considering quarter of the raft, $B = 4$ m, $L = 8$ m, and $z = 4$ m. Thus,

$$m = 4/4 = 1.0 \text{ and } n = 8/4 = 2.0$$

From Table 7.3, $I_2 = 0.1999$
Therefore, $\Delta\sigma_z = 4 \times 0.1999 \times 80 =$ **64.0 kN/m²**

Point *D*:
Considering upper half of the raft, $B = 4$ m, $L = 16$ m, and $z = 4$ m. Thus,

$$m = 4/4 = 1.0 \text{ and } n = 16/4 = 4.0$$

From Table 7.3, $I_2 = 0.2042$
Therefore, $\Delta\sigma_z = 2 \times 0.2042 \times 80 =$ **32.7 kN/m²** ■

Example 7.9

The flexible area shown in Figure 7.17 is uniformly loaded. Given that $q = 3000$ lb/ft², determine the vertical stress increase at point *A*.

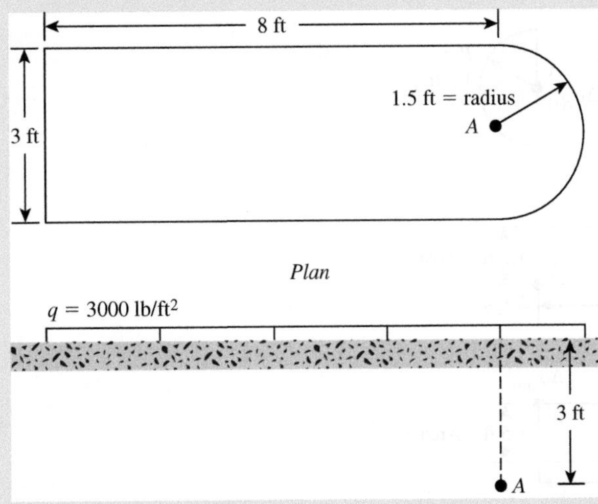

Figure 7.17 Uniformly loaded flexible area

Solution

The flexible area shown in Figure 7.17 is divided into three parts in Figure 7.18. At *A*,

$$\Delta\sigma_z = \Delta\sigma_{z(1)} + \Delta\sigma_{z(2)} + \Delta\sigma_{z(3)}$$

From Eq. (7.12),

$$\Delta\sigma_{z(1)} = \left(\frac{1}{2}\right)q\left\{1 - \frac{1}{[(R/z)^2 + 1]^{3/2}}\right\}$$

We know that $R = 1.5$ ft, $z = 3$ ft, and $q = 3000$ lb/ft^2, so

$$\Delta\sigma_{z(1)} = \frac{3000}{2}\left\{1 - \frac{1}{[(1.5/3)^2 + 1]^{3/2}}\right\} = 426.7 \text{ lb/ft}^2$$

We can see that $\Delta\sigma_{z(2)} = \Delta\sigma_{z(3)}$. From Eqs. (7.15) and (7.16),

$$m = \frac{1.5}{3} = 0.5$$

$$n = \frac{8}{3} = 2.67$$

From Table 7.3, for $m = 0.5$ and $n = 2.67$, the magnitude of $I_2 = 0.1264$. Thus, from Eq. (7.13),

$$\Delta\sigma_{z(2)} = \Delta\sigma_{z(3)} = qI_2 = (3000)(0.1264) = 379.2 \text{ lb/ft}^2$$

so

$$\Delta\sigma_z = 426.7 + 379.2 + 379.2 = \mathbf{1185.1\ lb/ft^2}$$

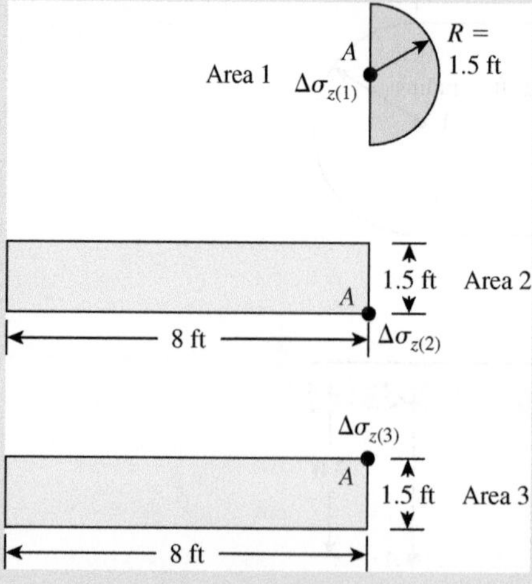

Figure 7.18 Division of uniformly loaded flexible area into three parts ■

7.7 Summary

In saturated soils, the normal stress at a point is carried partly by the soil grains as effective stress and the rest by the pore water in the form of pore water pressure. The total stress at a point is therefore the sum of these two components: effective stress and pore water pressure. This is known as Terzaghi's effective stress principle, which is valid in all directions.

When calculating the stresses beneath surface loads, for simplicity, we treat the soil as a continuum which has linear elastic properties. Charts and tables are provided in this chapter for computing vertical stress increases beneath point loads and selected shapes of surface loads such as rectangles and circles. These would become useful in computing settlements discussed in Chapters 8 and 12, where it is necessary to determine the vertical stress increase at specific depths due to a foundation load.

Problems

7.1 State whether the following are true or false.

a. In dry soils, total stress is the same as the effective stress.

b. Due to upward seepage, effective stress increases.

c. During upward seepage, the larger the hydraulic gradient, the larger pore water pressure.

d. Vertical stress increase due to surface load decays with depth.

e. At quick condition, the effective stress becomes zero.

7.2 through 7.3 Refer to Figure 7.19. Calculate σ, u, and σ' at A, B, C, and D for the following cases and plot the variations with depth. (*Note:* γ_d = dry unit weight, and γ_{sat} = saturated unit weight.)

	Details of soil layer		
Problem	**I**	**II**	**III**
7.2	$H_1 = 2$ ft $\gamma_d = 115$ lb/ft^3	$H_2 = 4$ ft $\gamma_{sat} = 118$ lb/ft^3	$H_3 = 6$ ft $\gamma_{sat} = 130$ lb/ft^3
7.3	$H_1 = 9$ ft $\gamma_d = 100$ lb/ft^3	$H_2 = 21$ ft $\gamma_{sat} = 122$ lb/ft^3	$H_3 = 36$ ft $\gamma_{sat} = 128$ lb/ft^3

7.4 Refer to the soil profile shown in Figure 7.20

a. Calculate and plot the variation of σ, u, and σ' with depth.

b. If the water table rises to the top of the ground surface, what is the change in the effective stress at the bottom of the clay layer?

c. How many feet must the groundwater table rise to decrease the effective stress by 300 lb/ft^2 at the bottom of the clay layer?

7.5 The soil profile at a site consists of 10 m of gravelly sand underlain by a soft clay layer. The water table lies 1 m below the ground level. The moist and saturated unit weights of the gravelly sand are 17.0 kN/m^3 and 20.0 kN/m^3, respectively. Due to some ongoing construction work, it is proposed to lower the water table to 3 m below the ground level. What will be the change in the effective stress on top of the soft clay layer?

7.6 The depth of water in a lake is 4 m. The soil at the bottom of the lake consists of sandy clay. The moisture content of the soil was determined to be 25.0%. The specific gravity of the soil

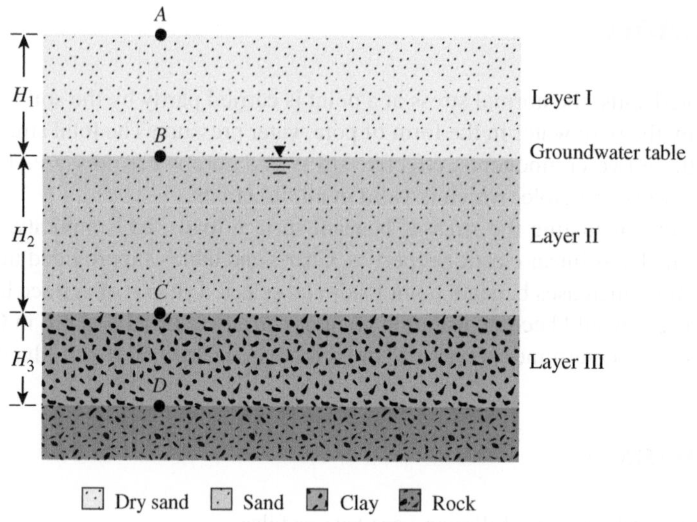

Figure 7.19

grains is 2.70. Determine the void ratio and the unit weight of the soil. What would be the total stress, effective stress, and pore water pressure at a 5 m depth into the bottom of the lake?

7.7 If the water level in the lake in Problem 7.6 rises by 2 m, what would be the total stress, pore water pressure, and the effective stress at the same location?

7.8 An exploratory drill hole was made in a stiff saturated clay (see Figure 7.21). The sand layer underlying the clay was observed to be under artesian pressure. Water in the drill hole rose to a height of 18 ft above the top of the sand layer. If an open excavation is to be made in the clay, how deep can the excavation proceed before the bottom heaves?

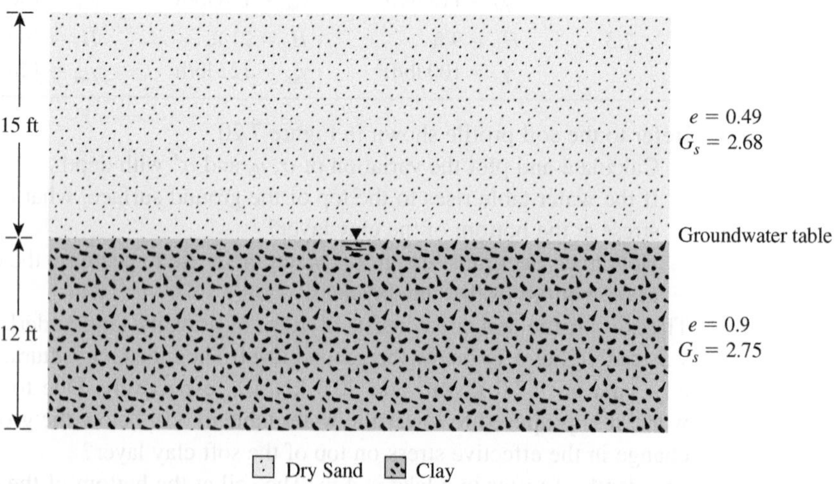

Figure 7.20

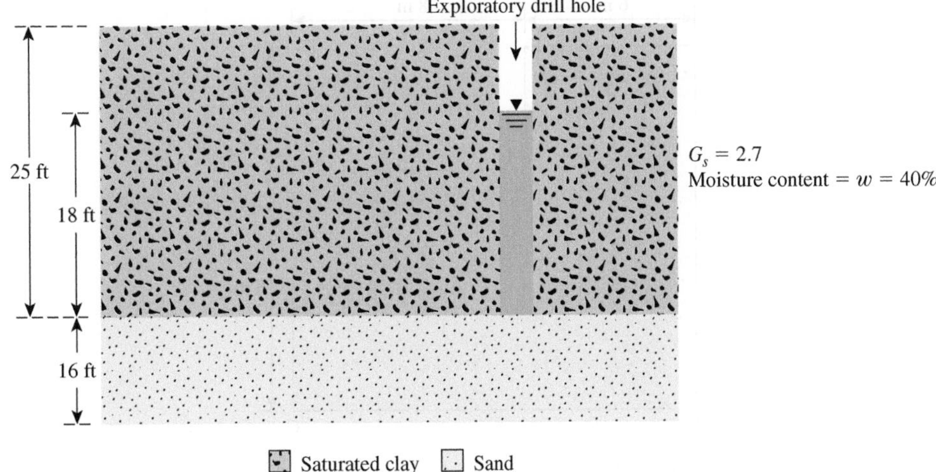

Figure 7.21

7.9 A sand has $G_s = 2.68$. Calculate the hydraulic gradient that will cause boiling for $e = 0.4, 0.5, 0.6$, and 0.7. Plot a graph for i_{cr} versus e.

7.10 A point load of 1000 kN is applied at the ground level. Plot the variation of the vertical stress increase $\Delta\sigma_z$ with depth at horizontal distance of 1 m, 2 m, and 4 m from the load.

7.11 A flexible circular footing of radius R carries a uniform pressure q. Find the depth (in terms of R) at which the vertical stress below the centre is 20% of q.

7.12 Consider a circularly loaded flexible area on the ground surface. Given that the radius of the circular area $R = 6$ ft and that the uniformly distributed load $q = 3500$ lb/ft^2, calculate the vertical stress increase, $\Delta\sigma_z$, at points 1.5, 3, 6, 9, and 12 ft below the ground surface (immediately below the center of the circular area).

7.13 The plan of a flexible rectangular loaded area is shown in Figure 7.22. The uniformly distributed load on the flexible area, q, is 1800 lb/ft^2. Determine the increase in the vertical stress, $\Delta\sigma_z$, at a depth of $z = 5$ ft below

a. Point A
b. Point B
c. Point C

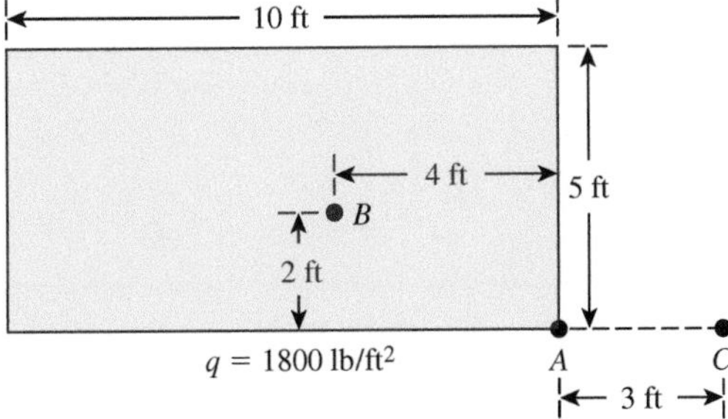

Figure 7.22

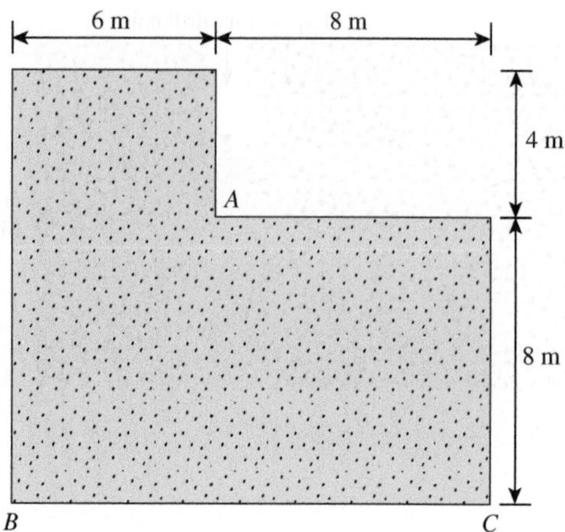

Figure 7.23

7.14 Refer to the flexible loaded rectangular area shown in Figure 7.22. Using Eq. (7.18), determine the vertical stress increase below the center of the area at a depth of 15 ft.

7.15 A flexible L-shaped raft shown in Figure 7.23 applies a uniform pressure of 60 kN/m^2 to the underlying ground. Find the vertical stress increase at 4 m below *A, B* and *C*.

References

Boussinesq, J. (1883). *Application des Potentials à L'Etude de L'Equilibre et du Mouvement des Solides Elastiques*, Gauthier–Villars, Paris.

Terzaghi, K. (1925). *Erdbaumechanik auf Bodenphysikalischer Grundlage*, Deuticke, Vienna.

Terzaghi, K. (1936). "Relation between Soil Mechanics and Foundation Engineering: Presidential Address," *Proceedings*, First International Conference on Soil Mechanics and Foundation Engineering, Boston, Vol. 3, 13–18.

8 Consolidation

8.1 Introduction

A stress increase caused by the construction of foundations, embankments, or other loads (e.g., surcharge) can compress the underlying soil layers, which results in settlement of the ground surface. The compression and subsequent settlement are caused by the displacement of soil grains and reduction in the void spaces that are filled with air and water. In general, the soil settlement caused by the surface loads can be divided into three broad categories.

1. *Elastic settlement*, which is caused by the elastic deformation of dry soil and of moist and saturated soils without any change in the moisture content. Elastic settlement calculations are generally based on equations derived from the theory of elasticity. It is discussed in Chapter 12.
2. *Primary consolidation settlement*, which is the result of a volume change in saturated cohesive soils because of the expulsion of water that occupies the void spaces.
3. *Secondary consolidation settlement*, which is observed in saturated cohesive soils and is the result of the plastic adjustment of soil fabrics. It follows the primary consolidation settlement under a constant effective stress.

This chapter presents the fundamental principles for estimating the consolidation settlement of soil layers under superimposed loadings. In this chapter, you will learn the following.

- The theory behind consolidation.
- The parameters governing consolidation.
- Rate of consolidation.
- Secondary consolidation.
- Laboratory tests for determining consolidation parameters.
- Calculation of consolidation settlements in clays.

8.2 Fundamentals of Consolidation

When a saturated soil layer is subjected to a stress increase, the pore water pressure suddenly increases. In sandy soils that are highly permeable, the drainage caused by the increase in the pore water pressure is completed immediately. Pore water drainage is accompanied by a reduction in the volume of the soil mass, resulting in settlement. Because of the rapid drainage of the pore water in sandy soils, elastic settlement and consolidation take place simultaneously. This is not the case, however, for clay soils, which have low hydraulic conductivity; hence, this consolidation process is very slow. The consolidation settlement is time dependent.

Keeping this in mind, we can analyze the strain of a saturated clay layer subjected to a stress increase (Figure 8.1a). A layer of saturated clay of thickness H is confined between two layers of sand

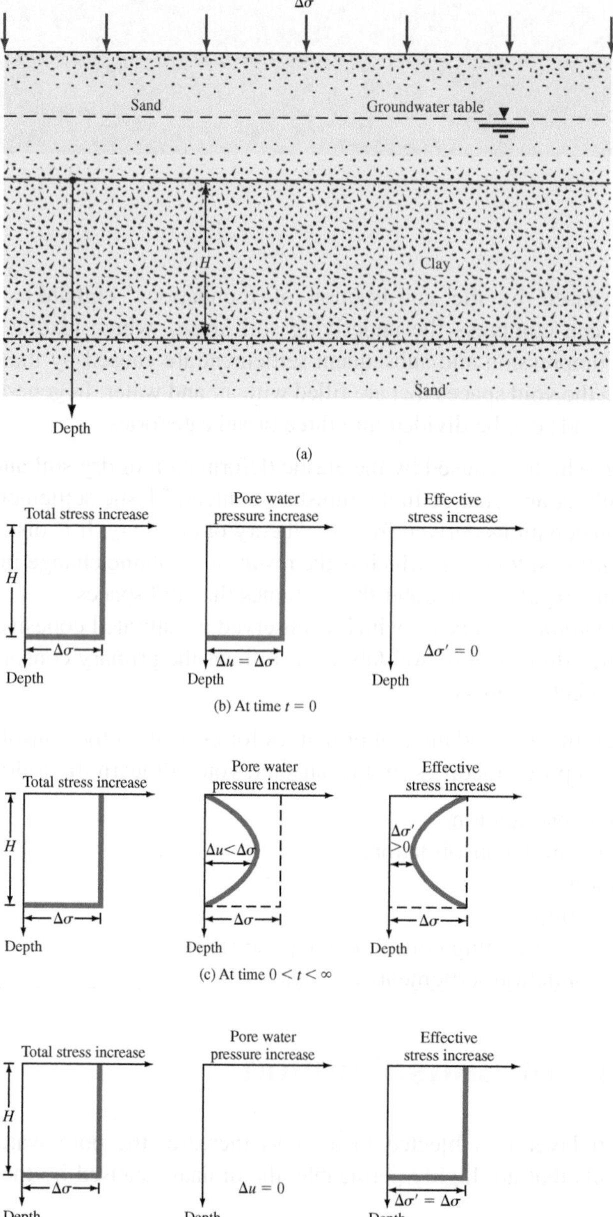

Figure 8.1 Variation of total stress, pore water pressure, and effective stress in a clay layer drained at top and bottom as the result of an added stress, $\Delta\sigma$

and is subjected to an instantaneous increase in *total stress* of $\Delta\sigma$. From Chapter 7, we know that

$$\Delta\sigma = \Delta\sigma' + \Delta u \tag{8.1}$$

where

$\Delta\sigma'$ = increase in the effective stress
Δu = increase in the pore water pressure

Since clay has very low hydraulic conductivity and water is incompressible compared with the soil skeleton, at time $t = 0$, the entire incremental stress, $\Delta\sigma$, will be carried by water ($\Delta\sigma = \Delta u$) at all depths (Figure 8.1b). None will be carried by the soil skeleton (that is, incremental effective stress, $\Delta\sigma' = 0$).

After the application of incremental stress, $\Delta\sigma$, to the clay layer, the water in the void spaces will begin to be squeezed out and will drain vertically in both directions into the sand layers. By this process, the excess pore water pressure at any depth on the clay layer will gradually decrease, and the stress carried by the soil grains (effective stress) will increase. Thus, at time $0 < t < \infty$,

$$\Delta\sigma = \Delta\sigma' + \Delta u \qquad (\Delta\sigma' > 0 \text{ and } \Delta u < \Delta\sigma)$$

However, the magnitudes of $\Delta\sigma'$ and Δu at various depths will change (Figure 8.1c), depending on the minimum distance of the drainage path to either the top or bottom sand layer.

Theoretically, at time $t = \infty$, the entire excess pore water pressure would dissipate by drainage from all points of the clay layer, thus giving $\Delta u = 0$. Then the total stress increase, $\Delta\sigma$, would be carried entirely by the soil structure (Figure 8.1d), so

$$\Delta\sigma = \Delta\sigma'$$

This gradual process of drainage under the application of an additional load and the associated transfer of excess pore water pressure to effective stress causes the time-dependent settlement (consolidation) in the clay soil layer. This process can take several months to a few years, depending on the thickness of the clay layer.

8.3 One-Dimensional Laboratory Consolidation Test

The one-dimensional consolidation testing procedure was first suggested by Terzaghi (1925). This test is performed in a consolidometer (sometimes referred to as an oedometer). Figure 8.2 is a

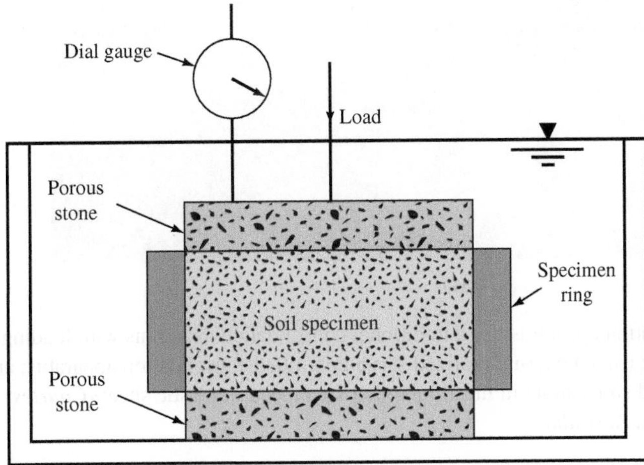

Figure 8.2 Consolidometer

schematic diagram of a consolidometer. The soil specimen is placed inside a metal ring with two porous stones, one at the top of the specimen and another at the bottom.

The specimens are usually 2.5 in. (63.5 mm) in diameter and 1.0 in. (25.4 mm) thick. Traditionally, the specimen is loaded using dead weights applied through a lever arm that magnifies the load acting on the specimen, and the change in the specimen height is measured by a dial gauge (Figure 8.3a). Nowadays, it is common to see motorized and servo-controlled machines such

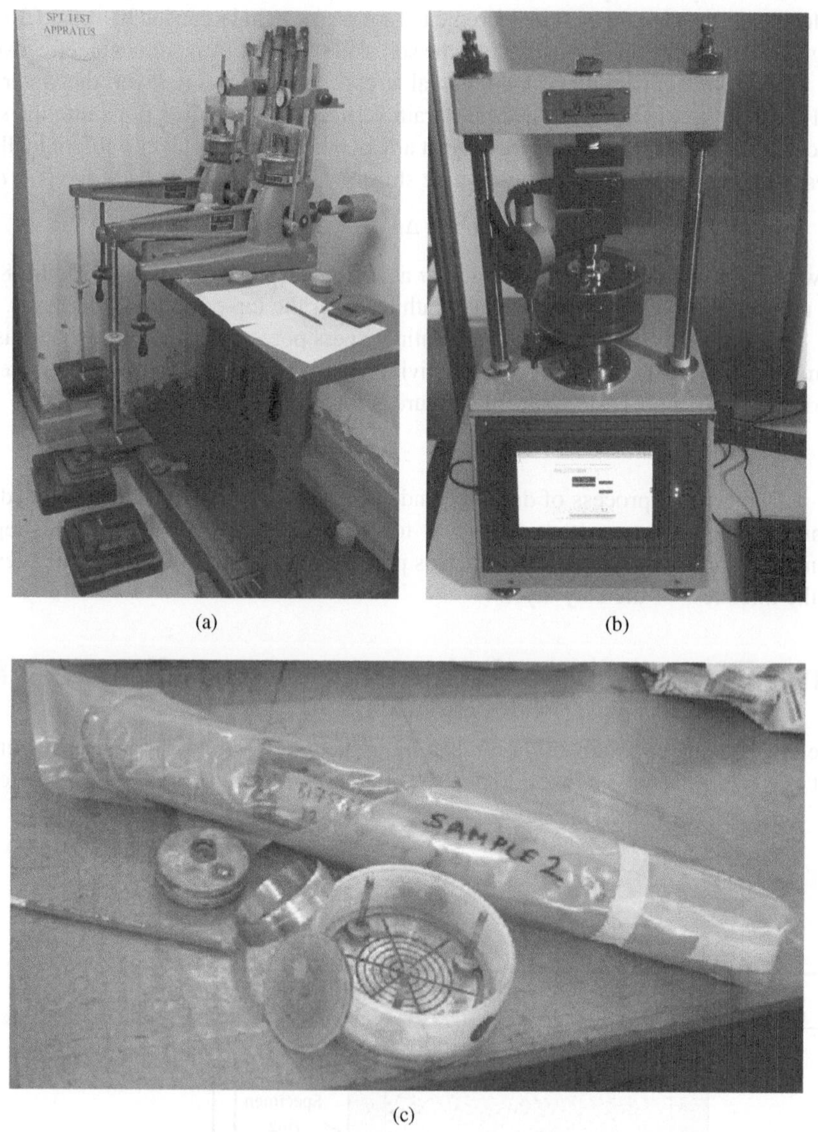

Figure 8.3 Consolidation apparatus: (a) a traditional consolidation apparatus with loading through dead weights and lever arm; (b) a modern motorized servo-controlled consolidation apparatus; and (c) oedoemeter ring, porous stone, and soil sample in metal tube that's wrapped in plastic sheet (*Courtesy of N. Sivakugan, James Cook University, Australia*)

as the one shown in Figure 8.3b. Figure 8.3b shows a consolidation test in progress with an S-type load cell measuring the load applied on to the specimen and a dial gauge measuring the reduction in the thickness of the specimen. Figure 8.3c shows the consolidation cell, the specimen ring, porous stone, and the soil sample contained in a metal tube wrapped in plastic sheets for preventing any loss of moisture. The specimen is kept under water during the test. Each load is usually kept for 24 hours. After that, the load is usually doubled, thus doubling the pressure on the specimen, and the compression measurement is continued. At the end of the test, the dry weight of the test specimen is determined.

The general shape of the plot of deformation of the specimen versus time for a given load increment is shown in Figure 8.4. From the plot, it can be observed that there are three distinct stages, which may be described as follows:

Stage I: Initial compression, which is mostly caused by preloading.

Stage II: Primary consolidation, during which excess pore water pressure is gradually transferred into effective stress by the expulsion of pore water.

Stage III: Secondary consolidation, which occurs after complete dissipation of the excess pore water pressure, when some deformation of the specimen takes place because of the plastic readjustment of soil fabric.

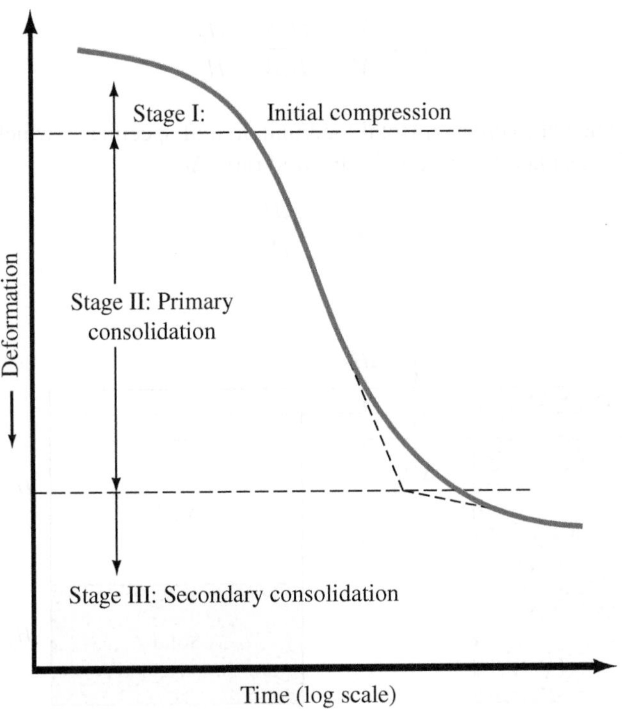

Figure 8.4 Time–deformation plot during consolidation for a given load increment

8.4 Void Ratio–Pressure Plots

After the time–deformation plots for various loadings are obtained in the laboratory, it is necessary to study the change in the void ratio of the specimen with pressure. Following is a step-by-step procedure:

1. Calculate the height of solids, H_s, in the soil specimen (Figure 8.5):

$$H_s = \frac{W_s}{AG_s\gamma_w} \tag{8.2}$$

where

W_s = dry weight of the specimen
A = area of the specimen
G_s = specific gravity of soil solids
γ_w = unit weight of water

2. Calculate the initial height of voids, H_v:

$$H_v = H - H_s \tag{8.3}$$

where H = initial height of the specimen.

3. Calculate the initial void ratio, e_0, of the specimen:

$$e_0 = \frac{V_v}{V_s} = \frac{H_vA}{H_sA} = \frac{H_v}{H_s} \tag{8.4}$$

4. For the first incremental loading σ_1 (total load/unit area of specimen), which causes deformation ΔH_1, calculate the change in the void ratio Δe_1:

$$\Delta e_1 = \frac{\Delta H_1}{H_s} \tag{8.5}$$

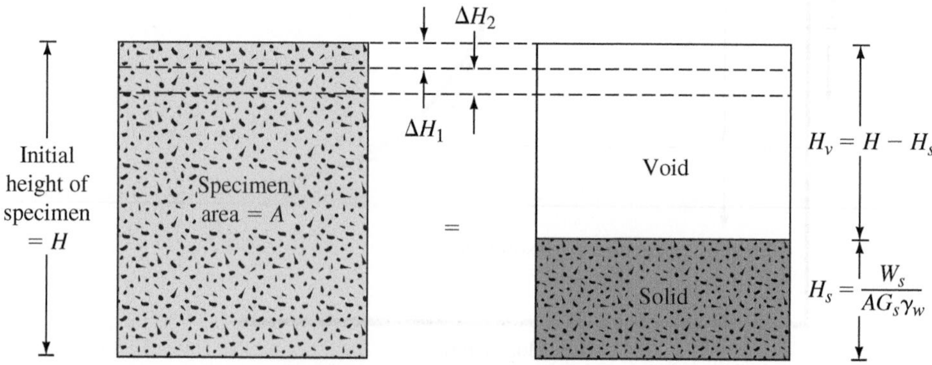

Figure 8.5 Change of height of specimen in one-dimensional consolidation test

ΔH_1 is obtained from the initial and the final dial readings for the loading. At this time, the effective pressure on the specimen is $\sigma' = \sigma_1 = \sigma_1'$.

5. Calculate the new void ratio, e_1, after consolidation caused by the pressure increment σ_1:

$$e_1 = e_0 - \Delta e_1 \tag{8.6}$$

For the next loading, σ_2 (*note*: σ_2 equals the cumulative load per unit area of specimen), which causes additional deformation ΔH_2, the void ratio e_2 at the end of consolidation can be calculated as

$$e_2 = e_1 - \frac{\Delta H_2}{H_s} \tag{8.7}$$

Note that, at this time, the effective pressure on the specimen is $\sigma' = \sigma_2 = \sigma_2'$.

Proceeding in a similar manner, we can obtain the void ratios at the end of the consolidation for all load increments.

The effective pressures ($\sigma = \sigma'$) and the corresponding void ratios (e) at the end of consolidation are plotted on semilogarithmic graph paper. The typical shape of such a plot is shown in Figure 8.6.

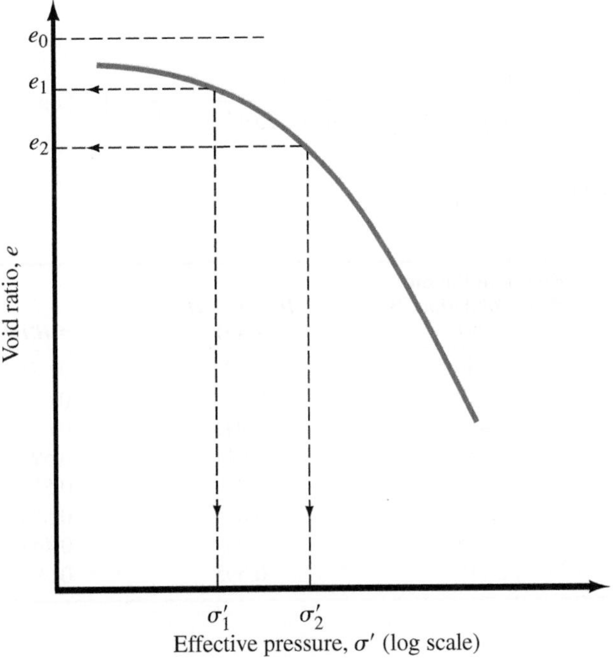

Figure 8.6 Typical plot of e versus log σ'

Example 8.1

Following are the results of a laboratory consolidation test on a soil specimen obtained from the field. Dry mass of specimen = 128 g, height of specimen at the beginning of the test = 2.54 cm, $G_s = 2.75$, and area of the specimen = 30.68 cm^2.

Pressure, σ' (ton/ft^2)	Final height of specimen at the end of consolidation (cm)
0	2.540
0.5	2.488
1	2.465
2	2.431
4	2.389
8	2.324
16	2.225
32	2.115

Make necessary calculations and draw an e versus log σ' curve.

Solution

Calculation of H_s

From Eq. (8.2)

$$H_s = \frac{W_s}{AG_s\gamma_w} = \frac{128 \text{ g}}{(30.68 \text{ cm}^2)(2.75)(1 \text{ g/cm}^3)} = 1.52 \text{ cm}$$

Now the following table can be prepared.

Pressure, σ' (ton/ft^2)	Height at the end of consolidation, H (cm)	$H_v = H - H_s$ (cm)	$e = H_v/H_s$
0	2.540	1.02	0.671
0.5	2.488	0.968	0.637
1	2.465	0.945	0.622
2	2.431	0.911	0.599
4	2.389	0.869	0.572
8	2.324	0.804	0.529
16	2.225	0.705	0.464
32	2.115	0.595	0.39

The e versus log σ' plot is shown in Figure 8.7

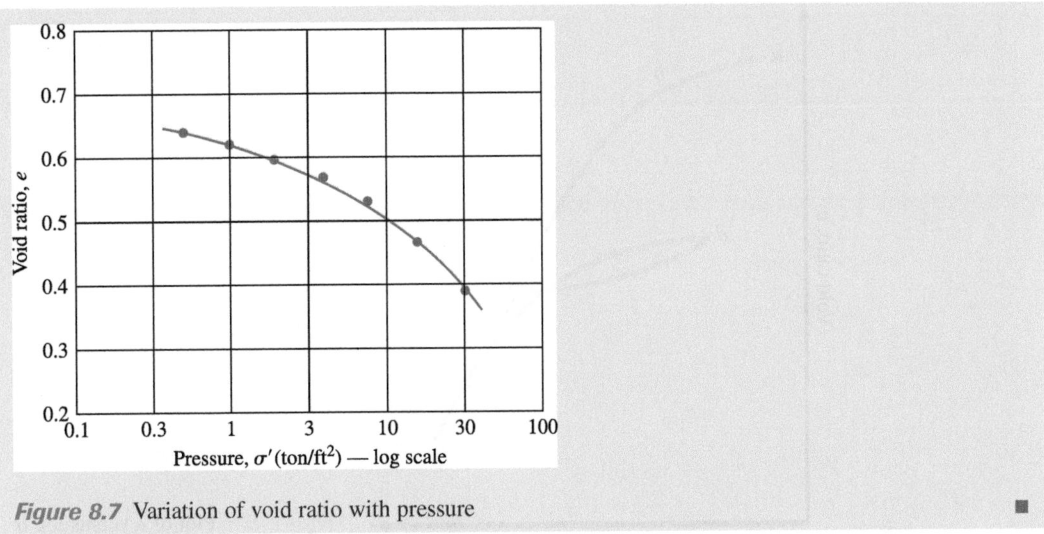

Figure 8.7 Variation of void ratio with pressure ■

8.5 Normally Consolidated and Overconsolidated Clays

Figure 8.6 showed that the upper part of the e–log σ' plot is somewhat curved with a flat slope, followed by a linear relationship for the void ratio, with log σ' having a steeper slope. This can be explained in the following manner.

A soil in the field at some depth has been subjected to a certain maximum effective past pressure in its geologic history. This maximum effective past pressure may be equal to or greater than the existing overburden pressure at the time of sampling. The reduction of pressure in the field may be caused by natural geologic processes or human processes. During the soil sampling, the existing effective overburden pressure is also released, resulting in some expansion. When this specimen is subjected to a consolidation test, a small amount of compression (that is, a small change in the void ratio) will occur when the total pressure applied is less than the maximum effective overburden pressure in the field to which the soil has been subjected in the past. When the total applied pressure on the specimen is greater than the maximum effective past pressure, the change in the void ratio is much larger, and the e–log σ' relationship is practically linear with a steeper slope.

This relationship can be verified in the laboratory by loading the specimen to exceed the maximum effective overburden pressure, and then unloading and reloading again. The e–log σ' plot for such cases is shown in Figure 8.8, in which *cd* represents unloading and *dfg* represents the reloading process.

This leads us to the two basic definitions of clay based on stress history:

1. *Normally consolidated*: The present effective overburden pressure is the maximum pressure to which the soil has been subjected in the past.
2. *Overconsolidated*: The present effective overburden pressure is less than that which the soil has experienced in the past. The maximum effective past pressure is called the *preconsolidation pressure*.

The past effective pressure cannot be determined explicitly because it is usually a function of geological processes and, consequently, it must be inferred from laboratory test results.

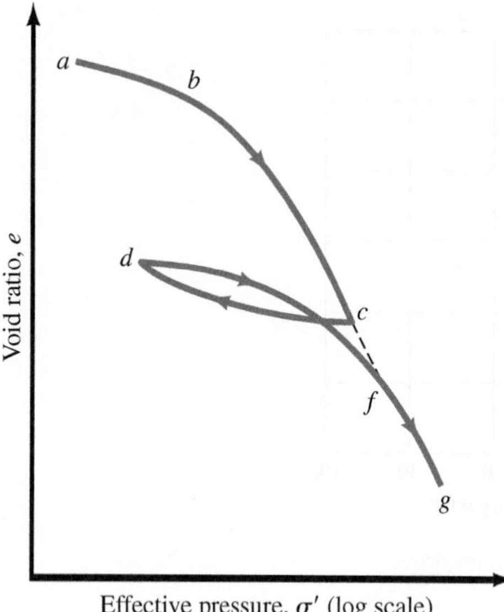

Figure 8.8 Plot of e versus log σ' showing loading, unloading, and reloading branches

Casagrande (1936) suggested a simple graphic construction to determine the preconsolidation pressure, σ'_c, from the laboratory e–log σ' plot. The procedure follows (see Figure 8.9):

1. By visual observation, establish point a at which the e–log σ' plot has a minimum radius of curvature.
2. Draw a horizontal line ab.
3. Draw the line ac tangent at a.
4. Draw the line ad, which is the bisector of the angle bac.

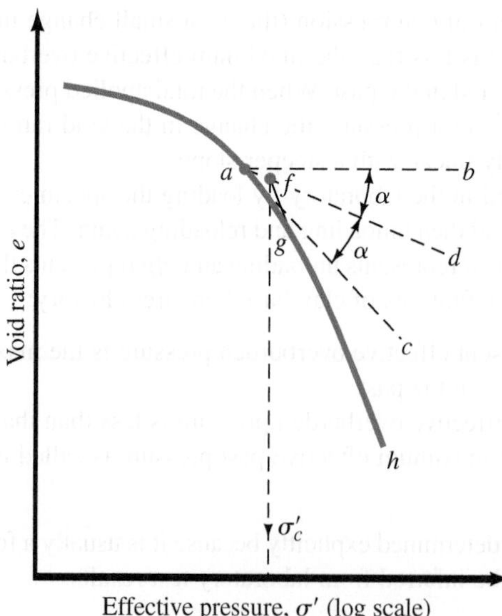

Figure 8.9 Graphic procedure for determining preconsolidation pressure

5. Project the straight-line portion gh of the e–log σ' plot back to intersect ad at f. The abscissa of point f is the preconsolidation pressure, σ'_c.

The overconsolidation ratio (OCR) for a soil can now be defined as

$$OCR = \frac{\sigma'_c}{\sigma'} \quad (8.8)$$

where

σ'_c = preconsolidation pressure of a specimen
σ' = present effective vertical pressure

8.6 Effect of Disturbance on Void Ratio–Pressure Relationship

A soil specimen will be remolded when it is subjected to some degree of disturbance. This will affect the void ratio–pressure relationship for the soil. For a normally consolidated clayey soil of low to medium sensitivity (Figure 8.10) under an effective overburden pressure of σ'_o and with a void ratio of e_0, the change in the void ratio with an increase of pressure in the field will be roughly as shown by curve 1. This is the *virgin consolidation line*, which is approximately a straight line on a semilogarithmic plot. However, the laboratory consolidation curve for a fairly undisturbed specimen of the same soil (curve 2) will be located to the left of curve 1. If the soil is completely remolded and a consolidation test is conducted on it, the general position of the e–log σ' plot will be represented by curve 3. Curves 1, 2, and 3 will intersect approximately at a void ratio of $e = 0.4e_0$ (Terzaghi and Peck, 1967).

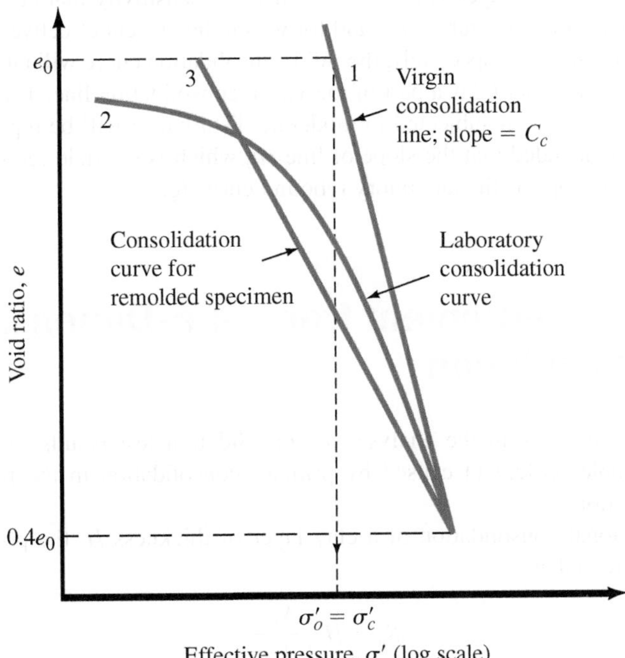

Figure 8.10 Consolidation characteristics of normally consolidated clay of low to medium sensitivity

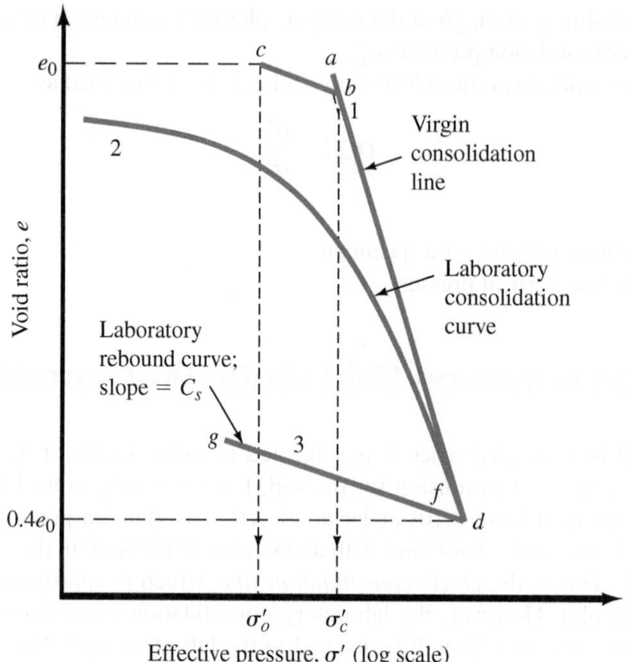

Figure 8.11 Consolidation characteristics of overconsolidated clay of low to medium sensitivity

For an overconsolidated clayey soil of low to medium sensitivity that has been subjected to a preconsolidation pressure of σ'_c (Figure 8.11) and for which the present effective overburden pressure and the void ratio are σ'_o and e_0, respectively, the field consolidation curve will take a path represented approximately by *cbd*. Note that *bd* is a part of the virgin consolidation line. The laboratory consolidation test results on a specimen subjected to moderate disturbance will be represented by curve 2. Schmertmann (1953) concluded that the slope of line *cb*, which is the field recompression path, has approximately the same slope as the laboratory rebound curve *fg*.

8.7 Calculation of Settlement from One-Dimensional Primary Consolidation

With the knowledge gained from the analysis of consolidation test results, we can now proceed to calculate the probable settlement caused by primary consolidation in the field, assuming one-dimensional consolidation.

For one-dimensional consolidation of a clay layer of thickness H, the primary consolidation settlement can be expressed as

$$S_p = H\frac{\Delta e}{1 + e_0} \tag{8.9}$$

where

e_0 = initial void ratio before the application of load
Δe = change in void ratio due to consolidation

For normally consolidated clays that exhibit a linear e–log σ' relationship (Figure 8.10) (*note*: $\Delta\sigma = \Delta\sigma'$ at the end of consolidation),

$$\Delta e = C_c[\log(\sigma'_o + \Delta\sigma') - \log\sigma'_o] \tag{8.10}$$

where C_c = slope of the e–log σ' plot and is defined as the *compression index* and σ'_o is the initial effective stress before the application of the load which is often called as *effective overburden stress*. Substituting Eq. (8.10) into Eq. (8.9) gives

$$S_p = \frac{C_c H}{1 + e_0} \log\left(\frac{\sigma'_o + \Delta\sigma'}{\sigma'_o}\right) \tag{8.11}$$

In overconsolidated clays (Figure 8.11), for $\sigma'_o + \Delta\sigma' \leq \sigma'_c$, field e–log σ' variation will be along the line *cb*, the slope of which will be approximately equal to the slope of the laboratory rebound curve. The slope of the rebound curve, C_s, is referred to as the *swell index*, so

$$\Delta e = C_s[\log(\sigma'_o + \Delta\sigma') - \log\sigma'_o] \tag{8.12}$$

From Eqs. (8.9) and (8.11), we have

$$S_p = \frac{C_s H}{1 + e_0} \log\left(\frac{\sigma'_o + \Delta\sigma'}{\sigma'_o}\right) \tag{8.13}$$

If $\sigma'_o + \Delta\sigma > \sigma'_c$, then

$$S_p = \frac{C_s H}{1 + e_0} \log\frac{\sigma'_c}{\sigma'_o} + \frac{C_c H}{1 + e_0} \log\left(\frac{\sigma'_o + \Delta\sigma'}{\sigma'_c}\right) \tag{8.14}$$

8.8 Compression Index (C_c) and Swell Index (C_s)

We can determine the compression index C_c from the e–log σ' plot generated in the laboratory consolidation test. As discussed in Section 8.6, the degree of disturbance to the clay specimen can distort the e–log σ' plot, which often leads to underestimation of C_c. It is suggested that the slope of the virgin consolidation line, established through the procedures suggested in Figures 8.10 and 8.11, be taken as C_c for computing the consolidation settlements in the field.

Terzaghi and Peck (1967) suggested empirical expressions for the compression index. For undisturbed clays:

$$C_c = 0.009(LL - 10) \tag{8.15}$$

For remolded clays:

$$C_c = 0.007(LL - 10) \tag{8.16}$$

where LL = liquid limit (%). In the absence of laboratory consolidation data, Eq. (8.15) often is used for an approximate calculation of primary consolidation in the field. Several other correlations for the compression index are also available now.

Based on observations on several natural clays, Rendon-Herrero (1983) gave the relationship for the compression index in the form

$$C_c = 0.141G_s^{1.2}\left(\frac{1 + e_0}{G_s}\right)^{2.38} \tag{8.17}$$

The swell index is appreciably smaller in magnitude than the compression index and can generally be determined from laboratory tests. In most cases,

$$C_s \approx \frac{1}{5} \text{ to } \frac{1}{10} C_c \tag{8.18}$$

Example 8.2

Refer to the e–log σ' curve shown in Figure 8.12.

a. Determine the preconsolidation pressure, σ'_c.
b. Find the compression index, C_c.

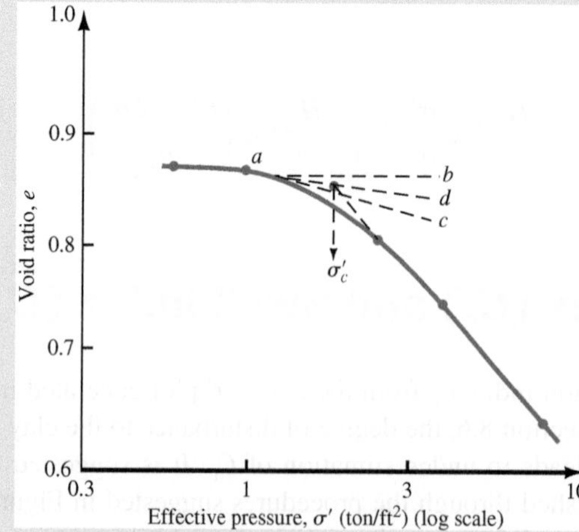

Figure 8.12

Solution

Part a

Using the procedure shown in Figure 8.9, we determine the preconsolidation pressure. From the plot, $\sigma'_c = \mathbf{1.6\ ton/ft^2}$.

Part b

From the e–log σ' plot, we find

$$\sigma'_1 = 4\ \text{ton/ft}^2 \qquad e_1 = 0.712$$

$$\sigma'_2 = 8\ \text{ton/ft}^2 \qquad e_2 = 0.627$$

So

$$C_c = \frac{e_1 - e_2}{\log(\sigma'_2/\sigma'_1)} = \frac{0.712 - 0.627}{\log(8/4)} = \mathbf{0.282}$$

■

Example 8.3

A soil profile is shown in Figure 8.13a. Laboratory consolidation tests were conducted on a sample collected from the middle of the clay layer. The field consolidation curve interpolated from the laboratory test results is shown in Figure 8.13b. Calculate the settlement in the field due to primary consolidation for a surcharge of 0.48 ton/ft² applied at the ground surface.

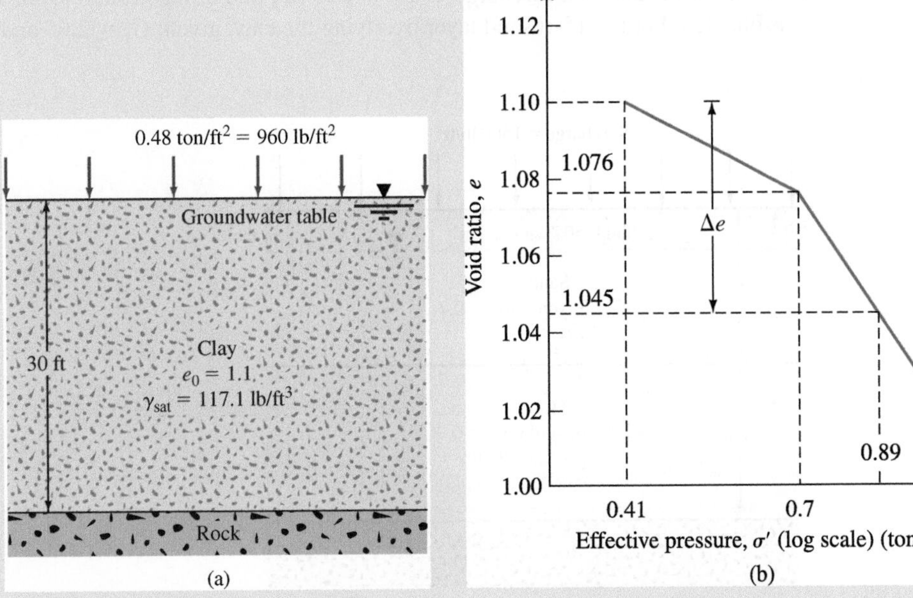

Figure 8.13 (a) Soil profile; (b) field consolidation curve

Solution

$$\sigma'_o = (15)(\gamma_{sat} - \gamma_w) = 15\,(117.1 - 62.4)$$
$$= 820.5 \text{ lb/ft}^2 = 0.41 \text{ ton/ft}^2$$
$$e_o = 1.1$$
$$\Delta\sigma' = 960 \text{ lb/ft}^2 = 0.48 \text{ ton/ft}^2$$
$$\sigma'_o + \Delta\sigma' = 0.41 + 0.48 = 0.89 \text{ ton/ft}^2$$

The void ratio corresponding to 0.89 ton/ft^2 (Figure 8.13b) is 1.045. Hence, $\Delta e = 1.1 - 1.045 = 0.055$.

$$\text{Settlement, } S_p = H\frac{\Delta e}{1 + e_O} \text{ [Eq. (8.9)]}$$

So

$$S_p = 30\,\frac{(0.055)}{1 + 1.1} = 0.786 \text{ ft} = \mathbf{9.43 \text{ in.}}$$ ■

Example 8.4

A soil profile is shown in Figure 8.14. Calculate the settlement due to primary consolidation for the 15-ft clay layer due to a surcharge of 1500 lb/ft^2 applied at the ground level. The clay is normally consolidated. For the 15-ft sand layer overlying the clay, given: $G_s = 2.65$ and $e = 0.7$.

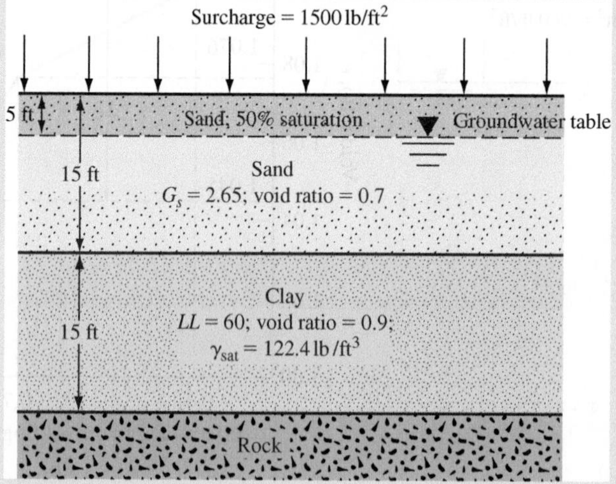

Figure 8.14

Solution

Calculation of Average Effective Overburden Pressure (σ'_o).

The moist unit weight of sand above the ground water table:

$$\gamma_{sand} = \frac{G_s\gamma_w + Se\gamma_w}{1+e} = \frac{[2.65 + (0.5 \times 0.7)]62.4}{1+0.7}$$

$$= 110.12 \text{ lb/ft}^3$$

Submerged unit weight of sand below the ground water table:

$$\gamma'_{sand} = \gamma_{sat(sand)} - \gamma_w$$

$$= \frac{G_s\gamma_w + e\gamma_w}{1+e} - \gamma_w = \frac{(G_s - 1)\gamma_w}{1+e}$$

$$= \frac{(2.65-1)62.4}{1+0.7} = 60.56 \text{ lb/ft}^3$$

Submerged unit weight of clay:

$$\gamma'_{clay} = \gamma_{sat(clay)} - \gamma_w = 122.4 - 62.4 = 60 \text{ lb/ft}^3$$

So,

$$\sigma'_o = 5\gamma_{sand} + 10\gamma'_{sand} + \frac{15}{2}\gamma'_{clay}$$

$$= 5(110.12) + 10(60.56) + 7.5(60)$$

$$= 1606.2 \text{ lb/ft}^2$$

Calculation of Compression Index [C_c]

$$C_c = 0.009(LL - 10) = 0.009(60 - 10) = 0.45$$

Calculation of Settlement

From Eq. (8.11),

$$S_p = \frac{C_cH}{1+e_O}\log\left(\frac{\sigma'_o + \Delta\sigma'}{\sigma'_o}\right)$$

$$= \frac{0.45(15 \times 12)}{1+0.9}\log\left(\frac{1606.2 + 1500}{1606.2}\right)$$

$$= \mathbf{12.21 \text{ in.}}$$

■

Example 8.5

In a normally consolidated clay specimen, the following data are given from the laboratory consolidation test.

- $e_1 = 1.10$
- $e_2 = 0.85$
- $\sigma'_1 = 65.0 \text{ kN/m}^2$
- $\sigma'_2 = 240.0 \text{ kN/m}^2$

What will be the void ratio when the next pressure increment raises the pressure to 460.0 kN/m^2?

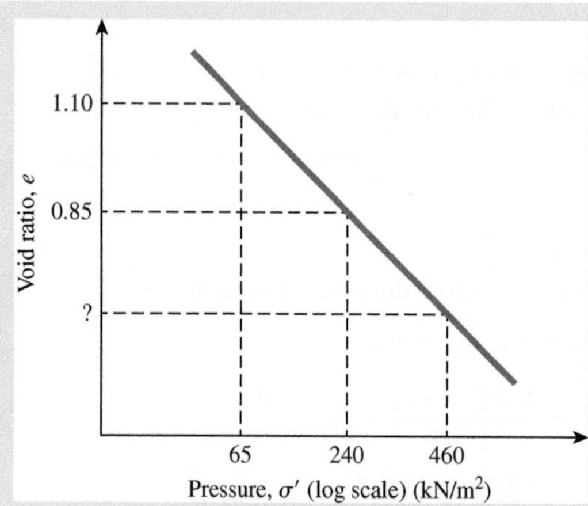

Figure 8.15

Solution

From Figure 8.15, the compression index C_c is given by

$$C_c = \frac{e_1 - e_2}{\log \sigma_2' - \log \sigma_1'} = \frac{1.10 - 0.85}{\log 240 - \log 65} = 0.441$$

Let's say the void ration at 460 kN/m^2 pressure is e_3.

$$e_1 - e_3 = C_c\,(\log 460 - \log 65) = 0.441 \times \log\left(\frac{460}{65}\right) = 0.375$$

Therefore, $e_3 = 1.10 - 0.375 = \mathbf{0.725}$. ■

Example 8.6

The soil profile at a site is shown is Figure 8.16. The moist and saturated unit weights of the sand are 17.0 kN/m^3 and 20.0 kN/m^3, respectively. A soil specimen was taken from the middle of the clay layer and subjected to a consolidation test, and the following properties are reported:

- Natural moisture content of the clay = 22.5%
- Specific gravity of the soil grains = 2.72
- Preconsolidation pressure = 110.0 kN/m^2
- Compression index = 0.52
- Swelling index = 0.06

Is the clay normally consolidated or overconsolidated?

If a 2-m-high compacted fill with a unit weight of 20.0 kN/m^3 is placed on the ground, what would be the final consolidation settlement?

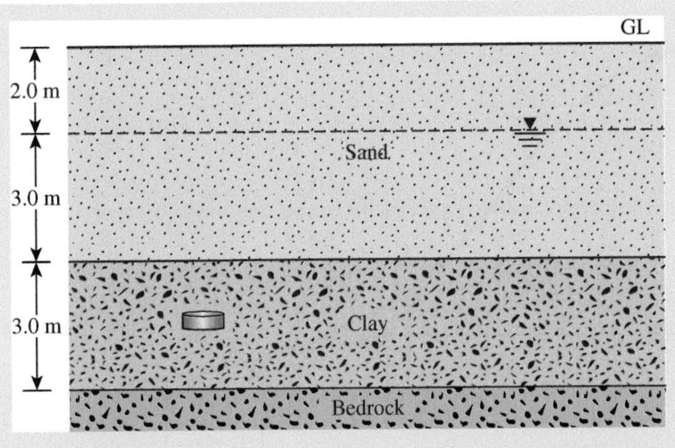

Figure 8.16

Solution

The clay is below the water table and hence is saturated. The initial void ratio e_0 can be determined as

$$e_0 = w\,G_s = 0.225 \times 2.72 = 0.612$$

The saturated unit weight is determined as

$$\gamma_{sat} = \frac{(G_s + e)\gamma_w}{1 + e} = \frac{(2.72 + 0.612) \times 9.81}{1 + 0.612} = 20.3 \text{ kN/m}^3$$

The effective overburden stress at the middle of the clay is

$$\sigma'_o = 2 \times 17.0 + 3(20.0 - 9.81) + 1.5(20.3 - 9.81) = 80.3 \text{ kN/m}^2 < 110.0 \text{ kN/m}^2$$

Since the preconsolidation pressure is greater than the current overburden pressure, the **clay is overconsolidated**. The overconsolidation ratio $OCR = 110.0/80.3 = 1.37$.

The 2-m-high compacted fill imposes a surcharge of $2 \times 20 = 40$ kN/m^2.

$$\text{i.e., } \Delta\sigma' = 40.0 \text{ kN/m}^2;\ \sigma'_o = 80.3 \text{ kN/m}^2;\text{ and } \sigma'_c = 110.0 \text{ kN/m}^2$$

Since $\sigma'_o + \Delta\sigma' > \sigma'_c$, the primary consolidation settlement can be computed from Eq. (8.14) as

$$S_p = \frac{C_s H}{1 + e_0}\log\left(\frac{\sigma'_c}{\sigma'_o}\right) + \frac{C_c H}{1 + e_0}\log\left(\frac{\sigma'_o + \Delta\sigma'}{\sigma'_c}\right)$$

$$S_p = \frac{0.06 \times 3000}{1 + 0.612}\log\left(\frac{110.0}{80.3}\right) + \frac{0.52 \times 3000}{1 + 0.612}\log\left(\frac{80.3 + 40.0}{110.0}\right) = \mathbf{52.9\ mm}$$ ■

8.9 Settlement from Secondary Consolidation

Section 8.3 showed that at the end of primary consolidation (that is, after complete dissipation of excess pore water pressure) some settlement is observed because of the plastic adjustment of soil fabrics, which is usually termed *creep*. This stage of consolidation is called *secondary consolidation*. During secondary consolidation, the plot of deformation versus the log of time is practically linear

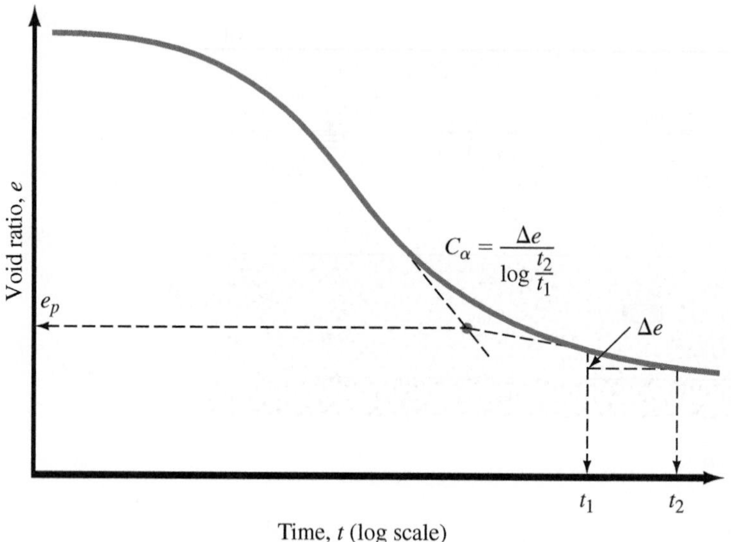

Figure 8.17 Variation of e with log t under a given load increment, and definition of secondary compression index

(Figure 8.4). The variation of the void ratio e with time t for a given load increment will be similar to that shown in Figure 8.4. This variation is illustrated in Figure 8.17.

The secondary compression index can be defined from Figure 8.17 as

$$C_\alpha = \frac{\Delta e}{\log t_2 - \log t_1} = \frac{\Delta e}{\log (t_2/t_1)} \tag{8.19}$$

where

C_α = secondary compression index
Δe = change of void ratio
t_1, t_2 = time

The magnitude of the secondary consolidation can be calculated as

$$S_s = C'_\alpha H \log\left(\frac{t_2}{t_1}\right) \tag{8.20}$$

where

$$C'_\alpha = \frac{C_\alpha}{1 + e_p} \tag{8.21}$$

and

e_p = void ratio at the end of primary consolidation (Figure 8.17)
H = thickness of clay layer

Example 8.7

Refer to Example 8.4. Assume that the primary consolidation will be complete in 3.5 years. Estimate the secondary consolidation that would occur from 3.5 years to 10 years after the load application. Given $C_\alpha = 0.022$. What is the total consolidation settlement after 10 years?

Solution

From Eq. (8.21),

$$C'_\alpha = \frac{C_\alpha}{1 + e_p}$$

The value of e_p can be calculated as

$$e_p = e_0 - \Delta e_{\text{primary}}$$

From Eq. (8.10),

$$\Delta e = C_c[\log(\sigma'_o + \Delta\sigma') - \log\sigma'_o]$$

Thus,

$$\begin{aligned} e_p &= e_0 - C_c[\log(\sigma'_o + \Delta\sigma') - \log\sigma'_o] \\ &= 0.9 - 0.45[\log(1606.2 + 1500) - \log(1606.2)] \\ &= 0.9 - 0.129 = 0.771 \end{aligned}$$

Hence,

$$C'_\alpha = \frac{0.022}{1 + 0.771} = 0.0124$$

Again, from Eq. (8.20),

$$S_s = C'_\alpha H \log\left(\frac{t_2}{t_1}\right) = (0.0124)(15 \times 12)\log\left(\frac{10}{3.5}\right) = \mathbf{1.02\ in.}$$

Total consolidation settlement = primary consolidation settlement (S_p) + secondary consolidation settlement (S_s). From Example 8.4, $S = 12.21$ in.

Total consolidation settlement = 12.21 + 1.02 = **13.23 in.** ■

Example 8.8

The piezometers installed at a site for monitoring pore water pressures show that the consolidation due to construction of an embankment is completed in the first two years. During the next three years, 12.5 mm of secondary compression settlement was recorded. What would be the additional secondary compression settlement that can be expected in the next 10 year?

Solution

From Eqs. (8.20) and (8.21), the secondary compression settlement is given by

$$S_s = \frac{C_\alpha H}{1 + e_p}\log\left(\frac{t_2}{t_1}\right)$$

During the first three years of secondary compression,

$$12.5 = \frac{C_\alpha H}{1 + e_p}\log\left(\frac{5}{2}\right)$$

Therefore, $\frac{C_\alpha H}{1 + e_p} = 31.4$ mm.

During the next 10 years, the secondary compression settlement is given by

$$S = \frac{C_\alpha H}{1 + e_p}\log\left(\frac{15}{5}\right) = 31.4\log\left(\frac{15}{5}\right) = \mathbf{15.0\ mm}$$

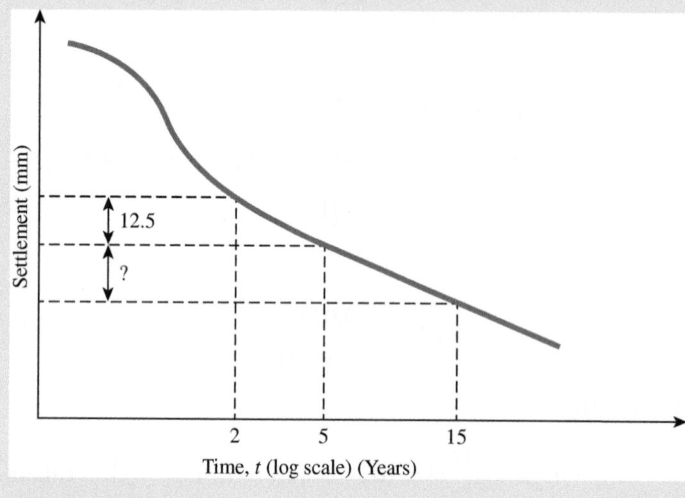

Figure 8.18 ■

8.10 Time Rate of Consolidation

The total settlement caused by primary consolidation resulting from an increase in the stress on a soil layer can be calculated by using one of the three equations [(8.11), (8.13), or (8.14)] given in Section 8.7. However, the equations do not provide any information regarding the rate of primary consolidation. Terzaghi (1925) proposed the first theory to consider the rate of one-dimensional consolidation for saturated clay soils. The mathematical derivations are based on the following assumptions.

1. The clay–water system is homogeneous.
2. Saturation is complete.
3. Compressibility of water is negligible.

4. Compressibility of soil grains is negligible (but soil grains rearrange).
5. The flow of water is in one direction only (that is, in the direction of compression).
6. Darcy's law is valid.

According to this theory, the average degree of consolidation can be defined as

$$U = \frac{S_t}{S_p} \tag{8.22}$$

where

S_t = consolidation settlement of a clay layer at time t after the beginning of consolidation
S_p = total primary consolidation of a clay layer

It was also established that

$$U \propto T_v \tag{8.23}$$

where T_v = time factor (nondimensional). The time factor is defined as

$$T_v = \frac{c_v t}{H_{\text{dr}}^2} \tag{8.24}$$

where

c_v = coefficient of consolidation [unit – (length)^2/time]
t = time after the beginning of consolidation
H_{dr} = maximum drainage path in the clay layer

The magnitude of H_{dr} can be determined as follows:

- If there is sand at the top of the clay layer and sand at the bottom of the clay layer, then $H_{\text{dr}} = H/2$ (H = thickness of clay layer). It is called *two-way drainage.*
- If there is sand at the top of the clay layer and rock at the bottom, then $H_{\text{dr}} = H$. It is called *one-way drainage*.
- If there is an impermeable layer at the top of the clay layer and sand at the bottom, then $H_{\text{dr}} = H$ *(one-way drainage).*

The variation in the average degree of consolidation with the nondimensional time factor, T_v, is given in Table 8.1. The values of the time factor and their corresponding average degrees of consolidation may also be approximated by the following simple relationships:

$$\text{For } U = 0 \text{ to } 60\%, \quad T_v = \frac{\pi}{4}\left(\frac{U\%}{100}\right)^2 \tag{8.25}$$

$$\text{For } U > 60\%, \quad T_v = 1.781 - 0.933 \log(100 - U\%) \tag{8.26}$$

8.11 Coefficient of Consolidation

The coefficient of consolidation, c_v, generally decreases as the liquid limit of soil increases. The range of variation of c_v for a given liquid limit of soil is rather wide.

For a given load increment on a specimen, there are two commonly used graphic methods for determining c_v from laboratory one-dimensional consolidation tests. One of them is the *logarithm-of-time*

Table 8.1 Variation of Time Factor with Degree of Consolidation

$U(\%)$	T_v	$U(\%)$	T_v	$U(\%)$	T_v
0	0	34	0.0907	68	0.377
1	0.00008	35	0.0962	69	0.390
2	0.0003	36	0.102	70	0.403
3	0.00071	37	0.107	71	0.417
4	0.00126	38	0.113	72	0.431
5	0.00196	39	0.119	73	0.446
6	0.00283	40	0.126	74	0.461
7	0.00385	41	0.132	75	0.477
8	0.00502	42	0.138	76	0.493
9	0.00636	43	0.145	77	0.511
10	0.00785	44	0.152	78	0.529
11	0.0095	45	0.159	79	0.547
12	0.0113	46	0.166	80	0.567
13	0.0133	47	0.173	81	0.588
14	0.0154	48	0.181	82	0.610
15	0.0177	49	0.188	83	0.633
16	0.0201	50	0.197	84	0.658
17	0.0227	51	0.204	85	0.684
18	0.0254	52	0.212	86	0.712
19	0.0283	53	0.221	87	0.742
20	0.0314	54	0.230	88	0.774
21	0.0346	55	0.239	89	0.809
22	0.0380	56	0.248	90	0.848
23	0.0415	57	0.257	91	0.891
24	0.0452	58	0.267	92	0.938
25	0.0491	59	0.276	93	0.993
26	0.0531	60	0.286	94	1.055
27	0.0572	61	0.297	95	1.129
28	0.0615	62	0.307	96	1.219
29	0.0660	63	0.318	97	1.336
30	0.0707	64	0.329	98	1.500
31	0.0754	65	0.304	99	1.781
32	0.0803	66	0.352	100	∞
33	0.0855	67	0.364		

method proposed by Casagrande and Fadum (1940), and the other is the *square-root-of-time method* suggested by Taylor (1942). The general procedures for obtaining c_v by the two methods are described next.

Logarithm-of-Time Method

For a given incremental loading of the laboratory test, the specimen deformation versus log-of-time plot is shown in Figure 8.19. The following constructions are needed to determine c_v:

1. Extend the straight-line portions of primary and secondary consolidations to intersect at *A*. The ordinate of *A* is represented by d_{100}—that is, the deformation at the end of 100% primary consolidation.

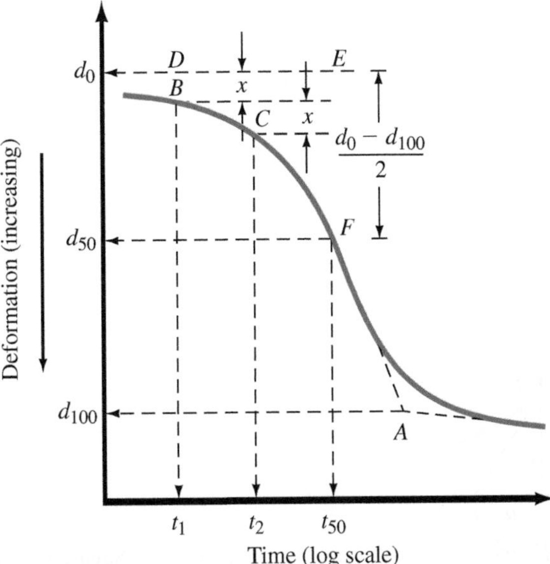

Figure 8.19 Logarithm-of-time method for determining coefficient of consolidation

2. The initial curved portion of the plot of deformation versus log t is approximated to be a parabola on the natural scale. Select times t_1 and t_2 on the curved portion such that $t_2 = 4t_1$. Let the difference of the specimen deformation during time $(t_2 - t_1)$ be equal to x.
3. Draw a horizontal line DE such that the vertical distance BD is equal to x. The deformation corresponding to the line DE is d_0 (that is, deformation at 0% consolidation).
4. The ordinate of point F on the consolidation curve represents the deformation at 50% primary consolidation, and its abscissa represents the corresponding time (t_{50}).
5. For 50% average degree of consolidation, $T_v = 0.197$ (Table 8.1);

$$T_{50} = \frac{c_v t_{50}}{H_{dr}^2}$$

or

$$c_v = \frac{0.197\, H_{dr}^2}{t_{50}} \tag{8.27}$$

where H_{dr} = average longest drainage path during consolidation.

For specimens drained at both top and bottom, H_{dr} equals one-half of the average height of the specimen during consolidation. For specimens drained on only one side, H_{dr} equals the average height of the specimen during consolidation.

Square-Root-of-Time Method

In this method, a plot of deformation versus the square root of time is drawn for the incremental loading (Figure 8.20). Other graphic constructions required are as follows:

1. Draw a line AB through the early portion of the curve.
2. Draw a line AC such that $\overline{OC} = 1.15\ \overline{OB}$. The abscissa of point D, which is the intersection of AC and the consolidation curve, gives the square root of time for 90% consolidation ($\sqrt{t_{90}}$).

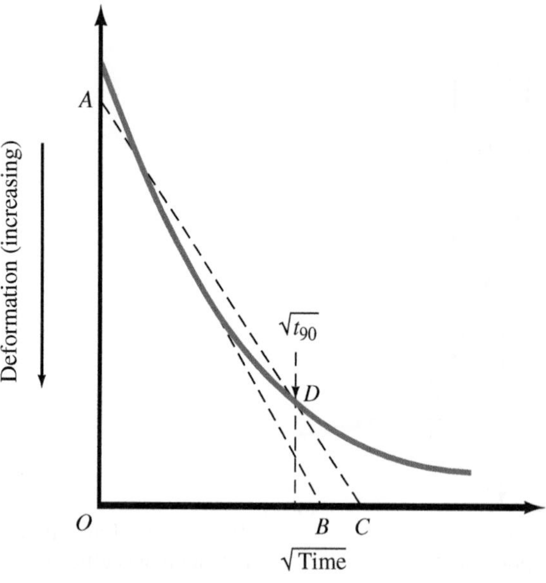

Figure 8.20 Square-root-of-time fitting method

3. For 90% consolidation, $T_{90} = 0.848$ (Table 8.1), so

$$T_{90} = 0.848 = \frac{C_v t_{90}}{H_{dr}^2}$$

or

$$c_v = \frac{0.848\ H_{dr}^2}{t_{90}} \qquad (8.28)$$

H_{dr} in Eq. (8.28) is determined in a manner similar to the logarithm-of-time method.

Example 8.9

For the problem in Example 8.4, answer the following:

a. What is the average degree of consolidation for the clay layer when the settlement is 3 in.?

b. If the average value of c_v for the pressure range is 0.003 cm^2/sec, how long will it take for 50% settlement to occur?

c. If the 15-ft clay layer is drained on both sides, how long will it take for 50% consolidation to occur?

Solution

Part a

$$U\% = \frac{\text{settlement at any time}}{\text{maximum settlement}} = \frac{3\text{ in.}}{12.21\text{ in.}} \times 100 = \mathbf{24.57\%}$$

Part b

$$U = 50\%; \text{ single drainage } T_{50} = \frac{c_v t_{50}}{H_{dr}^2}$$

From Table 8.1, for $U = 50\%$, $T_{50} = 0.197$. So

$$0.197 = \frac{0.003 \times t_{50}}{(15 \times 12 \times 2.54)^2}$$

or

$$t_{50} = \frac{0.197 \times (15 \times 12 \times 2.54)^2}{0.003 \times 60 \times 60 \times 24} = \mathbf{158.87\ days}$$

Part c

With double drainage, the maximum length drainage path = 15/2 = 7.5 ft

$$0.197 = \frac{0.003 \times t}{(7.5 \times 12 \times 2.54)^2}$$

or

$$t_{50} = \frac{0.197(7.5 \times 12 \times 2.54)^2}{0.003 \times 60 \times 60 \times 24} = \mathbf{39.72\ days}$$

■

Example 8.10

A 3-m thick layer (double drainage) of saturated clay under a surcharge loading underwent 90% primary consolidation in 75 days. Find the coefficient of consolidation of clay for the pressure range.

Solution

$$T_{90} = \frac{c_v t_{90}}{H_{dr}^2}$$

Since the clay layer has two-way drainage, $H_{dr} = 3\text{m}/2 = 1.5$ m; $T_{90} = 0.848$. So

$$0.848 = \frac{c_v(75 \times 24 \times 60 \times 60)}{(1.5 \times 100)^2}$$

$$c_v = \frac{0.848 \times 2.25 \times 10^4}{75 \times 24 \times 60 \times 60} = \mathbf{0.00294\ cm^2/sec}$$

■

Example 8.11

For a 30-mm thick undisturbed clay specimen as described in Example 8.10, how long will it take to undergo 90% consolidation in the laboratory for similar consolidation pressure range? The laboratory test specimen will have two-way drainage.

Solution

$$T_{90} = \frac{c_v t_{90(\text{field})}}{H^2_{\text{dr(field)}}} = \frac{c_v(75 \times 24 \times 60 \times 60)}{(1.5 \times 1000)^2}$$

and

$$T_{90} = \frac{c_v\, t_{90\,(\text{lab})}}{(30/2)^2}$$

So

$$\frac{4t_{90(\text{lab})}}{30^2} = \frac{75 \times 24 \times 60 \times 50}{2.25 \times 10^6}$$

or

$$t_{90(\text{lab})} = \frac{(75 \times 24 \times 60 \times 60)(9 \times 10^2)}{(2.25 \times 10^6) \times 4} = \mathbf{648\ s}$$

■

8.12 Summary

Consolidation is a process by which the load applied on a saturated soil is transferred from the pore water to the soil grains. Upon applying the load, the entire load is carried by the water, and with drainage taking place and dissipation of pore water pressure, the load is gradually transferred to the soil grains. When the consolidation is completed, the entire load is carried by the soil grains, where the effective stress increases. In granular soils, the consolidation process is almost instantaneous. In clays, the process can take from few months to several years. Secondary consolidation, also known as creep, takes place on the completion of primary consolidation.

The consolidation and secondary consolidation parameters can be determined from a one-dimensional laboratory consolidation test on a small clay specimen. The magnitude of the *final* consolidation settlement is governed by the loading at the ground level, and the consolidation parameters such as C_c, C_s, OCR, σ'_o, H, *and* e_0. The rate at which the consolidation occurs is governed by the coefficient of consolidation c_v.

This chapter provided the framework for computing the settlement of a clay layer under surface loads. The consolidation settlements and the secondary consolidation settlements have to be computed separately.

Problems

8.1 State whether the following are true or false.

a. During consolidation, the effective stress increases.

b. Larger the compression index, larger is the consolidation settlement.

c. Consolidation settlement is greater in overconsolidated clay than when it is normally consolidated.
d. Swelling index is always less than the compression index.
e. Larger the coefficient of consolidation, faster is the consolidation.

8.2 The results of a laboratory consolidation test on a clay sample are given below.

Pressure, σ' (ton/ft^2)	Void Ratio, e
0.24	1.112
0.48	1.105
0.96	1.080
1.92	0.985
3.84	0.850
7.68	0.731

a. Draw an e versus log σ' plot.
b. Determine the preconsolidation pressure, σ'_c.
c. Find the compression index, C_c.

8.3 A soil profile is shown in Figure 8.21. If a uniformly distributed load $\Delta\sigma$ is applied at the ground surface, what will be the settlement of the clay layer due to primary consolidation? Assume the sand above the ground water table to be dry. Given: $\Delta\sigma = 1000$ lb/ft^2, $H_1 = 23$ ft, $H_2 = 17$ ft, and $H_3 = 8$ ft.

- sand: $\gamma_{dry} = 110$ lb/ft^3
- $\gamma_{sat} = 115$ lb/ft^3
- clay: $\gamma_{sat} = 120$ lb/ft^3
- $LL = 50$, $e = 0.9$

(Assume the clay to be normally consolidated.)

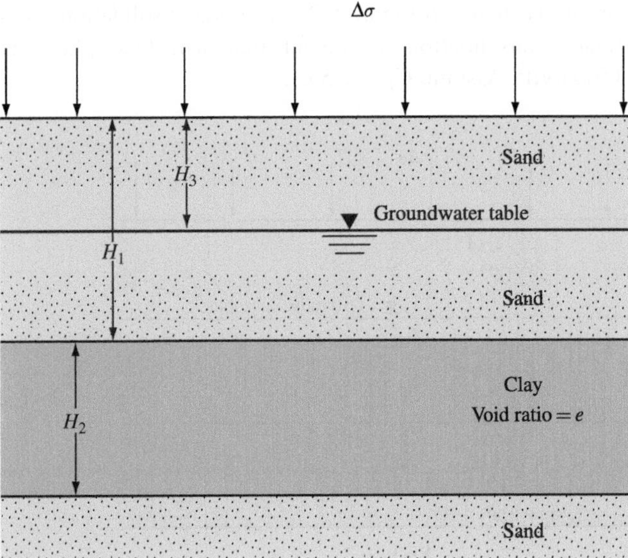

Figure 8.21

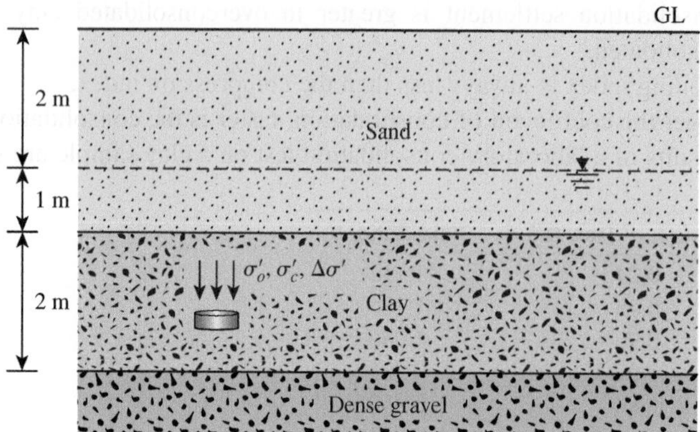

Figure 8.22

8.4 The soil profile at a site is shown in Figure 8.22. A clay specimen was taken from the middle of the clay layer, and the following properties were determined from a consolidation test. Given are the preconsolidation pressure is 80 kN/m^2, the compression index is 0.55, the swelling index is 0.07, and the initial in situ void ratio is 0.85.
The moist and saturated unit weight of the sand are 16.5 kN/m^3 and 19.0 kN/m^3, respectively. The saturated unit weight of the clay is 19.0 kN/m^3.

a. What is the overconsolidation ratio of the clay?

b. If a warehouse covering a large area is proposed to be built, imposing 50 kN/m^2 pressure at the ground level, what would be the consolidation settlement?

8.5 If the clay layer in Problem 8.3 has been preconsolidated and the average preconsolidation pressure is 2600 lb/ft^2, what will be the expected primary consolidation settlement due to the surcharge of 1000 lb/ft^2? Assume $C_s = 1/6\ C_c$.

8.6 A soil profile is shown in Figure 8.23. The preconsolidation pressure is 3400 lb/ft^2. Estimate the primary consolidation settlement that will take place as a result of a surcharge $\Delta\sigma = 1500$ lb/ft^2. Assume $C_s = 1/5\ C_c$.

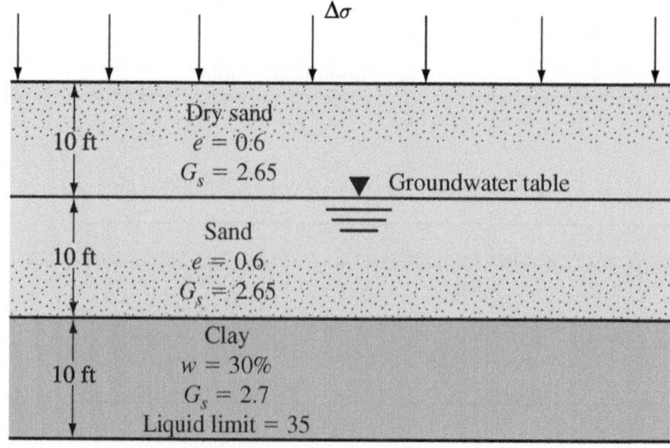

Figure 8.23

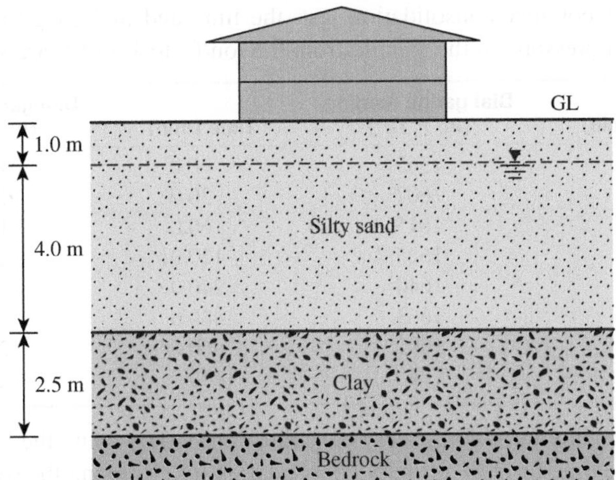

Figure 8.24

8.7 If the coefficient of consolidation for the clay layer in Problem 8.3 is 0.0018 cm^2/sec, how long will it take for 60% primary consolidation to take place? What will be the total compression due to consolidation at that time?

8.8 Figure 8.24 shows a soil profile where the moist and saturated unit weights of the sand are 17.0 kN/m^3 and 19.5 kN/m^3, respectively. The saturated unit weight of the clay 19.0 kN/m^3. The clay is normally consolidated with a compression index of 0.40. The natural moisture content of the clay is 25.0%, and the specific gravity of the clay grains is 2.70. The building imposes a stress increase of 30 kN/m^2 at the middle of the clay layer in addition to the effective overburden stresses due to the self-weights of the soils. Due to some proposed construction work, the water table will be lowered by 2 m and maintained at this new level for long time. What would be the additional consolidation settlement caused by the lowering of the water table?

8.9 The soil profile at a site consists of 2 m of sand at the ground level, underlain by 6 m of clay, and followed by a very stiff clay stratum that can be assumed to be impervious and incompressible. The water table is at 1.5 m below the ground level. The moist and saturated unit weights of the sand are 17.0 kN/m^3 and 18.5 kN/m^3, respectively. The clay has an initial void ratio of 0.810, saturated unit weight of 19.0 kN/m^3, and coefficient of consolidation of 0.0014 cm^2/s.

a. When the ground is surcharged with 3-m-high compacted fill with a moist unit weight of 19.0 kN/m^3, the settlement was 160 mm in the first year. What would be the consolidation settlement in the first two years?

b. If the clay is normally consolidated, what is the compression index of the clay?

8.10 Laboratory tests on a 25-mm thick clay sample drained at the top and bottom show that 50% consolidation takes place in 11 minutes.

a. How long will it take for a similar clay layer in the field, 4 m thick and drained at the top only, to undergo 50% consolidation?

b. Find the time required for the clay layer in the field as described in Part a to reach 70% consolidation.

8.11 A clay layer with two-way drainage reached 75% consolidation in *t* years. How long would it take for the same clay to consolidate 75% if it has one-way drainage?

8.12 During a laboratory consolidation test, the time and dial gauge readings obtained from an increase of pressure on the sample from 0.5 ton/ft^2 to 1 ton/ft^2 are given here.

Time (min)	Dial gauge reading (in. × 10^4)	Time (min)	Dial gauge reading (in. × 10^4)
0	1565	16.0	1800
0.1	1607	30.0	1865
0.25	1615	60.0	1938
0.5	1625	120.0	2000
1.0	1640	240.0	2050
2.0	1663	480.0	2080
4.0	1692	960.0	2100
8.0	1740	1440.0	2112

a. Find the time for 50% primary consolidation (t_{50}) by using the logarithm-of-time method.
b. Find the time for 90% primary consolidation (t_{90}) by using the square-root-of-time method.
c. If the average height of the sample during consolidation due to this incremental loading was 0.88 in., and it was drained at the top and bottom, calculate the coefficient of consolidation by using t_{50} and t_{90} obtained from Parts a and b.

8.13 The time for 50% consolidation of a 25-mm thick clay layer (drained at top and bottom) in the laboratory is 2 min, 20 sec. How long (in days) will it take for a 3-m thick layer of the same clay in the field under the same pressure increment to reach 50% consolidation? In the field, there is a rock layer at the bottom of the clay.

8.14 Refer to Problem 8.13. How long (in days) will it take in the field for 30% primary consolidation to occur?

8.15 The soil profile at a site consists of a 2.0-m-thick sand layer at the top, underlain by a 3.0-m-thick clay layer. The water table lies at a depth of 1.0 m below the ground level. The bulk and saturated unit weights of the sand are 16.0 kN/m^3 and 19.0 kN/m^3, respectively. The properties of the clay are moisture content of 45.0%, specific gravity of the soil grains at 2.70, compression index of 0.65, recompression index of 0.08 and overconsolidation ratio of 1.5.

a. The ground level is raised by placing a 1.5-m-high compacted fill with a unit weight of 20.0 kN/m^3. What is the consolidation settlement?
b. When the consolidation due to the fill is completed, it is proposed to construct a warehouse imposing a uniform pressure of 40.0 kN/m^2. What would be the consolidation settlement due to the construction of the warehouse?

References

Casagrande, A. (1936). "Determination of the Preconsolidation Load and Its Practical Significance," *Proceedings*, 1st International Conference on Soil Mechanics and Foundation Engineering, Cambridge, Mass., Vol. 3, 60–64.

Casagrande, A. and Fadum, R. E. (1940). "Notes on Soil Testing for Engineering Purposes," Harvard University Graduate School Engineering Publication No. 8.

Rendon-Herrero, O. (1980). "Universal Compression Index Equation," *Journal of the Geotechnical Engineering Division*, ASCE, Vol. 106, No. GT11, 1179–1200.

Schmertmann, J. H. (1953). "Undisturbed Consolidation Behavior of Clay," *Transactions*, ASCE, Vol. 120, 1201.

Taylor, D. W. (1942). "Research on Consolidation of Clays," *Serial No. 82*, Department of Civil and Sanitary Engineering, Massachusetts Institute of Technology, Cambridge, Massachusetts.

Terzaghi, K. (1925). *Erdbaumechanik auf Bodenphysikalischer Grundlage*, Deutichke, Vienna.

Terzaghi, K. and Peck, R. B. (1967). *Soil Mechanics in Engineering Practice*, 2nd ed., Wiley, New York.

9 Shear Strength of Soil

9.1 Introduction

The *shear strength* of a soil mass is the internal resistance per unit area that the soil mass can offer to resist failure and sliding along any plane inside it. Engineers must understand the nature of shearing resistance in order to analyze soil stability problems such as bearing capacity, slope stability, and lateral pressure on earth-retaining structures.

In this chapter, you will learn the following.

- Mohr-Coulomb failure criterion for shear strength of soils.
- Cohesive and frictional components of shear strength.
- Shear strength parameters—*cohesion* and *friction angle*.
- Laboratory tests for measuring cohesion and friction angle.
- Three types of triaxial tests.
- Unconfined compressive strength of a clay.

9.2 Mohr-Coulomb Failure Criterion

Mohr (1900) presented a theory for rupture in materials. This theory contended that a material fails because of a critical combination of normal stress and shear stress, and not from either maximum normal or shear stress alone. Thus, the functional relationship between normal stress and shear stress on a failure plane can be expressed in the form

$$\tau_f = f(\sigma) \tag{9.1}$$

where

τ_f = shear stress on the failure plane (i.e., shear strength)
σ = normal stress on the failure plane

The failure envelope defined by Eq. (9.1) is a curved line. For most soil mechanics problems, it is sufficient to approximate the shear stress on the failure plane as a linear function of the normal stress (Coulomb, 1776). This relation can be written as

$$\tau_f = c + \sigma \tan \phi \tag{9.2}$$

where

c = cohesion
ϕ = angle of internal friction

The preceding equation is called the *Mohr-Coulomb failure criterion.*

In saturated soil, the total normal stress at a point is the sum of the effective stress and the pore water pressure, or

$$\sigma = \sigma' + u$$

The effective stress, σ', is carried by the soil grains. So, to apply Eq. (9.2) to soil mechanics, we need to rewrite it as

$$\tau_f = c' + (\sigma - u)\tan\phi' = c' + \sigma'\tan\phi' \tag{9.3}$$

where

c' = effective stress cohesion
ϕ' = effective angle of friction

The value of c' for sand and inorganic silt is 0. For normally consolidated clays, c' can be approximated at 0. Overconsolidated clays have values of c' that are greater than 0. The angle of friction, ϕ', is sometimes referred to as the *drained angle of friction*. Typical values of ϕ' for some granular soils are given in Table 9.1.

For normally consolidated clays, the friction angle ϕ' generally ranges from 20° to 30°. For overconsolidated clays, the magnitude of ϕ' decreases. For natural noncemented, overconsolidated clays, the magnitude of c' is in the range of 100 to 300 lb/ft^2($\approx$ 5 to 15 kN/m^2).

Table 9.1 Typical values of drained angle of friction for sands and silts

Soil type	ϕ' (deg)
Sand: Rounded grains	
Loose	27–30
Medium	30–35
Dense	35–38
Sand: Angular grains	
Loose	30–35
Medium	35–40
Dense	40–45
Gravel with some sand	34–48
Silts	26–35

Laboratory Determination of Shear Strength Parameters

The shear strength parameters of a soil are determined in the laboratory primarily with two types of tests: direct shear test and triaxial test. The procedures for conducting each of these tests are explained in some detail in the following sections.

9.3 Direct Shear Test

This is the oldest and simplest form of shear test arrangement. A diagram of the direct shear test apparatus is shown in Figure 9.1. The test equipment consists of a metal shear box in which the soil specimen is placed. The soil specimens may be square or circular. The size of the specimens generally used is about 3 to 4 in.^{2} ($\approx$ 19 to 26 cm^2) across and 1 in. (25.4 mm) high. The box is split horizontally into halves. Normal force on the specimen is applied from the top of the shear box. The normal stress on the specimens can be as great as 150 lb/in^2 ($\approx$ 1035 kN/m^2). Shear force is applied by moving one half of the box relative to the other to cause failure in the soil specimen.

Depending on the equipment, the shear test can be either stress-controlled or strain-controlled. In stress-controlled tests, the shear force is applied in equal increments until the specimen fails. The failure takes place along the horizontal plane of split of the shear box. After the application of each incremental load, the shear displacement of the top half of the box is measured by a horizontal dial gauge. The change in the height of the specimen (and thus the volume change of the specimen) during the test can be obtained from the readings of a dial gauge that measures the vertical movement of the upper loading plate.

In strain-controlled tests, a constant rate of shear displacement is applied to one half of the box by a motor that acts through gears. The constant rate of shear displacement is measured by a

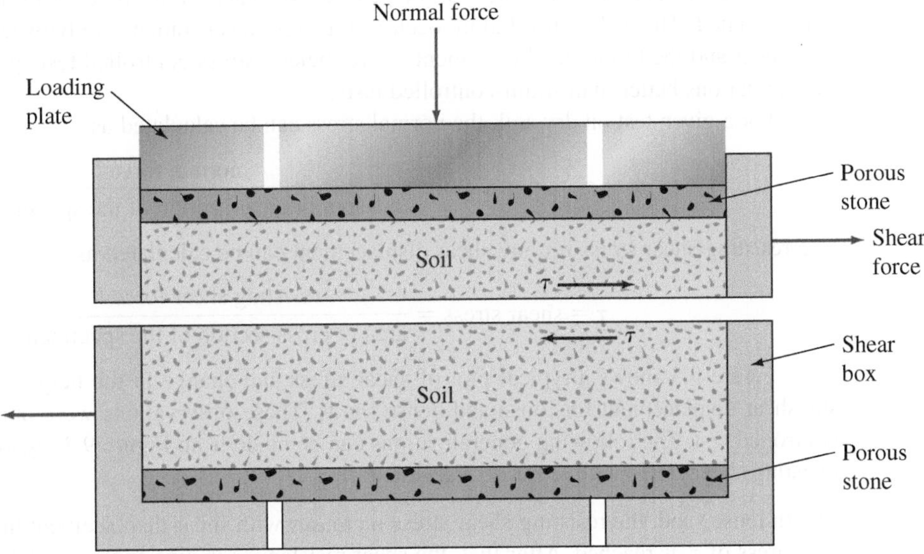

Figure 9.1 Diagram of direct shear test arrangement

(a)

(b)

Figure 9.2 (a) Direct shear apparatus for soils; (b) large direct shear apparatus for aggregates (*Courtesy of N. Sivakugan, James Cook University, Australia*)

horizontal dial gauge. The resisting shear force of the soil corresponding to any shear displacement can be measured by a horizontal proving ring or load cell. The volume change of the specimen during the test is obtained in a manner similar to the stress-controlled tests. Figure 9.2a is a photograph of direct shear equipment for testing soil specimens, where the normal stress is applied through dead weights and the horizontal shear load is applied through a loading ram. While testing aggregates, the specimen size has to be significantly larger, and the large direct shear test equipment shown in Figure 9.2b is used.

The advantage of the strain-controlled tests is that, in the case of dense sand, peak shear resistance (that is, at failure) as well as lesser shear resistance (that is, at a point after failure called *ultimate strength*) can be observed and plotted. In stress-controlled tests, only peak shear resistance can be observed and plotted. Note that the peak shear resistance in stress-controlled tests can only be approximated. This is because failure occurs at a stress level somewhere between the prefailure load increment and the failure load increment. Nevertheless, stress-controlled tests probably simulate real field situations better than strain-controlled tests.

For a given test on dry soil, the normal stress can be calculated as

$$\sigma = \sigma' = \text{normal stress} = \frac{\text{normal force}}{\text{area of cross section of the specimen}} \tag{9.4}$$

The resisting shear stress for any shear displacement can be calculated as

$$\tau = \text{shear stress} = \frac{\text{resisting shear force}}{\text{area of cross section of the specimen}} \tag{9.5}$$

Figure 9.3 shows a typical plot of shear stress and change in the height of the specimen versus shear displacement for loose and dense sands. These observations were obtained from a strain-controlled test. The following generalizations can be made from Figure 9.3 regarding the variation of resisting shear stress with shear displacement:

1. In loose sand, the resisting shear stress increases with shear displacement until a failure shear stress of τ_f is reached. After that, the shear resistance remains approximately constant with any further increase in the shear displacement.

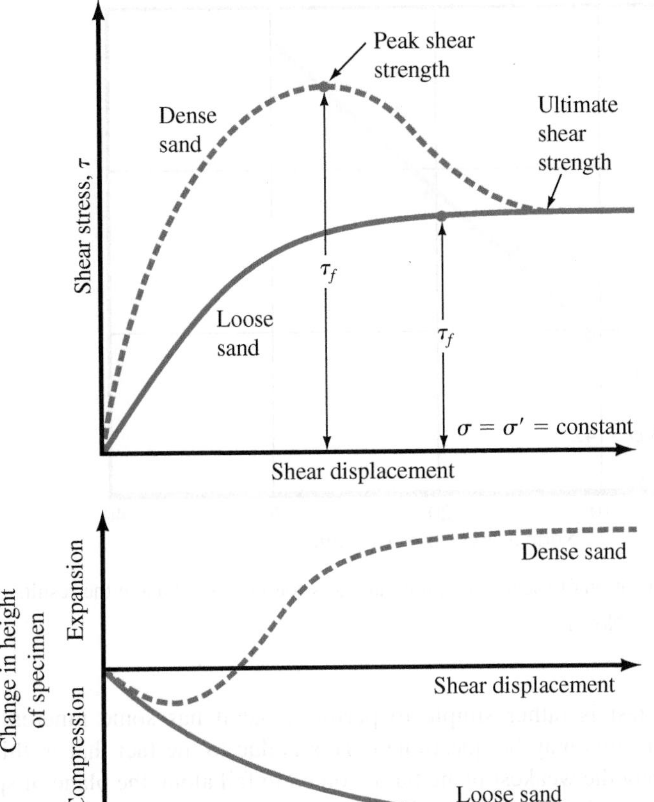

Figure 9.3 Plot of shear stress and change in height of specimen versus shear displacement for loose and dense dry sand (direct shear test)

2. In dense sand, the resisting shear stress increases with shear displacement until it reaches a failure stress of τ_f. This τ_f is called the *peak shear strength.* After failure stress is attained, the resisting shear stress gradually decreases as shear displacement increases until it finally reaches a constant value called the *ultimate shear strength.*

Direct shear tests are repeated on similar specimens at various normal stresses. The normal stresses and the corresponding values of τ_f obtained from a number of tests are plotted on a graph, from which the shear strength parameters are determined. Figure 9.4 shows such a plot for tests on a dry sand. The equation for the average line obtained from experimental results is

$$\tau_f = \sigma' \tan\phi' \tag{9.6}$$

(*Note*: $c' = 0$ for sand and $\sigma = \sigma'$ for dry conditions.) So the friction angle

$$\phi' = \tan^{-1}\left(\frac{\tau_f}{\sigma'}\right) \tag{9.7}$$

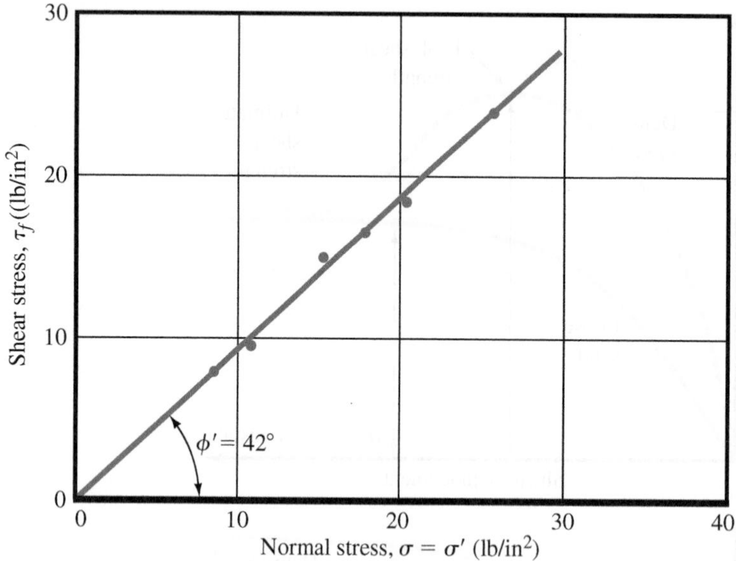

Figure 9.4 Determination of shear strength parameters for a dry sand using the results of direct shear tests (*Note:* 1 lb/in^2 = 6.9 kN/m^2)

The direct shear test is rather simple to perform, but it has some inherent shortcomings. The reliability of the results may be questioned. This is due to the fact that in this test the soil is not allowed to fail along the weakest plane but is forced to fail along the plane of split of the shear box. Also, the shear stress distribution over the shear surface of the specimen is not uniform. In spite of these shortcomings, the direct shear test is the simplest and most economical for a dry or saturated sandy soil.

9.4 Triaxial Shear Test

The triaxial shear test is one of the most reliable methods available for determining the shear strength parameters. It is widely used for both research and conventional testing. The test is considered reliable for the following reasons:

1. It provides information on the stress–strain behavior of the soil that the direct shear test does not.
2. It provides more uniform stress conditions than the direct shear test does with its stress concentration along the failure plane.
3. It provides more flexibility in terms of loading path.

A diagram of the triaxial test layout is shown in Figure 9.5.

In the triaxial shear test, a soil specimen about 1.5 in. (≈38 mm) in diameter and 3 in. (≈76 mm) long generally is used. The specimen is encased by a thin rubber membrane and placed inside a plastic cylindrical chamber that usually is filled with water or glycerine. The specimen is subjected to a confining pressure by compression of the fluid in the chamber. (Note that air is sometimes used as a

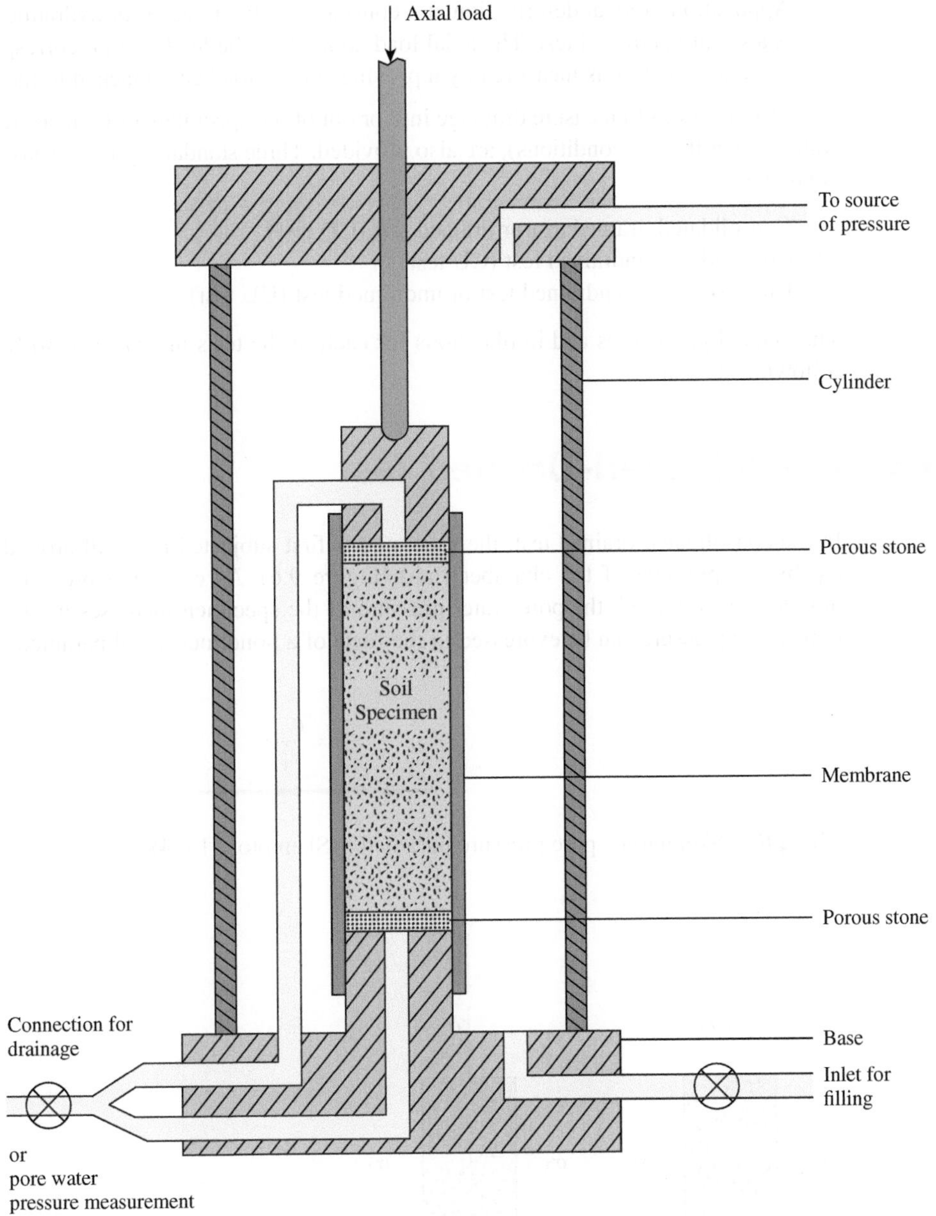

Figure 9.5 Schematic diagram of triaxial test equipment

compression medium.) To cause shear failure in the specimen, axial stress is applied through a vertical loading ram (sometimes called *deviator stress*). Stress is added in one of two ways:

1. Application of dead weights or hydraulic pressure in equal increments (i.e., stress controlled) until the specimen fails. (Axial deformation of the specimen resulting from the load applied through the ram is measured by a dial gauge.)

2. Application of axial deformation at a constant rate by a geared or hydraulic loading press. This is a strain-controlled test. The axial load applied by the loading ram corresponding to a given axial deformation is measured by a proving ring or load cell attached to the ram.

Connections to measure drainage into or out of the specimen, or to measure pressure in the pore water (as per the test conditions), are also provided. Three standard types of triaxial tests are generally conducted:

1. Consolidated-drained test or drained test (CD test)
2. Consolidated-undrained test (CU test)
3. Unconsolidated-undrained test or undrained test (UU test)

The general procedures and implications for each of the tests in *saturated soils* are described in the following sections.

9.5 Consolidated-Drained Test

In the consolidated-drained test, the specimen is first subjected to an all-around confining pressure, σ_3, by compression of the chamber fluid (Figure 9.6). As confining pressure is applied with the drainage valve closed, the pore water pressure of the specimen increases by u_c. This increase in the pore water pressure can be expressed in the form of a nondimensional parameter:

$$B = \frac{u_c}{\sigma_3} \tag{9.8}$$

where B = Skempton's pore pressure parameter (Skempton, 1954).

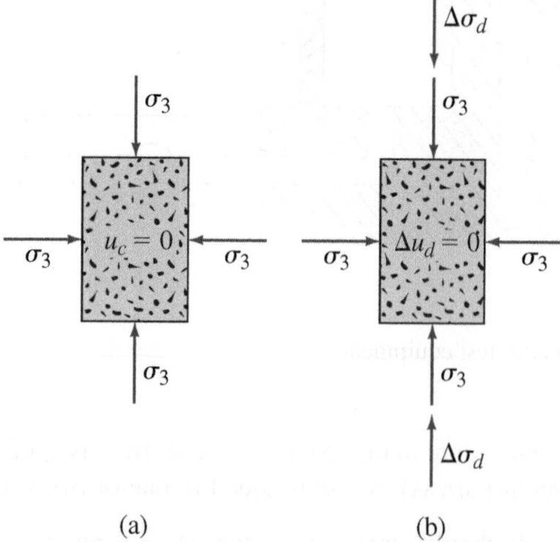

Figure 9.6 Consolidated-drained triaxial test: (a) specimen under chamber confining pressure; (b) deviator stress application

For saturated soft soils, B is approximately equal to 1; however, for saturated stiff soils, the magnitude of B can be less than 1. It is a common practice to measure B in a CD or CU triaxial test to check whether the specimen is saturated.

When the connection to drainage is kept open, dissipation of the excess pore water pressure, and thus consolidation, will occur. With time, u_c will become equal to 0 when the specimen is fully saturated. In saturated soil, the change in the volume of the specimen (ΔV_c) that takes place during consolidation can be obtained from the volume of pore water drained (Figure 9.7a). Then the deviator

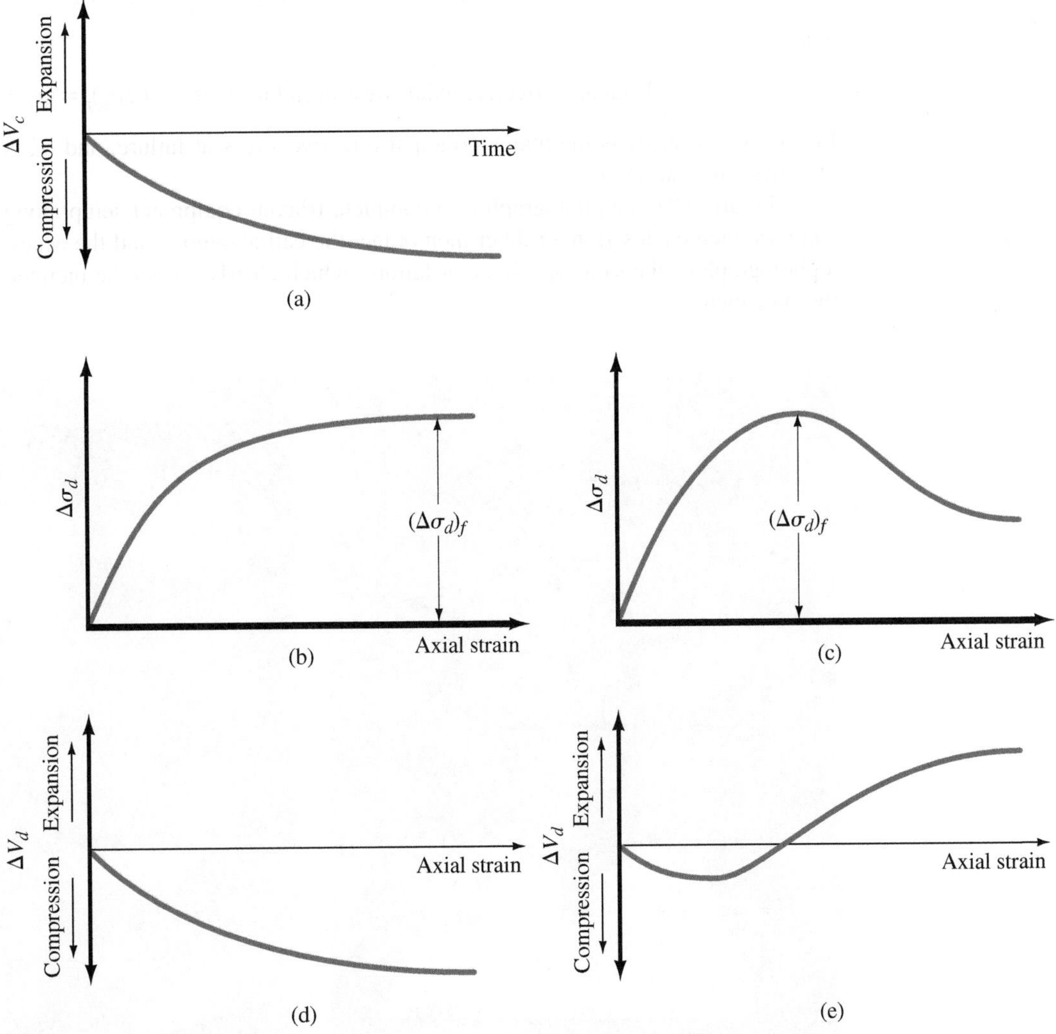

Figure 9.7 Consolidated-drained triaxial test: (a) volume change of specimen caused by chamber confining pressure; (b) plot of deviator stress against strain in the vertical direction for loose sand and normally consolidated clay; (c) plot of deviator stress against strain in the vertical direction for dense sand and overconsolidated clay; (d) volume change in loose sand and normally consolidated clay during deviator-stress application; (e) volume change in dense sand and overconsolidated clay during deviator-stress application

stress, $\Delta\sigma_d$, on the specimen is increased at a very slow rate (Figure 9.7b). The drainage connection is kept open, and the slow rate of deviator-stress application allows complete dissipation of any pore water pressure that developed as a result ($\Delta u_d = 0$).

A typical plot of the variation of deviator stress against strain in loose sand and normally consolidated clay is shown in Figure 9.7b. Figure 9.7c shows a similar plot for dense sand and overconsolidated clay. The volume change, ΔV_d, of specimens that occurs because of the application of deviator stress in various soils is also shown in Figures 9.7d and e.

Since the pore water pressure developed during the test is completely dissipated, we have

$$\text{Total and effective confining stress} = \sigma_3 = \sigma_3'$$

and

$$\text{Total and effective axial stress at failure} = \sigma_3 + (\Delta\sigma_d)_f = \sigma_1 = \sigma_1'$$

In a triaxial test, σ_1' is the major principal effective stress at failure, and σ_3' is the minor principal effective stress at failure.

Figure 9.8a is a photograph of a complete triaxial equipment setup along with the cylindrical soil specimen enclosed in a rubber membrane, the cell assembly, and the accessories. Figure 9.8b is a photograph of the same specimen at failure, which clearly shows the inclined failure plane within the specimen.

(a)

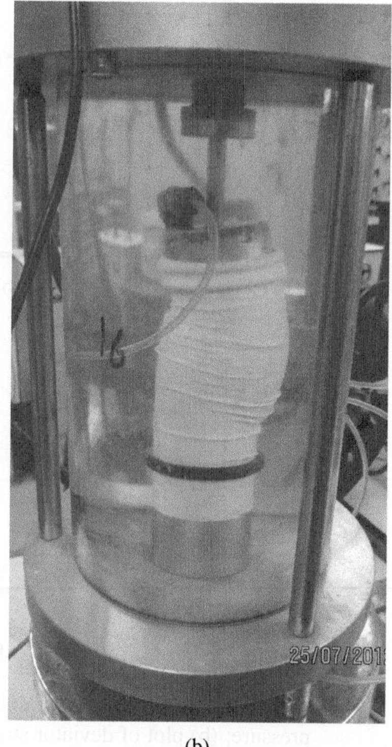

(b)

Figure 9.8 (a) Triaxial test in progress, and (b) a triaxial specimen at failure (*Courtesy of N. Sivakugan, James Cook University, Australia*)

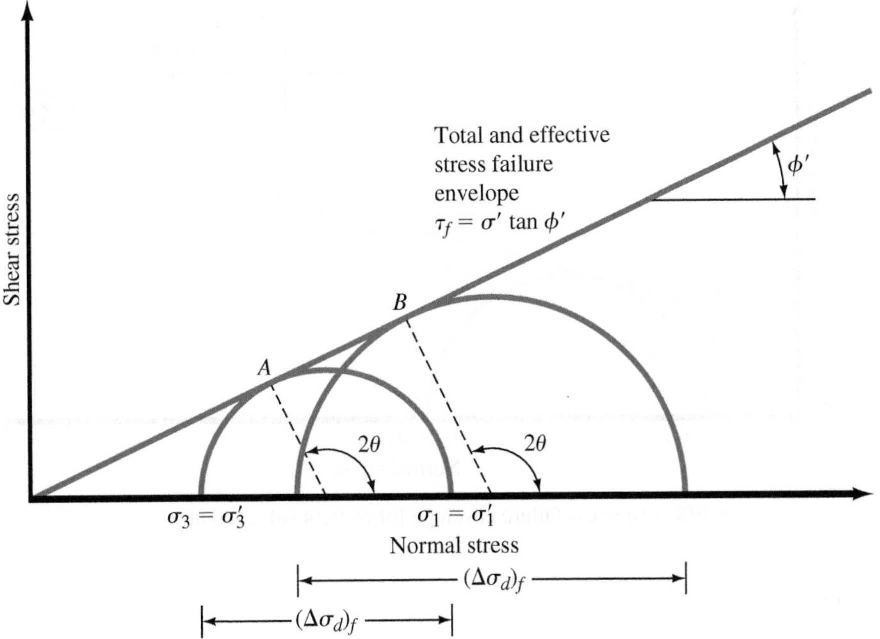

Figure 9.9 Effective stress failure envelope from drained tests in sand and normally consolidated clay

Several tests on similar specimens can be conducted by varying the confining pressure. With the major and minor principal stresses at failure for each test, the Mohr's circles can be drawn and the failure envelopes can be obtained. Figure 9.9 shows the type of effective stress failure envelope obtained for tests in sand and normally consolidated clay.

Overconsolidation results when a clay is initially consolidated under an all-around chamber pressure of $\sigma_c(= \sigma'_c)$ and is allowed to swell as the chamber pressure is reduced to $\sigma_3(= \sigma'_3)$. The failure envelope obtained from drained triaxial tests of such overconsolidated clay specimens shows two distinct branches (*ab* and *bc* in Figure 9.10). The portion *ab* has a flatter slope with a cohesion intercept, and the shear strength equation for this branch can be written as

$$\tau_f = c' + \sigma' \tan \phi'_1 \tag{9.9}$$

The portion *bc* of the failure envelope represents a normally consolidated stage of soil and follows the equation $\tau_f = \sigma' \tan \phi'$.

A consolidated-drained triaxial test on a clayey soil may take several days to complete. The time is needed to apply deviator stress at a very slow rate to ensure full drainage from the soil specimen. For that reason, the CD type of triaxial test is not commonly used.

It can also be shown that:

(a) For sand and normally consolidated clay (where $c' = 0$),

$$\sigma'_1 = \sigma'_3 \tan^2\left(45 + \frac{\phi'}{2}\right) \tag{9.10}$$

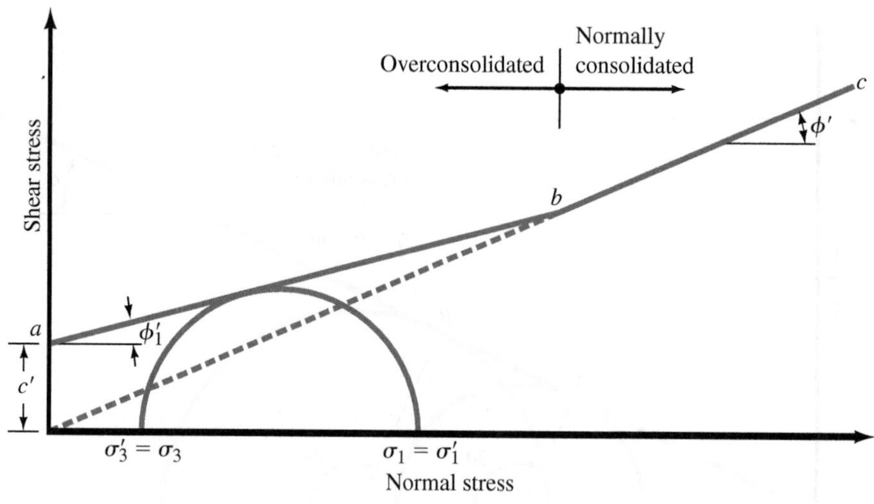

Figure 9.10 Effective stress failure envelope for overconsolidated clay

or

$$\phi' = 2\left[\tan^{-1}\left(\frac{\sigma'_1}{\sigma'_3}\right)^{0.5} - 45°\right] \tag{9.11}$$

(b) For overconsolidated clay where $c' \neq 0$,

$$\sigma'_1 = \sigma'_3 \tan^2\left(45 + \frac{\phi'}{2}\right) + 2c' \tan\left(45 + \frac{\phi'}{2}\right) \tag{9.12}$$

Example 9.1

A consolidated-drained triaxial test was conducted on a normally consolidated clay. The results are as follows:

- $\sigma_3 = 40$ lb/in.^{2}
- $(\Delta\sigma_d)_f = 40$ lb/in.^{2}

Determine the angle of friction, ϕ'.

Solution

For normally consolidated soil, the failure envelope equation is

$$\tau_f = \sigma' \tan \phi' \qquad \text{(since } c' = 0\text{)}$$

For the triaxial test, the effective major and minor principal stresses at failure are as follows:

$$\sigma'_1 = \sigma_1 = \sigma_3 + (\Delta\sigma_d)_f = 40 + 40 = 80 \text{ lb/in.}^2$$

and

$$\sigma'_3 = \sigma_3 = 40 \text{ lb/in.}^2$$

The Mohr's circle and the failure envelope are shown in Figure 9.11. From Eq. (9.11),

$$\phi' = 2\left[\tan^{-1}\left(\frac{\sigma'_1}{\sigma'_3}\right)^{0.5} - 45\right] = 2\left[\tan^{-1}\left(\frac{80}{40}\right)^{0.5} - 45\right] = \mathbf{19.47°}$$

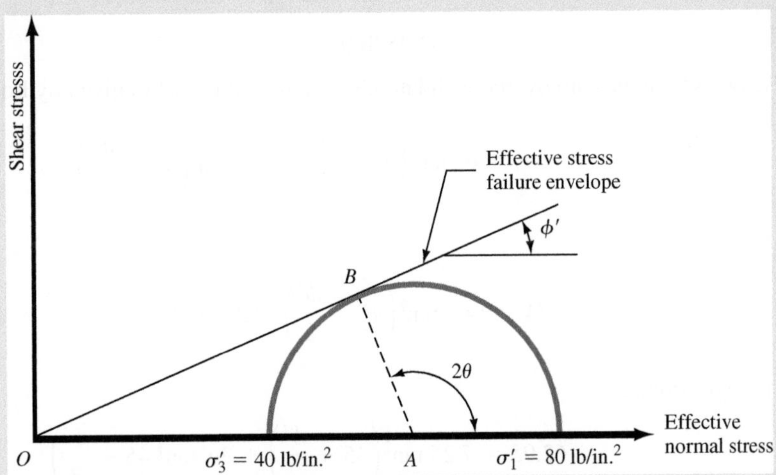

Figure 9.11

Example 9.2

Two similar clay soil specimens were preconsolidated in triaxial equipment under a chamber pressure of 85 lb/in.^{2}. Consolidated-drained triaxial tests were conducted on these two specimens. Following are the results of the tests:

- Specimen 1: $\sigma_3 = 14.5$ lb/in.^{2}
  $(\Delta\sigma_d)_f = 59.5$ lb/in.^{2}
- Specimen 2: $\sigma_3 = 7.25$ lb/in.^{2}
  $(\Delta\sigma_d)_f = 55.7$ lb/in.^{2}

Determine the shear strength parameters for the soil (see Figure 9.12).

Solution

For Specimen 1, the principal stresses at failure are

$$\sigma_3' = \sigma_3' = 14.5 \text{ lb/in.}^2$$

and

$$\sigma_1' = \sigma_1 = \sigma_3 + (\Delta\sigma_d)_f = 14.5 + 59.5 = 74 \text{ lb/in.}^2$$

Similarly, the principal stresses at failure for Specimen 2 are

$$\sigma_1' = \sigma_3 = 7.25 \text{ lb/in.}^2$$

and

$$\sigma_1' = \sigma_1 = \sigma_3 + (\Delta\sigma_d)_f = 7.25 + 55.7$$
$$= 62.95 \text{ lb/in.}^2$$

These two specimens are overconsolidated. Using the relationship given by Eq. (9.12)

$$\sigma_1' = \sigma_3' \tan^2\left(45 + \frac{\phi'}{2}\right) + 2c' \tan\left(45 + \frac{\phi'}{2}\right)$$

Thus, for Specimen 1

$$74 = 14.5 \tan^2\left(45 + \frac{\phi'}{2}\right) + 2c' \tan\left(45 + \frac{\phi'}{2}\right) \quad \text{(a)}$$

and for Specimen 2

$$62.95 = 7.25 \tan^2\left(45 + \frac{\phi'}{2}\right) + 2c' \tan\left(45 + \frac{\phi'}{2}\right) \quad \text{(b)}$$

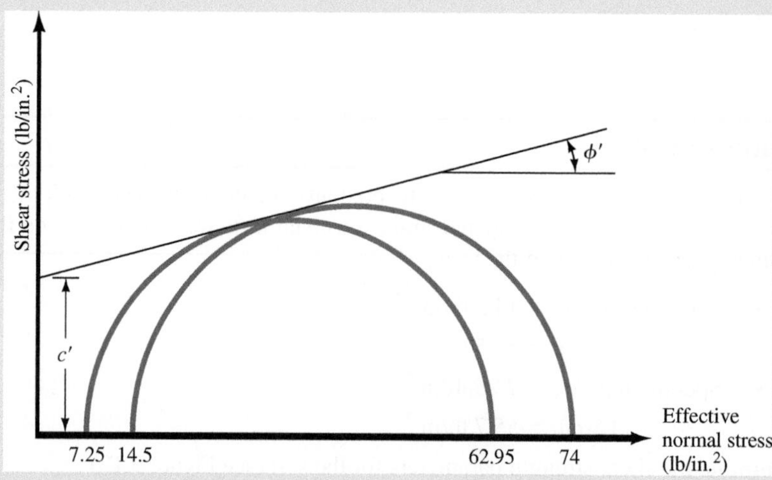

Figure 9.12

Subtracting Eq. (b) from Eq. (a)

$$11.05 = 7.25\tan^2\left(45 + \frac{\phi'}{2}\right)$$

$$45 + \frac{\phi'}{2} = \tan^{-1}\left[\sqrt{\frac{11.05}{7.25}}\right] = 51^\circ$$

or

$$\phi' = 12^\circ$$

Substituting $\phi' = 12^\circ$ in Eq. (a)

$$74 = 14.5\tan^2[45 + (12/2)] + 2c'\tan[45 + (12/2)]$$

or

$$74 = 22.1 + 2.47c'$$

So

$$c' = \mathbf{21\ lb/in.^2}$$

■

Example 9.3

A consolidated drained triaxial test is carried out on a sand specimen that is subjected to 100 kN/m^2 confining pressure. The vertical deviator stress was increased slowly such that there is no build-up pore water pressure within the specimen. The specimen failed when the deviator stress reached 260 kN/m^2. Find the friction angle of the sand.

Another identical sand specimen is subjected to 200 kN/m^2 confining pressure. What would be the deviator stress at failure?

Solution

$\sigma_3' = 100$ kN/m^2 and $(\Delta\sigma_d)_f = 260$ kN/m^2. Therefore, $\sigma_1' = \sigma_3' + (\Delta\sigma_d)_f = 360$ kN/m^2.

From Eq. (9.11),

$$\phi' = 2\left[\tan^{-1}\left(\frac{\sigma_1'}{\sigma_3'}\right)^{0.5} - 45\right] = 2\left[\tan^{-1}\left(\frac{360}{100}\right)^{0.5} - 45\right] = \mathbf{34.4^\circ}$$

For the second specimen, $\sigma_3' = 200$ kN/m^2.

From Eq. (9.10),

$$\sigma_1' = \sigma_3'\tan^2\left(45 + \frac{\phi'}{2}\right) = 200\tan^2\left(45 + \frac{34.4}{2}\right) = 719.5\ \text{kN/m}^2$$

Therefore, $(\Delta\sigma_d)_f = 719.5 - 200 = \mathbf{519.5\ kN/m^2}$

■

9.6 Consolidated-Undrained Test

The consolidated-undrained test is the most common type of triaxial test. In this test, the saturated soil specimen is first consolidated by an all-round chamber fluid pressure, σ_3, that results in drainage. After the pore water pressure generated by the application of confining pressure is completely dissipated (that is, $u_c = B\sigma_3 = 0$, see Figure 9.13a and b), the deviator stress, $\Delta\sigma_d$, on the specimen is increased to cause shear failure (Figure 9.13c). During this phase of the test, the drainage line from the specimen is kept closed. Since drainage is not permitted, the pore water pressure, Δu_d, will increase. During the test, measurements of $\Delta\sigma_d$ and Δu_d are made. The increase in the pore water pressure, Δu_d, can be expressed in a nondimensional form as

$$A = \frac{\Delta u_d}{\Delta\sigma_d} \tag{9.13}$$

where A = Skempton's pore pressure parameter (Skempton, 1954) which varies in the range of -0.5 to 1.0.

The general patterns of variation of $\Delta\sigma_d$ and Δu_d with axial strain for sand and clay soils are shown in Figures 9.13d, e, f, and g. In loose sand and normally consolidated clay, the pore water pressure increases with strain. In dense sand and overconsolidated clay, the pore water pressure increases with strain up to a certain limit, beyond which it decreases and becomes negative (with respect to the atmospheric pressure). This pattern is because the soil has a tendency to dilate.

Unlike in the consolidated-drained test, the total and effective principal stresses are not the same in the consolidated-undrained test where pore water pressure develops during undrained loading. Since the pore water pressure at failure is measured in this test, the principal stresses may be analyzed as follows:

- Major principal stress at failure (total):

$$\sigma_3 + (\Delta\sigma_d)_f = \sigma_1$$

- Major principal stress at failure (effective):

$$\sigma_1 - (\Delta u_d)_f = \sigma_1'$$

- Minor principal stress at failure (total):

$$\sigma_3$$

- Minor principal stress at failure (effective):

$$\sigma_3 - (\Delta u_d)_f = \sigma_3'$$

where $(\Delta u_d)_f$ = pore water pressure at failure. The preceding derivations show that

$$\sigma_1 - \sigma_3 = \sigma_1' - \sigma_3'$$

Tests on several similar specimens with varying confining pressures may be done to determine the shear strength parameters. Figure 9.14 shows the total and effective stress Mohr's circles at failure obtained from consolidated-undrained triaxial tests in sand and normally consolidated clay. Note that

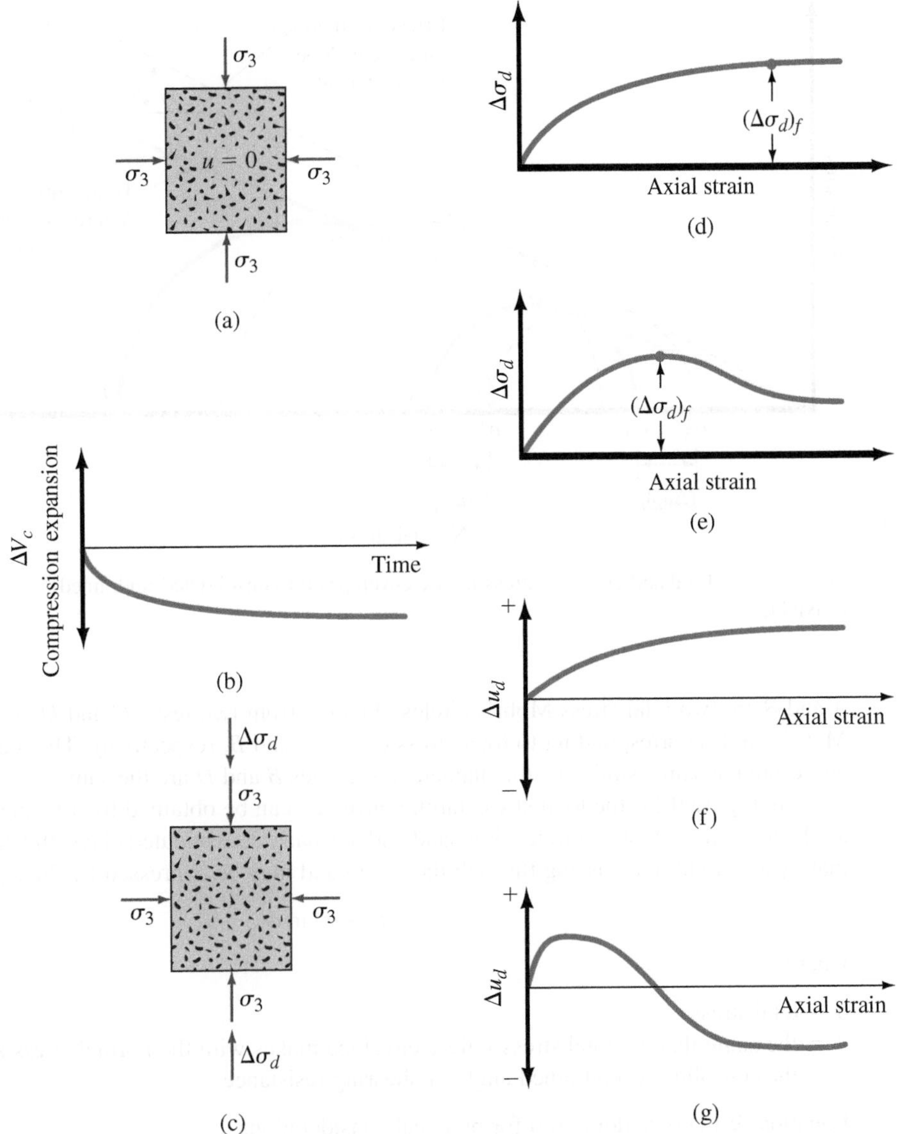

Figure 9.13 Consolidated-undrained test: (a) specimen under chamber confining pressure; (b) volume change in specimen caused by confining pressure; (c) deviator stress application; (d) deviator stress against axial strain for loose sand and normally consolidated clay; (e) deviator stress against axial strain for dense sand and overconsolidated clay; (f) variation of pore water pressure with axial strain for loose sand and normally consolidated clay; (g) variation of pore water pressure with axial strain for dense sand and overconsolidated clay

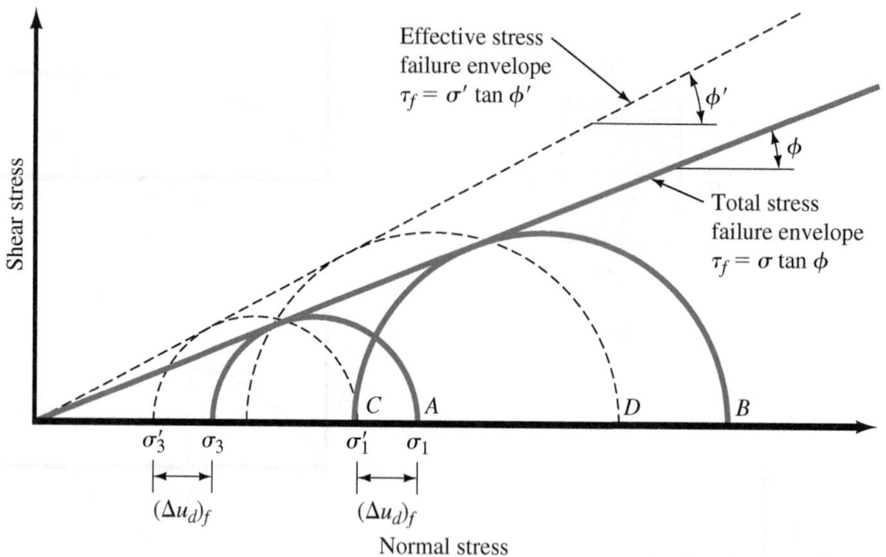

Figure 9.14 Total and effective stress failure envelopes for consolidated-undrained triaxial tests.

A and *B* are two total stress Mohr's circles obtained from two tests. *C* and *D* are the effective stress Mohr's circles corresponding to total stress circles *A* and *B*, respectively. The diameters of circles *A* and *C* are the same; similarly, the diameters of circles *B* and *D* are the same.

In Figure 9.14, the total stress failure envelope can be obtained by drawing a line that touches all the total stress Mohr's circles. For sand and normally consolidated clays, this line will be approximately a straight line passing through the origin and may be expressed by the equation

$$\tau_f = \sigma \tan \phi \tag{9.14}$$

where

σ = total stress
ϕ = the angle that the total stress failure envelope makes with the normal stress axis, also known as the consolidated-undrained angle of shearing resistance

Equation (9.14) is seldom used for practical considerations.

Again referring to Figure 9.14, we see that the failure envelope that is tangent to all the effective stress Mohr's circles can be represented by the equation $\tau_f = \sigma' \tan \phi'$, which is the same as the failure envelope obtained from consolidated-drained tests (see Figure 9.9).

In overconsolidated clays, the total stress failure envelope obtained from consolidated-undrained tests takes the shape shown in Figure 9.15. The straight line $a'\ b'$ is represented by the equation

$$\tau_f = c + \sigma \tan \phi_1 \tag{9.15}$$

and the straight line $b'\ c'$ follows the relationship given by Eq. (9.14). The effective stress failure envelope drawn from the effective stress Mohr's circles is similar to that shown in Figure 9.15.

Consolidated-drained tests on clay soils take considerable time. For that reason, consolidated-undrained tests can be conducted on such soils with pore pressure measurements to obtain the drained

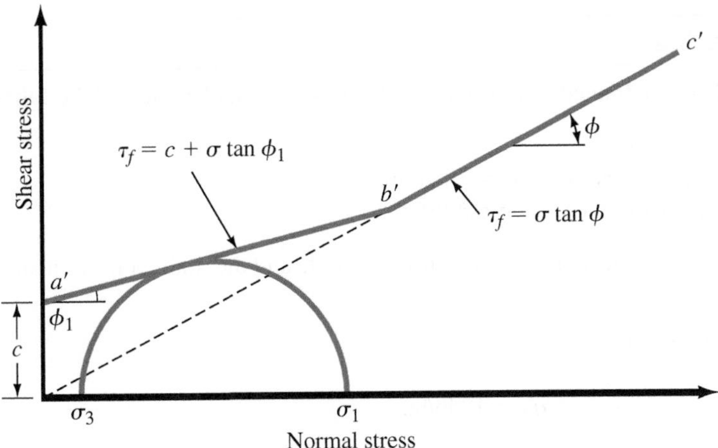

Figure 9.15 Total stress failure envelope obtained from consolidated-undrained tests in overconsolidated clay

shear strength parameters. Since drainage is not allowed in these tests during the application of deviator stress, the tests can be performed rather quickly.

It is important to note that, for the Mohr's circles shown in Figure 9.14 (sand or normally consolidated clay),

$$\phi' = 2\left[\tan^{-1}\left(\frac{\sigma_1'}{\sigma_3'}\right)^{0.5} - 45^\circ\right]$$

and

$$\phi = 2\left[\tan^{-1}\left(\frac{\sigma_1}{\sigma_3}\right)^{0.5} - 45^\circ\right]$$

For overconsolidated clays ($c' \neq 0$), σ_1' and σ_3' are related by Eq. (9.12).

When analyzing a geotechnical problem such as foundation or embankment in terms of effective stresses, it is necessary to use the effective stress shear strength parameters c' and ϕ'. These parameters can be determined from either a consolidated drained or a consolidated undrained triaxial test. When the soil is fully drained or when the pore water pressures are known, it is possible to separate the effective stresses and pore water pressures and hence to carry out the analysis in terms of effective stresses.

In situations where the soil is undrained (e.g., short-term stability of a footing in clay), we do not bother separating the effective stresses and pore water pressure but carry out the analysis in terms of total stresses. Here, we use *undrained shear strength* c_u ($\phi = 0$), which is derived from an unconsolidated undrained test discussed in Section 9.7, and carry out the analysis in terms of total stresses.

The direct shear test discussed in Section 9.3 can be carried out as drained (i.e., very slow loading with no pore water pressure development) or undrained (i.e., quick loading with no drainage) to determine c', ϕ', or c_u.

Example 9.4

A consolidated-undrained test on a normally consolidated clay yielded the following results:

- $\sigma_3 = 12 \text{ lb/in}^2$.
- Deviator stress, $(\Delta\sigma_d)_f = 9.1 \text{ lb/in.}^2$
- Pore pressure, $(\Delta u_d)_f = 6.8 \text{ lb/in.}^2$

Calculate the consolidated-undrained friction angle and the drained friction angle.

Solution

Refer to Figure 9.16.

$$\sigma_3 = 12 \text{ lb/in.}^2$$

$$\sigma_1 = \sigma_3 + (\Delta\sigma_d)_f = 12 + 9.1 = 21.1 \text{ lb/in.}^2$$

$$\sigma_1 = \sigma_3 \tan^2\left(45 + \frac{\phi}{2}\right)$$

$$21.1 = 12 \tan^2\left(45 + \frac{\phi}{2}\right)$$

$$\phi = 2\left[\tan^{-1}\left(\frac{21.1}{12}\right)^{0.5} - 45\right] = \mathbf{16°}$$

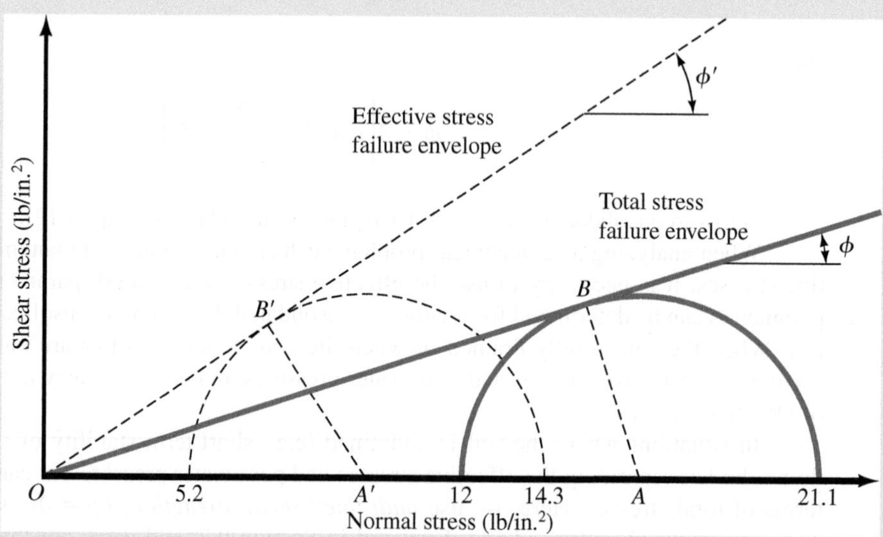

Figure 9.16

Again,

$$\sigma'_3 = \sigma_3 - (\Delta u_d)_f = 12 - 6.8 = 5.2 \text{ lb/in.}^2$$

$$\sigma'_1 = \sigma_1 - (\Delta u_d)_f = 21.1 - 6.8 = 14.3 \text{ lb/in.}^2$$

$$\sigma'_1 = \sigma'_3 \tan^2\left(45 + \frac{\phi'}{2}\right)$$

$$14.3 = 5.2 \tan^2\left(45 + \frac{\phi'}{2}\right)$$

$$\phi' = 2\left[\tan^{-1}\left(\frac{14.3}{5.2}\right)^{0.5} - 45\right] = \mathbf{27.8°}$$

■

Example 9.5

A consolidated-drained triaxial test was carried out on a normally consolidated clay specimen, and the following results were recorded: $\sigma'_3 = 150 \text{ kN/m}^2$ and $(\Delta\sigma_d)_f = 260 \text{ kN/m}^2$. An identical specimen from the same clay was subjected to a consolidated-undrained test with a confining pressure of 150 kN/m^2, and the deviator stress at failure was 115 kN/m^2. What is the pore water pressure at failure in this second specimen? What is Skempton's A-parameter at failure?

Solution

In normally consolidated clay, $c' = 0$. For the first specimen (consolidated-drained test), using Eq. (9.11), we have

$$\phi' = 2\left[\tan^{-1}\left(\frac{\sigma'_1}{\sigma'_3}\right)^{0.5} - 45\right] = 2\left[\tan^{-1}\left(\frac{410}{150}\right)^{0.5} - 45\right] = 27.7°$$

In the second specimen (consolidated-undrained test), applying the same value of ϕ' in Eq, (9.10), we have

$$\sigma_3 = 150 \text{ kN/m}^2 \text{ and } (\Delta\sigma_d)_f = 115 \text{ kN/m}^2$$

$$\sigma_1 = \sigma_3 + (\Delta\sigma_d)_f = 265 \text{ kN/m}^2$$

$$\sigma'_3 = 150 - \Delta u_d \text{ and } \sigma'_1 = 265 - \Delta u_d$$

$$\sigma'_1 = \sigma'_3 \tan^2\left(45 + \frac{\phi'}{2}\right) = \sigma'_3 \tan^2\left(45 + \frac{27.7}{2}\right) = 2.737\sigma'_3$$

$$265 - \Delta u_d = (150 - \Delta u_d) \times 2.737$$

$$\Delta u_d = 83.8 \text{ kN/m}^2$$

From Eq. (9.13),

$$A = \frac{\Delta u_d}{\Delta\sigma_d} = \frac{83.8}{115} = \mathbf{0.73}$$

■

9.7 Unconsolidated-Undrained Test

In unconsolidated-undrained tests, drainage from the soil specimen is not permitted during the application of chamber pressure, σ_3. The test specimen is sheared to failure by the application of deviator stress, $\Delta\sigma_d$, with no drainage allowed. Since drainage is not allowed at any stage, the test can be performed very quickly. Because of the application of chamber confining pressure, σ_3, the pore water pressure in the soil specimen will increase by u_c. There will be a further increase in the pore water pressure, Δu_d, because of the deviator stress application. Hence, the total pore water pressure, u, in the specimen at any stage of deviator stress application can be given as

$$u = u_c + \Delta u_d \tag{9.16}$$

The unconsolidated-undrained test is usually conducted on clay specimens and depends on a very important strength concept for saturated cohesive soils. The added axial stress at failure $(\Delta\sigma_d)_f$ is practically the same regardless of the chamber confining pressure. This result is shown in Figure 9.17. The failure envelope for the total stress Mohr's circles becomes a horizontal line and hence is called a $\phi = 0$ condition, and

$$\tau_f = c_u \tag{9.17}$$

where c_u is the *undrained shear strength* and is equal to the radius of the Mohr's circles.

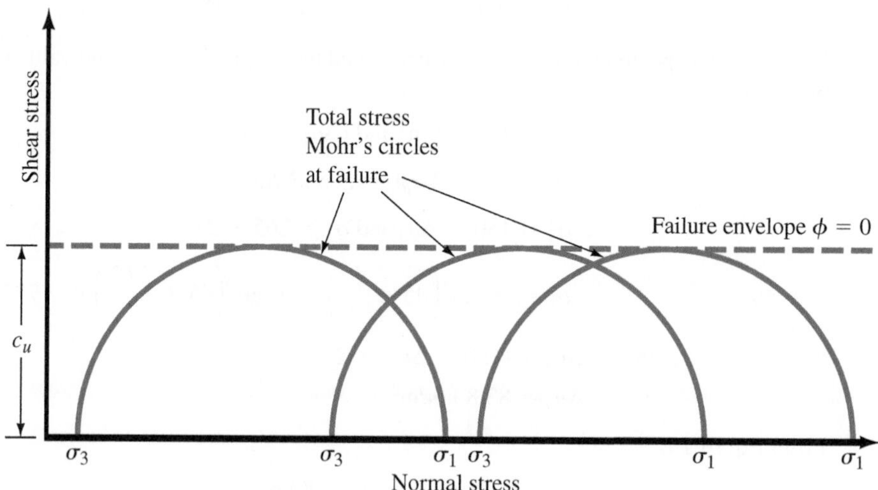

Figure 9.17 Total stress Mohr's circles and failure envelope ($\phi = 0$) obtained from unconsolidated-undrained triaxial tests

9.8 Unconfined Compression Test on Saturated Clay

The unconfined compression test is a special type of unconsolidated-undrained test that is commonly used for clay specimens. In this test, the confining pressure σ_3 is 0. An axial load is rapidly applied to the specimen to cause failure. At failure, the total minor principal stress is 0 and the total major principal stress is σ_1 (Figure 9.18). Since the undrained shear strength is independent of the confining pressure, we have

$$\tau_f = \frac{\sigma_1}{2} = \frac{q_u}{2} = c_u \tag{9.18}$$

where q_u is the *unconfined compression strength*. Table 9.2 gives the approximate consistencies of clays based on their unconfined compression strengths. A photograph of unconfined compression test equipment is shown in Figure 9.19.

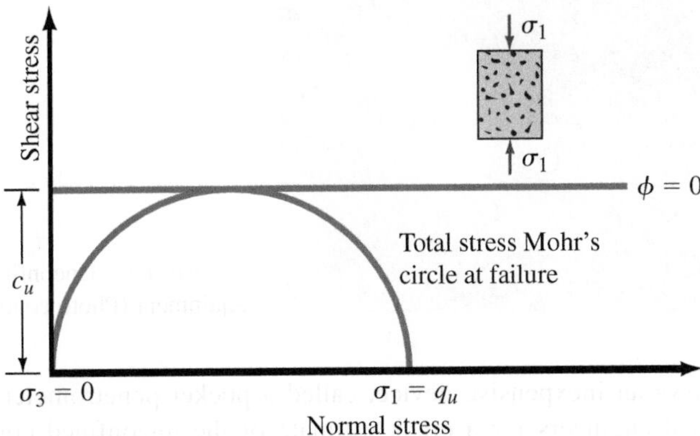

Figure 9.18 Unconfined compression test

Table 9.2 General relationship of consistency and unconfined compression strength of clays

	q_u	
Consistency	**(lb/ft²)**	**kN/m²**
Very soft	0–500	0–24
Soft	500–1000	24–48
Medium	1000–2000	48–96
Stiff	2000–4000	96–192
Very stiff	4000–8000	192–384
Hard	>8000	>384

Figure 9.19 Unconfined compression test equipment (Photo courtesy of Braja Das)

Figure 9.20 shows an inexpensive device, called a pocket penetrometer, that is commonly carried by geotechnical engineers for a quick estimate of the unconfined compressive strength. By pushing the spring-loaded piston into the clay sample or wall of an excavation, the unconfined compressive strength can be read from the scale. While the estimates cost literally nothing, they are very approximate.

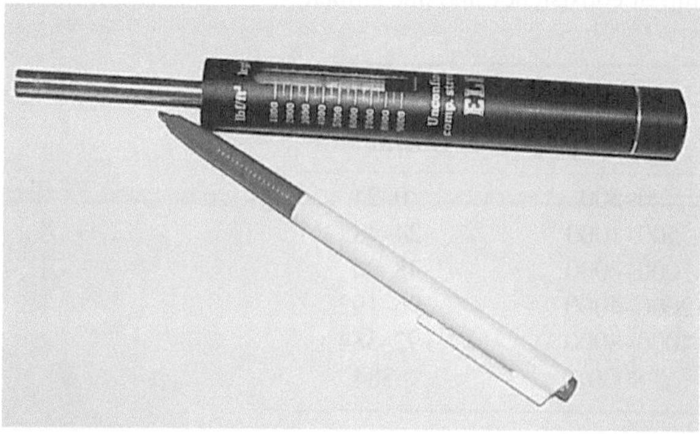

Figure 9.20 Pocket penetrometer

9.9 Summary

Soils always fail in shear where the shear stress along the failure surface becomes equal to or exceeds the shear strength. Shear failure in soils is governed by the Mohr-Coulomb failure criterion given by $\tau_f = c + \sigma \tan \phi$. The shear strength ($\tau_f$) consists of two separate components: one cohesive and the other frictional. The frictional shear resistance is proportional to the normal stress acting on the shear plane. The two shear-strength parameters cohesion (c), and friction angle (ϕ) can be measured in the laboratory using triaxial or direct shear tests.

Consolidated drained (CD), consolidated undrained (CU), and unconsolidated undrained (UU) are the three types of trial tests carried out in the laboratory to simulate drained and undrained loading and thus derive the appropriate c and ϕ. From CD or CU tests, it is possible to determine c' and ϕ'. The UU test gives $c_u(\phi = 0)$. The unconfined compression test is a special type of UU test that gives the unconfined compressive strength $q_u\,(=2c_u)$.

Problems

9.1 State whether the following are true or false.

a. Shear strength increases with the normal stress.

b. In normally consolidated clays and granular soils, $c' = 0$.

c. Undrained shear strength c_u can be derived from a consolidated-undrained triaxial test.

d. In a consolidated-drained triaxial test, the larger the confining pressure, the larger is the deviator stress.

e. Undrained shear strength is twice the unconfined compressive strength.

9.2 A direct shear test was conducted on a specimen of dry sand with a normal stress of 27 lb/in.^{2}. Failure occurred at a shear stress of 17 lb/in.^{2}. The size of the sample tested was 2 in. $\times$ 2 in. $\times$ 1 in. (height). Determine the angle of friction, ϕ'. For a normal stress of 20 lb/in.^{2}, what shear force would be required to cause failure in the sample?

9.3 The angle of friction of a compacted dry sand is 41°. In a direct shear test on this sand, a normal stress of 15 lb/in.^{2} was applied. The size of the specimen was 2 in. $\times$ 2 in. $\times$ 1.2 in. (height). What shear force (in lbs) will cause failure?

9.4 A direct shear test is conducted on a 60 mm $\times$ 60 mm normally consolidated clay specimen. The loading was very slow, ensuring that there is no pore water pressure development within the specimen (i.e., drained loading). The following data were recorded.

Normal load (N)	Shear load (N)	σ (kN/m^2)	τ (kN/m^2)
178	102	49.4	28.3
362	174	100.6	48.3
537	256	149.2	71.1
719	332	199.7	92.2

Determine the shear strength parameters c' and ϕ'.

9.5 Following are the results of four drained direct shear tests on a normally consolidated clay:
Sample size: diameter of sample = 2 in.
height of sample = 1 in.

Test No.	Normal force (lb)	Shear force at failure (lb)
1	60	27
2	91	38
3	107	46
4	122	55

Draw a graph for shear stress at failure against normal stress. Determine the drained angle of friction from the graph.

9.6 The equation of the effective stress failure envelope for a loose sandy soil was obtained from a direct shear test as $\tau_f = \sigma' \tan 30°$. A drained triaxial test was conducted with the same soil at a chamber confining pressure of 10 lb/in^2. Calculate the deviator stress at failure.

9.7 A consolidated-drained triaxial test was carried out on some normally consolidated clay specimens. The average ratio of the principal stresses (σ_1'/σ_3') at failure is 2.4. Determine the drained friction angle of the soil.

9.8 For a normally consolidated clay, the results of a drained triaxial test are as follows:

- Chamber confining pressure = 20 lb/in^2.
- Deviator stress at failure = 37.6 lb/in^2.

Determine the soil friction angle, ϕ'

9.9 The results of two drained triaxial tests on a saturated clay are given below:

- Specimen I: Chamber confining pressure = 10 lb/in^2
  Deviator stress at failure = 31 lb/in^2
- Specimen II: Chamber confining pressure = 17.4 lb/in^2
  Deviator stress at failure = 37.5 lb/in^2

Calculate the shear strength parameters of the soil.

9.10 If a specimen of clay described in Problem 9.9 is tested in a triaxial apparatus with a chamber confining pressure of 29 lb/in^2, what will be the major principal stress at failure? Assume full drained condition during the test.

9.11 The specimens obtained from a clay layer at a site gave the following shear strength parameters from a consolidated-drained triaxial test: $c' = 10$ kN/m^2 and $\phi' = 26°$. A consolidated-undrained triaxial test is carried out on this soil, where a specimen is consolidated under confining pressure of 100 kN/m^2 and loaded under undrained conditions. The specimen failed under a deviator stress of 107.0 kN/m^2. What is the pore water pressure within the specimen?

9.12 A sandy soil has a drained angle of friction of 35°. In a drained triaxial test on the same soil, the deviator stress at failure is 2.69 ton/ft^2. What is the chamber confining pressure?

9.13 A consolidated-undrained test on a normally consolidated clay yielded the following results:

- $\sigma_3 = 12$ lb/in.^{2}
- Deviator stress: $(\Delta\sigma_d)_f = 9.14$ lb/in.^{2}
- Pore pressure: $(\Delta u_d)_f = 6.83$ lb/in.^{2}

Calculate the consolidated-undrained friction angle and the drained friction angle.

9.14 The following are the results of a consolidated-undrained triaxial test in a clay:

Sample No.	σ_3 (lb/in^2).	σ_1 at failure (lb/in^2).
1	27.8	54.5
2	55.6	92.2

Draw the total stress Mohr's circles and determine the shear strength parameters for consolidated-undrained conditions.

9.15 The data from a series of consolidated-undrained triaxial tests are summarized here. Draw the three Mohr's circles, plot the failure envelope in terms of effective stresses, and find c' and ϕ'.

Sample number	Cell pressure (kN/m^2)	Deviator stress at failure (kN/m^2)	Pore water pressure at failure (kN/m^2)
1	100	88.2	57.4
2	200	138.5	123.7
3	300	232.1	208.8

9.16 The unconsolidated-undrained test results of a saturated clay specimen are as follows:

- $\sigma_3 = 2000$ lb/ft^2
- σ_1 at failure $= 3900$ lb/ft^2

What will be the axial stress at failure if a similar sample is subjected to an unconfined compression test?

9.17 Steel plates with a mass of 1500 g each were stacked on top of a 75-mm-diameter and 150-mm-high clay specimen, as shown in Figure 9.21. If the undrained shear strength of the specimen is 45.0 kN/m^2, how many plates can be stacked before the specimen fails? What is the consistency term for this clay?

9.18 A consolidated-drained triaxial test was carried out on a normally consolidated clay specimen with confining pressure of 75.0 kN/m^2. The deviator stress at failure was 96.0 kN/m^2. An identical specimen of the same clay was subjected to a consolidated-undrained triaxial test, where the confining pressure is 150.0 kN/m^2. The deviator stress at failure was 115.0 kN/m^2. What would be the pore water pressure at failure?

9.19 Consolidated-undrained triaxial tests were carried out on three clay specimens, and the corresponding Mohr's circles are shown in Figure 9.22. Find the shear strength parameters c, ϕ, c', and ϕ'.

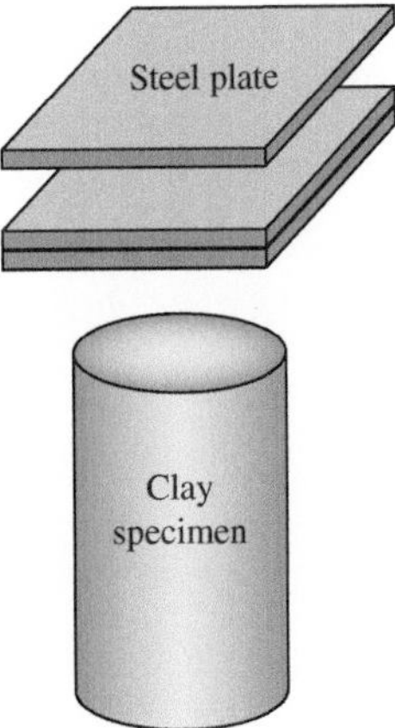

Figure 9.21

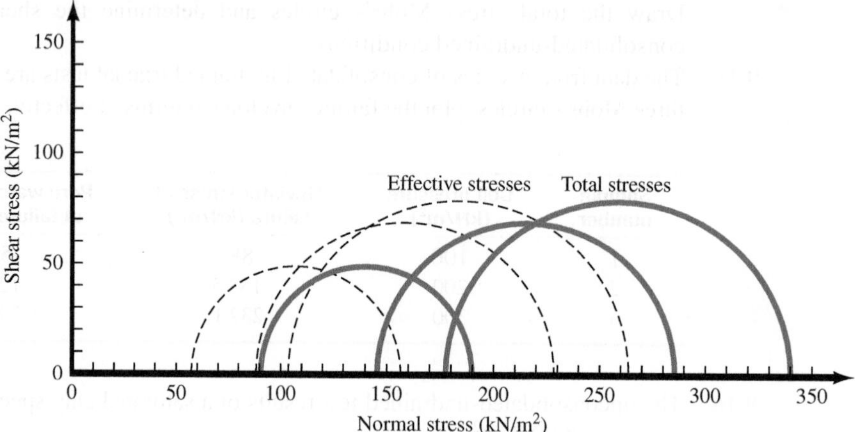

Figure 9.22

References

Coulomb, C. A. (1776). "Essai sur une application des regles de Maximums et Minimis á quelques Problèmes de Statique, relatifs á l'Architecture," *Memoires de Mathematique et de Physique*, Présentés, á l'Academie Royale des Sciences, Paris, Vol. 3, 38.

Mohr, O. (1900). "Welche Umstände Bedingen die Elastizitätsgrenze und den Bruch eines Materiales?," *Zeitschrift des Vereines Deutscher Ingenieure*, Vol. 44, 1524–1530, 1572–1577.

Skempton, A. W. (1954). "The Pore Water Coefficients *A* and *B*," *Geotechnique*, Vol. 4, 143–147.

10 Subsurface Exploration

10.1 Introduction

The process of identifying the layers of deposits that underlie a proposed structure and their physical characteristics is generally referred to as *subsurface exploration*. The purpose of subsurface exploration is to obtain information that will aid the geotechnical engineer in these tasks.

1. Selecting the type and depth of foundation suitable for a given structure
2. Evaluating the load-bearing capacity of the foundation
3. Estimating the probable settlement of a structure
4. Determining potential foundation problems (for example, expansive soil, collapsible soil, sanitary landfill, and so on)
5. Determining the location of the water table
6. Predicting lateral earth pressure for structures such as retaining walls, sheet pile bulkheads, and braced cuts
7. Establishing construction methods for changing subsoil conditions

Subsurface exploration is also necessary for underground construction and excavation. It may be required when additions or alterations to existing structures are contemplated. In this chapter, you will learn the following.

- Boreholes and trial pits
- Soil sampling
- Standard penetration test
- Vane shear test
- Cone penetration test
- Rock cores
- Bore logs and subsoil exploration report

10.2 Subsurface Exploration Program

Subsurface exploration comprises several steps, including collection of preliminary information, reconnaissance, and site investigation.

Collection of Preliminary Information

Information must be obtained regarding the type of structure to be built and its general use. For the construction of buildings, the approximate column loads and their spacing and the local building-code and basement requirements should be known. The construction of bridges requires determining span length and the loading on piers and abutments.

A general idea of the topography and the type of soil to be encountered near and around the proposed site can be obtained from the following sources:

1. Google Earth
2. U.S. Geological Survey maps
3. State government geological survey maps
4. U.S. Department of Agriculture's Soil Conservation Service county soil reports
5. Agronomy maps published by the agriculture departments of various states
6. Hydrological information published by the U.S. Corps of Engineers, including the records of stream flow, high flood levels, tidal records, and so on
7. Highway department soils manuals published by several states

The information collected from these sources can be extremely helpful to those planning a site investigation. In some cases, substantial savings are realized by anticipating problems that may be encountered later in the exploration program.

Reconnaissance

The engineer should always make a visual inspection of the site with a camera to obtain information about these features:

1. The general topography of the site and the possible existence of drainage ditches, abandoned dumps of debris, or other materials. Also, evidence of creep of slopes and deep, wide shrinkage cracks at regularly spaced intervals may be indicative of expansive soils.
2. Soil stratification from deep cuts, such as those made for construction of nearby highways and railroads.
3. Type of vegetation at the site, which may indicate the nature of the soil. For example, a mesquite cover in central Texas may indicate the existence of expansive clays that can cause possible foundation problems.
4. High-water marks on nearby buildings and bridge abutments.
5. Groundwater levels, which can be determined by checking nearby wells.
6. Types of construction nearby and existence of any cracks in walls or other problems.

The nature of stratification and physical properties of the soil nearby can also be obtained from any available soil-exploration reports for nearby existing structures.

Site Investigation

The site investigation phase of the exploration program consists of planning, making test boreholes, and collecting soil samples at desired intervals for subsequent observation and laboratory tests. The approximate required minimum depth of the borings should be predetermined; however, the depth can be changed during the drilling operation, depending on the subsoil encountered. Table 10.1 gives approximate depths of boring for a building with a width of 100 ft. ($\approx$ 30 m).

Sometimes subsoil conditions require that the foundation load be transmitted to bedrock. The minimum depth of core boring into the bedrock is about 10 ft. ($\approx$ 3 m). If the bedrock is irregular or weathered, the core borings may have to be deeper.

Table 10.1 Approximate Depths of Borings for Buildings with a Width of 100 ft (≈ 30 m)

No. of stories	Boring depth	
	ft	m
1	10	3
2	20	6
3	30	9
4	50	15
5	75	23

Table 10.2 Approximate Spacing of Boreholes

Type of project	Spacing	
	ft	m
Multistory building	30–90	9–27
One-story industrial plants	60–120	18–36
Highways	750–1500	230–460
Residential subdivision	750–1500	230–460
Dams and dikes	125–250	38–76

There are no hard and fast rules for borehole spacing. Table 10.2 gives some general guidelines. The spacing can be increased or decreased, depending on the subsoil condition. If various soil strata are more or less uniform and predictable, fewer boreholes are needed than in nonhomogeneous soil strata.

The engineer should also take into account the ultimate cost of the structure when making decisions regarding the extent of field exploration. The exploration cost generally should be 0.1% to 0.5% of the cost of the structure.

10.3 Exploratory Borings in the Field

Soil borings can be made by several methods, including auger boring, wash boring, percussion drilling, and rotary drilling.

Auger boring is the simplest method of making exploratory boreholes. Figure 10.1a and b show two types of hand auger: the *post hole auger* and the *helical auger*. Figure 10.1c shows a power drivers continuous flight auger. Hand augers cannot be used for advancing holes to depths exceeding 10 to 15 ft (≈ 3 to 4.5 m); however, they can be used for soil exploration work for some highways and small structures. *Portable power-driven helical augers* [3 to 12 in. (≈ 76 to 305 mm) in diameter] are available for making deeper boreholes. The soil samples obtained from such borings are highly disturbed. In some noncohesive soils or soils that have low cohesion, the walls of the boreholes will not stand unsupported. In such circumstances, a metal pipe is used as a *casing* to prevent the soil from caving in.

When power is available, *continuous-flight augers* are probably the most common method used for advancing a borehole. The power for drilling is delivered by truck- or tractor-mounted drilling rigs. Boreholes up to about 230 ft (≈ 70 m) can be made easily by this method. Continuous-flight augers are available in sections of about 3 to 5 ft (≈ 0.9 to 1.5 m) with either a solid or hollow stem. Some of the commonly used solid stem augers have outside diameters of $2\frac{5}{8}$ in., $3\frac{1}{4}$ in., 4 in., and $4\frac{1}{2}$ in. (67 mm, 83 mm, 101 mm, 114 mm). Hollow stem augers commercially available have dimensions of 2.5 in. inside diameter (ID) and 6.25 in. outside diameter (OD) [≈ 64 mm (ID), 159 mm (OD)], 2.75 in. ID and 7 in. OD [≈ 70 mm (ID), 178 mm (OD)], 3 in. ID and 8 in. OD [≈ 76 mm (ID), 203 mm (OD)], and 3.25 in. ID and 9 in. OD [≈ 83 mm (ID), 229 mm (OD)].

The tip of the auger is attached to a cutter head (Figure 10.1c). During the drilling operation (Figure 10.2), section after section of auger can be added and the hole extended downward. The flights of the augers bring the loose soil from the bottom of the hole to the surface. The driller can detect

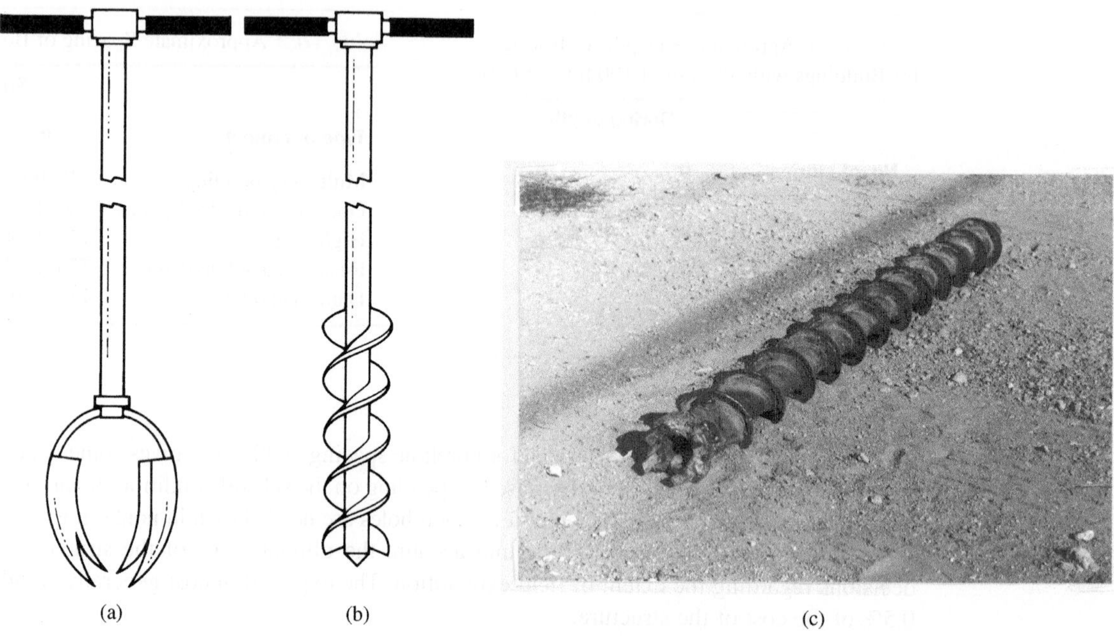

Figure 10.1 Hand tools: (a) post hole auger; (b) helical auger; (c) continuous flight auger (*Courtesy of Braja Das*)

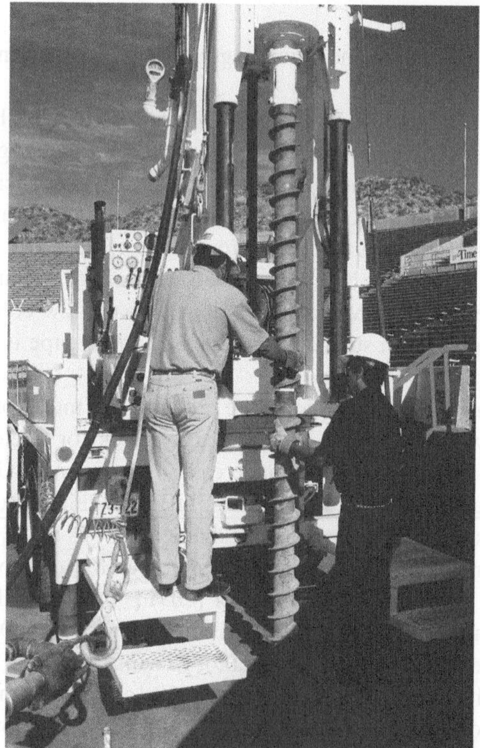

Figure 10.2 Drilling with a continuous-flight auger (*Courtesy of Braja Das*)

changes in soil type by noting changes in the speed and sound of drilling. When solid stem augers are used, the auger must be withdrawn at regular intervals to obtain soil samples and also to conduct other operations such as standard penetration tests. Hollow stem augers have a distinct advantage over solid stem augers in that they do not have to be removed frequently for sampling or other tests. As shown schematically (Figure 10.3), the outside of the hollow stem auger acts like a casing.

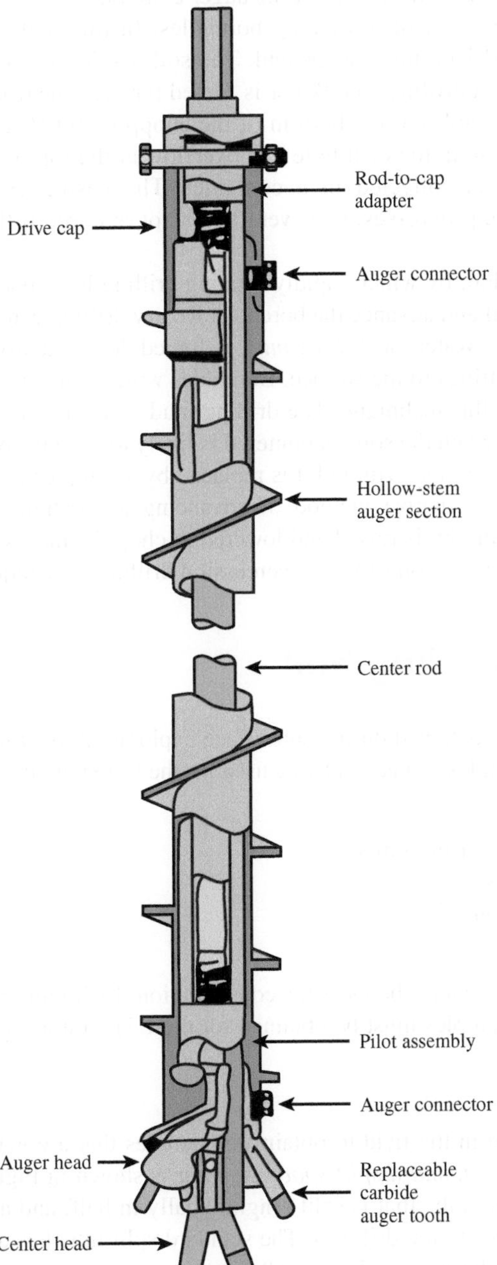

Figure 10.3 Hollow-stem auger components (After ASTM, 2001) (Copyright ASTM International. Reprinted with permission.)

The hollow-stem auger system includes the following:

Outer component: (a) hollow auger sections, (b) hollow auger cap, and (c) drive cap
Inner component: (a) pilot assembly, (b) center rod column, and (c) rod-to-cap adapter

The auger head contains replaceable carbide teeth (see Figure 10.1c). During drilling, if soil samples are to be collected at a certain depth, the pilot assembly and the center rod are removed. The soil sampler is then inserted through the hollow stem of the auger column.

Wash boring is another method of advancing boreholes. In this method, a casing about 6 to 10 ft ($\approx$ 1.8 to 3 m) long is driven into the ground. The soil inside the casing is then removed using a chopping bit attached to a drilling rod. Water is forced through the drilling rod and exits at a very high velocity through the holes at the bottom of the chopping bit (Figure 10.4). The water and the chopped soil particles rise in the drill hole and overflow at the top of the casing through a T connection. The washwater is collected in a container. The casing can be extended with additional pieces as the borehole progresses; however, that is not required if the borehole will stay open and not cave in.

Rotary drilling is a procedure by which rapidly rotating drilling bits attached to the bottom of drilling rods cut and grind the soil and advance the borehole. Rotary drilling can be used in sand, clay, and rocks (unless badly fissured). Water, or *drilling mud*, is forced down the drilling rods to the bits, and the return flow forces the cuttings to the surface. Boreholes with diameters of 2 to 8 in. ($\approx$ 51 to 203 mm) can be made easily by this technique. The drilling mud is a slurry of water and bentonite. Generally, rotary drilling is used when the soil encountered is likely to cave in. When soil samples are needed, the drilling rod is raised and the drilling bit is replaced by a sampler.

Percussion drilling is an alternative method of advancing a borehole, particularly through hard soil and rock. A heavy drilling bit is raised and lowered to chop the hard soil. The chopped soil particles are brought up by the circulation of water. Percussion drilling may require casing.

10.4 Procedures for Sampling Soil

Two types of soil samples can be obtained during subsurface exploration: *disturbed* and *undisturbed*. Disturbed, but representative, samples can generally be used for the following types of laboratory tests:

1. Grain-size analysis
2. Determination of liquid and plastic limits
3. Specific gravity of soil solids
4. Organic content determination
5. Classification of soil

Disturbed soil samples, however, cannot be used for consolidation, hydraulic conductivity, or shear strength tests. Undisturbed soil samples must be obtained for these laboratory tests.

Split-Spoon Sampling

Split-spoon samplers can be used in the field to obtain soil samples that are generally disturbed but still representative. A section of a *standard split-spoon sampler* is shown in Figure 10.5a. It consists of a tool-steel driving shoe, a steel tube that is split longitudinally in half, and a coupling at the top. The coupling connects the sampler to the drill rod. The standard split tube has an inside diameter of $1\frac{3}{8}$ in. (34.93 mm) and an outside diameter of 2 in. (50.8 mm); however, samplers that have inside and

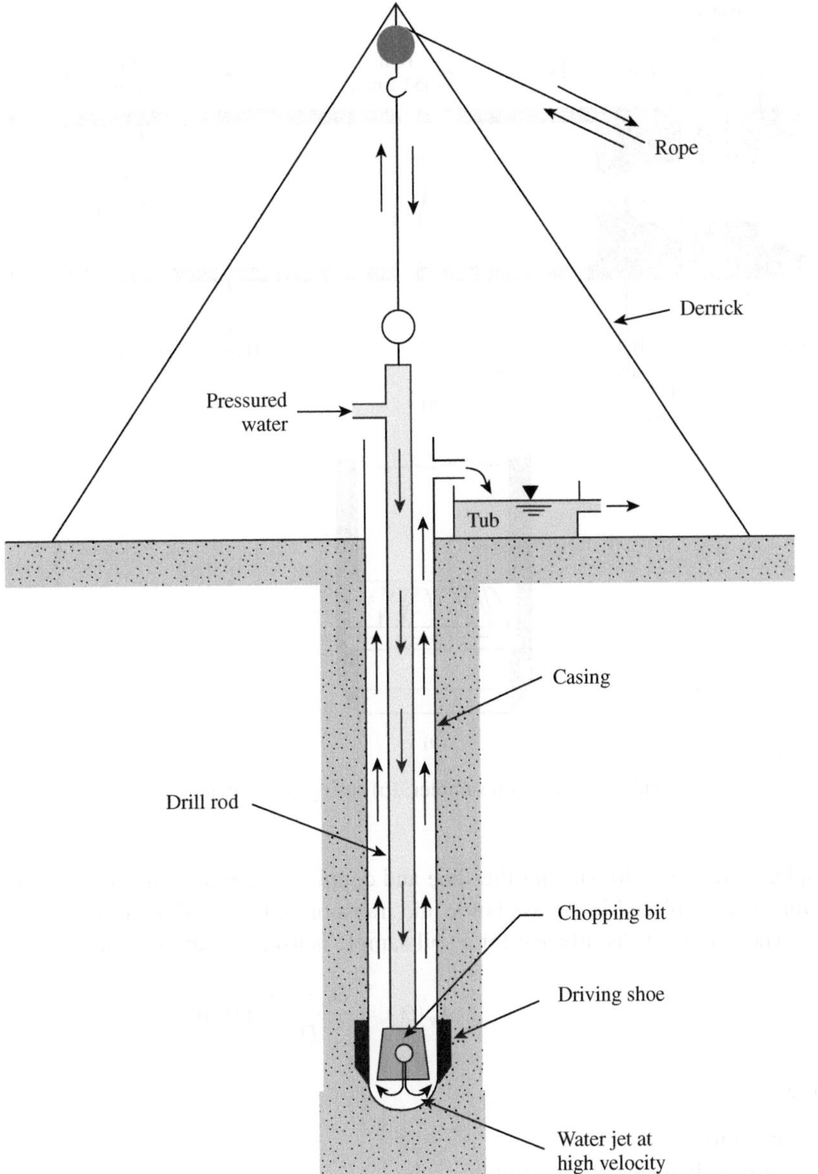

Figure 10.4 Wash boring

outside diameters up to $2\frac{1}{2}$ in. (63.5 mm) and 3 in. (76.2 mm), respectively, are also available. When a borehole is extended to a predetermined depth, the drill tools are removed and the sampler is lowered to the bottom of the borehole. The sampler is driven into the soil by hammer blows to the top of the drill rod. The standard weight of the hammer is 140 lb (mass = 63.5 kg) and, for each blow, the hammer drops a distance of 30 in. (762 mm). The number of blows required for spoon penetration of three 6 in. (152.4 mm) intervals is recorded. The numbers of blows required for the last two intervals are added to give the *standard penetration number, N*, at that depth. This number is generally referred to as the *N value* (American Society for Testing and Materials, 2013, Designation D-1586). The

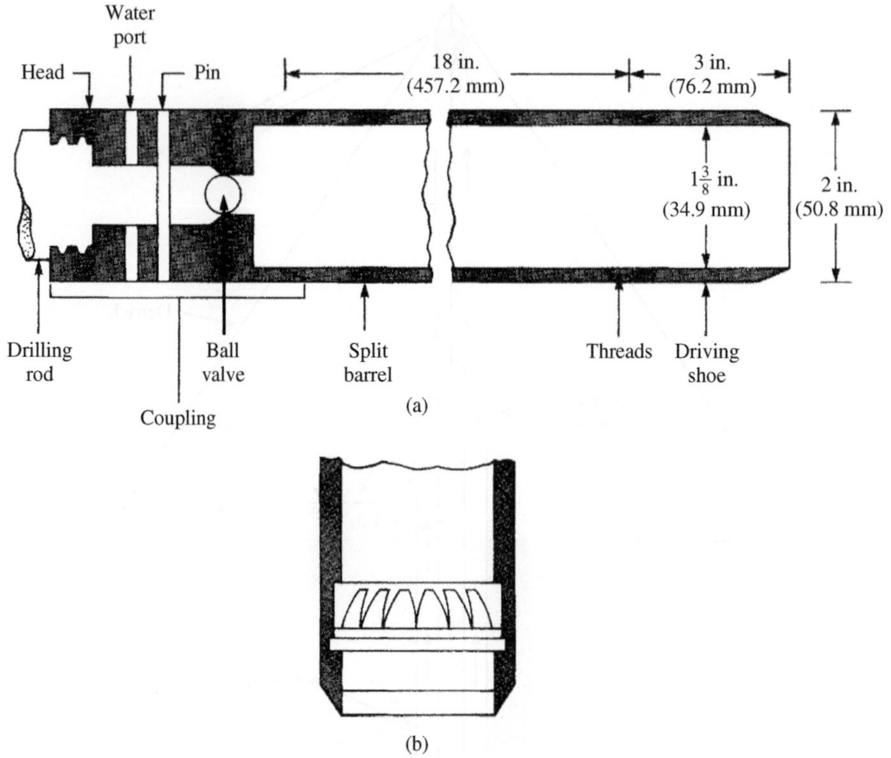

Figure 10.5 (a) Standard split-spoon sampler; (b) spring core catcher

sampler is then withdrawn, and the shoe and coupling are removed. The soil sample recovered from the tube is then placed in a glass bottle and transported to the laboratory.

The degree of disturbance for a soil sample is usually expressed as

$$A_R(\%) = \frac{D_o^2 - D_i^2}{D_i^2}(100) \tag{10.1}$$

where

A_R = area ratio
D_o = outside diameter of the sampling tube
D_i = inside diameter of the sampling tube

When the area ratio is 10% or less, the sample is generally considered to be undisturbed.

Split-spoon samples generally are taken at intervals of about 5 ft. (≈1.5 m). When the material encountered in the field is sand (particularly fine sand below the water table), sample recovery by a split-spoon sampler may be difficult. In that case, a device such as a *spring core catcher* (Figure 10.5b) may have to be placed inside the split spoon.

At this point, it is important to point out that there are several factors that will contribute to the variation of the standard penetration number N at a given depth for similar soil profiles. These factors include standard penetration test (SPT) hammer efficiency, borehole diameter, sampling method, and

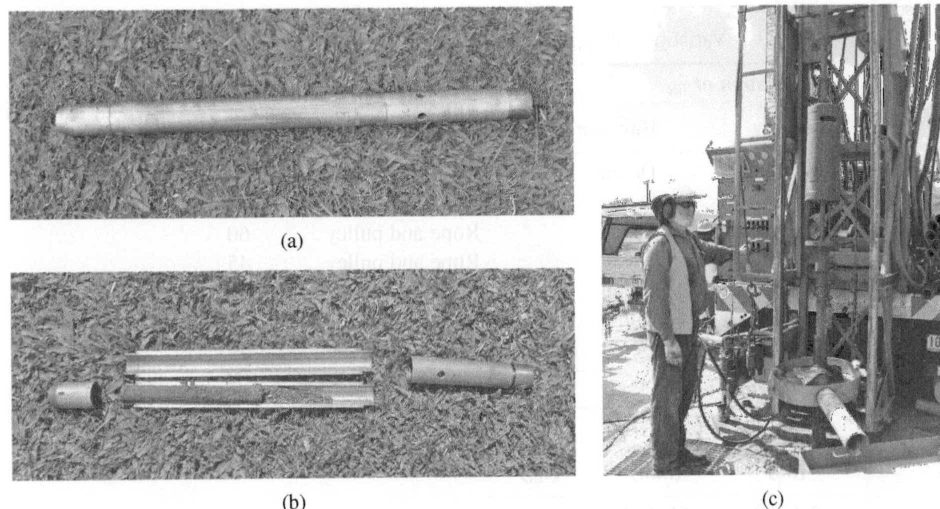

(a)

(b)

(c)

Figure 10.6 Split spoon sampler for standard penetration test: (a) closed and (b) open positions; (c) standard penetration test setup with safety hammer (*Courtesy of Dr. N. Sivakugan, James Cook University, Australia*)

rod-length factor (Seed et al., 1985; Skempton, 1986). The two most common types of SPT hammers used in the field are the *safety hammer* and *donut hammer*. They are commonly dropped by a rope with *two wraps around a pulley*.

Figure 10.6a shows a standard split-spoon sampler with the driving shoe on one end and the connection to the drill rod at the other end. Figure 10.6b shows the same sampler with the middle section split longitudinally, showing the soil sample collected while driving into the borehole. Figure 10.6c shows the standard penetration test rig with the 140 lb (mass = 63.5 kg) hammer falling on the anvil at the top of the drill rod.

Based on field observations it appears reasonable to standardize the field standard penetration number based on the input driving energy and its dissipation around the sampler into the surrounding soil, borehole diameter, drill rod length, and whether or not a liner is used within the borehole. The corrected penetration number is given by

$$N_{60} = \frac{N\eta_H\eta_B\eta_S\eta_R}{60} \tag{10.2}$$

where

N_{60} = standard penetration number corrected for field conditions
N = measured penetration number
η_H = hammer efficiency (%)
η_B = correction for borehole diameter
η_S = sampler correction
η_R = correction for rod length

Based on the recommendations of Seed et al. (1985) and Skempton (1986), the variations of η_H, η_B, η_S, and η_R are summarized in Table 10.3.

Table 10.3 Variations of η_H, η_B, η_S, and η_R [Eq. (10.2)]

1. Variation of η_H

Country	Hammer type	Hammer release	η_H (%)
Japan	Donut	Free fall	78
	Donut	Rope and pulley	67
United States	Safety	Rope and pulley	60
	Donut	Rope and pulley	45
Argentina	Donut	Rope and pulley	45
China	Donut	Free fall	60
	Donut	Rope and pulley	50

2. Variation of η_B

Diameter		
(in.)	(mm)	(η_B)
2.4–4.8	61–122	1
6	152	1.05
8	203	1.15

3. Variation of η_S

Variable	η_S
Standard sampler	1.0
With liner for dense sand and clay	0.8
With liner for loose sand	0.9

4. Variation of η_R

Rod length		
(ft)	(m)	(η_R)
>30	>9	1.0
20–30	6–9	0.95
10–20	3–6	0.85
0–10	0–3	0.75

Besides obtaining soil samples, standard penetration tests provide several useful correlations. For example, the consistency of clayey soils can often be estimated from the standard penetration number, N_{60}, as shown in Table 10.4. However, correlations for clays require tests to verify that the relationships are valid for the clay deposit being examined. The standard prenetration test is unreliable in clays due to pore water pressure developed during driving, which alters the effective stress.

In granular soils, the N_{60} value is affected by the effective overburden pressure, σ'_o at the test depth. For that reason, the N_{60} value obtained from field exploration under different effective overburden pressures should be adjusted to correspond to a standard value of σ'_o; that is,

$$(N_1)_{60} = C_N N_{60} \tag{10.3}$$

Table 10.4 Consistency of Clays and Approximate Correlation to the Standard Penetration Number, N_{60}

Standard penetration number, N_{60}	Consistency	Unconfined compression strength, q_u (lb/ft²)	Unconfined compression strength, q_u (kN/m²)
0–2	Very soft	0–500	0–24
2–5	Soft	500–1000	24–48
5–10	Medium stiff	1000–2000	48–96
10–20	Stiff	2000–4000	96–192
20–30	Very stiff	4000–8500	192–407
>30	Hard	>8500	>407

where

$(N_1)_{60}$ = corrected N value to a standard value of σ'_o [(2000 lb/ft² (≈ 100 kN/m²)]
C_N = correction factor
N_{60} = N value obtained from the field corrected by Eq. (10.2)

A number of empirical relationships have been proposed for C_N. The most commonly cited relationships are those given by Liao and Whitman (1986) and Skempton (1986). They are given in Table 10.5.

An approximate relationship between the corrected standard penetration number and the relative density of sand is given in Table 10.6. These values are approximate, primarily because the effective overburden pressure and the stress history of the soil significantly influence the N_{60} values of sand.

Schmertmann (1975) provided a correlation between N_{60}, σ'_o, and ϕ', which is shown in Figure 10.7. The correlation can be approximated as (Kulhawy and Mayne, 1990).

$$\phi' = \tan^{-1}\left[\frac{N_{60}}{12.2 + 20.3\left(\dfrac{\sigma'_o}{p_a}\right)}\right]^{0.34} \tag{10.4}$$

Table 10.5 Empirical Relationships for C_N^*

Source	C_N
Liao and Whitman (1986)	$\left(\dfrac{p_a}{\sigma'_o}\right)^{0.5}$
Skempton (1986)	$\dfrac{2}{1 + \left(\dfrac{\sigma'_o}{p_a}\right)}$

**Note:* p_a = atmospheric pressure [(2000 lb/ft² (≈ 100 kN/m²)]

Table 10.6 Relation Between the Corrected N Values and the Relative Density in Sands

Standard penetration number, N_{60}	Approximate relative density, D_r (%)
0–5	0–5
5–10	5–30
10–30	30–60
30–50	60–95

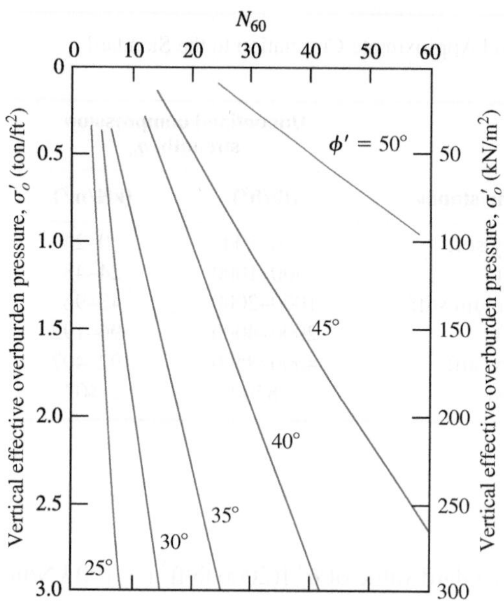

Figure 10.7 Schmertmann's (1975) correlation between N_{60}, σ'_o and ϕ' for granular soils

where

N_{60} = field standard penetration number corrected by Eq. (10.2)
σ'_o = effective overburden pressure
p_a = atmospheric pressure in the same unit as σ'_o
ϕ' = soil friction angle (effective)

When the standard penetration resistance values are used in the preceding correlations to estimate soil parameters, the following qualifications should be noted:

1. The equations are approximate and largely empirical.
2. Because the soil is not homogeneous, the N_{60} values obtained from a given borehole vary widely.
3. In soil deposits that contain large boulders and gravel, standard penetration numbers may be erratic and unreliable.

Although the correlations are approximate, with correct interpretation the standard penetration test provides a good evaluation of soil properties. The primary sources of errors in standard penetration tests are inadequate cleaning of the borehole, careless measurement of the blow count, eccentric hammer strikes on the drill rod and inadequate maintenance of water head in the borehole.

Scraper Bucket

When the soil deposit is sand mixed with pebbles, obtaining samples by split spoon with a spring core catcher may not be possible because the pebbles may prevent the springs from closing. In such cases a scraper bucket may be used to obtain disturbed representative samples (Figure 10.8). The scraper

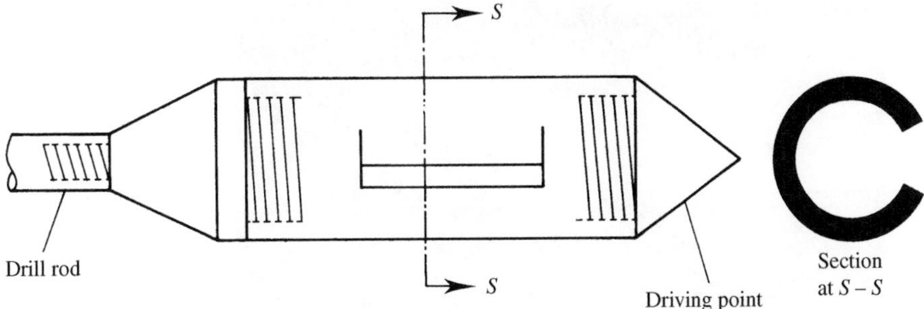

Figure 10.8 Scraper bucket

bucket has a driving point and can be attached to a drilling rod. The sampler is driven down into the soil and rotated, and the scrapings from the side fall into the bucket.

Thin-Wall Tube

Thin-wall tubes are sometimes called *Shelby tubes*. They are made of seamless steel and commonly are used to obtain undisturbed clayey soils. The commonly used thin wall tube samplers have outside diameters of 2 in. (50.8 mm) and 3 in. (76.2 mm). The bottom end of the tube is sharpened. The tubes can be attached to drilling rods (Figure 10.9). The drilling rod with the sampler attached is lowered to the bottom of the borehole, and the sampler is pushed into the soil. The soil sample inside the tube is then pulled out. The two ends of the sampler are sealed, and it is sent to the laboratory for testing.

Samples obtained in this manner may be used for consolidation or shear tests. A thin-wall tube with a 2 in. (50.8 mm) outside diameter has an inside diameter of about 1.875 in. (47.6 mm). The area ratio is

$$A_R(\%) = \frac{D_o^2 - D_i^2}{D_i^2}(100) = \frac{(2)^2 - (1.875)^2}{(1.875)^2}(100) = 13.75\%$$

Increasing the diameters of samples increases the cost of obtaining them.

Figure 10.10 shows soil samples within a plastic liner that are collected from a borehole. These are stacked within a sample tray and are numbered to reflect the sample depth. The sample trays are taken to the soil laboratory for further tests.

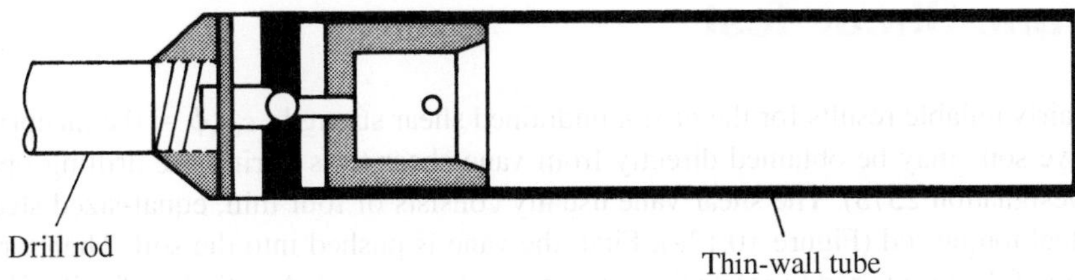

Figure 10.9 Thin-wall tube

Figure 10.10 Soil samples recovered from borehole placed in a sample tray (*Courtesy of N. Sivakugan, James Cook University, Australia*)

10.5 Observation of Water Levels

The presence of a water table near a foundation significantly affects a foundation's load-bearing capacity and settlement. The water level will change seasonally. In many cases, establishing the highest and lowest possible levels of water during the life of a project may become necessary.

If water is encountered in a borehole during a field exploration, that fact should be recorded. In soils with high hydraulic conductivity, the level of water in a borehole will stabilize about 24 hours after completion of the boring. The depth of the water table can then be recorded by lowering a chain or tape into the borehole.

In highly impermeable layers, the water level in a borehole may not stabilize for several weeks. In such cases, if accurate water-level measurements are required, a *piezometer* can be used.

The simplest piezometer (Figure 10.11a) is a standpipe or Casagrande-type piezometer. It consists of a riser pipe joined to a filter tip that is placed in sand. A bentonite seal is placed above the sand to isolate the pore water pressure at the filter tip. The annular space between the riser pipe and the borehole is backfilled with bentonite-cement grout to prevent vertical migration of water. This allows periodic checking until the water level stabilizes. Figure 10.11b shows a vibrating-wire piezometer.

10.6 Vane Shear Test

Fairly reliable results for the *in situ* undrained shear strength, c_u ($\phi = 0$ concept) of soft plastic cohesive soils may be obtained directly from vane shear tests during the drilling operation (ASTM Test Designation 2573). The shear vane usually consists of four thin, equal-sized steel plates welded to a steel torque rod (Figure 10.12a). First, the vane is pushed into the soil. Then torque is applied at the top of the torque rod to rotate the vane at a uniform speed. A cylinder of soil of height h and diameter d will resist the torque until the soil fails. The undrained shear strength of the soil can be calculated as follows.

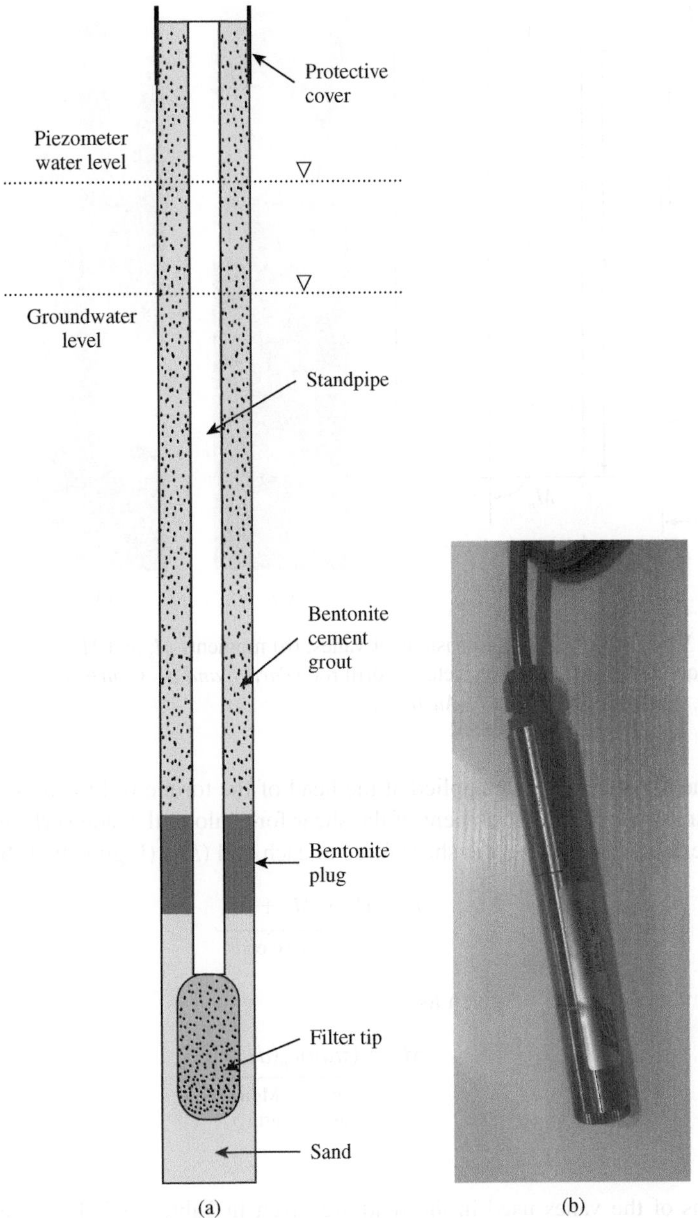

Figure 10.11 (a) Casagrande-type piezometer; (b) vibrating wire piezometer (*Courtesy of N. Sivakugan, James Cook University, Australia*)

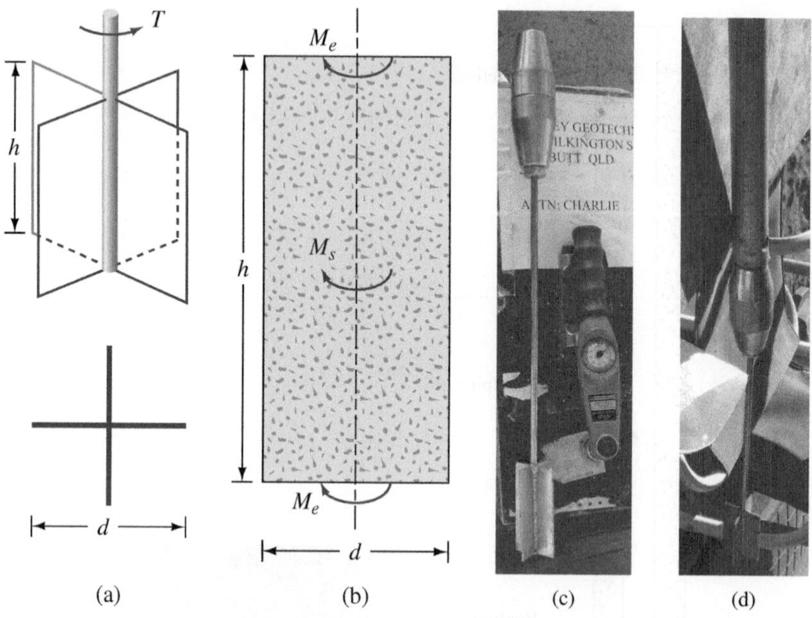

Figure 10.12 Vane Shear test: (a) dimensions of vanes; (b) moments M_s and M_e; (c) vane and torque meter; (d) vane connected to drill rod (*Parts c and d—Courtesy of N. Sivakugan, James Cook University, Australia*)

If T is the maximum torque applied at the head of the torque rod to cause failure, it should be equal to the sum of the resisting moment of the shear force along the side surface of the soil cylinder (M_s) and the resisting moment of the shear force at each end (M_e) (Figure 10.12b):

$$T = M_s + \underbrace{M_e + M_e}_{\text{Two ends}} \tag{10.5}$$

The resisting moment M_s can be given as

$$M_S = \underbrace{(\pi dh)}_{\substack{\text{Surface}\\ \text{area}}} c_u \underbrace{(d/2)}_{\substack{\text{Moment}\\ \text{arm}}} \tag{10.5a}$$

The dimensions of the vanes used in the field are given in Table 10.7. The standard rate of torque application is 0.1°/sec. The maximum torque T applied to cause failure can be given as

$$T = f(c_u, h, \text{and } d) \tag{10.6}$$

or

$$c_u = \frac{T}{K} \tag{10.7}$$

where K = a constant with a magnitude depending on the dimension and shape of the vane.

Table 10.7 Recommended Dimensions of Field Vanes*

	Diameter, *d*		Height, *h*		Thickness of blade		Diameter of rod	
Casting size	**(in.)**	**(mm)**	**(in.)**	**(mm)**	**(in.)**	**(mm)**	**(in.)**	**(mm)**
AX	$1\frac{1}{2}$	38.1	3.0	76.2	$\frac{1}{16}$	1.59	$\frac{1}{2}$	12.7
BX	2.0	50.8	4.0	101.6	$\frac{1}{16}$	1.59	$\frac{1}{2}$	12.7
NX	$2\frac{1}{2}$	63.5	5.0	127.0	$\frac{1}{8}$	3.18	$\frac{1}{2}$	12.7
101.6 mm†	$3\frac{5}{8}$	92.1	$7\frac{1}{4}$	184.2	$\frac{1}{8}$	3.18	$\frac{1}{2}$	12.7

*Selection of the vane size is directly related to the consistency of the soil being tested; that is, the softer the soil, the larger the vane diameter should be.
†Inside diameter.

For rectangular vanes,

$$K = \frac{\pi d^2}{2}\left(h + \frac{d}{3}\right) \tag{10.8}$$

If $h/d = 2$,

$$K = \frac{7\pi d^3}{6}$$

Thus,

$$c_u = \frac{6T}{7\pi d^3} \tag{10.9}$$

Field vane shear tests are moderately rapid and economical and are used extensively in field soil-exploration programs. The test gives good results in soft and medium-stiff clays and gives excellent results in determining the peak and residual undrained shear strengths of sensitive clays.

Sources of significant error in the field vane shear test are poor calibration of torque measurement and damaged vanes. Other errors may be introduced if the rate of rotation of the vane is not controlled properly.

10.7 Cone Penetration Test

The cone penetration test (CPT), originally known as the Dutch cone penetration test, is a versatile sounding method that can be used to determine the materials in a soil profile and estimate their engineering properties. This test is also called the *static penetration test*, and no boreholes are necessary to perform it. In the original version, a 60° cone with a base area of 10 cm^2 was pushed into the ground at a steady rate of about 20 mm/sec, and the resistance to penetration (called the point resistance) was measured. Figure 10.13 shows a piezocone, which is a modified version of the Dutch cone that can also measure pore water pressure through the porous stone between the cone and the shaft. These days, piezocones are quite common for their ability to give pore water pressure measurements as well.

Figure 10.13 Piezocone: a cone penetrometer with pore pressure measurement facility (*Courtesy of N. Sivakugan, James Cook University, Australia*)

The cone penetrometers in use at present measure (a) the *cone resistance,* q_c, to penetration developed by the cone, which is equal to the vertical force applied to the cone divided by its horizontally projected area; and (b) the *frictional resistance,* f_c, which is the resistance measured by a sleeve located above the cone with the local soil surrounding it. The frictional resistance is equal to the vertical force applied to the sleeve divided by its surface area—actually, the sum of friction and adhesion.

Generally, two types of penetrometers are used to measure q_c and f_c:

1. *Mechanical friction-cone penetrometer* (Figure 10.14). In this case, the penetrometer tip is connected to an inner set of rods. The tip is first advanced about 40 mm, thus giving the cone resistance. With further thrusting, the tip engages the friction sleeve. As the inner rod advances, the rod force is equal to the sum of the vertical forces on the cone and the sleeve. Subtracting the force on the cone gives the side resistance.
2. *Electric friction-cone penetrometer* (Figure 10.15). In this case, the tip is attached to a string of steel rods. The tip is pushed into the ground at the rate of 20 mm/sec. Wires from the transducers are threaded through the center of the rods and continuously give the cone and side resistances.

Figure 10.16 shows the results of penetrometer tests in a soil profile with friction measurement by an electric friction-cone penetrometer.

Several correlations that are useful in estimating the properties of soils encountered during an exploration program have been developed for the cone resistance, q_c, and the friction ratio, F_r, obtained from the cone penetration tests. The friction ratio, F_r, is defined as

$$F_r\,(\%) = \frac{\text{frictional resistance}}{\text{cone resistance}} = \frac{f_c}{q_c} \times 100 \qquad (10.10)$$

It varies in the range of 0 to 10%, where the lower end of the range applies to granular soils, and the higher end applies to clays.

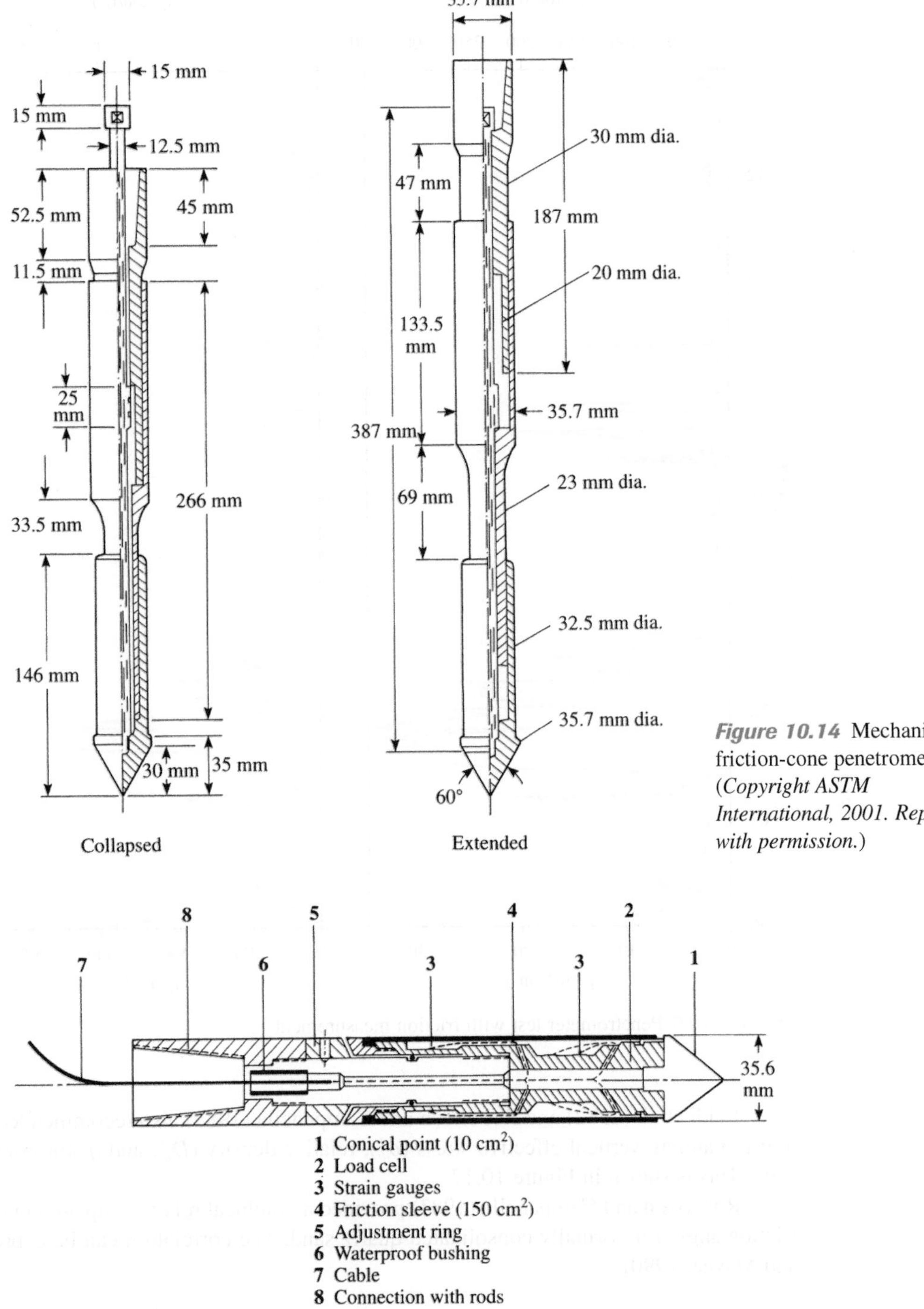

Figure 10.14 Mechanical friction-cone penetrometer (*Copyright ASTM International, 2001. Reprinted with permission.*)

Figure 10.15 Electric friction-cone penetrometer (*Copyright ASTM International, 1997. Reprinted with permission.*)

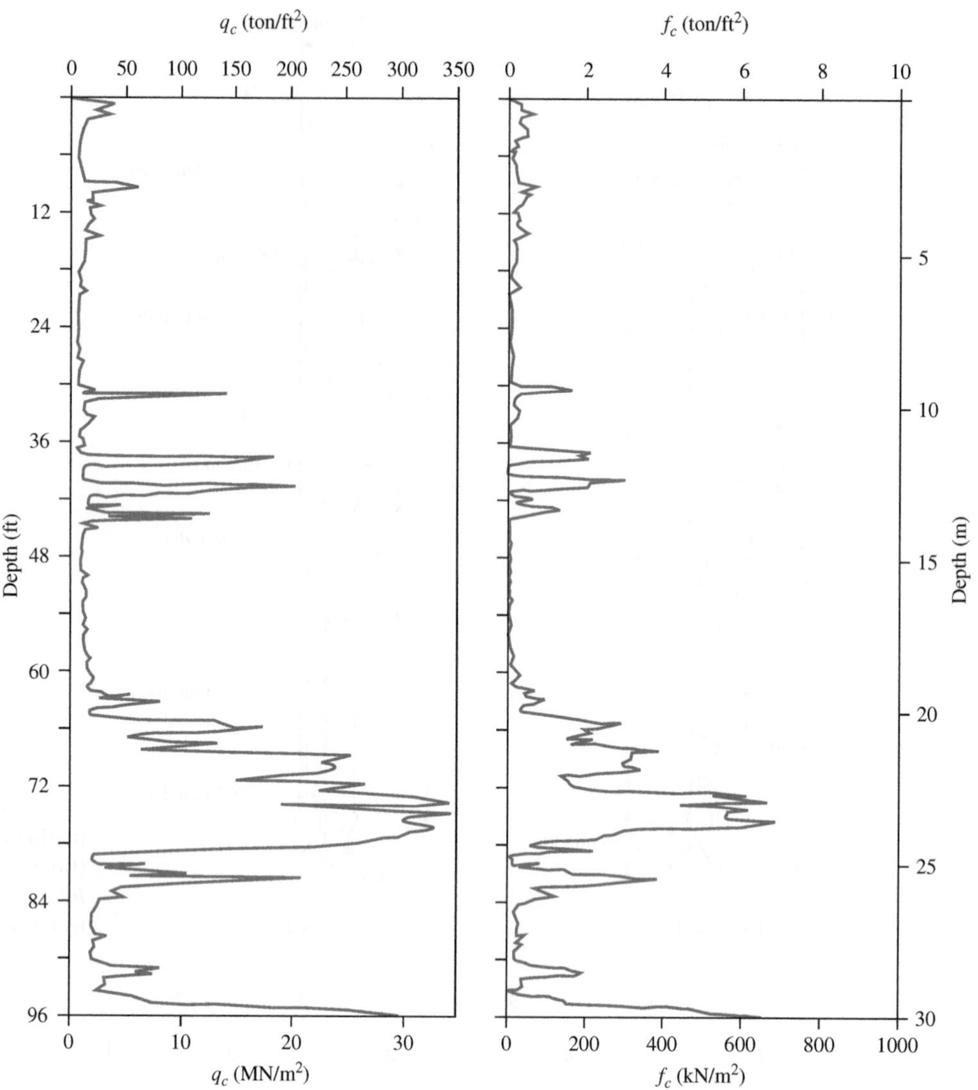

Figure 10.16 Penetrometer test with friction measurement

Baldi et al. (1982) and Robertson and Campanella (1983) also recommended an empirical relationship among vertical effective stress (σ'_o), relative density (D_r), and q_c for *normally consolidated sand*. This is shown in Figure 10.17.

Robertson and Campanella (1983) provided a graphical relationship among σ'_o, q_c, and the peak friction angle for normally consolidated quartz sand. The correlation can be expressed as (Kulhawy and Mayne, 1990)

$$\phi' = \tan^{-1}\left[0.1 + 0.38 \log\left(\frac{q_c}{\sigma'_o}\right)\right] \tag{10.11}$$

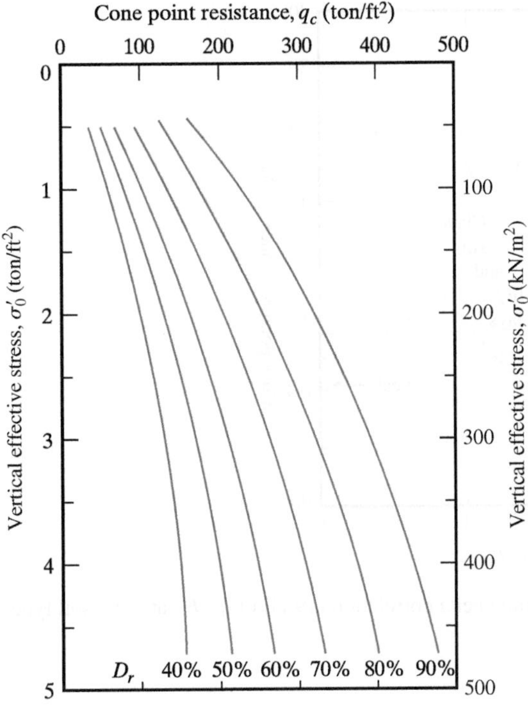

Figure 10.17 Variation of q_c, σ'_o, and D_r for normally consolidated quartz sand (*Based on Baldi et al., 1982, and Robertson and Campanella, 1983*)

Robertson and Campanella (1983) also found a general correlation among q_c, friction ratio F_r, and the type of soil encountered in the field (Figure 10.18). The pair of values of q_c and F_c is used to identify the soil type at a certain depth.

According to Mayne and Kemper (1988), in clayey soil the undrained shear strength c_u can be correlated as

$$\frac{c_u}{\sigma'_o} = \left(\frac{q_c - \sigma_o}{\sigma'_o}\right)\frac{1}{N_K} \tag{10.12}$$

or

$$c_u = \frac{q_c - \sigma_o}{N_K} \tag{10.13}$$

where

N_K = bearing capacity factor ($N_K = 15$ for electric cone and $N_K = 20$ for mechanical cone)
σ_o = *total* vertical stress
σ'_o = effective vertical stress

Consistent units of c_u, σ_o, σ'_o, and q_c should be used with Eq. (10.12).

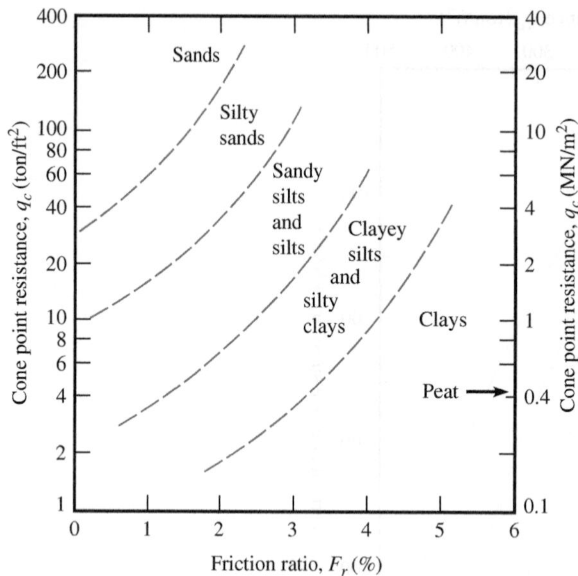

Figure 10.18 Robertson and Campanella correlation (1983) of q_c, F_r, and the soil type

10.8 Coring of Rocks

When a rock layer is encountered during a drilling operation, rock coring may be necessary. To core rocks, a *core barrel* is attached to a drilling rod. A *coring bit* is attached to the bottom of the barrel (Fig. 10.19). The cutting elements may be diamond, tungsten, or carbide of high hardness. Table 10.8 summarizes the various types of core barrel and their sizes, as well as the compatible drill rods commonly used for exploring foundations. The coring is advanced by rotary drilling. Water is circulated through the drilling rod during coring, and the cutting is washed out.

Two types of core barrel are available: the *single-tube core barrel* (Figure 10.19a) and the *double-tube core barrel* (Figure 10.19b). Rock cores obtained by single-tube core barrels can be highly disturbed and fractured because of torsion. Rock cores smaller than the BX size tend to fracture during the coring process.

Table 10.8 Standard Size and Designation of Casing, Core Barrel, and Compatible Drill Rod

Casing and core barrel designation	Outside diameter of core barrel bit (in.)	Outside diameter of core barrel bit (mm)	Drill rod designation	Outside diameter of drill rod (in.)	Outside diameter of drill rod (mm)	Diameter of borehole (in.)	Diameter of borehole (mm)	Diameter of core sample (in.)	Diameter of core sample (mm)
EX	$1\frac{7}{16}$	36.5	E	$1\frac{5}{16}$	33.3	$1\frac{1}{2}$	38.1	$\frac{7}{8}$	22.2
AX	$1\frac{7}{8}$	47.6	A	$1\frac{5}{8}$	41.3	2	50.8	$1\frac{1}{8}$	28.6
BX	$2\frac{5}{16}$	58.7	B	$1\frac{7}{8}$	47.6	$2\frac{1}{2}$	63.5	$1\frac{5}{8}$	41.3
NX	$2\frac{15}{16}$	74.6	N	$2\frac{3}{8}$	60.3	3	76.2	$2\frac{1}{8}$	54.0

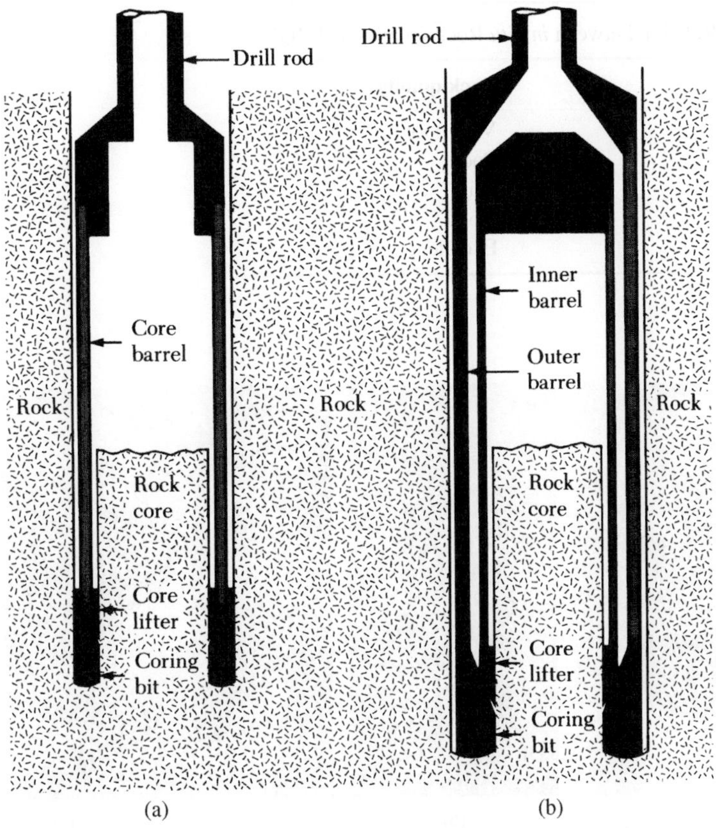

Figure 10.19 Rock coring: (a) single-tube core barrel; (b) double-tube core barrel

Figures 10.20a and b show photographs of a single-tube and a double-tube core barrel, respectively. Figure 19.20c shows a diamond-tipped coring bit.

When the core samples are recovered, the depth of recovery should be properly recorded for further evaluation in the laboratory. Based on the length of the rock core recovered from each run, the following quantities may be calculated for a general evaluation of the rock quality encountered:

$$\text{Recovery ratio} = \frac{\text{length of core recovered}}{\text{theoretical length of rock cored}} \tag{10.14}$$

Rock quality designation (RQD)

$$= \frac{\Sigma \text{ length of recovered pieces equal to or larger than 4 in. (101.6 mm)}}{\text{theoretical length of rock cored}} \tag{10.15}$$

A recovery ratio of unity indicates the presence of intact rock; for highly fractured rocks, the recovery ratio may be 0.5 or smaller. Table 10.9 presents the general relationship (Deere, 1963) between the RQD and the *in situ* rock quality.

Table 10.9 Relation between *in situ* Rock Quality and RQD

RQD	Rock quality
0–0.25	Very poor
0.25–0.5	Poor
0.5–0.75	Fair
0.75–0.9	Good
0.9–1	Excellent

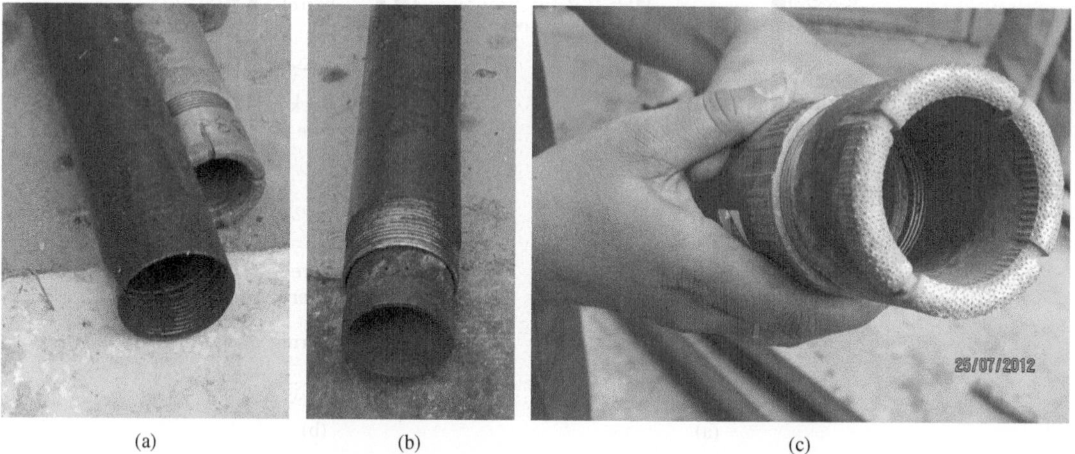

(a) (b) (c)

Figure 10.20 Photographs of: (a) single-tube core barrel; (b) double tube core barrel; (c) diamond tipped coring bit (*Courtesy of N. Sivakugan, James Cook University, Australia*)

10.9 Preparation of Boring Logs

The detailed information gathered from each borehole is presented in a graphical form called the *boring log*. As a borehole is advanced downward, the driller generally should record the following information in a standard log:

1. Name and address of the drilling company
2. Driller's name
3. Job description and number
4. Number, type, and location of boring
5. Date of boring
6. Subsurface stratification, which can be obtained by visual observation of the soil brought out by auger, split-spoon sampler, and thin-walled Shelby tube sampler
7. Elevation of water table and date observed, use of casing and mud losses, and so on
8. Standard penetration resistance and the depth of SPT

9. Number, type, and depth of soil sample collected
10. In case of rock coring, type of core barrel used and, for each run, the actual length of coring, length of core recovery, and RQD

This information should never be left to memory, because doing so often results in erroneous boring logs.

After completion of the necessary laboratory tests, the geotechnical engineer prepares a finished log that includes notes from the driller's field log and the results of tests conducted in the laboratory. Figure 10.21 shows a typical boring log.

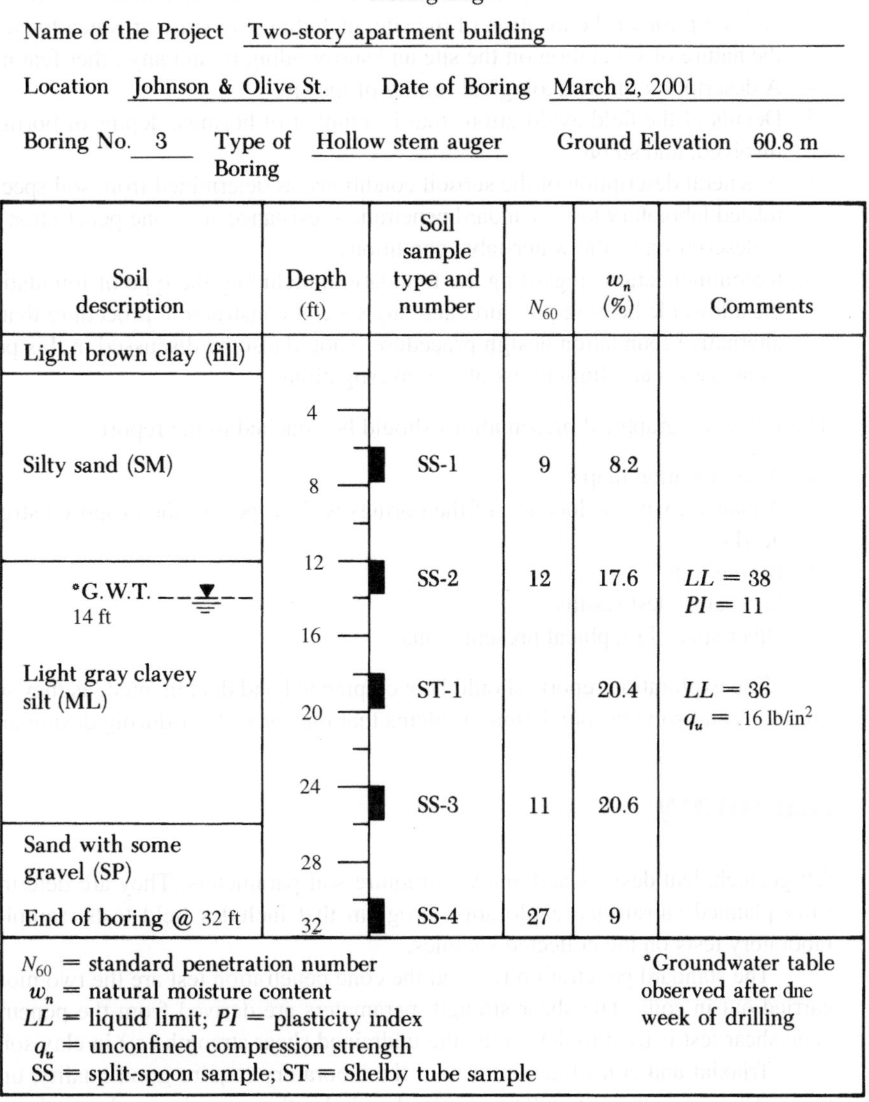

Boring Log

Name of the Project Two-story apartment building

Location Johnson & Olive St. Date of Boring March 2, 2001

Boring No. 3 Type of Boring Hollow stem auger Ground Elevation 60.8 m

Soil description	Depth (ft)	Soil sample type and number	N_{60}	w_n (%)	Comments
Light brown clay (fill)					
Silty sand (SM)	4, 8	SS-1	9	8.2	
°G.W.T. 14 ft	12	SS-2	12	17.6	$LL = 38$, $PI = 11$
Light gray clayey silt (ML)	16, 20	ST-1		20.4	$LL = 36$, $q_u = 16$ lb/in^2
	24	SS-3	11	20.6	
Sand with some gravel (SP)	28				
End of boring @ 32 ft	32	SS-4	27	9	

N_{60} = standard penetration number
w_n = natural moisture content
LL = liquid limit; PI = plasticity index
q_u = unconfined compression strength
SS = split-spoon sample; ST = Shelby tube sample

°Groundwater table observed after one week of drilling

Figure 10.21 A typical boring log

10.10 Subsoil Exploration Report

At the end of all soil exploration programs, the soil and rock specimens collected in the field are subject to visual observation and appropriate laboratory testing. After all the required information has been compiled, a soil exploration report is prepared for use by the design office and for reference during future construction work. Although the details and sequence of information in such reports may vary to some degree, depending on the structure under consideration and the person compiling the report, each report should include the following items:

1. A description of the scope of the investigation
2. A description of the proposed structure for which the subsoil exploration has been conducted
3. A description of the location of the site, including any structures nearby, drainage conditions, the nature of vegetation on the site and surrounding it, and any other features unique to the site
4. A description of the geological setting of the site
5. Details of the field exploration—that is, number of borings, depths of borings, types of borings involved, and so on
6. A general description of the subsoil conditions, as determined from soil specimens and from related laboratory tests, standard penetration resistance and cone penetration resistance, and so on
7. A description of the water-table conditions
8. Recommendations regarding the foundation, including the type of foundation recommended, the allowable bearing pressure, and any special construction procedure that may be needed; alternative foundation design procedures should also be discussed in this portion of the report
9. Conclusions and limitations of the investigations

The following graphical presentations should be attached to the report:

1. A site location map
2. A plan view of the location of the borings with respect to the proposed structures and those nearby
3. Boring logs
4. Laboratory test results
5. Other special graphical presentations

The exploration reports should be well planned and documented, as they will help in answering questions and solving foundation problems that may arise later during design and construction.

10.11 Summary

All geotechnical designs and analyses require soil parameters. They are determined through a carefully planned subsurface exploration program that includes field tests, sample collection, and the laboratory tests on the collected samples.

The standard penetration test and the cone penetration test are the two most common field tests carried out in soils. The shear strength parameters are derived from the penetration resistance. The vane shear test is used to determine the undrained shear strength (c_u) in clay soils.

Triaxial and consolidation tests in the laboratory require good quality, undisturbed soil specimens. They are generally collected from boreholes that are advanced into the ground through auger boring, wash boring, or rotary drilling. Thin-walled samplers where the area ratio is less than 10% will produce good quality samples.

The data recorded from the boreholes, including the *in situ* test results and the soil profile thus determined, are summarized as boring logs. All boring logs are collated and presented in the subsoil exploration report, which includes the laboratory test results as well.

Problems

10.1 State whether the following are true or false.

a. The larger the area ratio of a sampling tube, the lower is the sample disturbance.

b. In a standard penetration test in sands, the larger the penetration number, the larger is the friction angle.

c. The vane shear test is commonly carried out in sands.

d. The friction ratio in a cone penetration test is generally higher for sands than clays.

e. The higher the RQD, the lesser is the degree of fragmentation.

10.2 Determine the area ratio of a Shelby tube sampler having inside and outside diameters of 4.37 in. and 4.49 in., respectively.

10.3 A clay specimen was collected from the split-spoon sampler in a standard penetration test. The sampler has an inner diameter of 34.9 mm and an outer diameter of 50.8 mm. Determine the area ratio. Would you use this specimen in a consolidated undrained triaxial test?

10.4 It is required to fabricate a thin-walled sampling tube to obtain 75-mm-diameter good quality, undisturbed clay samples. What is the maximum possible wall thickness?

10.5 The field standard penetration numbers for a deposit of dry sand are given below. For the sand, given $\gamma = 116$ lb/ft^3. Determine the variation of $(N_1)_{60}$ with depth. Use Skempton's correction factor given in Table 10.5.

Depth (ft)	N_{60}
5	9
10	9
15	12
20	12
25	16

10.6 Standard penetration test is carried out at a site consisting of sandy soils and the following data were recorded. Determine the $(N_1)_{60}$ values, using the correction factors proposed by (a) Liao and Whitman (1986) and (b) Skempton (1986). Round off the $(N_1)_{60}$ values.

Depth (m)	Unit weight (kN/m^3)	N_{60}
3.5	18.0	12
8.8	18.5	8
12.4	18.5	15
18.5	18.5	19
23.6	19.0	21
26.9	19.0	26

10.7 Standard penetration tests were carried out at a site at different times, using different test rigs. In the first time, the hammer efficiency was 55% and the measured N value was 22 at a specific depth. The second time, the hammer efficiency was 75% and the N value was measured as 16 at the same depth. What is the approxiamate N_{60} value at this depth? Assume η_B η_s, and η_R are unity.

10.8 For the sand deposit described in Problem 10.5, estimate the average friction angle, ϕ'. Use Eq. (10.4).

10.9 The N_{60} value at a 15 m depth in a sand is 21. The water table is at a 5 m depth below the ground level, and the moist and saturated unit weights of the sand are 17.0 kN/m^3 and 19.5 kN/m^3, respectively. Determine the friction angle of the sand using Eq. (10.4) and Figure 10.7.

10.10 A soil profile is shown in Figure 10.22 along with the variation in the standard penetration numbers (N_{60}). Use Liao and Whitman's correction factor (C_N) and Eq. (10.3) to determine the variation of $(N_1)_{60}$.

10.11 The table gives the standard penetration numbers determined from a sandy soil deposit in the field:

Depth (ft)	Unit weight of soil (lb/ft^3)	N_{60}
10	106	7
15	106	9
20	106	11
25	118	16
30	118	18
35	118	20
40	118	22

Using Eq. (10.4), determine the variation of the peak soil friction angle, ϕ'. Estimate an average value of ϕ' for the design of a shallow foundation. *Note*: For depth greater than 20 ft, the unit weight of soil is 118 lb/ft^3.

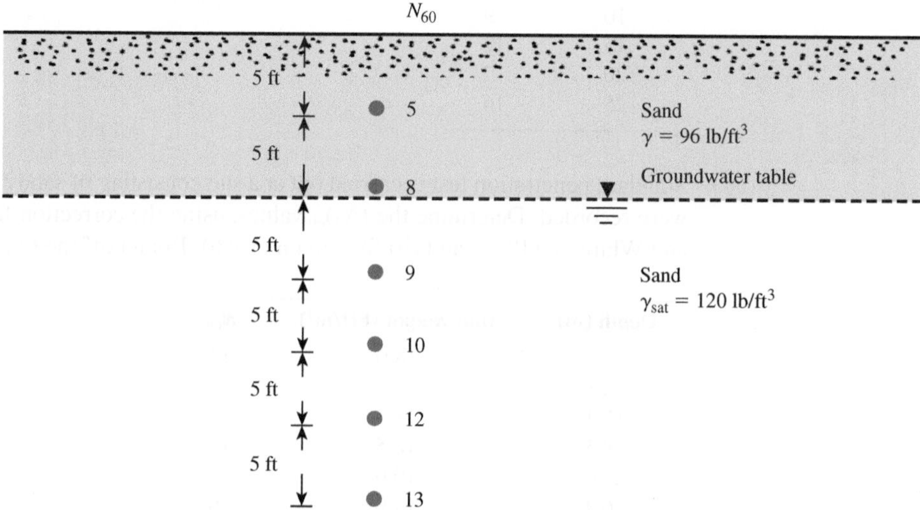

Figure 10.22

10.12 A vane shear test was conducted in a saturated clay. The height and diameter of the vane were 4 in. and 2 in., respectively. During the test, the maximum torque applied was 25 lb·ft. Determine the undrained shear strength of the clay.

10.13 A vane shear test was conducted in a saturated soft clay, using a 100 mm × 200 mm vane. When the vane was rotated at the standard rate of 0.1°/s, the torque measured in the torque meter increased to 60 N·m with further rotation reduced to 35 N·m. Determine the peak and ultimate undrained shear strength of the clay.

10.14 In a deposit of normally consolidated dry sand, a cone penetration test was conducted. The table gives the results:

Depth (ft)	Point resistance of cone, q_c (ton/ft^2)
5	41
10	82
15	121
20	163
25	202
30	232

Assume the dry unit weight of sand is 102 lb/ft^3.

a. Estimate the average peak friction angle, ϕ', of the sand. Use Eq. (10.11).

b. Estimate the average relative density of the sand. Use Figure 10.17.

10.15 In a clay layer, the groundwater table is located at a depth of 10 ft below the ground surface. The unit weights of the soil above and below the groundwater table are 113 lb/ft^3 and 125 lb/ft^3, respectively. The cone penetration resistance (electric friction cone) at a depth of 30 ft below the ground surface is 8.2 ton/ft^2. Determine the undrained cohesion, c_u.

10.16 In the cone penetrometer data shown in Figure 10.16, what is the soil encountered at 96 ft (29.3 m) depth? The water table is at a 15 ft depth (4.57 m), and the average moist and saturated unit weights of the soil are 108.1 1b/ft^3 (17.0 kN/m^3) and 124.0 1b/ft^3 (19.5 kN/m^3), respectively. Determine the relative density and friction angle of this soil.

10.17 The following measurements were recorded from a cone penetration test. Identify the soil type at each depth.

Depth (m)	q_c (MN/m^2)	f_c (kN/m^2)
5.0	2.0	10.0
10.0	1.5	75.5
15.0	0.2	7.2
20.0	1.1	38.5
25.0	15.0	150.0

10.18 During a field exploration, coring of rock was required. The core barrel was advanced 5 ft during the coring. The length of the core recovered was 3.2 ft. What was the recovery ratio?

References

AMERICAN SOCIETY for TESTING and MATERIALS (2014). *Annual Book of ASTM Standards*, Vol. 04.09, West Conshohocken, PA.

AMERICAN SOCIETY for TESTING and MATERIALS (2001). *Annual Book of ASTM Standards*, Vol. 04.08, West Conshohocken, PA.

AMERICAN SOCIETY for TESTING and MATERIALS (1997). *Annual Book of ASTM Standards*, Vol. 04.08, West Conshohocken, PA.

BALDI, G., BELLOTTI, R., GHIONNA, V., and JAMIOLKOWSKI, M. (1982). "Design Parameters for Sands from CPT," *Proceedings*, Second European Symposium on Penetration Testing, Amsterdam, Vol. 2, 425–438.

DEERE, D. U. (1963). "Technical Description of Rock Cores for Engineering Purposes," *Felsmechanik und Ingenieurgeologie*, Vol. 1, No. 1, 16–22.

KULHAWY, F. H. and MAYNE, P. W. (1990). *Manual on Estimating Soil Properties for Foundation Design* Electric Power Research Institute, Palo Alto, CA.

LIAO, S. S. C. and WHITMAN, R. V. (1986). "Overburden Correction Factors for SPT in Sand." *Journal of Geotechnical Engineering*, American Society of Civil Engineers, Vol. 112, No. 3, 373–377.

MAYNE, P. W. and KEMPER, J. B. (1988). "Profiling OCR in Stiff Clays by CPT and SPT," *Geotechnical Testing Journal*, ASTM, Vol. 11, No. 2, 139–147.

ROBERTSON, P. K. and CAMPANELLA, R. G. (1983). "Interpretation of Cone Penetration Tests. Part I: Sand," *Canadian Geotechnical Journal*, Vol. 20, No. 4, 718–733.

SCHMERTMANN, J. H. (1975). "Measurement of *In Situ* Shear Strength," *Proceedings*, Specialty Conference on *In Situ* Measurement of Soil Properties, ASCE, Vol. 2, 57–138.

SEED, H. B., TOKIMATSU, K., HARDER, L. F., and CHUNG, R. M. (1985). "Influence of SPT Procedures in Soil Liquefaction Resistance Evaluations," *Journal of Geotechnical Engineering*, ASCE, Vol. 111, No. 12, 1425–1445.

SKEMPTON, A. W. (1986). "Standard Penetration Test Procedures and the Effect in Sands of Overburden Pressure, Relative Density, Particle Size, Aging and Overconsolidation," *Geotechnique*, Vol. 36, No. 3, 425–447.

11 Lateral Earth Pressure: At-Rest, Rankine, and Coulomb

11.1 Introduction

Retaining structures such as retaining walls, basement walls, and bulkheads commonly are encountered in foundation engineering as they support slopes of earth masses. Proper design and construction of these structures require a thorough knowledge of the lateral forces that act between the retaining structures and the soil masses being retained. These lateral forces are caused by lateral earth pressure. Figure 11.1 shows some earth retaining structures. This chapter is devoted to the study of the various earth pressure theories. In this chapter, you will learn the following:

- Earth pressure at rest.
- Active and passive earth pressures.
- Rankine's earth pressure theories.
- Earth pressure coefficients K_o, K_a and K_p.
- Coulomb's earth pressure theories.
- Computing lateral loads on earth retaining structures.

11.2 At-Rest, Active, and Passive Pressures

Consider a mass of soil shown in Figure. 11.2a. The mass is bounded by a *frictionless wall AB* of height H. A soil element located at a depth z is subjected to a vertical effective pressure, σ_o', and a horizontal effective pressure, σ_h'. There are no shear stresses on the vertical and horizontal planes of the soil element. Let us define the ratio of σ_h' to σ_o' as a nondimensional quantity K, or

$$K = \frac{\sigma_h'}{\sigma_o'} \tag{11.1}$$

Figure 11.1 Earth retaining structures: (a) retaining wall; (b) crib wall; (c) basement wall (*Courtesy of N. Sivakugan, James Cook University, Australia*)

Now, three possible cases may arise concerning the retaining wall, and they are described here.

Case 1: If the wall *AB* is static—that is, if it does not move either to the right or to the left of its initial position—the soil mass will be in a state of *static equilibrium*. In that case, σ'_h is referred to as the *at-rest earth pressure*, or

$$K = K_o = \frac{\sigma'_h}{\sigma'_o} \tag{11.2}$$

where K_o = at-rest earth pressure coefficient that is the same at all locations in a homogeneous soil.

Case 2: If the frictionless wall rotates sufficiently about its bottom to a position of $A'B$ (Figure 11.2b), then a triangular soil mass ABC' adjacent to the wall will reach a state of *plastic equilibrium* and will fail sliding down the plane BC'. At this time, the horizontal effective stress, $\sigma'_h = \sigma'_a$, will be referred to as *active pressure*. Now,

$$K = K_a = \frac{\sigma'_h}{\sigma'_o} = \frac{\sigma'_a}{\sigma'_o} \tag{11.3}$$

where K_a = active earth pressure coefficient.

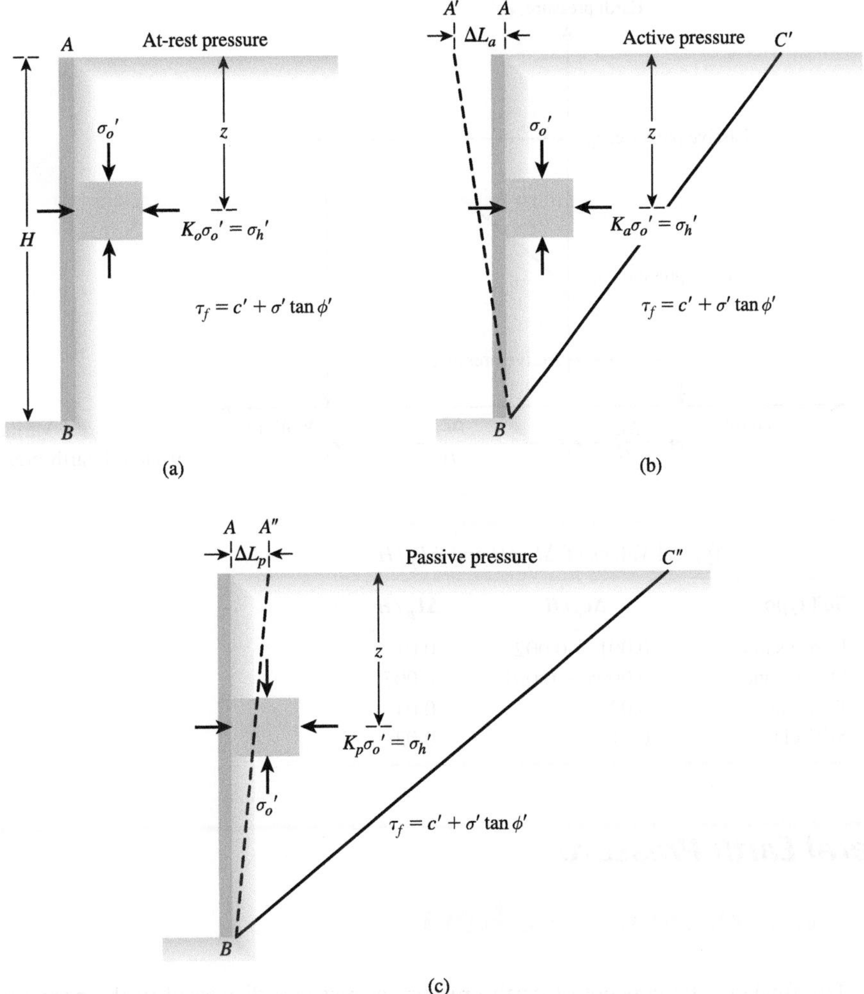

Figure 11.2 Definition of at-rest, active, and passive pressures (*Note*: Wall AB is frictionless)

Case 3: If the frictionless wall rotates sufficiently about its bottom to a position $A''B$ (Figure 11.2c), then a triangular soil mass ABC'' will reach a state of *plastic equilibrium* and will fail sliding upward along the plane BC''. The horizontal effective stress at this time will be $\sigma_h' = \sigma_p'$, the so-called *passive pressure*. In this case,

$$K = K_p = \frac{\sigma_h'}{\sigma_o'} = \frac{\sigma_p'}{\sigma_o'} \tag{11.4}$$

where K_p = passive earth pressure coefficient

Figure 11.3 shows the nature of variation of lateral earth pressure with the wall tilt. Typical values of $\Delta L_a/H$ ($\Delta L_a = A'A$ in Figure 11.2b) and $\Delta L_p/H$ ($\Delta L_p = A''A$ in Figure 11.2c) for attaining the active and passive states in various soils are given in Table 11.1. It can be seen that the wall tilt required to reach the active state is less than that for the passive state.

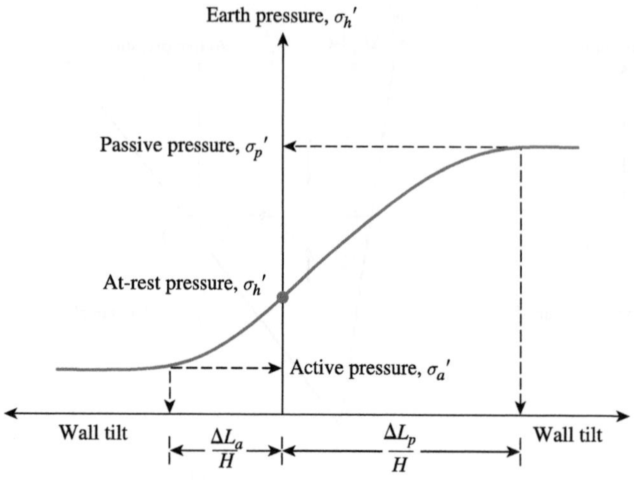

Figure 11.3 Variation of the magnitude of lateral earth pressure with wall tilt

Table 11.1 Typical Values of $\Delta L_a/H$ and $\Delta L_p/H$

Soil type	$\Delta L_a/H$	$\Delta L_p/H$
Loose sand	0.001 – 0.002	0.01
Dense sand	0.0005 – 0.001	0.005
Soft clay	0.02	0.04
Stiff clay	0.01	0.02

At-Rest Lateral Earth Pressure

11.3 Earth Pressure At-Rest

The fundamental concept of earth pressure at-rest was discussed in the preceding section. In order to define the earth pressure coefficient K_o at-rest, we refer to Figure 11.4, which shows a wall AB retaining a dry soil with a unit weight of γ. The wall is static. At a depth z,

$$\text{Vertical effective stress} = \sigma_o' = \gamma z$$

$$\text{Horizontal effective stress} = \sigma_h' = K_o \gamma z$$

So,

$$K_o = \frac{\sigma_h'}{\sigma_o'} = \text{at-rest earth pressure coefficient}$$

For coarse-grained soils, the coefficient of earth pressure at-rest can be estimated by using the empirical relationship (Jaky, 1944)

$$K_o = 1 - \sin \phi' \tag{11.5}$$

where ϕ' = drained friction angle.

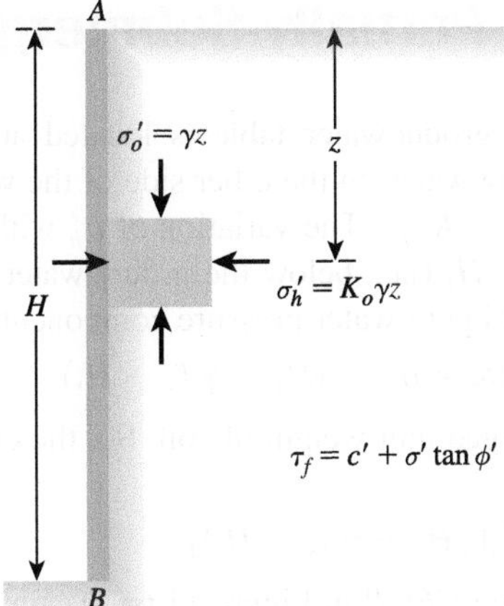

Figure 11.4 Earth pressure at-rest

For fine-grained, normally consolidated soils, Massarsch (1979) suggested the following equation for K_o:

$$K_o = 0.44 + 0.42\left[\frac{PI(\%)}{100}\right] \tag{11.5a}$$

For overconsolidated clays, the coefficient of earth pressure at-rest can be approximated as

$$K_{o\,(\text{overconsolidated})} = K_{o\,(\text{normally consolidated})}\sqrt{OCR} \tag{11.6}$$

Figure 11.5 shows the distribution of lateral earth pressure at-rest on a wall of height H retaining a dry soil having a unit weight of γ. The total force per unit length of the wall, P_o, is equal to the area of the pressure diagram, and hence,

$$P_o = \tfrac{1}{2}K_o\gamma H^2 \tag{11.7}$$

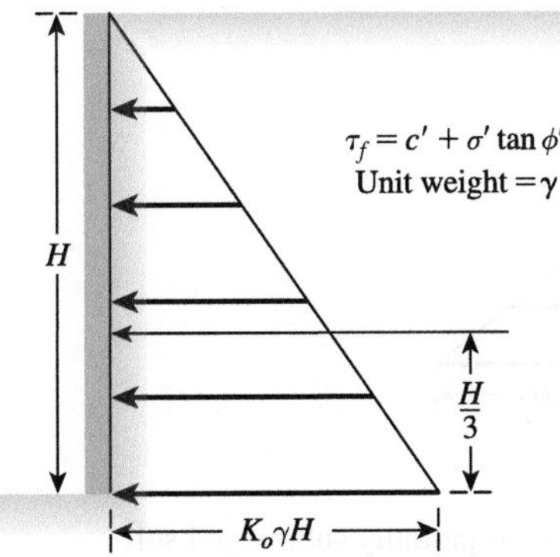

Figure 11.5 Distribution of lateral earth pressure at-rest on a wall

11.4 Earth Pressure At-Rest for Partially Submerged Soil

Figure 11.6a shows a wall of height H. The groundwater table is located at a depth H_1 below the ground surface, and there is no compensating water on the other side of the wall. For $z \leq H_1$, the lateral earth pressure at-rest can be given as $\sigma'_h = K_o \gamma z$. The variation of σ'_h with depth is shown by triangle ACE in Figure 11.6a. However, for $z \geq H_1$ (i.e., below the groundwater table), the pressure on the wall is found from the effective stress and pore water pressure components via the equation

$$\text{Effective vertical pressure} = \sigma'_o = \gamma H_1 + \gamma'(z - H_1)$$

where $\gamma' = \gamma_{sat} - \gamma_w$ = the effective or submerged unit weight of soil. So, the effective lateral pressure at rest is

$$\sigma'_h = K_o \sigma'_o = K_o[\gamma H_1 + \gamma'(z - H_1)]$$

The variation of σ'_h with depth is shown by $CEGB$ in Figure 11.6a. Again, the lateral pressure from pore water is

$$u = \gamma_w(z - H_1)$$

The variation of u with depth is shown in Figure 11.6b.

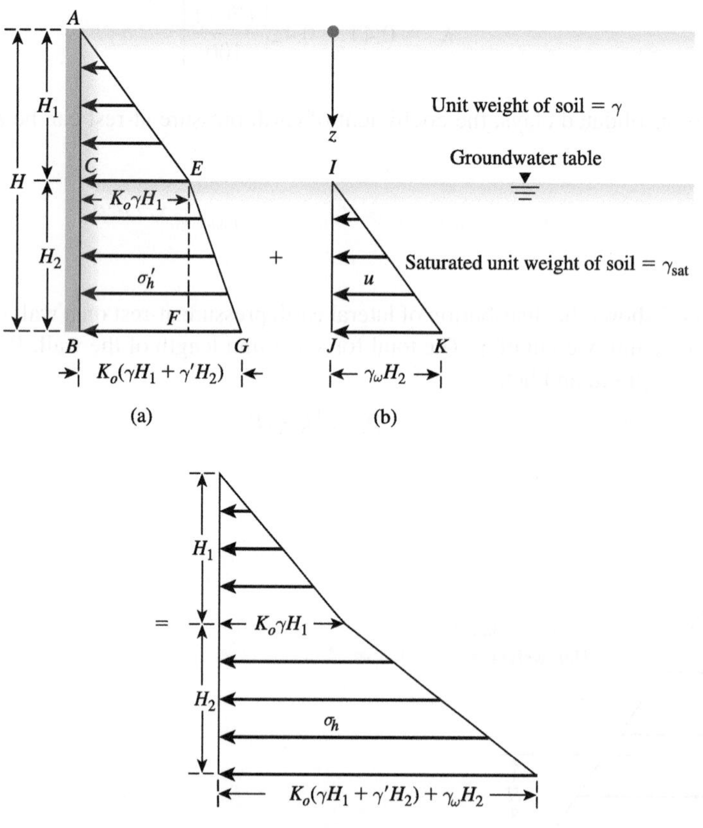

Figure 11.6 Distribution of earth pressure at-rest for partially submerged soil

Hence, the total lateral pressure from earth and water at any depth $z \geq H_1$ is equal to

$$\begin{aligned}\sigma_h &= \sigma'_h + u \\ &= K_o[\gamma H_1 + \gamma'(z - H_1)] + \gamma_w(z - H_1)\end{aligned} \tag{11.8}$$

The force per unit length of the wall can be found from the sum of the areas of the pressure diagrams in Figures 11.6a and 11.6b and is equal to (Figure 11.6c)

$$P_o = \underbrace{\tfrac{1}{2} K_o \gamma H_1^2}_{\substack{\text{Area}\\ ACE}} + \underbrace{K_o \gamma H_1 H_2}_{\substack{\text{Area}\\ CEFB}} + \underbrace{\tfrac{1}{2} (K_o \gamma' + \gamma_w) H_2^2}_{\substack{\text{Areas}\\ EFG \text{ and } IJK}} \tag{11.9}$$

Example 11.1

If the retaining wall shown in Figure 11.7a is restrained from moving, what will be the lateral force per unit length of the wall?

Solution

If the wall is restrained from moving, the backfill will exert at-rest earth pressure. Thus,

$$\sigma'_h = K_o \sigma'_o = K_o \gamma z \qquad \text{[Eq. (11.2)]}$$
$$K_o = 1 - \sin \phi' \qquad \text{[Eq. (11.5)]}$$

or

$$K_o = 1 - \sin 30^\circ = 0.5$$

and at $z = 0$, $\sigma'_h = 0$; at $z = 15$ ft, $\sigma'_h = (0.5)(15)(100) = 750$ lb/ft^2.

The pressure distribution diagram is shown in Figure 11.7b.

$$P_o = \frac{1}{2}(15)(750) = \mathbf{5625\ lb/ft}$$

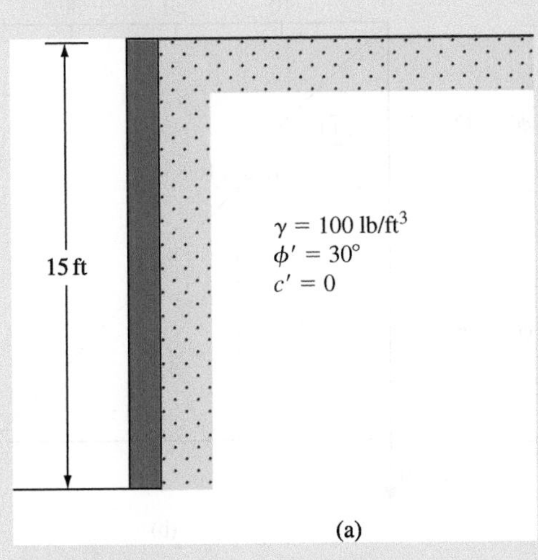

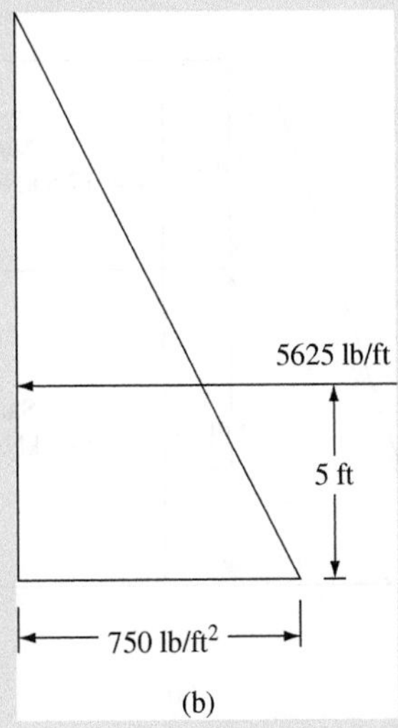

Figure 11.7 ■

Example 11.2

Refer to the retaining wall shown in Figure 11.7a. Assume that the groundwater level rises to 5 ft below the top of the retaining wall. Given: $\gamma = 100$ lb/ft^3, $\gamma_{sat} = 122.4$ lb/ft^3, $\phi' = 30°$, and $c' = 0$. If the wall is restrained from moving, what will be the lateral force per unit length of the wall?

Solution

Refer to Figure 11.6a. For this problem,

$$H_1 = 5 \text{ ft}, H_2 = 10 \text{ ft}, \phi' = 30°, \text{and } \gamma = 100 \text{ lb/ft}^3.$$

So

$$\gamma' = \gamma_{sat} - \gamma_w = 122.4 - 62.4 = 60 \text{ lb/ft}^3$$

Hence,

$$K_o = 1 - \sin\phi' = 1 - \sin 30 = 0.5$$

From Eq. (11.9),

$$\begin{aligned} P_o &= \tfrac{1}{2}K_o\gamma H_1^2 + K_o\gamma H_1 H_2 + \tfrac{1}{2}(K_o\gamma' + \gamma_w)H_2^2 \\ &= (0.5)(0.5)(100)(5)^2 + (0.5)(100)(5)(10) + (0.5)[(0.5)(60) + 62.4](10)^2 \\ &= 625 + 2500 + 4620 = \mathbf{7745\ lb/ft} \end{aligned}$$

■

Example 11.3

The backfill retained by a gravity retaining wall shown in Figure 11.8a consists of two sand layers, compacted at different densities. The properties of the sand are shown in the figure. Assuming that the gravity wall does not move laterally, determine the magnitude and location of the thrust on the wall.

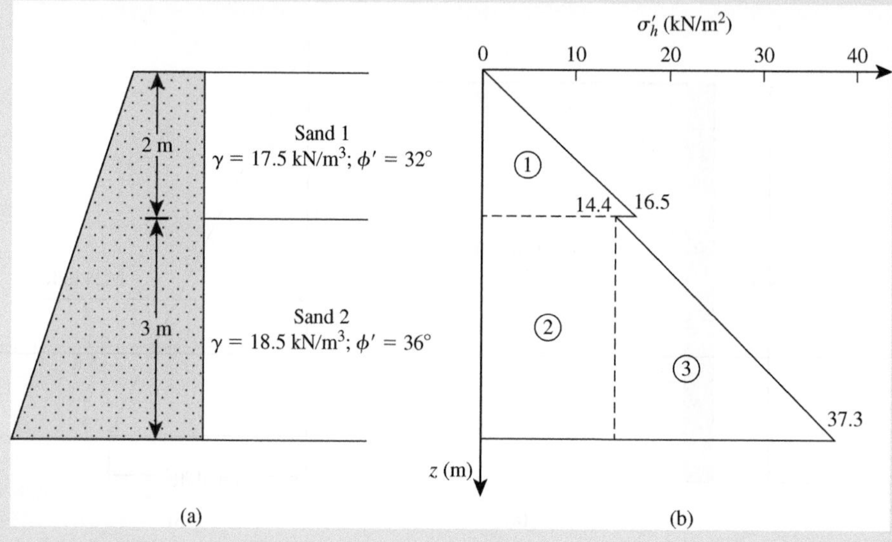

Figure 11.8

Solution

$K_{o,\text{sand1}} = 1 - \sin 32 = 0.470$, and $K_{o,\text{sand2}} = 1 - \sin 36 = 0.412$. At top of sand 1, the vertical effective pressure $\sigma'_o = 0$; hence, $\sigma'_h = 0$. At the bottom of sand 1, $\sigma'_o = 2 \times 17.5 = 35.0 \text{ kN/m}^2$; hence, $\sigma'_h = 0.470 \times 35.0 = 16.5 \text{ kN/m}^2$.

At the top of sand 2, $\sigma'_o = 2 \times 17.5 = 35.0 \text{ kN/m}^2$ and $\sigma'_h = 0.412 \times 35.0 = 14.4 \text{ kN/m}^2$. Yes, there is break in the σ'_h value at the 2 m depth due to different K_o values. At the bottom of sand 2, $\sigma'_o = 2 \times 17.5 + 3 \times 18.5 = 90.5 \text{ kN/m}^2$ and $\sigma'_h = 0.412 \times 90.5 = 37.3 \text{ kN/m}^2$. The variation of σ'_h with depth is shown in Figure 11.8b.

The lateral force on the wall per unit length is simply the area of the diagram in Figure 11.8b. It is divided into rectangles and triangles for convenience and the following table can be prepared.

Zone Col. 1	Horizontal load (kN/m) Col. 2	Centroidal height From base (m) Col. 3	Moment (kN · m/m) Col. 2 × Col. 3
1	0.5 × 2 × 16.5 = 16.5	3.67	60.6
2	3 × 14.4 = 43.2	1.5	64.8
3	0.5 × 3 × (37.3−14.4) = 34.4	2.0	68.8
SUM	**94.1**		**194.2**

The lateral load per unit length P_o is 94.1 kN/m. It acts horizontally at a height of 194.2/94.1 (= **2.06 m**) above the base of the wall. ■

Rankine Earth-Pressure Theory

11.5 Rankine Active Earth Pressure

Figure 11.9 shows a *frictionless* retaining wall of height H with a backfill whose shear strength, τ_f, can be given by the relation

$$\tau_f = c' + \sigma' \tan \phi'$$

If the wall tilts far enough away from the backfill and the soil reaches a state of plastic equilibrium, it is referred to as Rankine active state (Rankine, 1857). In that case, a soil wedge ABC will fail and slide to the left. The line BC will make an angle of $45 + \phi'/2$ with the horizontal. According to Rankine's active pressure theory, the lateral active earth pressure (σ'_a) at a depth z can be expressed as

$$\sigma'_a = \sigma'_o K_a - 2c'\sqrt{K_a} \tag{11.10}$$

where

K_a = Rankine active earth-pressure coefficient
σ'_o = effective vertical stress at a depth z

The active earth-pressure coefficient is given by the relation

$$K_a = \tan^2(45 - \phi'/2) \tag{11.11}$$

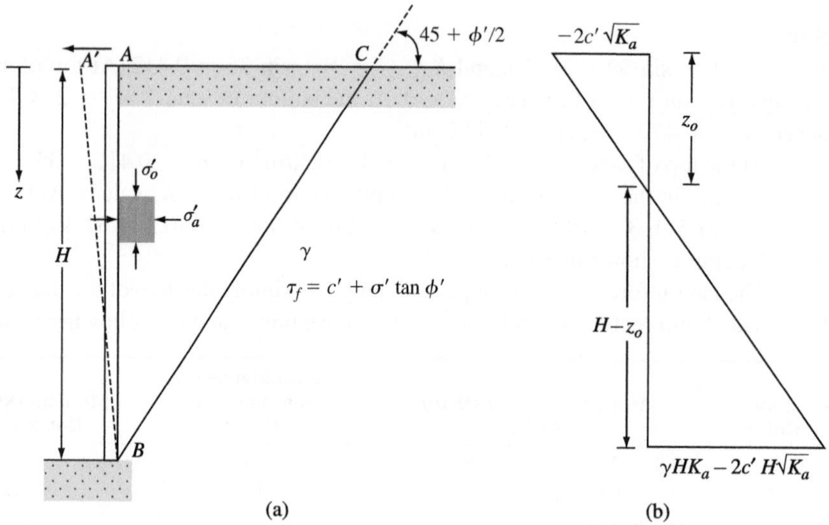

Figure 11.9 Rankine active earth pressure

For the case shown in Figure 11.9a, $\sigma'_o = \gamma z$. If the distribution of the active pressure is plotted with depth, it will be of the type shown in Figure 11.9b. It is important to note the following.

a. For $z = 0$ to z_o, the magnitude of σ'_a is negative.
b. From $z = z_o$ to H, the magnitude of σ'_a is positive.
c. The magnitude of z_o can be calculated as follows. At $z = z_o$

$$\sigma'_a = 0 = \sigma'_o K_a - 2c'\sqrt{K_a} = \gamma z_o K_a - 2c'\sqrt{K_a}$$

or

$$z_o = \frac{2c'}{\gamma\sqrt{K_a}} \tag{11.12}$$

d. For undrained condition, $\phi = 0$ and $c = c_u$, so

$$z_o = \frac{2c_u}{\gamma} \tag{11.13}$$

e. The active force per unit length of wall can be given by the area of the pressure diagram shown in Figure 11.9b, or

$$P_a = \tfrac{1}{2}K_a\gamma H^2 - 2\sqrt{K_a}c'H \tag{11.14}$$

Since the active pressure from $z = 0$ to $z = z_o$ is negative (that is, directed towards the backfill), a tensile crack occurs over time. This means that there will be no contact between the wall and backfill from $z = 0$ to z_o which is the depth of the tensile crack. After the occurrence of the tensile crack, the active force on the wall (per unit length) can be given by the area of the pressure diagram between $z = z_o$ to $z = H$. Or

$$P_a = \frac{1}{2}(K_a\gamma H - 2c'\sqrt{K_a})\left(H - \frac{2c'}{\gamma\sqrt{K_a}}\right) = \frac{1}{2}K_a\gamma H^2 - 2\sqrt{K_a}\,c'H + \frac{2c'^2}{\gamma} \tag{11.15}$$

For undrained condition (that is, $\phi = 0$ and $c = c_u$),

$$P_a = \frac{1}{2}\gamma H^2 - 2c_u H + \frac{2c_u^2}{\gamma} \tag{11.15a}$$

11.6 Rankine Active Pressure—Partially Submerged Cohesionless Soil ($c' = 0$) with Backfill Supporting a Surcharge

Figure 11.10a shows a frictionless retaining wall of height H and a backfill of cohensionless soil. The groundwater table is located at a depth of H_1 below the ground surface, and the backfill is supporting a surcharge pressure of q per unit area. From Eq. (11.10), the effective active earth pressure at any depth can be given by

$$\sigma'_a = K_a\sigma'_o \tag{11.16}$$

where σ'_o and σ'_a = the effective vertical pressure and lateral pressure, respectively. At $z = 0$,

$$\sigma_o = \sigma'_o = q \tag{11.17}$$

and

$$\sigma'_a = K_a q \tag{11.18}$$

At depth $z = H_1$,

$$\sigma'_o = (q + \gamma H_1) \tag{11.19}$$

and

$$\sigma'_a = K_a(q + \gamma H_1) \tag{11.20}$$

At depth $z = H$,

$$\sigma'_o = (q + \gamma H_1 + \gamma' H_2) \tag{11.21}$$

and

$$\sigma'_a = K_a(q + \gamma H_1 + \gamma' H_2) \tag{11.22}$$

where $\gamma' = \gamma_{sat} - \gamma_w$. The variation of σ'_a with depth is shown in Figure 11.10b.

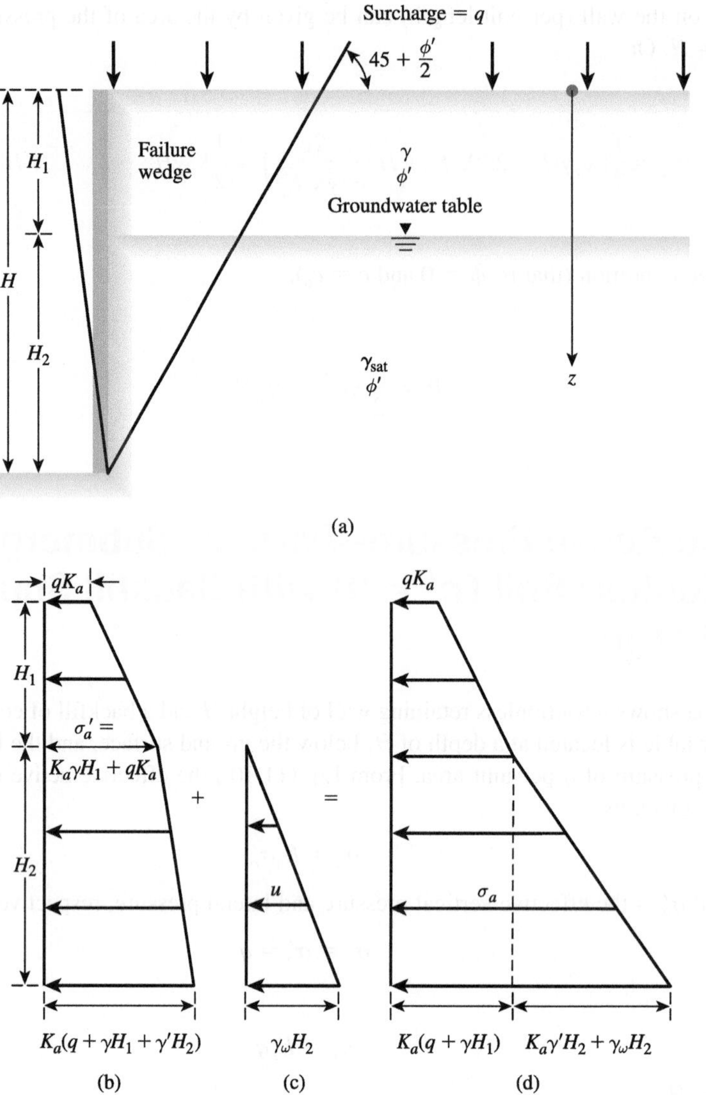

Figure 11.10 Rankine's active earth-pressure distribution against a retaining wall with partially submerged cohensionless soil backfill supporting a surcharge

The lateral pressure on the wall from the pore water between $z = 0$ and H_1 is 0, and for $z > H_1$, it increases linearly with depth (Figure 11.10c). At $z = H$,

$$u = \gamma_w H_2$$

The total lateral-pressure diagram (Figure 11.10d) is the sum of the pressure diagrams shown in Figures 11.10b and 11.10c. The total active force per unit length of the wall is the area of the total pressure diagram. Thus,

$$P_a = K_a q H + \tfrac{1}{2} K_a \gamma H_1^2 + K_a \gamma H_1 H_2 + \tfrac{1}{2}(K_a \gamma' + \gamma_w) H_2^2 \tag{11.23}$$

11.7 Rankine Active Pressure with Inclined Granular ($c' = 0$) Backfill

Figure 11.11 shows a frictionless retaining wall with a vertical back. The granular backfill is inclined at an angle α with the horizontal. The Rankine active pressure, σ'_a, for this condition can be given as

$$\sigma'_a = \gamma z K_a \tag{11.24}$$

where

$$K_a = \cos\alpha \frac{\cos\alpha - \sqrt{\cos^2\alpha - \cos^2\phi'}}{\cos\alpha + \sqrt{\cos^2\alpha - \cos^2\phi'}} \tag{11.25}$$

The direction of the pressure σ'_a is inclined at an angle α with the horizontal. The Rankine active force, P_a, per unit length of the wall is

$$P_a = \tfrac{1}{2}\gamma H^2 K_a \tag{11.26}$$

Note that the resultant force P_a will be inclined at an angle α with the horizontal and will intersect the wall at a distance of $H/3$ measured from the bottom of the wall. Table 11.2 gives the variation of K_a with α and ϕ'.

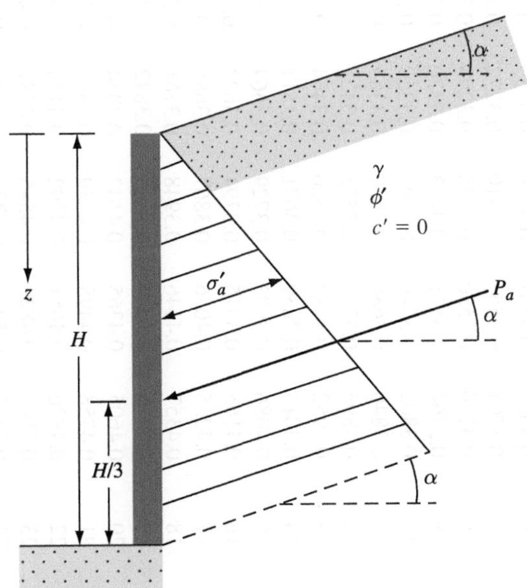

Figure 11.11 Notations for active pressure—Eqs. (11.24), (11.25), (11.26)

Table 11.2 Values of K_a [Eq. (11.25)]

α (deg) ↓	φ′ (deg) → 28	29	30	31	32	33	34	35	36	37	38	39	40
0	0.3610	0.3470	0.3333	0.3201	0.3073	0.2948	0.2827	0.2710	0.2596	0.2486	0.2379	0.2275	0.2174
1	0.3612	0.3471	0.3335	0.3202	0.3074	0.2949	0.2828	0.2711	0.2597	0.2487	0.2380	0.2276	0.2175
2	0.3618	0.3476	0.3339	0.3207	0.3078	0.2953	0.2832	0.2714	0.2600	0.2489	0.2382	0.2278	0.2177
3	0.3627	0.3485	0.3347	0.3214	0.3084	0.2959	0.2837	0.2719	0.2605	0.2494	0.2386	0.2282	0.2181
4	0.3639	0.3496	0.3358	0.3224	0.3094	0.2967	0.2845	0.2726	0.2611	0.2500	0.2392	0.2287	0.2186
5	0.3656	0.3512	0.3372	0.3237	0.3105	0.2978	0.2855	0.2736	0.2620	0.2508	0.2399	0.2294	0.2192
6	0.3676	0.3531	0.3389	0.3253	0.3120	0.2992	0.2868	0.2747	0.2631	0.2518	0.2409	0.2303	0.2200
7	0.3701	0.3553	0.3410	0.3272	0.3138	0.3008	0.2883	0.2761	0.2644	0.2530	0.2420	0.2313	0.2209
8	0.3730	0.3580	0.3435	0.3294	0.3159	0.3027	0.2900	0.2778	0.2659	0.2544	0.2432	0.2325	0.2220
9	0.3764	0.3611	0.3463	0.3320	0.3182	0.3049	0.2921	0.2796	0.2676	0.2560	0.2447	0.2338	0.2233
10	0.3802	0.3646	0.3495	0.3350	0.3210	0.3074	0.2944	0.2818	0.2696	0.2578	0.2464	0.2354	0.2247
11	0.3846	0.3686	0.3532	0.3383	0.3241	0.3103	0.2970	0.2841	0.2718	0.2598	0.2482	0.2371	0.2263
12	0.3896	0.3731	0.3573	0.3421	0.3275	0.3134	0.2999	0.2868	0.2742	0.2621	0.2503	0.2390	0.2281
13	0.3952	0.3782	0.3620	0.3464	0.3314	0.3170	0.3031	0.2898	0.2770	0.2646	0.2527	0.2412	0.2301
14	0.4015	0.3839	0.3671	0.3511	0.3357	0.3209	0.3068	0.2931	0.2800	0.2674	0.2552	0.2435	0.2322
15	0.4086	0.3903	0.3729	0.3564	0.3405	0.3253	0.3108	0.2968	0.2834	0.2705	0.2581	0.2461	0.2346
16	0.4165	0.3975	0.3794	0.3622	0.3458	0.3302	0.3152	0.3008	0.2871	0.2739	0.2612	0.2490	0.2373
17	0.4255	0.4056	0.3867	0.3688	0.3518	0.3356	0.3201	0.3053	0.2911	0.2776	0.2646	0.2521	0.2401
18	0.4357	0.4146	0.3948	0.3761	0.3584	0.3415	0.3255	0.3102	0.2956	0.2817	0.2683	0.2555	0.2433
19	0.4473	0.4249	0.4039	0.3842	0.3657	0.3481	0.3315	0.3156	0.3006	0.2862	0.2724	0.2593	0.2467
20	0.4605	0.4365	0.4142	0.3934	0.3739	0.3555	0.3381	0.3216	0.3060	0.2911	0.2769	0.2634	0.2504
21	0.4758	0.4498	0.4259	0.4037	0.3830	0.3637	0.3455	0.3283	0.3120	0.2965	0.2818	0.2678	0.2545
22	0.4936	0.4651	0.4392	0.4154	0.3934	0.3729	0.3537	0.3356	0.3186	0.3025	0.2872	0.2727	0.2590
23	0.5147	0.4829	0.4545	0.4287	0.4050	0.3832	0.3628	0.3438	0.3259	0.3091	0.2932	0.2781	0.2638
24	0.5404	0.5041	0.4724	0.4440	0.4183	0.3948	0.3731	0.3529	0.3341	0.3164	0.2997	0.2840	0.2692
25	0.5727	0.5299	0.4936	0.4619	0.4336	0.4081	0.3847	0.3631	0.3431	0.3245	0.3070	0.2905	0.2750

Example 11.4

Calculate the Rankine active force per unit length of the wall shown in Figure 11.12a, and also determine the location of the resultant.

Solution

Since $c' = 0$,

$$\sigma'_a = K_a\sigma'_o = K_a\gamma z$$

$$K_a = \tan^2\left(45 - \frac{\phi'}{2}\right) = \tan^2\left(45 - \frac{30}{2}\right) = \frac{1}{3}$$

At $z = 0$, $\sigma'_a = 0$; at $z = 15$ ft, $\sigma'_a = (1/3)(100)(15) = 500$ lb/ft^2.

The active pressure distribution diagram is shown in Figure 11.12b.

$$\text{Active force } P_a = \frac{1}{2}(15)(500)$$

$$= \mathbf{3750\ lb/ft}$$

The pressure distribution is triangular, and so P_a will act at a distance of $15/3 = 5$ ft above the bottom of the wall.

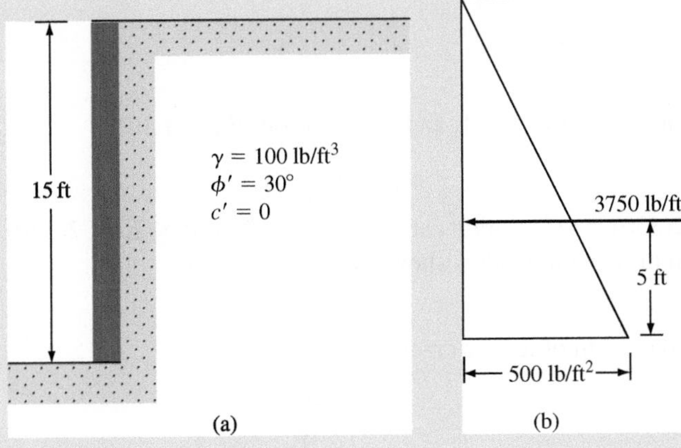

Figure 11.12 ■

Example 11.5

A retaining wall, having a soft, saturated clay backfill, is shown in Figure 11.13a. For undrained condition ($\phi = 0$) of the backfill, determine:

a. The maximum depth of the tensile crack
b. P_a before the tensile crack occurs
c. P_a after the tensile crack occurs

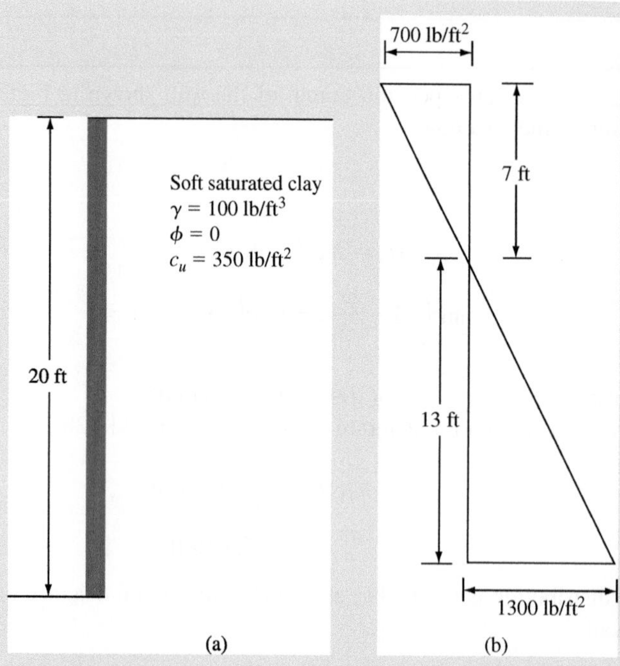

Figure 11.13

Solution

Since $\phi = 0$, then $K_a = \tan^2(45°) = 1$, and $c = c_u$. From Eq. (11.10)

$$\sigma_a = \gamma z - 2c_u$$

At $z = 0$, $\sigma_a = -2(350) = -700$ lb/ft²; at $z = 20$ ft, $\sigma_a = 100 \times 20 - 2(350) = 1300$ lb/ft². The variation of σ_a with depth is shown in Figure 11.13b.

Part a

From Eq. (11.13), the depth of tensile crack equals

$$z_o = \frac{2c_u}{\gamma}$$

or

$$z_o = \frac{2 \times 350}{100} = \mathbf{7\ ft}$$

Part b

Before the tensile crack occurs,

$$P_a = \tfrac{1}{2}\gamma H^2 - 2c_u H \qquad \text{[Eq. (11.14)]}$$

or

$$P_a = \tfrac{1}{2}(100)(20)^2 - 2(350)(20)$$
$$= 20{,}000 - 14{,}000 = \mathbf{6000\ lb/ft}$$

Part c

After the tensile crack has occurred,

$$P_a = \tfrac{1}{2}(20 - 7)(1300) = \mathbf{8450\ lb/ft}$$ ■

Example 11.6

A smooth vertical wall retains a clay soil backfill with $c' = 10\ \text{kN/m}^2$, $\phi' = 25°$, and $\gamma = 19.0\ \text{kN/m}^3$. If the soil in the backfill is in an active state, determine the following.

a. The maximum tensile stress within the clay.
b. The depth of the tensile cracks.

Solution

Part a

$$K_a = \tan^2\left(45 - \frac{\phi'}{2}\right) = \tan^2\left(45 - \frac{25}{2}\right) = 0.406$$

The maximum tensile stress occurs at the top of the clay, and is given by

$$\sigma'_a = -2c'\sqrt{K_a} = -2 \times 10 \times \sqrt{0.406} = \mathbf{-12.7\ kN/m^2}$$

Part b

The depth of the tensile crack is given by

$$z_o = \frac{2c'}{\gamma\sqrt{K_a}} = \frac{2 \times 10}{19.0\sqrt{0.406}} = \mathbf{1.65\ m}$$ ■

Example 11.7

A retaining wall with a granular backfill ($c' = 0$) is shown in Figure 11.14. Determine the force per unit length of the wall at Rankine active state. Also determine the location of the resultant force.

Solution

$$K_a = \tan^2\left(45 - \frac{\phi'}{2}\right) = \tan^2\left(45 - \frac{35}{2}\right) = 0.271$$

At $z = 0$,

$$\sigma'_a = \sigma'_o K_a = (1000)(0.271) = 271\ \text{lb/ft}^2$$
$$u = 0$$

At $z = 5$ ft,

$$\sigma'_a = \sigma'_o K_a = [1000 + (5)(100)](0.271) = 406.5\ \text{lb/ft}^2$$
$$u = 0$$

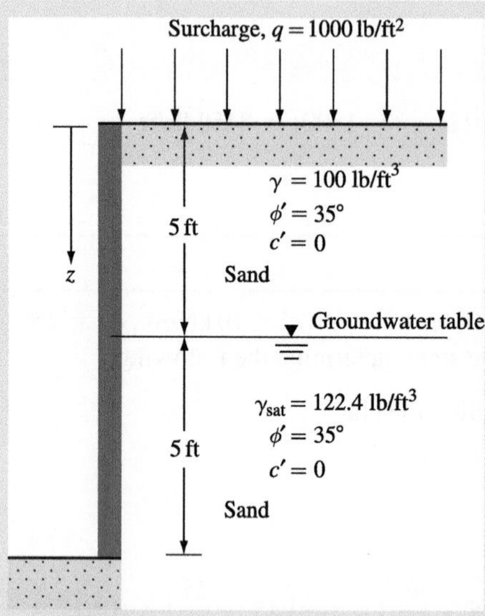

Figure 11.14

At $z = 10$ ft,

$$\sigma'_a = \sigma'_o K_a = [1000 + (5)(100) + (5)(122.4 - 62.4)](0.271) = 487.8 \text{ lb/ft}^2$$

$$u = 5\gamma_w = (5)(62.4) = 312 \text{ lb/ft}^2$$

The pressure diagram is shown in Figure 11.15.

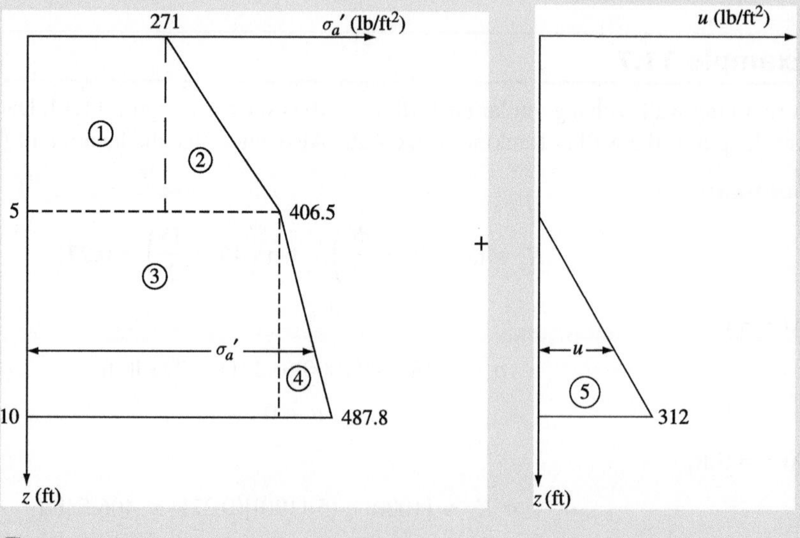

Figure 11.15

$$P_a = \text{Area 1} + \text{Area 2} + \text{Area 3} + \text{Area 4} + \text{Area 5}$$
$$= (271)(5) + (\tfrac{1}{2})(5)(406.5 - 271) + (5)(406.5) + (\tfrac{1}{2})(5)(487.8 - 406.50) + (\tfrac{1}{2})(5)(312)$$
$$= 1355 + 338.75 + 2032.5 + 203.25 + 780 = \mathbf{4709.5\ lb/ft}$$

Location of P_a: Taking the moment of the forces about the bottom of the wall,

$$\bar{z} = \frac{(1355)\left(5 + \frac{5}{2}\right) + (338.75)\left(5 + \frac{5}{3}\right) + (2032.5)\left(\frac{5}{2}\right) + (203.25)\left(\frac{5}{3}\right) + (780)\left(\frac{5}{3}\right)}{4709.5}$$

or

$$\bar{z} = \frac{10{,}162.5 + 2258.3 + 5081.25 + 338.75 + 1300}{4709.5}$$
$$= \mathbf{4.06\ ft\ (measured\ from\ the\ bottom\ of\ the\ wall)}$$ ■

11.8 Rankine Passive Earth Pressure

Figure 11.16a shows a *frictionless* retaining wall of height H. If the wall is pushed enough into the soil mass and the soil reaches a state of plastic equilibrium, it is referred to as *Rankine passive state*. In that state, a soil wedge ABC will be pushed into the soil mass and up. The line BC will make an

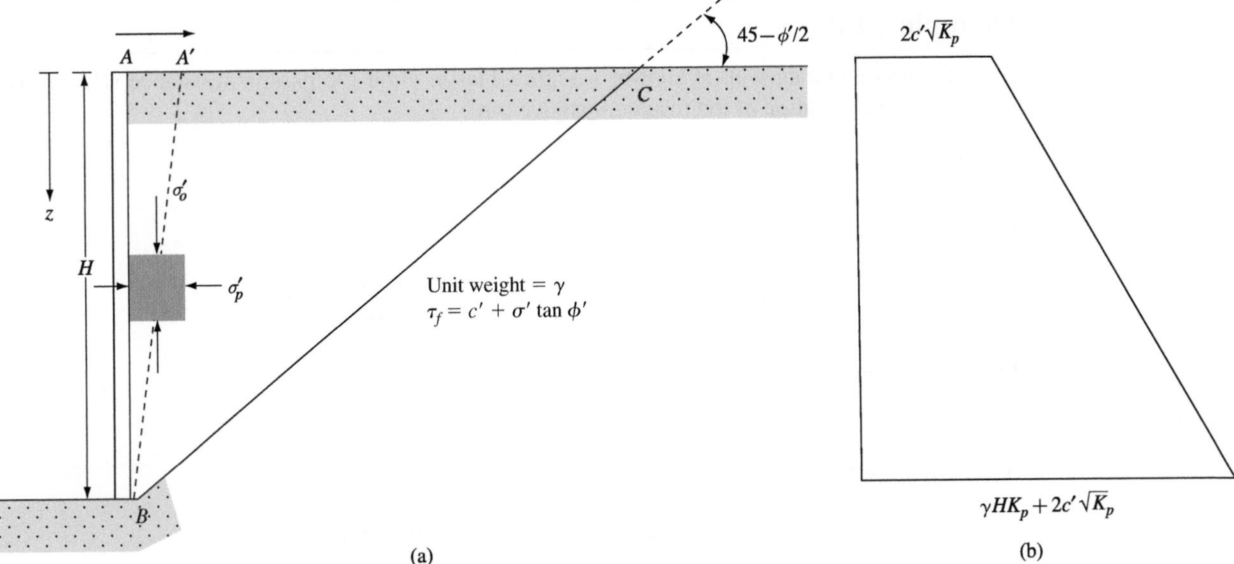

Figure 11.16 Rankine passive earth pressure

angle of $45 - \phi'/2$ with the horizontal. The lateral earth pressure σ'_p at a depth z measured from the top of the wall can be expressed as

$$\sigma'_p = \sigma'_o K_p + 2c'\sqrt{K_p} \tag{11.27}$$

where

σ'_o = effective vertical stress at a depth z
K_p = Rankine passive earth-pressure coefficient

The coefficient K_p is given by the relation

$$K_p = \tan^2(45 + \phi'/2) \tag{11.28}$$

Based on Eq. (11.27), the variation of σ'_p with depth is shown in Figure 11.16b. Note that:
At $z = 0$,

$$\sigma'_p = 2c'\sqrt{K_p}$$

At $z = H$,

$$\sigma'_p = \gamma z K_p + 2c'\sqrt{K_p}$$

The passive force, P_p, per unit length of the wall can be obtained by finding the area of the pressure diagram shown in Figure 11.16b. Or,

$$P_p = 2c'\sqrt{K_p}H + \frac{1}{2}\gamma H^2 K_p \tag{11.29}$$

Example 11.8

A frictionless retaining wall is shown in Figure 11.17a. Find the passive resistance (P_p) on the backfill, and the location of the resultant passive force.

Solution

Passive Resistance
Given: $\phi' = 26°$.

$$K_p = \frac{1 + \sin\phi'}{1 - \sin\phi'} = \frac{1 + \sin 26°}{1 - \sin 26°} = \frac{1.4384}{0.5616} = 2.56$$

From Eq. (11.27),

$$\sigma'_p = K_p \sigma'_o + 2c'\sqrt{K_p}$$

At $z = 0$, $\sigma'_o = 200\ \text{lb/ft}^2$

$$\sigma'_p = (2.56)(200) + (2)(170)\sqrt{2.56}$$
$$= 512 + 544 = 1056\ \text{lb/ft}^2$$

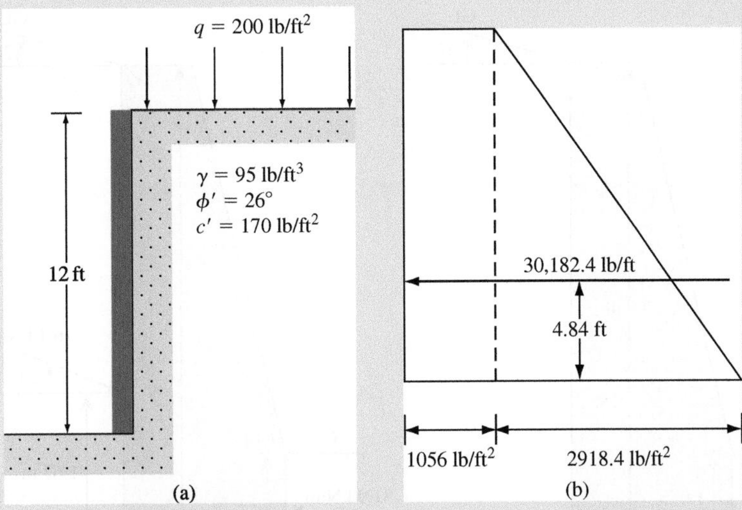

Figure 11.17

Again, at $z = 12$ ft, $\sigma'_o = (200 + 12 \times 95) = 1340$ lb/ft². So

$$\sigma'_p = (2.56)(1340) + (2)(170)\sqrt{2.56}$$
$$= 3974.4 \text{ lb/ft}^2$$

The pressure distribution is shown in Figure 11.17b. The passive resistance per unit length of wall

$$P_p = (1056)(12) + \frac{1}{2}(12)(2918.4)$$
$$= 12{,}672 + 17{,}510.4 = \mathbf{30{,}182.4\ lb/ft}$$

Location of Resultant

Taking the moment of the pressure diagram about the bottom of the wall

$$\bar{z} = \frac{(12{,}672)\left(\dfrac{12}{2}\right) + (17{,}510.4)\left(\dfrac{12}{3}\right)}{30{,}182.4} = \mathbf{4.84\ ft}$$

■

Example 11.9

Figure 11.18a shows a gravity retaining wall with sand ($\gamma = 18.0$ kN/m³ and $\phi' = 34°$) on the right and left sides in active and passive states, respectively. Show the direction, magnitude, and location of P_a and P_p on the same wall in Figure 11.18b.

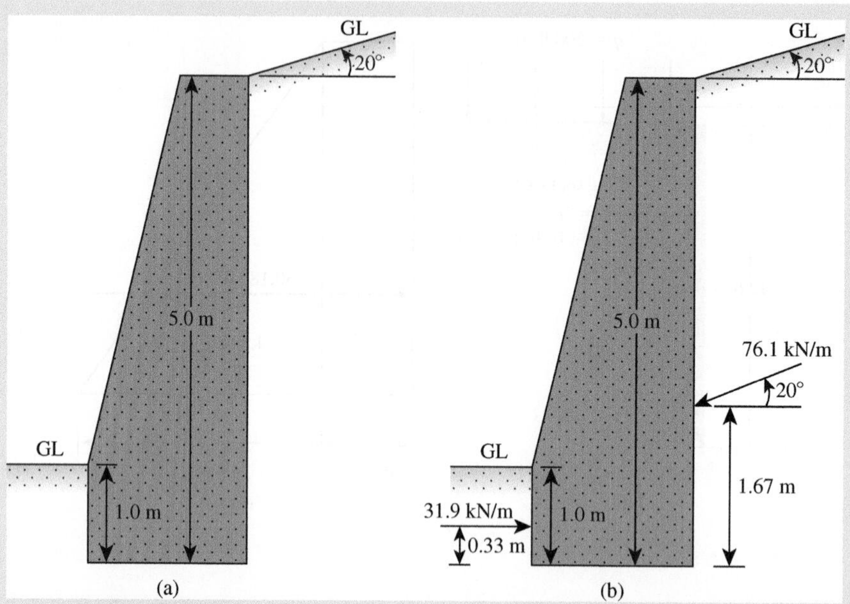

Figure 11.18

Solution

Since the backfill on the active side of the sand is inclined with $\alpha = 20°$ and $\phi' = 34°$, using Eq. (11.25) or Table 11.2 gives $K_a = 0.3381$

For the passive side, $K_p = \tan^2\left(45 + \frac{\phi'}{2}\right) = \tan^2\left(45 + \frac{34}{2}\right) = 3.54.$

From Eq. (11.26),

$P_a = \frac{1}{2}\gamma H^2 K_a = \frac{1}{2} \times 18 \times 5^2 \times 0.338 =$ **76.1 kN/m, acting at 1.67 m height from the bottom, inclined at 20° to horizontal.**

From Eq. (11.29) with $c' = 0$,

$P_p = \frac{1}{2}\gamma H^2 K_p = \frac{1}{2} \times 18 \times 1^2 \times 3.54 =$ **31.9kN/m, acting horizontally at a height of 0.33 m above the base.**

The two forces P_a and P_p are shown in Figure 11.18b. ■

Coulomb's Earth-Pressure Theory

So far in our study of active and passive earth pressures, we have considered the case of frictionless walls. In reality retaining walls are rough, and shear forces develop between the face of the wall and the backfill. More than 200 years ago, Coulomb (1776) presented a theory for active and passive earth pressures against retaining walls. In this theory, Coulomb assumed that the failure surface is a plane. The *wall friction* was taken into consideration. The following sections discuss the general principles of the derivation of Coulomb's earth-pressure theory for a cohesionless backfill (shear strength defined by the equation $\tau_f = \sigma' \tan \phi'$).

11.9 Coulomb's Active Pressure

Let AB (Figure 11.19a) be the back face of a retaining wall supporting a granular soil; the surface of which is constantly sloping at an angle α with the horizontal. BC is a trial failure surface. In the stability consideration of the probable failure wedge ABC, the following forces are involved (per unit length of the wall):

1. W—the weight of the soil wedge.
2. F—the resultant of the shear and normal forces on the surface of failure, BC. This is inclined at an angle of ϕ' to the normal drawn to the plane BC.
3. P_a—the active force per unit length of the wall. The direction of P_a is inclined at an angle δ' to the normal drawn to the face of the wall that supports the soil. δ' is the angle of friction between the soil and the wall.

The force triangle for the wedge is shown in Figure 11.19b. From the law of sines, we have

$$\frac{W}{\sin(90 + \theta + \delta' - \beta + \phi')} = \frac{P_a}{\sin(\beta - \phi')} \tag{11.30}$$

or

$$P_a = \frac{\sin(\beta - \phi')}{\sin(90 + \theta + \delta' - \beta + \phi')} W \tag{11.31}$$

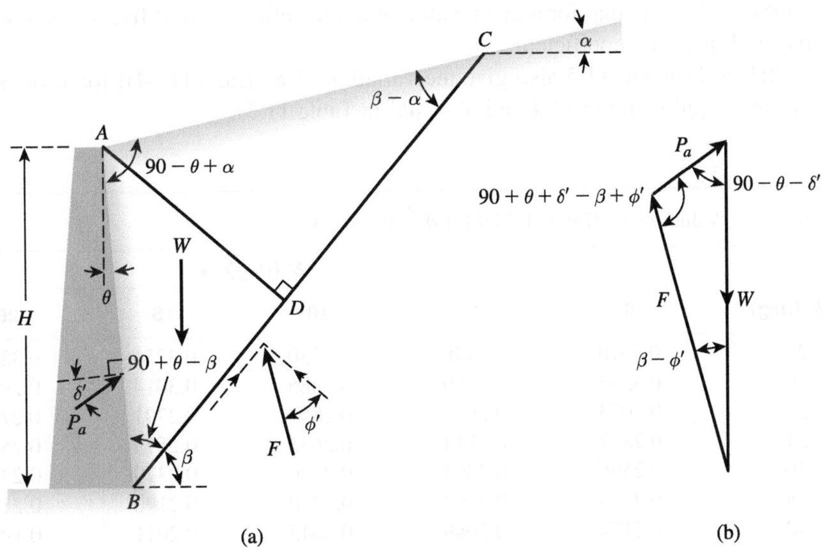

Figure 11.19 Coulomb's active pressure: (a) trial failure wedge; (b) force polygon

The preceding equation can be written in the form

$$P_a = \frac{1}{2}\gamma H^2\left[\frac{\cos(\theta-\beta)\cos(\theta-\alpha)\sin(\beta-\phi')}{\cos^2\theta\sin(\beta-\alpha)\sin(90+\theta+\delta'-\beta+\phi')}\right] \tag{11.32}$$

where γ = unit weight of the backfill. The values of γ, H, θ, α, ϕ', and δ' are constants, and β is the only variable.

Note that the relationship for P_a in Eq. (11.32) is only for the trial wedge *ABC*. If the value of β is varied, the wedge which will give the maximum value of P_a is the critical wedge. For this case, we can write

$$P_a = \tfrac{1}{2}K_a\gamma H^2 \tag{11.33}$$

where K_a is Coulomb's active earth-pressure coefficient and is given by

$$K_a = \frac{\cos^2(\phi'-\theta)}{\cos^2\theta\cos(\delta'+\theta)\left[1+\sqrt{\dfrac{\sin(\delta'+\phi')\sin(\phi'-\alpha)}{\cos(\delta'+\theta)\cos(\theta-\alpha)}}\right]^2} \tag{11.34}$$

P_a acts at a height of $H/3$ from the bottom of the wall. The variation of the values of K_a for retaining walls with a vertical back ($\theta = 0°$) and horizontal backfill ($\alpha = 0°$) is given in Table 11.3. From this table, note that for a given value of ϕ', the effect of wall friction is to reduce somewhat the active earth-pressure coefficient.

Tables 11.4 and 11.5 also give the variation of K_a [Eq. (11.34)] for various values of α, ϕ', θ, and δ' ($\delta' = \frac{2}{3}\phi'$ in Table 11.4 and $\delta' = \frac{1}{2}\phi'$ in Table 11.5).

Table 11.3 Values of K_a [Eq. (11.34)] for $\theta = 0°$, $\alpha = 0°$

	δ' (deg) →					
↓ ϕ' (deg)	0	5	10	15	20	25
28	0.3610	0.3448	0.3330	0.3251	0.3203	0.3186
30	0.3333	0.3189	0.3085	0.3014	0.2973	0.2956
32	0.3073	0.2945	0.2853	0.2791	0.2755	0.2745
34	0.2827	0.2714	0.2633	0.2579	0.2549	0.2542
36	0.2596	0.2497	0.2426	0.2379	0.2354	0.2350
38	0.2379	0.2292	0.2230	0.2190	0.2169	0.2167
40	0.2174	0.2089	0.2045	0.2011	0.1994	0.1995
42	0.1982	0.1916	0.1870	0.1841	0.1828	0.1831

Table 11.4 Values of K_a [Eq. (11.34)] (Note: $\delta' = \frac{2}{3}\phi'$)

α (deg)	ϕ' (deg)	θ (deg)					
		0	**5**	**10**	**15**	**20**	**25**
0	28	0.3213	0.3588	0.4007	0.4481	0.5026	0.5662
	29	0.3091	0.3467	0.3886	0.4362	0.4908	0.5547
	30	0.2973	0.3349	0.3769	0.4245	0.4794	0.5435
	31	0.2860	0.3235	0.3655	0.4133	0.4682	0.5326
	32	0.2750	0.3125	0.3545	0.4023	0.4574	0.5220
	33	0.2645	0.3019	0.3439	0.3917	0.4469	0.5117
	34	0.2543	0.2916	0.3335	0.3813	0.4367	0.5017
	35	0.2444	0.2816	0.3235	0.3713	0.4267	0.4919
	36	0.2349	0.2719	0.3137	0.3615	0.4170	0.4824
	37	0.2257	0.2626	0.3042	0.3520	0.4075	0.4732
	38	0.2168	0.2535	0.2950	0.3427	0.3983	0.4641
	39	0.2082	0.2447	0.2861	0.3337	0.3894	0.4553
	40	0.1998	0.2361	0.2774	0.3249	0.3806	0.4468
	41	0.1918	0.2278	0.2689	0.3164	0.3721	0.4384
	42	0.1840	0.2197	0.2606	0.3080	0.3637	0.4302
5	28	0.3431	0.3845	0.4311	0.4843	0.5461	0.6190
	29	0.3295	0.3709	0.4175	0.4707	0.5325	0.6056
	30	0.3165	0.3578	0.4043	0.4575	0.5194	0.5926
	31	0.3039	0.3451	0.3916	0.4447	0.5067	0.5800
	32	0.2919	0.3329	0.3792	0.4324	0.4943	0.5677
	33	0.2803	0.3211	0.3673	0.4204	0.4823	0.5558
	34	0.2691	0.3097	0.3558	0.4088	0.4707	0.5443
	35	0.2583	0.2987	0.3446	0.3975	0.4594	0.5330
	36	0.2479	0.2881	0.3338	0.3866	0.4484	0.5221
	37	0.2379	0.2778	0.3233	0.3759	0.4377	0.5115
	38	0.2282	0.2679	0.3131	0.3656	0.4273	0.5012
	39	0.2188	0.2582	0.3033	0.3556	0.4172	0.4911
	40	0.2098	0.2489	0.2937	0.3458	0.4074	0.4813
	41	0.2011	0.2398	0.2844	0.3363	0.3978	0.4718
	42	0.1927	0.2311	0.2753	0.3271	0.3884	0.4625
10	28	0.3702	0.4164	0.4686	0.5287	0.5992	0.6834
	29	0.3548	0.4007	0.4528	0.5128	0.5831	0.6672
	30	0.3400	0.3857	0.4376	0.4974	0.5676	0.6516
	31	0.3259	0.3713	0.4230	0.4826	0.5526	0.6365
	32	0.3123	0.3575	0.4089	0.4683	0.5382	0.6219
	33	0.2993	0.3442	0.3953	0.4545	0.5242	0.6078
	34	0.2868	0.3314	0.3822	0.4412	0.5107	0.5942
	35	0.2748	0.3190	0.3696	0.4283	0.4976	0.5810
	36	0.2633	0.3072	0.3574	0.4158	0.4849	0.5682
	37	0.2522	0.2957	0.3456	0.4037	0.4726	0.5558
	38	0.2415	0.2846	0.3342	0.3920	0.4607	0.5437
	39	0.2313	0.2740	0.3231	0.3807	0.4491	0.5321
	40	0.2214	0.2636	0.3125	0.3697	0.4379	0.5207
	41	0.2119	0.2537	0.3021	0.3590	0.4270	0.5097
	42	0.2027	0.2441	0.2921	0.3487	0.4164	0.4990

(continued)

Table 11.4 (*continued*)

α (deg)	φ′ (deg)	θ (deg) 0	5	10	15	20	25
15	28	0.4065	0.4585	0.5179	0.5868	0.6685	0.7670
	29	0.3881	0.4397	0.4987	0.5672	0.6483	0.7463
	30	0.3707	0.4219	0.4804	0.5484	0.6291	0.7265
	31	0.3541	0.4049	0.4629	0.5305	0.6106	0.7076
	32	0.3384	0.3887	0.4462	0.5133	0.5930	0.6895
	33	0.3234	0.3732	0.4303	0.4969	0.5761	0.6721
	34	0.3091	0.3583	0.4150	0.4811	0.5598	0.6554
	35	0.2954	0.3442	0.4003	0.4659	0.5442	0.6393
	36	0.2823	0.3306	0.3862	0.4513	0.5291	0.6238
	37	0.2698	0.3175	0.3726	0.4373	0.5146	0.6089
	38	0.2578	0.3050	0.3595	0.4237	0.5006	0.5945
	39	0.2463	0.2929	0.3470	0.4106	0.4871	0.5805
	40	0.2353	0.2813	0.3348	0.3980	0.4740	0.5671
	41	0.2247	0.2702	0.3231	0.3858	0.4613	0.5541
	42	0.2146	0.2594	0.3118	0.3740	0.4491	0.5415
20	28	0.4602	0.5205	0.5900	0.6714	0.7689	0.8880
	29	0.4364	0.4958	0.5642	0.6445	0.7406	0.8581
	30	0.4142	0.4728	0.5403	0.6195	0.7144	0.8303
	31	0.3935	0.4513	0.5179	0.5961	0.6898	0.8043
	32	0.3742	0.4311	0.4968	0.5741	0.6666	0.7799
	33	0.3559	0.4121	0.4769	0.5532	0.6448	0.7569
	34	0.3388	0.3941	0.4581	0.5335	0.6241	0.7351
	35	0.3225	0.3771	0.4402	0.5148	0.6044	0.7144
	36	0.3071	0.3609	0.4233	0.4969	0.5856	0.6947
	37	0.2925	0.3455	0.4071	0.4799	0.5677	0.6759
	38	0.2787	0.3308	0.3916	0.4636	0.5506	0.6579
	39	0.2654	0.3168	0.3768	0.4480	0.5342	0.6407
	40	0.2529	0.3034	0.3626	0.4331	0.5185	0.6242
	41	0.2408	0.2906	0.3490	0.4187	0.5033	0.6083
	42	0.2294	0.2784	0.3360	0.4049	0.4888	0.5930

Table 11.5 Values of K_a [Eq. (11.34)] (Note: $\delta' = \phi'/2$)

α (deg)	φ′ (deg)	θ (deg) 0	5	10	15	20	25
0	28	0.3264	0.3629	0.4034	0.4490	0.5011	0.5616
	29	0.3137	0.3502	0.3907	0.4363	0.4886	0.5492
	30	0.3014	0.3379	0.3784	0.4241	0.4764	0.5371
	31	0.2896	0.3260	0.3665	0.4121	0.4645	0.5253
	32	0.2782	0.3145	0.3549	0.4005	0.4529	0.5137
	33	0.2671	0.3033	0.3436	0.3892	0.4415	0.5025
	34	0.2564	0.2925	0.3327	0.3782	0.4305	0.4915
	35	0.2461	0.2820	0.3221	0.3675	0.4197	0.4807

α (deg)	ϕ' (deg)	θ (deg) 0	5	10	15	20	25
	36	0.2362	0.2718	0.3118	0.3571	0.4092	0.4702
	37	0.2265	0.2620	0.3017	0.3469	0.3990	0.4599
	38	0.2172	0.2524	0.2920	0.3370	0.3890	0.4498
	39	0.2081	0.2431	0.2825	0.3273	0.3792	0.4400
	40	0.1994	0.2341	0.2732	0.3179	0.3696	0.4304
	41	0.1909	0.2253	0.2642	0.3087	0.3602	0.4209
	42	0.1828	0.2168	0.2554	0.2997	0.3511	0.4117
5	28	0.3477	0.3879	0.4327	0.4837	0.5425	0.6115
	29	0.3337	0.3737	0.4185	0.4694	0.5282	0.5972
	30	0.3202	0.3601	0.4048	0.4556	0.5144	0.5833
	31	0.3072	0.3470	0.3915	0.4422	0.5009	0.5698
	32	0.2946	0.3342	0.3787	0.4292	0.4878	0.5566
	33	0.2825	0.3219	0.3662	0.4166	0.4750	0.5437
	34	0.2709	0.3101	0.3541	0.4043	0.4626	0.5312
	35	0.2596	0.2986	0.3424	0.3924	0.4505	0.5190
	36	0.2488	0.2874	0.3310	0.3808	0.4387	0.5070
	37	0.2383	0.2767	0.3199	0.3695	0.4272	0.4954
	38	0.2282	0.2662	0.3092	0.3585	0.4160	0.4840
	39	0.2185	0.2561	0.2988	0.3478	0.4050	0.4729
	40	0.2090	0.2463	0.2887	0.3374	0.3944	0.4620
	41	0.1999	0.2368	0.2788	0.3273	0.3840	0.4514
	42	0.1911	0.2276	0.2693	0.3174	0.3738	0.4410
10	28	0.3743	0.4187	0.4688	0.5261	0.5928	0.6719
	29	0.3584	0.4026	0.4525	0.5096	0.5761	0.6549
	30	0.3432	0.3872	0.4368	0.4936	0.5599	0.6385
	31	0.3286	0.3723	0.4217	0.4782	0.5442	0.6225
	32	0.3145	0.3580	0.4071	0.4633	0.5290	0.6071
	33	0.3011	0.3442	0.3930	0.4489	0.5143	0.5920
	34	0.2881	0.3309	0.3793	0.4350	0.5000	0.5775
	35	0.2757	0.3181	0.3662	0.4215	0.4862	0.5633
	36	0.2637	0.3058	0.3534	0.4084	0.4727	0.5495
	37	0.2522	0.2938	0.3411	0.3957	0.4597	0.5361
	38	0.2412	0.2823	0.3292	0.3833	0.4470	0.5230
	39	0.2305	0.2712	0.3176	0.3714	0.4346	0.5103
	40	0.2202	0.2604	0.3064	0.3597	0.4226	0.4979
	41	0.2103	0.2500	0.2956	0.3484	0.4109	0.4858
	42	0.2007	0.2400	0.2850	0.3375	0.3995	0.4740
15	28	0.4095	0.4594	0.5159	0.5812	0.6579	0.7498
	29	0.3908	0.4402	0.4964	0.5611	0.6373	0.7284
	30	0.3730	0.4220	0.4777	0.5419	0.6175	0.7080
	31	0.3560	0.4046	0.4598	0.5235	0.5985	0.6884
	32	0.3398	0.3880	0.4427	0.5059	0.5803	0.6695
	33	0.3244	0.3721	0.4262	0.4889	0.5627	0.6513
	34	0.3097	0.3568	0.4105	0.4726	0.5458	0.6338
	35	0.2956	0.3422	0.3953	0.4569	0.5295	0.6168
	36	0.2821	0.3282	0.3807	0.4417	0.5138	0.6004
	37	0.2692	0.3147	0.3667	0.4271	0.4985	0.5846

(continued)

Table 11.5 (*continued*)

		θ (deg)					
α (deg)	ϕ' (deg)	0	5	10	15	20	25
	38	0.2569	0.3017	0.3531	0.4130	0.4838	0.5692
	39	0.2450	0.2893	0.3401	0.3993	0.4695	0.5543
	40	0.2336	0.2773	0.3275	0.3861	0.4557	0.5399
	41	0.2227	0.2657	0.3153	0.3733	0.4423	0.5258
	42	0.2122	0.2546	0.3035	0.3609	0.4293	0.5122
20	28	0.4614	0.5188	0.5844	0.6608	0.7514	0.8613
	29	0.4374	0.4940	0.5586	0.6339	0.7232	0.8313
	30	0.4150	0.4708	0.5345	0.6087	0.6968	0.8034
	31	0.3941	0.4491	0.5119	0.5851	0.6720	0.7772
	32	0.3744	0.4286	0.4906	0.5628	0.6486	0.7524
	33	0.3559	0.4093	0.4704	0.5417	0.6264	0.7289
	34	0.3384	0.3910	0.4513	0.5216	0.6052	0.7066
	35	0.3218	0.3736	0.4331	0.5025	0.5851	0.6853
	36	0.3061	0.3571	0.4157	0.4842	0.5658	0.6649
	37	0.2911	0.3413	0.3991	0.4668	0.5474	0.6453
	38	0.2769	0.3263	0.3833	0.4500	0.5297	0.6266
	39	0.2633	0.3120	0.3681	0.4340	0.5127	0.6085
	40	0.2504	0.2982	0.3535	0.4185	0.4963	0.5912
	41	0.2381	0.2851	0.3395	0.4037	0.4805	0.5744
	42	0.2263	0.2725	0.3261	0.3894	0.4653	0.5582

Example 11.10

Refer to Figure 11.19. Given: $\alpha = 10°$; $\theta = 5°$; $H = 12$ ft; unit weight of soil, $\gamma = 100$ lb/ft^3; soil friction angle, $\phi' = 30°$; and $\delta' = 15°$: Estimate the active force, P_a, per unit length of the wall. Also, state the direction and location of the resultant force, P_a.

Solution

From Eq. (11.33)

$$P_a = \frac{1}{2}\gamma H^2 K_a$$

For $\phi' = 30°$; $\delta' = 15°$ - that is, $\frac{\delta'}{\phi'} = \frac{15}{30} = \frac{1}{2}$; $\alpha = 10°$; and $\theta = 5°$, the magnitude of K_a is 0.3872 (Table 11.5). So,

$$P_a = \frac{1}{2}(100)(12)^2(0.3872) = \mathbf{2787.8\ lb/ft}$$

The resultant will act at a vertical distance equal to $H/3 = 12/3 = 4$ ft above the bottom of the wall and will be inclined at an angle of 15° (= δ') to be back face of the wall. ■

Example 11.11

For a smooth vertical wall ($\theta = 0$ and $\delta' = 0$) retaining a horizontal ($\alpha = 0$) granular ($c' = 0$) backfill, find the K_a values using Rankine's and Coulomb's expressions for $\phi' = 30°, 34°, 38°$, and $42°$.

Solution

ϕ'(°)	Rankine's K_a [Eq. (11.11)]	Coulomb's K_a (Table 11.3)
30	0.3333	0.3333
34	0.2827	0.2827
38	0.2379	0.2379
42	0.1982	0.1982

For smooth vertical walls retaining horizontal backfills, both methods give the same values for K_a. ■

11.10 Coulomb's Passive Pressure

Figure 11.20a shows a retaining wall with a sloping cohesionless backfill similar to that considered in Figure 11.19a. The force polygon for equilibrium of the wedge *ABC* for the passive state is shown in Figure 11.20b. P_p is the notation for the passive force. Other notations used are the same as those for the active case (Section 11.9). In a procedure similar to the one that we followed in the active case [Eq. (11.33)], we get

$$P_p = \tfrac{1}{2}K_p\gamma H^2 \tag{11.35}$$

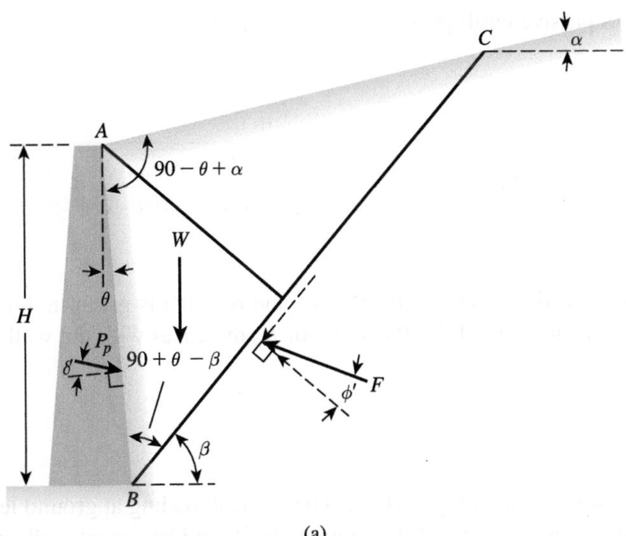

Figure 11.20 Coulomb's passive pressure: (a) trial failure wedge

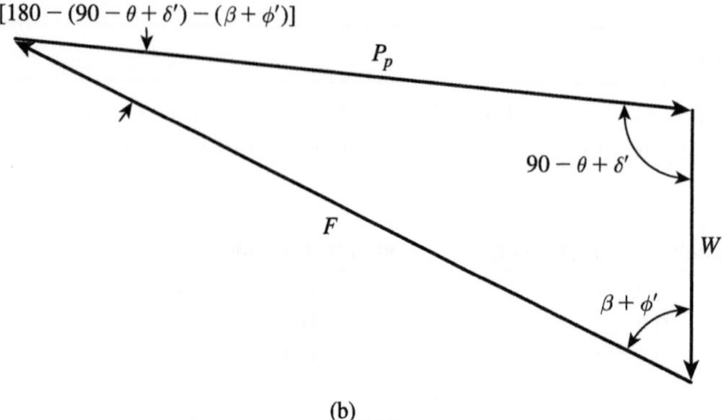

(b)

Figure 11.20 (*continued*) (b) force polygon

Table 11.6 Values of K_p [Eq. 11.36] for $\theta = 0°$, $\alpha = 0°$

↓ φ′ (deg)	δ′ (deg) → 0	5	10	15	20
15	1.698	1.900	2.130	2.405	2.735
20	2.040	2.313	2.636	3.030	3.525
25	2.464	2.830	3.286	3.855	4.597
30	3.000	3.506	4.143	4.977	6.105
35	3.690	4.390	5.310	6.854	8.324
40	4.600	5.590	6.946	8.870	11.772

where K_p = Coulomb's passive earth pressure coefficient, or

$$K_p = \frac{\cos^2(\phi' + \theta)}{\cos^2\theta\cos(\delta' - \phi)\left[1 - \sqrt{\frac{\sin(\phi' + \delta')\sin(\phi' + \alpha)}{\cos(\delta' - \phi)\cos(\alpha - \theta)}}\right]^2} \qquad (11.36)$$

The variation of K_p with ϕ' and δ' (for $\theta = 0°$ and $\alpha = 0°$) is given in Table 11.6. We can see from this table that for given value of ϕ', the value of K_p increases with the wall friction.

11.11 Summary

The stresses within the soils are not only vertical. The vertical loading at ground level induces stresses in all directions within the underlying soil. In retaining walls and basement walls in contact with soils, braced excavations, and sheet piles, it is often necessary to determine the lateral earth pressures acting on them.

Active and passive states are two different failure states in the soil. In the active state, the horizontal stress is reduced until failure occurs. In the passive state, the stress increases until failure occurs. The lateral earth pressure coefficients at active and passive states are denoted by K_a and K_p, respectively. Soils at rest are very stable, where the earth pressure coefficient is K_o.

The lateral earth pressures can be determined by Rankine's or Coulomb's earth pressure theories. Rankine's theory is simpler than Coulomb's. Rankine's theory assumes that the wall is smooth and vertical. Coulomb's theory is more realistic and allows for an inclined wall and for friction along the wall.

Problems

11.1 State whether the following are true or false.

a. The higher the friction angle, the higher the value of K_o.

b. K_o is greater for normally consolidated clays than for overconsolidated clays.

c. The active earth-pressure coefficient is greater than the passive one.

d. The larger the cohesion, the larger is the depth of the tensile cracks in clays in the active state.

e. Lateral earth pressures increase linearly with depth.

11.2 Figure 11.21 shows a retaining wall with cohesionless soil backfill. For cases a through c, determine the following: total active force per unit length of the wall for Rankine's state, the location of the resultant, and the variation of the active pressure with depth.

a. $H = 10$ ft, $\gamma = 110$ lb/ft^3, $\phi' = 32°$

b. $H = 12$ ft, $\gamma = 98$ lb/ft^3, $\phi' = 28°$

c. $H = 18$ ft, $\gamma = 115$ lb/ft^3, $\phi' = 40°$

11.3 Assuming that the wall shown in Figure 11.21 is restrained from yielding, find the magnitude and location of the resultant lateral force per unit length of the wall for the following cases:

a. $H = 8$ ft, $\gamma = 105$ lb/ft^3, $\phi' = 34°$

b. $H = 14$ ft, $\gamma = 108$ lb/ft^3, $\phi' = 36°$

11.4 Referring to Figure 11.21, determine the passive force, P_p, per unit length of the wall for Rankine's state. Also state the Rankine passive pressure at the bottom of the wall. Given the following cases:

a. $H = 10$ ft, $\gamma = 110$ lb/ft^3, $\phi' = 30°$

b. $H = 14$ ft, $\gamma = 120$ lb/ft^3, $\phi' = 36°$

11.5 Referring to Figure 11.21 and given that $H = 5$ m, $\gamma = 18.5$ kN/m^3, and $\phi' = 34°$, determine the location and magnitude of the thrust on the wall for (a) active, (b) at-rest, and (c) passive states.

11.6 A 4-m-high smooth vertical wall retains a granular backfill with $\gamma = 18.0$ kN/m^3 and $\phi' = 33°$, Using Rankine's earth pressure coefficients, determine the active force per unit length P_a for the following cases.

a. Horizontal backfill

b. Inclined backfill with $\alpha = 10°$

c. Inclined backfill with $\alpha = 20°$

11.7 A retaining wall is shown in Figure 11.22. Determine the Rankine active force, P_a, per unit length of the wall and the location of the resultant for each of the following cases:

a. $H = 12$ ft, $H_1 = 4$ ft, $\gamma_1 = 105$ lb/ft^3, $\gamma_2 = 122$ lb/ft^3, $\phi_1' = 30°$, $\phi_2' = 30°$, $q = 0$

b. $H = 20$ ft, $H_1 = 6$ ft, $\gamma_1 = 110$ lb/ft^3, $\gamma_2 = 126$ lb/ft^3, $\phi_1' = 34°$, $\phi_2' = 34°$, $q = 300$ lb/ft^2

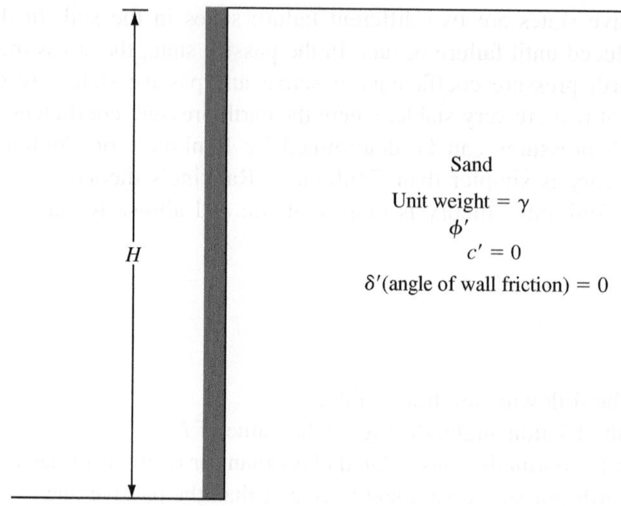

Figure 11.21

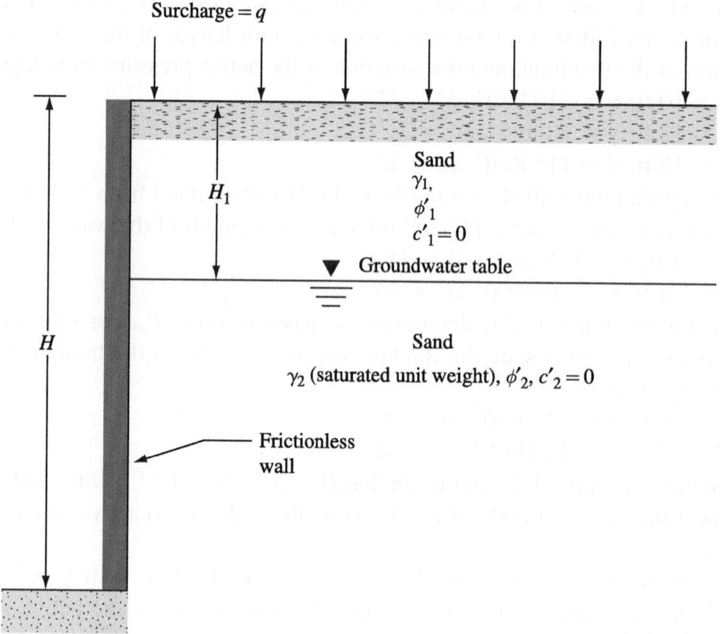

Figure 11.22

11.8 Refer to Figure 11.22. Determine the Rankine passive force, P_p, per unit length of the wall for the following cases. Also find the location of the resultant for each case.

a. $H = 12$ ft, $H_1 = 4$ ft, $\gamma_1 = 105$ lb/ft^3, $\gamma_2 = 122$ lb/ft^3, $\phi_1' = 30°$, $\phi_2' = 30°$, $q = 0$

b. $H = 20$ ft, $H_1 = 6$ ft, $\gamma_1 = 110$ lb/ft^3, $\gamma_2 = 126$ lb/ft^3, $\phi_1' = 34°$, $\phi_2' = 34°$, $q = 300$ lb/ft^2

11.9 Figure 11.23 shows a smooth vertical wall retaining a sandy backfill underlain by clay. Assuming the entire soil is in the active state, determine the magnitude and location of the resultant active force on the wall.

11.10 A retaining wall 14 ft high with a vertical back face retains a homogeneous saturated soft clay. The saturated unit weight of the clay is 124.5 lb/ft^3. Laboratory tests showed that the undrained shear strength, c_u, of the clay is equal to 400 lb/ft^2.

a. Make necessary calculations and draw the variation of Rankine's active pressure on the wall with depth.

b. Find the depth up to which a tensile crack can occur.

c. Determine the total active force per unit length of the wall before the tensile crack occurs.

d. Determine the total active force per unit length of the wall after the occurrence of the tensile crack. Also find the location of the resultant.

11.11 A smooth vertical wall retains a horizontal granular backfill. Determine the Rankine's and Coulomb's active passive pressure coefficients for $\phi' = 30°$, 35°, and 40°.

11.12 Redo Problem 11.10a, b, c, and d assuming that the backfill is supporting a surcharge of 150 lb/ft^2.

11.13 A retaining wall 18 ft high with a vertical back face has a $c' - \phi'$ soil for backfill. For the backfill, given: $\gamma = 118$ lb/ft^3, $c' = 520$ lb/ft, $\phi' = 16°$. Taking the existence of the tensile crack into consideration, determine the active force, P_a, on the wall for Rankine's active state.

11.14 Figure 11.24 shows a gravity retaining wall for a granular backfill ($\gamma = 19.0$ kN/m^3 and $\phi' = 36°$). Assuming the backfill is in the active state, find the magnitude, location, and direction of the resultant active force P_a based on (a) Rankine's and (b) Coulomb's (assume $\delta' = \frac{2}{3}\phi'$) earth pressure theories. Discuss the differences.

11.15 Determine the resultant active thrust on the retaining wall shown in Figure 11.25. The properties of the sandy backfill are $\gamma = 19.5$ kN/m^3, $\phi' = 36°$, and $\delta' = \frac{2}{3}\phi'$. What is the inclination of the resultant active force P_a to horizontal?

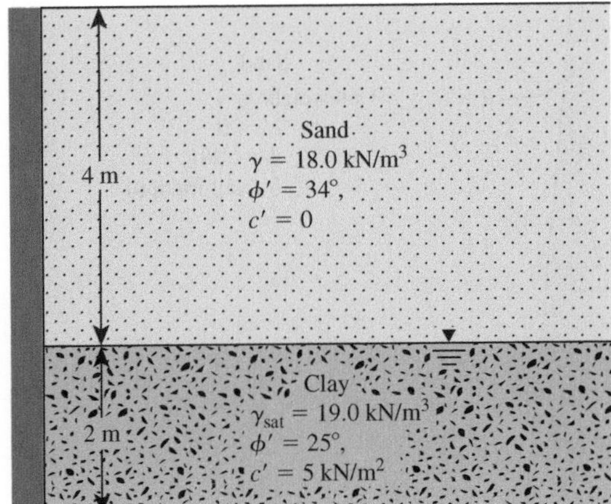

Figure 11.23

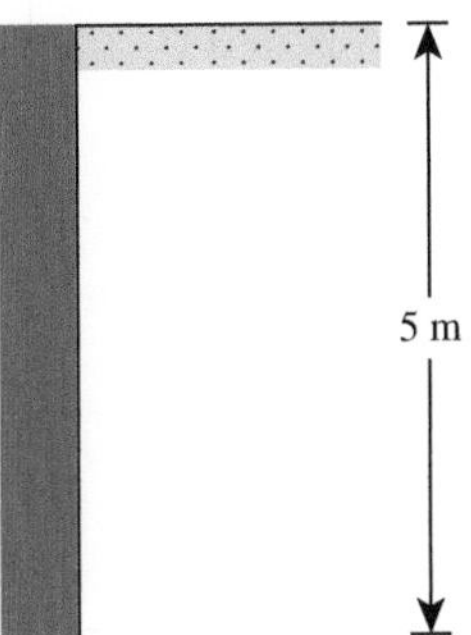

Figure 11.24

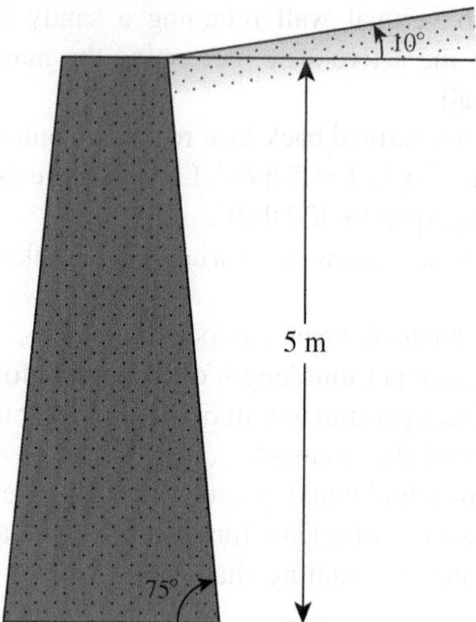

Figure 11.25

11.16 For the wall described in Problem 11.13, determine the passive force, P_p, for Rankine's passive state.

11.17 A retaining wall is shown in Figure 11.26. Given: height of the wall is equal to 16 ft and unit weight of the sand backfill is 114 lb/ft^3; calculate the active force, P_a, on the wall using Coulomb's equation [Eqs. (11.33) and (11.34)] for the following values of the angle of wall friction.

a. $\delta' = 0°$
b. $\delta' = 10°$
c. $\delta' = 20°$

Comment on the direction and location of the resultant.

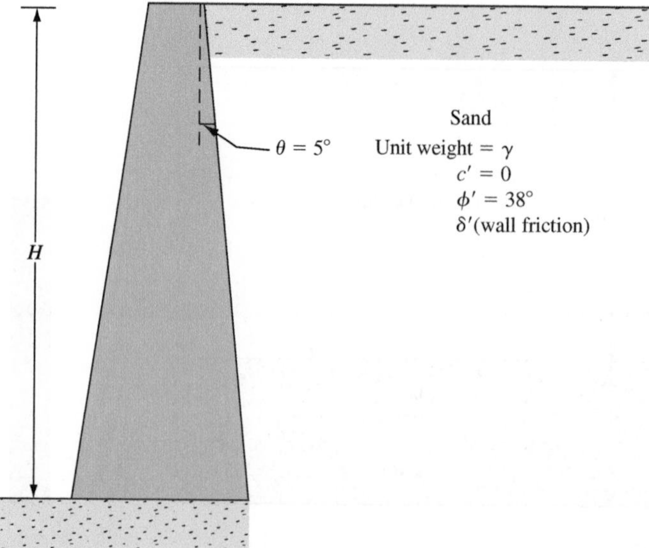

Figure 11.26

11.18 For the retaining wall described in Problem 11.17, determine the passive force, P_p, by using Coulomb's equation [Eqs. (11.35) and (11.36)] for the following values of the angle of wall friction.

a. $\delta' = 0°$
b. $\delta' = 10°$
c. $\delta' = 20°$

References

Coulomb, C. A. (1776). "Essai sur une Application des Règles de Maximis et Minimis à quelques Problèmes de Statique, relatifs a l'Architecture," *Mem. Roy. des Sciences*, Paris, Vol. 3, 38.

Jaky, J. (1944). "The Coefficient of Earth Pressure at Rest," *Journal of the Society of Hungarian Architects and Engineers*, Vol. 7, 355–358.

Rankine, W. M. J. (1857). "On Stability on Loose Earth," *Philosophic Transactions of Royal Society*, London, Part I, 9–27.

12 Shallow Foundations—Bearing Capacity and Settlement

12.1 Introduction

The lowest part of a structure is generally referred to as the *foundation*. Its function is to transfer the load of the structure to the soil on which it is resting. A properly designed foundation is one that transfers the load throughout the soil without overstressing the soil. Overstressing the soil can result in either excessive settlement or shear failure of the soil, both of which cause damage to the structure. Thus, geotechnical and structural engineers who design foundations must evaluate the bearing capacity of soils.

Depending on the structure and soil encountered, various types of foundations are used. A *spread footing* is simply an enlargement of a load-bearing wall or column that makes it possible to spread the load of the structure over a larger area of the soil. In soil with low load-bearing capacity, the size of the spread footings required is impracticably large. In that case, it is more economical to construct the entire structure over a concrete pad. This is called a *mat foundation.*

Pile and *drilled shaft foundations* are used for heavier structures when great depth is required for supporting the load. Piles are structural members made of timber, concrete, or steel that transmit the load of the superstructure to the lower layers of the soil. According to how they transmit their load into the subsoil, piles can be divided into two categories: friction piles and end-bearing piles. In the case of friction piles, the superstructure load is resisted by the shear stresses generated along the surface of the pile. In the end-bearing pile, the load carried by the pile is transmitted at its tip to a firm stratum.

In the case of drilled shafts, a shaft is drilled into the subsoil and is then filled with concrete. A metal casing may be used while the shaft is being drilled. The casing may be left in place or withdrawn during the placing of concrete. Generally, the diameter of a drilled shaft is much larger than that of a pile. The distinction between piles and drilled shafts becomes hazy at an approximate diameter of 3 ft (900 mm), and then the definitions and nomenclature are inaccurate.

Spread footings and mat foundations are generally referred to as shallow foundations, and pile and drilled shaft foundations are classified as deep foundations. In a more general sense, shallow foundations are those foundations that have a depth-of-embedment-to-width ratio of approximately less than four. When the depth-of-embedment-to-width ratio of a foundation is greater than four, it may be classified as a deep foundation.

In this chapter, we discuss the soil-bearing capacity for shallow foundations. As mentioned before, for a foundation to function properly, (1) the settlement of soil caused by the load must be within the tolerable limit, and (2) shear failure of the soil supporting the foundation must not occur.

This chapter introduces the load-carrying capacity of shallow foundations based on the criterion of shear failure in soil; and also the settlement. In this chapter, you will learn the following.

- Difference between shallow and deep foundations.
- Bearing capacity theory for shallow foundations.
- Settlement of a shallow foundation.
- Plate load test, simulating a footing under load.
- Mat foundations.

Ultimate Bearing Capacity of Shallow Foundations

12.2 General Concepts

Consider a strip (i.e., theoretically length is infinity) foundation resting on the surface of a dense sand or stiff cohesive soil, as shown in Figure 12.1a, with a width of B. Now, if load is gradually applied to the foundation, settlement will increase. The variation of the load per unit area on the foundation, q, with the foundation settlement is also shown in Figure 12.1a. At a certain point—when the load per unit area equals q_u—a sudden failure in the soil supporting the foundation will take place, and the failure surface in the soil will extend to the ground surface. This load per unit area, q_u, is usually referred to as the *ultimate bearing capacity of the foundation*. When this type of sudden failure in soil takes place, it is called *general shear failure*.

If the foundation under consideration rests on sand or clayey soil of medium compaction (Figure 12.1b), an increase of load on the foundation will also be accompanied by an increase of settlement. However, in this case the failure surface in the soil will gradually extend outward from the foundation, as shown by the solid lines in Figure 12.1b. When the load per unit area on the foundation equals $q_{u(1)}$, the foundation movement will be accompanied by sudden jerks. A considerable movement of the foundation is then required for the failure surface in soil to extend to the ground surface (as shown by the broken lines in Figure 12.1b). The load per unit area at which this happens is the *ultimate bearing capacity*, q_u. Beyond this point, an increase of load will be accompanied by a large increase of foundation settlement. The load per unit area of the foundation, $q_{u(1)}$, is referred to as the *first failure load* (Vesic, 1963). Note that a peak value of q is not realized in this type of failure, which is called *local shear failure* in soil.

If the foundation is supported by a fairly loose soil, the load–settlement plot will be like the one in Figure 12.1c. In this case, the failure surface in soil will not extend to the ground surface. Beyond the ultimate failure load, q_u, the load-settlement plot will be steep and practically linear. This type of failure in soil is called *punching shear failure*.

Based on experimental results, Vesic (1963) proposed a relationship for the mode of bearing capacity failure of foundations resting on sands. Figure 12.2 shows this relationship, which involves the following notation:

D_r = relative density of sand
D_f = depth of foundation measured from the ground surface
B = width of foundation
L = length of foundation

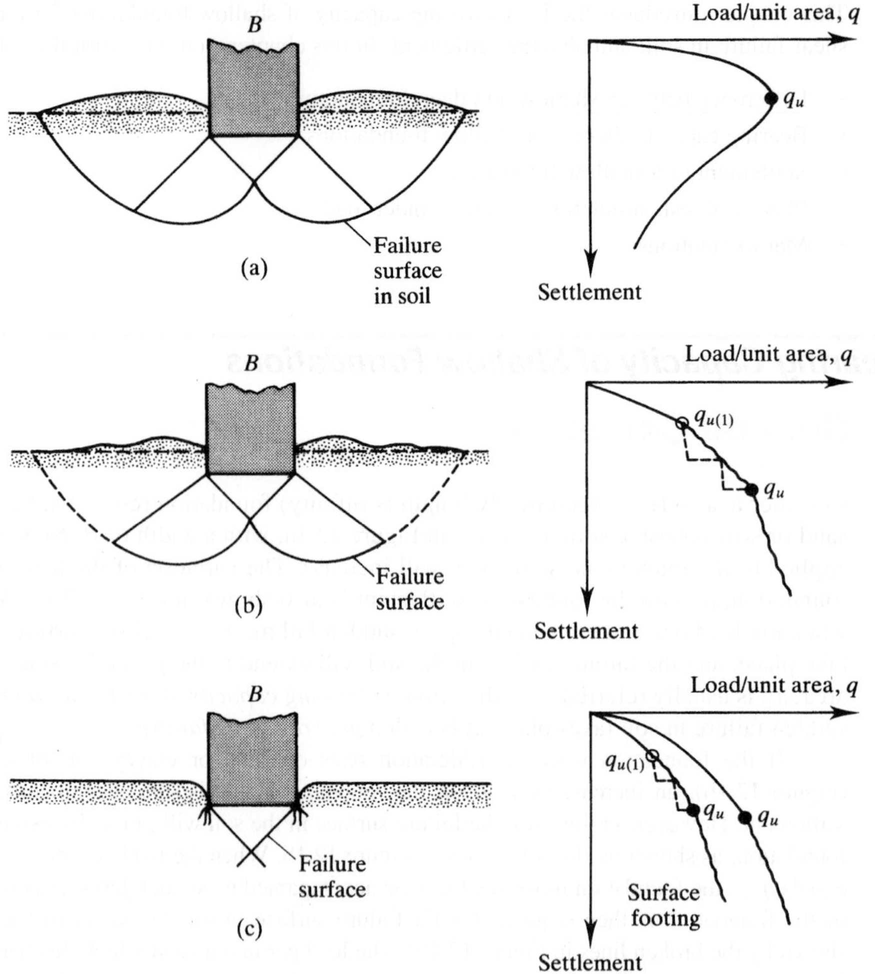

Figure 12.1 Nature of bearing capacity failure in soil: (a) general shear failure; (b) local shear failure; (c) punching shear failure

From Figure 12.2, it can be seen that

$$\text{Nature of failure in soil} = f\left(D_r, \frac{D_f}{B}, \frac{B}{L}\right) \tag{12.1}$$

For foundations at a shallow depth (that is, small D_f/B^*), the ultimate load may occur at a foundation settlement of 4% to 10% of B. This condition occurs with general shear failure in soil; however, with local or punching shear failure, the ultimate load may occur in settlements of 15% to 25% of the width of foundation (B). Note that

$$B^* = \frac{2BL}{B + L} \tag{12.2}$$

where B^* varies between B and $2B$, depending on the B/L ratio.

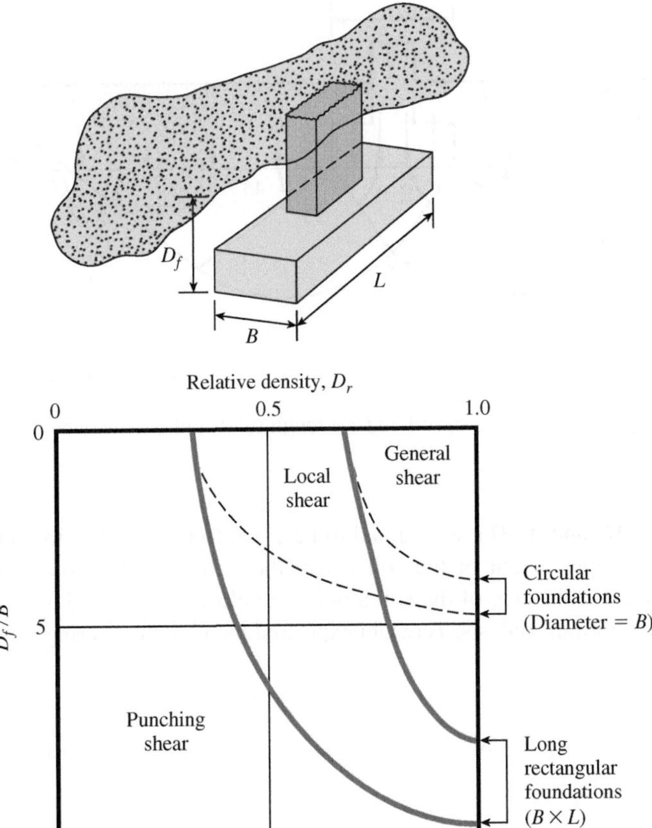

Figure 12.2 Vesic's (1963) test results for modes of foundation failure in sand

12.3 Ultimate Bearing Capacity Theory

Terzaghi (1943) was the first to present a comprehensive theory for evaluating the ultimate bearing capacity of rough shallow foundations. According to this theory, a foundation is *shallow* if the depth, D_f (Figure 12.3), of the foundation is less than or equal to the width of the foundation. Later investigators, however, have suggested that foundations with D_f equal to 3 to 4 times the width of the foundation may be defined as *shallow foundations*.

Terzaghi suggested that for a *continuous*, or *strip, foundation* (that is, the width-to-length ratio of the foundation approaches 0), the failure surface in soil at ultimate load may be assumed to be similar to that shown in Figure 12.3. (Note that this is the case of general shear failure as defined in Figure 12.1a.) The effect of soil above the bottom of the foundation may also be assumed to be replaced by an equivalent surcharge, $q = \gamma D_f$ (where γ = unit weight of soil). The failure zone under the foundation can be separated into three parts (see Figure 12.3):

1. *The triangular zone ACD* immediately under the foundation
2. The *radial shear zones ADF* and *CDE*, with the curves *DE* and *DF* being arcs of a logarithmic spiral
3. Two *triangular Rankine passive zones AFH* and *CEG*

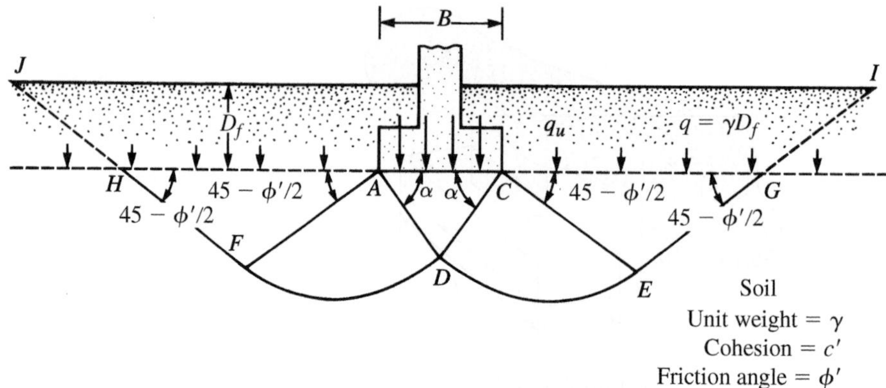

Figure 12.3 Bearing capacity failure in soil under a rough rigid continuous foundation

The angles *CAD* and *ACD* are assumed to be equal to the soil friction angle (that is, $\alpha = \phi'$). Note that, with the replacement of the soil above the bottom of the foundation by an equivalent surcharge q, the shear resistance of the soil along the failure surfaces *GI* and *HJ* was neglected.

Using the equilibrium analysis, Terzaghi expressed the ultimate bearing capacity in the form

$$q_u = c'N_c + qN_q + \frac{1}{2}\gamma BN_\gamma \qquad \text{(strip foundation)} \tag{12.3}$$

where

c' = cohesion of soil
γ = unit weight of soil
$q = \gamma D_f$
N_c, N_q, N_γ = bearing capacity factors that are nondimensional and are only functions of the soil friction angle, ϕ'

Based on laboratory and field studies of bearing capacity, the basic nature of the failure surface in soil suggested by Terzaghi now appears to be correct (Vesic, 1973). However, the angle α shown in Figure 12.3 is closer to $45 + \phi'/2$ than to ϕ', as was originally assumed by Terzaghi. With $\alpha = 45 + \phi'/2$, the relations for N_c and N_q can be derived as

$$N_q = \tan^2\left(45 + \frac{\phi'}{2}\right)e^{\pi \tan \phi} \tag{12.4}$$

$$N_c = (N_q - 1)\cot \phi' \tag{12.5}$$

The equation for N_c given by Eq. (12.5) was originally derived by Prandtl (1921), and the relation for N_q [Eq. (12.4)] was presented by Reissner (1924). Caquot and Kerisel (1953) and Vesic (1973) gave the relation for N_γ as

$$N_\gamma = 2(N_q + 1)\tan\phi' \tag{12.6}$$

Table 12.1 shows the variation of the preceding bearing capacity factors with soil friction angles.

The ultimate bearing capacity expression presented in Eq. (12.3) is for a continuous foundation only. It does not apply to the case of rectangular foundations. Also, the equation does not take into account the shearing resistance along the failure surface in soil above the bottom of the foundation (portion of the failure surface marked as *GI* and *HJ* in Figure 12.3). In addition, the load on the foundation may be inclined. To account for all these shortcomings, Meyerhof (1963) suggested the following form of the general bearing capacity equation:

$$q_u = c'N_cF_{cs}F_{cd}F_{ci} + qN_qF_{qs}F_{qd}F_{qi} + \frac{1}{2}\gamma BN_\gamma F_{\gamma s}F_{\gamma d}F_{\gamma i} \tag{12.7}$$

Table 12.1 Bearing Capacity Factors [Eqs. (12.4), (12.5), and (12.6)]

ϕ'	N_c	N_q	N_γ	ϕ'	N_c	N_q	N_γ
0	5.14	1.00	0.00	23	18.05	8.66	8.20
1	5.38	1.09	0.07	24	19.32	9.60	9.44
2	5.63	1.20	0.15	25	20.72	10.66	10.88
3	5.90	1.31	0.24	26	22.25	11.85	12.54
4	6.19	1.43	0.34	27	23.94	13.20	14.47
5	6.49	1.57	0.45	28	25.80	14.72	16.72
6	6.81	1.72	0.57	29	27.86	16.44	19.34
7	7.16	1.88	0.71	30	30.14	18.40	22.40
8	7.53	2.06	0.86	31	32.67	20.63	25.99
9	7.92	2.25	1.03	32	35.49	23.18	30.22
10	8.35	2.47	1.22	33	38.64	26.09	35.19
11	8.80	2.71	1.44	34	42.16	29.44	41.06
12	9.28	2.97	1.69	35	46.12	33.30	48.03
13	9.81	3.26	1.97	36	50.59	37.75	56.31
14	10.37	3.59	2.29	37	55.63	42.92	66.19
15	10.98	3.94	2.65	38	61.35	48.93	78.03
16	11.63	4.34	3.06	39	67.87	55.96	92.25
17	12.34	4.77	3.53	40	75.31	64.20	109.41
18	13.10	5.26	4.07	41	83.86	73.90	130.22
19	13.93	5.80	4.68	42	93.71	85.38	155.55
20	14.83	6.40	5.39	43	105.11	99.02	186.54
21	15.82	7.07	6.20	44	118.37	115.31	224.64
22	16.88	7.82	7.13	45	133.88	134.88	271.76

where

c' = cohesion
q = effective stress at the level of the bottom of the foundation
γ = unit weight of soil
B = width of foundation (= diameter for a circular foundation)
$F_{cs}, F_{qs}, F_{\gamma s}$ = shape factors
$F_{cd}, F_{qd}, F_{\gamma d}$ = depth factors
$F_{ci}, F_{qi}, F_{\gamma i}$ = load inclination factors
N_c, N_q, N_γ = bearing capacity factors

The relationships for the shape factors, depth factors, and inclination factors *recommended for use* are given in Table 12.2. The bearing capacity equation [Eq. (12.7)] also can be applied to undrained clays with $\phi = 0$.

Table 12.2 Shape, Depth, and Inclination Factors Recommended for Use

Factor	Relationship	Source
Shape*	$F_{cs} = 1 + \frac{B}{L}\frac{N_q}{N_c}$ $F_{qs} = 1 + \frac{B}{L}\tan\phi'$ $F_{\gamma s} = 1 - 0.4\frac{B}{L}$ where L = length of the foundation ($L > B$)	De Beer (1970)
Depth†	*Condition (a)*: $D_f/B \le 1$ $F_{cd} = 1 + 0.4\frac{D_f}{B}$ $F_{qd} = 1 + 2\tan\phi'(1 - \sin\phi')^2\frac{D_f}{B}$ $F_{\gamma d} = 1$ *Condition (b)*: $D_f/B > 1$ $F_{cd} = 1 + (0.4)\tan^{-1}\left(\frac{D_f}{B}\right)$ $F_{qd} = 1 + 2\tan\phi'(1 - \sin\phi')^2\tan^{-1}\left(\frac{D_f}{B}\right)$ $F_{\gamma d} = 1$	Hansen (1970)
Inclination	$F_{ci} = F_{qi} = \left(1 - \frac{\beta^\circ}{90^\circ}\right)^2$ $F_{\gamma i} = \left(1 - \frac{\beta}{\phi'}\right)^2$ where β = inclination of the load on the foundation with respect to the vertical	Meyerhof (1963); Hanna and Meyerhof (1981)

*These shape factors are empirical relations based on extensive laboratory tests.
†The factor $\tan^{-1}(D_f/B)$ is in radians.

Net Ultimate Bearing Capacity

The net ultimate bearing capacity is defined as the ultimate pressure per unit area of the foundation that can be supported by the soil in excess of the pressure caused by the surrounding soil at the foundation level. If the difference between the unit weight of concrete used in the foundation and the unit weight of soil surrounding the foundation is assumed to be negligible, then

$$q_{\text{net}(u)} = q_u - q \tag{12.8}$$

where $q_{\text{net}(u)}$ = net ultimate bearing capacity.

The expressions given in Table 12.2 are also valid for $\phi = 0$ (undrained clays).

12.4 Modification of Bearing Capacity Equations for Water Table

Equation (12.7) was developed for determining the ultimate bearing capacity based on the assumption that the water table is located well below the foundation. However, if the water table is close to the foundation, some modifications of the bearing capacity equation are necessary, depending on the location of the water table (see Figure 12.4).

Case 1: If the water table is located so that $0 \leq D_1 \leq D_f$, the factor q in the bearing capacity equations takes the form

$$q = \text{effective surcharge} = D_1\gamma + D_2(\gamma_{\text{sat}} - \gamma_w) \tag{12.9}$$

where

γ_{sat} = saturated unit weight of soil
γ_w = unit weight of water

Also, the value of γ in the last term of the equations has to be replaced by $\gamma' = \gamma_{\text{sat}} - \gamma_w$.

Case 2: For a water table located so that $0 \leq d \leq B$,

$$q = \gamma D_f \tag{12.10}$$

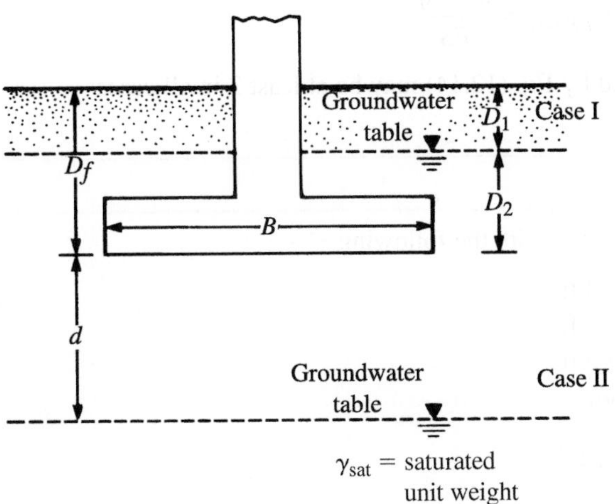

Figure 12.4 Modification of bearing capacity equations for water table

The factor γ in the last term of the bearing capacity equations must be replaced by the factor

$$\bar{\gamma} = \gamma' + \frac{d}{B}(\gamma - \gamma') \tag{12.11}$$

The preceding modifications are based on the assumption that there is no seepage force in the soil.

Case 3: When the water table is located so that $d \geq B$, the water will have no effect on the ultimate bearing capacity.

12.5 The Factor of Safety

Calculating the gross allowable load-bearing capacity of shallow foundations requires the application of a factor of safety FS to the gross ultimate bearing capacity, or

$$q_{\text{all}} = \frac{q_u}{\text{FS}} \tag{12.12}$$

However, some practicing engineers prefer to use a factor of safety of

$$\text{Net stress increase on soil} = \frac{\text{net ultimate bearing capacity}}{\text{FS}} \tag{12.13}$$

The net ultimate bearing capacity was defined in Eq. (12.8) as

$$q_{\text{net}(u)} = q_u - q$$

Substituting this equation into Eq. (12.13) yields

$$\begin{aligned}
&\text{Net stress increase on soil} \\
&\quad = \text{load from the superstructure per unit area of the foundation} \\
&\quad = q_{\text{all(net)}} = \frac{q_u - q}{\text{FS}}
\end{aligned} \tag{12.14}$$

The factor of safety defined by Eq. (12.14) may be at least 3 in all cases.

Example 12.1

Consider a shallow foundation with the following:

- Foundation: $B = 3$ ft
  $L = 5$ ft
  $D_f = 3$ ft
- Soil: $\gamma = 110\ \text{lb/ft}^3$
  $\phi' = 25°$
  $c' = 400\ \text{lb/ft}^2$

Determine the gross allowable load the foundation can carry. Use FS = 4.

Solution

From Eq. (12.7), noting that $F_{ci} = F_{qi} = F_{\gamma i} = 1$

$$q_u = c'N_cF_{cs}F_{cd} + qN_qF_{qs}F_{qd} + \frac{1}{2}\gamma BN_\gamma F_{\gamma s}F_{\gamma q}$$

For $\phi' = 25°$, from Table 12.1,

$N_c = 20.72$
$N_q = 10.66$
$N_\gamma = 10.88$
$q = \gamma D_f = (110)(3) = 330 \text{ lb/ft}^2$

From Table 12.2,

$$F_{cs} = 1 + \frac{B\,N_q}{L\,N_c} = 1 + \left(\frac{3}{5}\right)\left(\frac{10.66}{20.72}\right) = 1.309$$

$$F_{qs} = 1 + \frac{B}{L}\tan\phi' = 1 + \left(\frac{3}{5}\right)\tan 25 = 1.28$$

$$F_{\gamma s} = 1 - 0.4\frac{B}{L} = 1 - (0.4)\left(\frac{3}{5}\right) = 0.76$$

$$F_{cd} = 1 + 0.4\frac{D_f}{B} = 1 + (0.4)\left(\frac{3}{3}\right) = 1.4$$

$$F_{qd} = 1 + 2\tan\phi'(1 - \sin\phi')^2\frac{D_f}{B} = 1 + 2\tan 25(1 - \sin 25)^2\left(\frac{3}{3}\right) = 1.311$$

$$F_{\gamma d} = 1$$

$$q_u = (400)(20.72)(1.309)(1.4) + (330)(10.66)(1.28)(1.311) + \tfrac{1}{2}(110)(3)(10.88)(0.76)(1)$$

$$= 15{,}188.6 + 5903.1 + 1364.4 = 22{,}456.1 \text{ lb/ft}^2$$

$$q_{all} = \frac{q_u}{FS} = \frac{22{,}456.1}{4} \approx 5614 \text{ lb}$$

$$Q_{all} = (q_{all})(BL) = (5614)(3 \times 5) = 84{,}210 \text{ lb} \approx \mathbf{84.2\ kip}$$

■

Example 12.2

A 2.0 m wide strip foundation carries a wall load of 350 kN/m in a soil where $\gamma = 19.0$ kN/m^3, $c' = 5$ kN/m^2 and $\phi' = 23°$. The foundation depth is 1.5 m. Determine the factor of safety of this foundation.

Solution

For $\phi' = 23°$ from Table 12.1, $N_c = 18.05$, $N_q = 8.66$, and $N_\gamma = 8.20$. Given: $B = 2.0$ m and $D_f =$ 1.5 m. For strip foundation, $L \gg B$. Hence $B/L \approx 0$.

Shape Factors

$$F_{cs} = 1 + \frac{B}{L}\frac{N_q}{N_c} \approx 1$$

$$F_{qs} = 1 + \frac{B}{L}\tan\phi' \approx 1$$

$$F_{\gamma s} = 1 - 0.4\frac{B}{L} \approx 1$$

Note that all three shape factors are 1.0 for all strip foundations.

Depth Factors

$$F_{cd} = 1 + 0.4\frac{D_f}{B} = 1 + 0.4\frac{1.5}{2.0} = 1.30$$

$$F_{qd} = 1 + 2\tan\phi'(1 - \sin\phi')^2\frac{D_f}{B} = 1 + 2\tan 25(1 - \sin 25)^2\frac{1.5}{2} = 1.23$$

$$F_{\gamma d} = 1.00$$

Since there is no inclination in the load, all three inclination factors (F_{ci}, F_{qi}, and $F_{\gamma i}$) are unity.

From Eq. (12.7), the ultimate bearing capacity is given by

$$q_u = c'N_cF_{cs}F_{cd}F_{ci} + qN_qF_{qs}F_{qd}F_{qi} + 0.5\gamma B N_\gamma F_{\gamma s}F_{\gamma d}F_{\gamma i}$$

$$= (5.0)(18.05)(1)(1.30)(1) + (1.5 \times 19.0)(8.66)(1)(1.23)(1) + 0.5(19.0)(2)(8.20)(1)(1)(1)$$

$$= 576.7 \text{ kN/m}^2$$

The pressure applied on the soil is $350/2 = 175.0$ kN/m^2. Therefore, the factor of safety is

$$\text{FS} = \frac{576.7}{175} = \mathbf{3.30}$$

■

Example 12.3

A square column foundation to be constructed on a sandy soil has to carry a gross allowable inclined load of 59.6 kip. The depth of the foundation will be 2.3 ft. The load will be inclined at an angle of 20° to the vertical (Figure 12.5).

Assume that the unit weight of the soil is 114.5 lb/ft^3. Determine the width of the foundation, *B*. Use a factor of safety of 3.

Solution

With $c' = 0$, the ultimate bearing capacity [Eq. (12.7)] becomes

$$q_u = qN_qF_{qs}F_{qd}F_{qi} + \frac{1}{2}\gamma BN_\gamma F_{\gamma s}F_{\gamma d}F_{\gamma i}$$

$$q = (2.3)(114.5) = 263.35 \text{ lb/ft}^2$$

$$\gamma = 114.5 \text{ lb/ft}^3$$

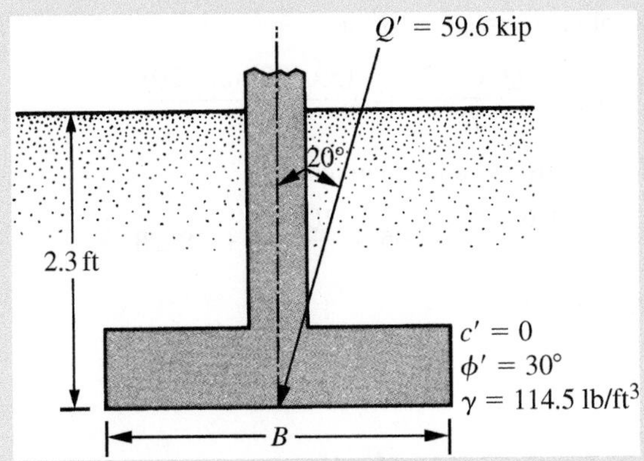

Figure 12.5

From Table 12.1, for $\phi' = 30°$, we find

$$N_q = 18.4$$
$$N_\gamma = 22.4$$

From Table 12.2,

$$F_{qs} = 1 + \left(\frac{B}{L}\right)\tan\phi' = 1 + 0.577 = 1.577$$

$$F_{\gamma s} = 1 - 0.4\left(\frac{B}{L}\right) = 0.6$$

$$F_{qd} = 1 + 2\tan\phi'(1-\sin\phi')^2\frac{D_f}{B} = 1 + \frac{(0.289)(2.3)}{B} = 1 + \frac{0.665}{B}$$

$$F_{\gamma d} = 1$$

$$F_{qi} = \left(1 - \frac{\beta°}{90°}\right)^2 = \left(1 - \frac{20}{90}\right)^2 = 0.605$$

$$F_{\gamma i} = \left(1 - \frac{\beta°}{\phi'}\right)^2 = \left(1 - \frac{20}{30}\right)^2 = 0.11$$

Hence,

$$q_u = (263.35)(18.4)(1.577)\left(1 + \frac{0.665}{B}\right)(0.605) + (0.5)(114.5)(B)(22.4)(0.6)(1)(0.11)$$

$$= 4623.15 + \frac{3074.4}{B} + 84.6B \qquad \textbf{(a)}$$

Thus,

$$q_{\text{all}} = \frac{q_u}{3} = 1541.05 + \frac{1024.8}{B} + 28.2B \qquad \textbf{(b)}$$

For Q = total vertical allowable load = $q_{all} \times B^2$ or

$$q_{all} = \frac{Q' \cos 20}{B^2} = \frac{(59{,}600)(\cos 20)}{B^2} \qquad \textbf{(c)}$$

Equating the right-hand sides of Eqs. (b) and (c) gives

$$\frac{56{,}006}{B^2} = 1541.05 + \frac{1024.8}{B} + 28.2B$$

By trial and error, we find $B \approx$ **5.5 ft** ■

12.6 Eccentrically Loaded Foundations

As with the base of a retaining wall, there are several instances in which foundations are subjected to moments in addition to the vertical load, as shown in Figure 12.6a. In such cases, the distribution of pressure by the foundation on the soil is not uniform. It is assumed to linearly vary along the width from q_{max} to q_{min}– the maximum and minimum values respectively. The distribution of nominal pressure is

$$q_{max} = \frac{Q}{BL} + \frac{6M}{B^2L} \qquad (12.15)$$

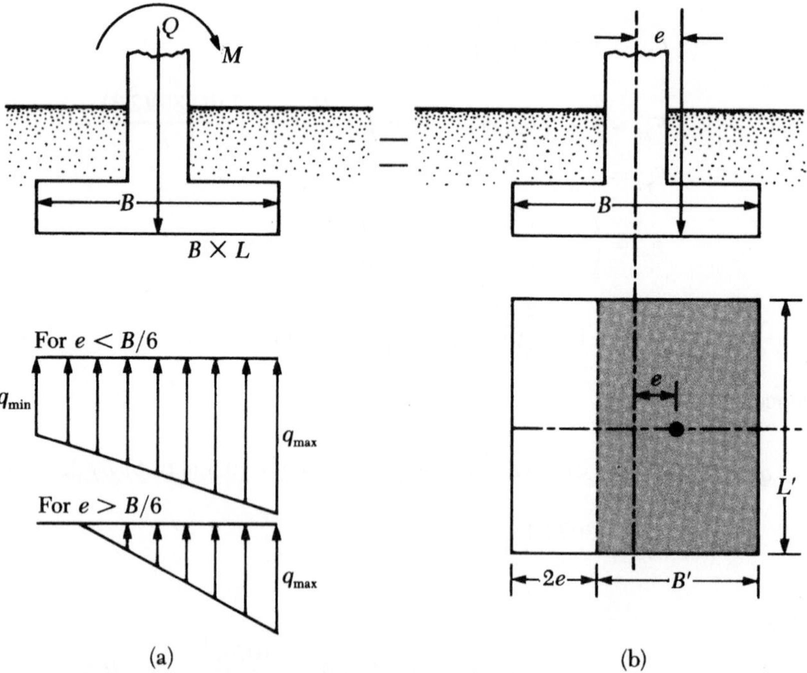

Figure 12.6 Eccentrically loaded foundations

and

$$q_{min} = \frac{Q}{BL} - \frac{6M}{B^2L} \tag{12.16}$$

where

Q = total vertical load
M = moment on the foundation

The exact distribution of pressure is difficult to estimate.

The factor of safety for such types of loading against bearing capacity failure can be evaluated using the procedure suggested by Meyerhof (1953), which is generally referred to as the *effective area* method. The following is Meyerhof's step-by-step procedure for determining the ultimate load that the soil can support and the factor of safety against bearing capacity failure.

Step 1: Figure 12.6b shows a force system equivalent to that shown in Figure 12.6a. The distance e is the eccentricity, or

$$e = \frac{M}{Q} \tag{12.17}$$

Substituting Eq. (12.17) in Eqs. (12.15) and (12.16) gives

$$q_{max} = \frac{Q}{BL}\left(1 + \frac{6e}{B}\right) \tag{12.18}$$

and

$$q_{min} = \frac{Q}{BL}\left(1 - \frac{6e}{B}\right) \tag{12.19}$$

Note that, in these equations, when the eccentricity e becomes $B/6$, q_{min} is 0. For $e > B/6$, q_{min} will be negative, which means that tension will develop. Because soil cannot take any tension, there will be a separation between the foundation and the soil underlying it. A redistribution in the contact pressure takes place. The nature of the pressure distribution on the soil will be as shown in Figure 12.6a. The value of q_{max} is difficult to determine when $e > B/6$.

Step 2: Determine the effective dimensions of the foundation as

$$B' = \text{effective width} = B - 2e$$
$$L' = \text{effective length} = L$$

Note that, if the eccentricity were in the direction of the length of the foundation, then the value of L' would be equal to $L - 2e$. The value of B' would equal B. The smaller of the two dimensions (that is, L' and B') is the effective width of the foundation.

Step 3: Use Eq. (12.7) for the ultimate bearing capacity as

$$q'_u = c'N_c F_{cs} F_{cd} F_{ci} + qN_q F_{qs} F_{qd} F_{qi} + \frac{1}{2}\gamma B' N_\gamma F_{\gamma s} F_{\gamma d} F_{\gamma i} \tag{12.20}$$

To evaluate F_{cs}, F_{qs}, and $F_{\gamma s}$, use Table 12.2 with *effective length* and *effective width* dimensions instead of L and B, respectively. To determine F_{cd}, F_{qd}, and $F_{\gamma d}$, use Table 12.2 (*do not* replace B with B').

Step 4: The total ultimate load that the foundation can sustain is

$$Q_{\text{ult}} = q'_u \overbrace{(B')(L')}^{A'} \tag{12.21}$$

where A' = effective area.

Step 5: The factor of safety against bearing capacity failure is

$$\text{FS} = \frac{Q_{\text{ult}}}{Q} \tag{12.22}$$

Example 12.4

A continuous foundation is shown in Figure 12.7. If the load eccentricity is 0.5 ft, determine the ultimate load, Q_{ult}, per unit length of the foundation.

Solution

For $c' = 0$, Eq. (12.20) gives

$$q'_u = qN_q F_{qs} F_{qd} F_{qi} + \frac{1}{2}\gamma B' N_\gamma F_{\gamma s} F_{\gamma d} F_{\gamma i}$$

$$q = (110)(4) = 440 \text{ lb/ft}^2$$

For $\phi' = 35°$, from Table 12.1, $N_q = 33.3$ and $N_\gamma = 48.03$

$$B' = 6 - (2)(0.5) = 5 \text{ ft}$$

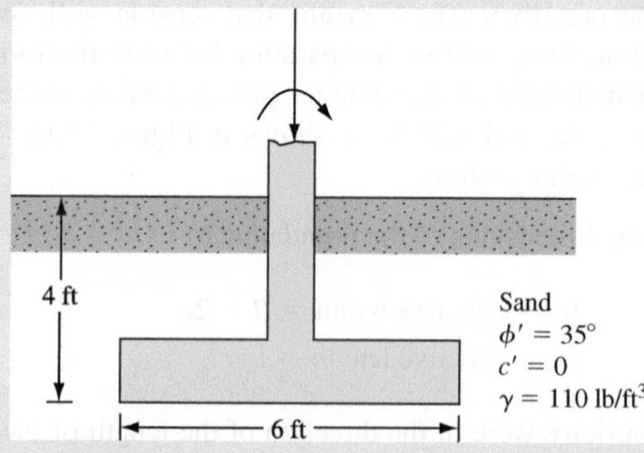

Figure 12.7

Because it is a strip foundation, B'/L' is zero. Hence $F_{qs} = 1$ and $F_{\gamma s} = 1$, and

$$F_{qi} = F_{\gamma i} = 1$$

From Table 12.2,

$$F_{qd} = 1 + 2\tan\phi'(1 - \sin\phi')^2 \frac{D_f}{B} = 1 + 0.255\left(\frac{4}{6}\right) = 1.17$$

$$F_{\gamma d} = 1$$

$$q'_u = (440)(33.3)(1)(1.17)(1) + \left(\frac{1}{2}\right)(110)(5)(48.03)(1)(1)(1) = 30{,}351 \text{ lb/ft}^2$$

Hence,

$$Q_{\text{ult}} = (B')(1)(q'_u) = (5)(1)(30{,}351) = 151{,}755 \text{ lb/ft} = \mathbf{75.88 \text{ ton/ft}}$$

■

Example 12.5

A 2 m × 3 m spread foundation placed at a depth of 2 m carries a vertical load of 3000 kN and moment of 300 kN·m, as shown in Figure 12.8, Determine the factor of safety.

Solution

From Eq. (12.17), eccentricity $e = \frac{M}{Q} = \frac{300}{3000} = 0.10$ m. Therefore, $B' = B - 2e = 2.00 - 2 \times 0.1 = 1.80$ m. Also $L' = L = 3$ m.

For $\phi' = 32°$, from Table 12.1, $N_c = 35.49$, $N_q = 23.18$, and $N_\gamma = 30.22$.

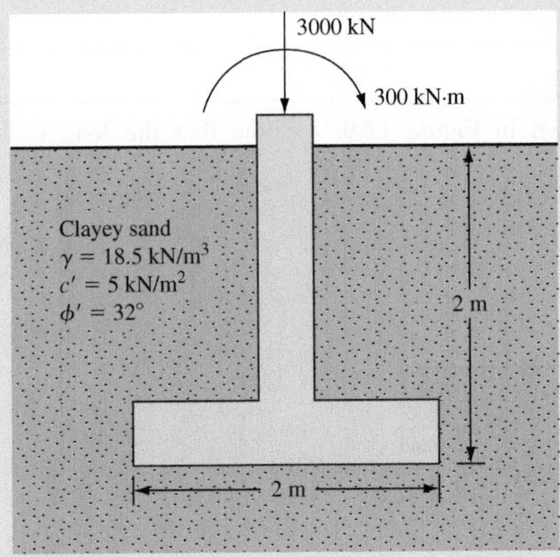

Figure 12.8

Shape Factors

$$F_{cs} = 1 + \frac{B'}{L}\frac{N_q}{N_c} = 1 + \frac{1.8}{3} \times \frac{23.18}{35.49} = 1.39$$

$$F_{qs} = 1 + \frac{B'}{L}\tan\phi' = 1 + \frac{1.8}{3}\tan 32 = 1.37$$

$$F_{\gamma s} = 1 - 0.4\frac{B'}{L} = 1 - 0.4 \times \frac{1.8}{3} = 0.76$$

Depth Factors

$$F_{cd} = 1 + 0.4\frac{D_f}{B} = 1 + 0.4 \times \frac{2.0}{2.0} = 1.40$$

$$F_{qd} = 1 + 2\tan\phi'(1 - \sin\phi')^2\frac{D_f}{B} = 1 + 2\tan 32(1 - \sin 32)^2 \times \frac{2}{2} = 1.28$$

$$F_{qi} = 1.00$$

Inclination factors are $F_{ci} = 1$, $F_{qi} = 1$, and $F_{\gamma i} = 1$.

From Eq. (12.20), the ultimate bearing capacity is given by

$$\begin{aligned} q'_u &= c'N_c F_{cs} F_{cd} F_{ci} + qN_q F_{qs} F_{qd} F_{qi} + 0.5\,\gamma B' N_\gamma F_{\gamma s} F_{\gamma d} F_{\gamma i} \\ &= (5)(35.49)(1.39)(1.40)(1) + (2 \times 18.5)(23.18)(1.37)(1.28)(1) \\ &\quad + 0.5(18.5 \times 1.8)(30.22)(0.76)(1.0)(1) \\ &= 2231.17 \text{ kN/m}^2 \end{aligned}$$

From Eq. (12.21), $Q_{ult} = q'_u \times B'L = 2231.7 \times 1.8 \times 3 = 12951.2$ kN

$$FS = 12951.9/3000 = \mathbf{4.01}$$

■

Example 12.6

A square foundation is shown in Figure 12.9. Assume that the load eccentricity $e = 0.5$ ft. Determine the ultimate load, Q_{ult}.

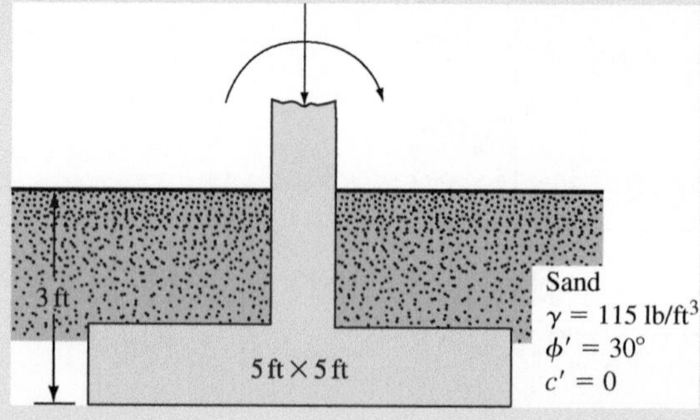

Figure 12.9

Solution

With $c' = 0$, Eq. 12.20 becomes

$$q'_u = qN_q F_{qs} F_{qd} F_{qi} + \frac{1}{2}\gamma B' N_\gamma F_{\gamma s} F_{\gamma a} F_{\gamma i}$$

$$q = (3)(115) = 345 \text{ lb/ft}^2$$

For $\phi' = 30°$, from Table 12.1, $N_q = 18.4$ and $N_\gamma = 22.4$.

$$B' = 5 - 2(0.5) = 4 \text{ ft}$$
$$L' = 5 \text{ ft}$$

From Table 12.2

$$F_{qs} = 1 + \frac{B'}{L'} \tan\phi' = 1 + \left(\frac{4}{5}\right)\tan 30° = 1.462$$

$$F_{qd} = 1 + 2\tan\phi'(1 - \sin\phi')^2 \frac{D_f}{B} = 1 + \frac{(0.289)(3)}{5} = 1.173$$

$$F_{\gamma s} = 1 - 0.4\left(\frac{B'}{L'}\right) = 1 - 0.4\left(\frac{4}{5}\right) = 0.68$$

$$F_{\gamma d} = 1$$

So

$$q'_u = (345)(18.4)(1.462)(1.173) + \frac{1}{2}(115)(4)(22.4)(0.68)(1)$$
$$= 10{,}886.4 + 3503.4 \approx 14{,}390 \text{ lb/ft}^2$$

Hence,

$$Q_{\text{ult}} = B'L'(q'_u) = (4)(5)(14{,}390) = \mathbf{287{,}800\ lb}$$
$$\approx \mathbf{287.8\ kip}$$

■

Settlement of Shallow Foundations

12.7 Types of Foundation Settlement

The settlement of a foundation consists of two parts. They are

1. Elastic settlement (S_e)
2. Consolidation settlement (S_c)

The general principles of consolidation settlement calculation were presented in Chapter 8 and will be revisited in this chapter. Elastic settlement is caused by the elastic deformation of dry soil and of moist and saturated soils without any change in moisture content.

It is important to point out that, theoretically at least, a foundation could be considered fully flexible or fully rigid. A uniformly loaded, perfectly flexible foundation resting on an elastic material

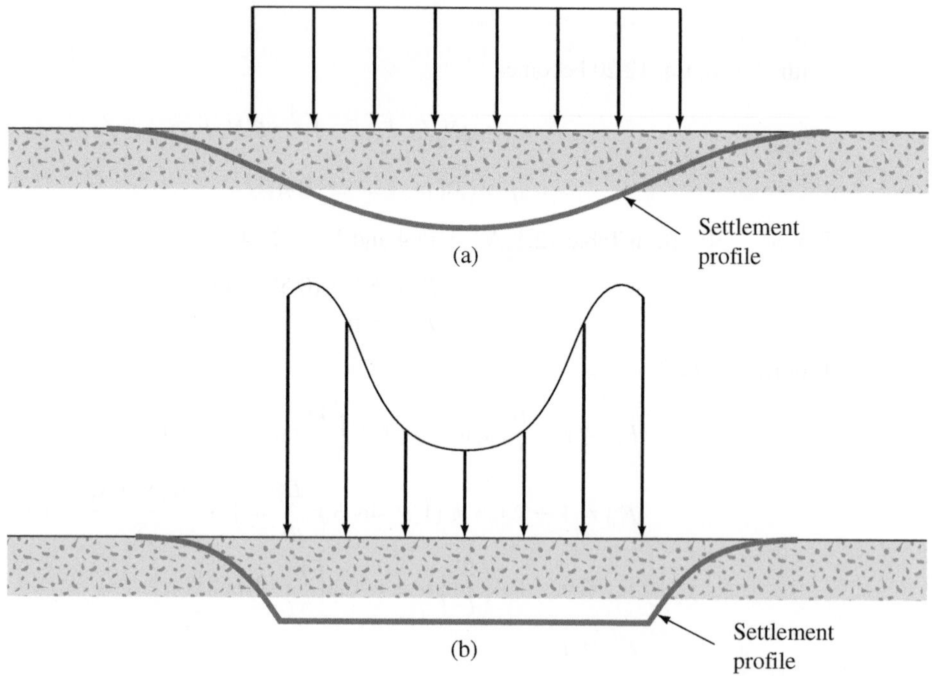

Figure 12.10 Elastic settlement profile and contact pressure in clay: (a) flexible foundation; (b) rigid foundation

such as saturated clay will have a sagging profile, as shown in Figure 12.10a, because of elastic settlement. However, if the foundation is rigid and is resting on an elastic material such as clay, it will undergo uniform settlement and the contact pressure will be redistributed (Figure 12.10b).

12.8 Elastic Settlement

Figure 12.11 shows a shallow foundation subjected to a net force per unit area equal to q_o. Let the Poisson's ratio and the modulus of elasticity of the soil supporting it be μ_s and E_s, respectively. Theoretically, if the foundation is perfectly flexible, the settlement may be expressed as

$$S_e = q_o(\alpha B')\frac{1 - \mu_s^2}{E_s} I_s I_f \tag{12.23}$$

where

q_o = net applied pressure on the foundation
μ_s = Poisson's ratio of soil
E_s = average modulus of elasticity of the soil under the foundation measured from $z = 0$ to about $z = 5B$
B' = $B/2$ for center of foundation
= B for corner of foundation

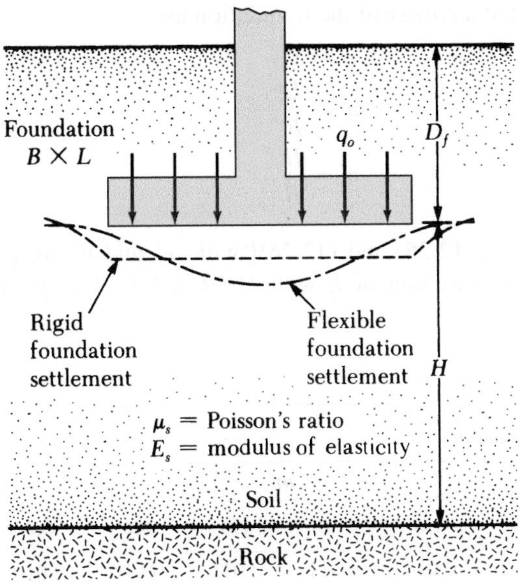

Figure 12.11 Elastic settlement of flexible and rigid foundations

I_s = shape factor (Steinbrenner, 1934)

$$= F_1 + \frac{1 - 2\mu_s}{1 - \mu_s} F_2 \tag{12.24}$$

$$F_1 = \frac{1}{\pi}(A_0 + A_1) \tag{12.25}$$

$$F_2 = \frac{n'}{2\pi} \tan^{-1} A_2 \tag{12.26}$$

$$A_0 = m' \ln \frac{(1 + \sqrt{m'^2 + 1})\sqrt{m'^2 + n'^2}}{m'(1 + \sqrt{m'^2 + n'^2 + 1})} \tag{12.27}$$

$$A_1 = \ln \frac{(m' + \sqrt{m'^2 + 1})\sqrt{1 + n'^2}}{m' + \sqrt{m'^2 + n'^2 + 1}} \tag{12.28}$$

$$A_2 = \frac{m'}{n'\sqrt{m'^2 + n'^2 + 1}} \tag{12.29}$$

$$I_f = \text{depth factor (Fox, 1948)} = f\left(\frac{D_f}{B}, \mu_s, \text{and } \frac{L}{B}\right) \tag{12.30}$$

α = factor that depends on the location on the foundation where settlement is being calculated

- For calculation of settlement at the *center* of the foundation use

$$\alpha = 4$$

$$m' = \frac{L}{B}$$

$$n' = \frac{H}{\left(\frac{B}{2}\right)}$$

- For calculation of settlement at a *corner* of the foundation use

$$\alpha = 1$$

$$m' = \frac{L}{B}$$

$$n' = \frac{H}{B}$$

The variations of F_1 and F_2 [Eqs. (12.25) and (12.26)] with m' and n' are given in Tables 12.3 and 12.4, respectively. Also, the variation of I_f with D_f/B and L/B is given in Figures 12.12, 12.13, and 12.14.

Table 12.3 Variation of F_1 with m' and n'

	m'									
n'	1.0	1.2	1.4	1.6	1.8	2.0	2.5	3.0	3.5	4.0
0.25	0.014	0.013	0.012	0.011	0.011	0.011	0.010	0.010	0.010	0.010
0.50	0.049	0.046	0.044	0.042	0.041	0.040	0.038	0.038	0.037	0.037
0.75	0.095	0.090	0.087	0.084	0.082	0.080	0.077	0.076	0.074	0.074
1.00	0.142	0.138	0.134	0.130	0.127	0.125	0.121	0.118	0.116	0.115
1.25	0.186	0.183	0.179	0.176	0.173	0.170	0.165	0.161	0.158	0.157
1.50	0.224	0.224	0.222	0.219	0.216	0.213	0.207	0.203	0.199	0.197
1.75	0.257	0.259	0.259	0.258	0.255	0.253	0.247	0.242	0.238	0.235
2.00	0.285	0.290	0.292	0.292	0.291	0.289	0.284	0.279	0.275	0.271
2.25	0.309	0.317	0.321	0.323	0.323	0.322	0.317	0.313	0.308	0.305
2.50	0.330	0.341	0.347	0.350	0.351	0.351	0.348	0.344	0.340	0.336
2.75	0.348	0.361	0.369	0.374	0.377	0.378	0.377	0.373	0.369	0.365
3.00	0.363	0.379	0.389	0.396	0.400	0.402	0.402	0.400	0.396	0.392
3.25	0.376	0.394	0.406	0.415	0.420	0.423	0.426	0.424	0.421	0.418
3.50	0.388	0.408	0.422	0.431	0.438	0.442	0.447	0.447	0.444	0.441
3.75	0.399	0.420	0.436	0.447	0.454	0.460	0.467	0.458	0.466	0.464
4.00	0.408	0.431	0.448	0.460	0.469	0.476	0.484	0.487	0.486	0.484
4.25	0.417	0.440	0.458	0.472	0.481	0.484	0.495	0.514	0.515	0.515
4.50	0.424	0.450	0.469	0.484	0.495	0.503	0.516	0.521	0.522	0.522
4.75	0.431	0.458	0.478	0.494	0.506	0.515	0.530	0.536	0.539	0.539
5.00	0.437	0.465	0.487	0.503	0.516	0.526	0.543	0.551	0.554	0.554
5.25	0.443	0.472	0.494	0.512	0.526	0.537	0.555	0.564	0.568	0.569
5.50	0.448	0.478	0.501	0.520	0.534	0.546	0.566	0.576	0.581	0.584
5.75	0.453	0.483	0.508	0.527	0.542	0.555	0.576	0.588	0.594	0.597
6.00	0.457	0.489	0.514	0.534	0.550	0.563	0.585	0.598	0.606	0.609
6.25	0.461	0.493	0.519	0.540	0.557	0.570	0.594	0.609	0.617	0.621
6.50	0.465	0.498	0.524	0.546	0.563	0.577	0.603	0.618	0.627	0.632
6.75	0.468	0.502	0.529	0.551	0.569	0.584	0.610	0.627	0.637	0.643
7.00	0.471	0.506	0.533	0.556	0.575	0.590	0.618	0.635	0.646	0.653
7.25	0.474	0.509	0.538	0.561	0.580	0.596	0.625	0.643	0.655	0.662
7.50	0.477	0.513	0.541	0.565	0.585	0.601	0.631	0.650	0.663	0.671
7.75	0.480	0.516	0.545	0.569	0.589	0.606	0.637	0.658	0.671	0.680

n'	m' 1.0	1.2	1.4	1.6	1.8	2.0	2.5	3.0	3.5	4.0
8.00	0.482	0.519	0.549	0.573	0.594	0.611	0.643	0.664	0.678	0.688
8.25	0.485	0.522	0.552	0.577	0.598	0.615	0.648	0.670	0.685	0.695
8.50	0.487	0.524	0.555	0.580	0.601	0.619	0.653	0.676	0.692	0.703
8.75	0.489	0.527	0.558	0.583	0.605	0.623	0.658	0.682	0.698	0.710
9.00	0.491	0.529	0.560	0.587	0.609	0.627	0.663	0.687	0.705	0.716
9.25	0.493	0.531	0.563	0.589	0.612	0.631	0.667	0.693	0.710	0.723
9.50	0.495	0.533	0.565	0.592	0.615	0.634	0.671	0.697	0.716	0.729
9.75	0.496	0.536	0.568	0.595	0.618	0.638	0.675	0.702	0.721	0.735
10.00	0.498	0.537	0.570	0.597	0.621	0.641	0.679	0.707	0.726	0.740
20.00	0.529	0.575	0.614	0.647	0.677	0.702	0.756	0.797	0.830	0.858
50.00	0.548	0.598	0.640	0.678	0.711	0.740	0.803	0.853	0.895	0.931
100.00	0.555	0.605	0.649	0.688	0.722	0.753	0.819	0.872	0.918	0.956

n'	m' 4.5	5.0	6.0	7.0	8.0	9.0	10.0	25.0	50.0	100.0
0.25	0.010	0.010	0.010	0.010	0.010	0.010	0.010	0.010	0.010	0.010
0.50	0.036	0.036	0.036	0.036	0.036	0.036	0.036	0.036	0.036	0.036
0.75	0.073	0.073	0.072	0.072	0.072	0.072	0.071	0.071	0.071	0.071
1.00	0.114	0.113	0.112	0.112	0.112	0.111	0.111	0.110	0.110	0.110
1.25	0.155	0.154	0.153	0.152	0.152	0.151	0.151	0.150	0.150	0.150
1.50	0.195	0.194	0.192	0.191	0.190	0.190	0.189	0.188	0.188	0.188
1.75	0.233	0.232	0.229	0.228	0.227	0.226	0.225	0.223	0.223	0.223
2.00	0.269	0.267	0.264	0.262	0.261	0.260	0.259	0.257	0.256	0.256
2.25	0.302	0.300	0.296	0.294	0.293	0.291	0.291	0.287	0.287	0.287
2.50	0.333	0.331	0.327	0.324	0.322	0.321	0.320	0.316	0.315	0.315
2.75	0.362	0.359	0.355	0.352	0.350	0.348	0.347	0.343	0.342	0.342
3.00	0.389	0.386	0.382	0.378	0.376	0.374	0.373	0.368	0.367	0.367
3.25	0.415	0.412	0.407	0.403	0.401	0.399	0.397	0.391	0.390	0.390
3.50	0.438	0.435	0.430	0.427	0.424	0.421	0.420	0.413	0.412	0.411
3.75	0.461	0.458	0.453	0.449	0.446	0.443	0.441	0.433	0.432	0.432
4.00	0.482	0.479	0.474	0.470	0.466	0.464	0.462	0.453	0.451	0.451
4.25	0.516	0.496	0.484	0.473	0.471	0.471	0.470	0.468	0.462	0.460
4.50	0.520	0.517	0.513	0.508	0.505	0.502	0.499	0.489	0.487	0.487
4.75	0.537	0.535	0.530	0.526	0.523	0.519	0.517	0.506	0.504	0.503
5.00	0.554	0.552	0.548	0.543	0.540	0.536	0.534	0.522	0.519	0.519
5.25	0.569	0.568	0.564	0.560	0.556	0.553	0.550	0.537	0.534	0.534
5.50	0.584	0.583	0.579	0.575	0.571	0.568	0.585	0.551	0.549	0.548
5.75	0.597	0.597	0.594	0.590	0.586	0.583	0.580	0.565	0.583	0.562
6.00	0.611	0.610	0.608	0.604	0.601	0.598	0.595	0.579	0.576	0.575
6.25	0.623	0.623	0.621	0.618	0.615	0.611	0.608	0.592	0.589	0.588
6.50	0.635	0.635	0.634	0.631	0.628	0.625	0.622	0.605	0.601	0.600
6.75	0.646	0.647	0.646	0.644	0.641	0.637	0.634	0.617	0.613	0.612
7.00	0.656	0.658	0.658	0.656	0.653	0.650	0.647	0.628	0.624	0.623
7.25	0.666	0.669	0.669	0.668	0.665	0.662	0.659	0.640	0.635	0.634
7.50	0.676	0.679	0.680	0.679	0.676	0.673	0.670	0.651	0.646	0.645
7.75	0.685	0.688	0.690	0.689	0.687	0.684	0.681	0.661	0.656	0.655

(continued)

Table 12.3 *(continued)*

	m'									
n'	4.5	5.0	6.0	7.0	8.0	9.0	10.0	25.0	50.0	100.0
8.00	0.694	0.697	0.700	0.700	0.698	0.695	0.692	0.672	0.666	0.665
8.25	0.702	0.706	0.710	0.710	0.708	0.705	0.703	0.682	0.676	0.675
8.50	0.710	0.714	0.719	0.719	0.718	0.715	0.713	0.692	0.686	0.684
8.75	0.717	0.722	0.727	0.728	0.727	0.725	0.723	0.701	0.695	0.693
9.00	0.725	0.730	0.736	0.737	0.736	0.735	0.732	0.710	0.704	0.702
9.25	0.731	0.737	0.744	0.746	0.745	0.744	0.742	0.719	0.713	0.711
9.50	0.738	0.744	0.752	0.754	0.754	0.753	0.751	0.728	0.721	0.719
9.75	0.744	0.751	0.759	0.762	0.762	0.761	0.759	0.737	0.729	0.727
10.00	0.750	0.758	0.766	0.770	0.770	0.770	0.768	0.745	0.738	0.735
20.00	0.878	0.896	0.925	0.945	0.959	0.969	0.977	0.982	0.965	0.957
50.00	0.962	0.989	1.034	1.070	1.100	1.125	1.146	1.265	1.279	1.261
100.00	0.990	1.020	1.072	1.114	1.150	1.182	1.209	1.408	1.489	1.499

Table 12.4 Variation of F_2 with m' and n'

	m'									
n'	1.0	1.2	1.4	1.6	1.8	2.0	2.5	3.0	3.5	4.0
0.25	0.049	0.050	0.051	0.051	0.051	0.052	0.052	0.052	0.052	0.052
0.50	0.074	0.077	0.080	0.081	0.083	0.084	0.086	0.086	0.0878	0.087
0.75	0.083	0.089	0.093	0.097	0.099	0.101	0.104	0.106	0.107	0.108
1.00	0.083	0.091	0.098	0.102	0.106	0.109	0.114	0.117	0.119	0.120
1.25	0.080	0.089	0.096	0.102	0.107	0.111	0.118	0.122	0.125	0.127
1.50	0.075	0.084	0.093	0.099	0.105	0.110	0.118	0.124	0.128	0.130
1.75	0.069	0.079	0.088	0.095	0.101	0.107	0.117	0.123	0.128	0.131
2.00	0.064	0.074	0.083	0.090	0.097	0.102	0.114	0.121	0.127	0.131
2.25	0.059	0.069	0.077	0.085	0.092	0.098	0.110	0.119	0.125	0.130
2.50	0.055	0.064	0.073	0.080	0.087	0.093	0.106	0.115	0.122	0.127
2.75	0.051	0.060	0.068	0.076	0.082	0.089	0.102	0.111	0.119	0.125
3.00	0.048	0.056	0.064	0.071	0.078	0.084	0.097	0.108	0.116	0.122
3.25	0.045	0.053	0.060	0.067	0.074	0.080	0.093	0.104	0.112	0.119
3.50	0.042	0.050	0.057	0.0.64	0.070	0.076	0.089	0.100	0.109	0.116
3.75	0.040	0.047	0.054	0.060	0.067	0.073	0.086	0.096	0.105	0.113
4.00	0.037	0.044	0.051	0.057	0.063	0.069	0.082	0.093	0.102	0.110
4.25	0.036	0.042	0.049	0.055	0.061	0.066	0.079	0.090	0.099	0.107
4.50	0.034	0.040	0.046	0.052	0.058	0.063	0.076	0.086	0.096	0.104
4.75	0.032	0.038	0.044	0.050	0.055	0.061	0.073	0.083	0.093	0.101
5.00	0.031	0.036	0.042	0.048	0.053	0.058	0.070	0.080	0.090	0.098
5.25	0.029	0.035	0.040	0.046	0.051	0.056	0.067	0.078	0.087	0.095
5.50	0.028	0.033	0.039	0.044	0.049	0.054	0.065	0.075	0.084	0.092
5.75	0.027	0.032	0.037	0.042	0.047	0.052	0.063	0.073	0.082	0.090
6.00	0.026	0.031	0.036	0.040	0.045	0.050	0.060	0.070	0.079	0.087
6.25	0.025	0.030	0.034	0.039	0.044	0.048	0.058	0.068	0.077	0.085
6.50	0.024	0.029	0.033	0.038	0.042	0.046	0.056	0.066	0.075	0.083
6.75	0.023	0.028	0.032	0.036	0.041	0.045	0.055	0.064	0.073	0.080

n'	m' 1.0	1.2	1.4	1.6	1.8	2.0	2.5	3.0	3.5	4.0
7.00	0.022	0.027	0.031	0.035	0.039	0.043	0.053	0.062	0.071	0.078
7.25	0.022	0.026	0.030	0.034	0.038	0.042	0.051	0.060	0.069	0.076
7.50	0.021	0.025	0.029	0.033	0.037	0.041	0.050	0.059	0.067	0.074
7.75	0.020	0.024	0.028	0.032	0.036	0.039	0.048	0.057	0.065	0.072
8.00	0.020	0.023	0.027	0.031	0.035	0.038	0.047	0.055	0.063	0.071
8.25	0.019	0.023	0.026	0.030	0.034	0.037	0.046	0.054	0.062	0.069
8.50	0.018	0.022	0.026	0.029	0.033	0.036	0.045	0.053	0.060	0.067
8.75	0.018	0.021	0.025	0.028	0.032	0.035	0.043	0.051	0.059	0.066
9.00	0.017	0.021	0.024	0.028	0.031	0.034	0.042	0.050	0.057	0.064
9.25	0.017	0.020	0.024	0.027	0.030	0.033	0.041	0.049	0.056	0.063
9.50	0.017	0.020	0.023	0.026	0.029	0.033	0.040	0.048	0.055	0.061
9.75	0.016	0.019	0.023	0.026	0.029	0.032	0.039	0.047	0.054	0.060
10.00	0.016	0.019	0.022	0.025	0.028	0.031	0.038	0.046	0.052	0.059
20.00	0.008	0.010	0.011	0.013	0.014	0.016	0.020	0.024	0.027	0.031
50.00	0.003	0.004	0.004	0.005	0.006	0.006	0.008	0.010	0.011	0.013
100.00	0.002	0.002	0.002	0.003	0.003	0.003	0.004	0.005	0.006	0.006

n'	m' 4.5	5.0	6.0	7.0	8.0	9.0	10.0	25.0	50.0	100.0
0.25	0.053	0.053	0.053	0.053	0.053	0.053	0.053	0.053	0.053	0.053
0.50	0.087	0.087	0.088	0.088	0.088	0.088	0.088	0.088	0.088	0.088
0.75	0.109	0.109	0.109	0.110	0.110	0.110	0.110	0.111	0.111	0.111
1.00	0.121	0.122	0.123	0.123	0.124	0.124	0.124	0.125	0.125	0.125
1.25	0.128	0.130	0.131	0.132	0.132	0.133	0.133	0.134	0.134	0.134
1.50	0.132	0.134	0.136	0.137	0.138	0.138	0.139	0.140	0.140	0.140
1.75	0.134	0.136	0.138	0.140	0.141	0.142	0.142	0.144	0.144	0.145
2.00	0.134	0.136	0.139	0.141	0.143	0.144	0.145	0.147	0.147	0.148
2.25	0.133	0.136	0.140	0.142	0.144	0.145	0.146	0.149	0.150	0.150
2.50	0.132	0.135	0.139	0.142	0.144	0.146	0.147	0.151	0.151	0.151
2.75	0.130	0.133	0.138	0.142	0.144	0.146	0.147	0.152	0.152	0.153
3.00	0.127	0.131	0.137	0.141	0.144	0.145	0.147	0.152	0.153	0.154
3.25	0.125	0.129	0.135	0.140	0.143	0.145	0.147	0.153	0.154	0.154
3.50	0.122	0.126	0.133	0.138	0.142	0.144	0.146	0.153	0.155	0.155
3.75	0.119	0.124	0.131	0.137	0.141	0.143	0.145	0.154	0.155	0.155
4.00	0.116	0.121	0.129	0.135	0.139	0.142	0.145	0.154	0.155	0.156
4.25	0.113	0.119	0.127	0.133	0.138	0.141	0.144	0.154	0.156	0.156
4.50	0.110	0.116	0.125	0.131	0.136	0.140	0.143	0.154	0.156	0.156
4.75	0.107	0.113	0.123	0.130	0.135	0.139	0.142	0.154	0.156	0.157
5.00	0.105	0.111	0.120	0.128	0.133	0.137	0.140	0.154	0.156	0.157
5.25	0.102	0.108	0.118	0.126	0.131	0.136	0.139	0.154	0.156	0.157
5.50	0.099	0.106	0.116	0.124	0.130	0.134	0.138	0.154	0.156	0.157
5.75	0.097	0.103	0.113	0.122	0.128	0.133	0.136	0.154	0.157	0.157
6.00	0.094	0.101	0.111	0.120	0.126	0.131	0.135	0.153	0.157	0.157
6.25	0.092	0.098	0.109	0.118	0.124	0.129	0.134	0.153	0.157	0.158
6.50	0.090	0.096	0.107	0.116	0.122	0.128	0.132	0.153	0.157	0.158
6.75	0.087	0.094	0.105	0.114	0.121	0.126	0.131	0.153	0.157	0.158

(continued)

Table 12.4 (*continued*)

	m'									
n'	4.5	5.0	6.0	7.0	8.0	9.0	10.0	25.0	50.0	100.0
7.00	0.085	0.092	0.103	0.112	0.119	0.125	0.129	0.152	0.157	0.158
7.25	0.083	0.090	0.101	0.110	0.117	0.123	0.128	0.152	0.157	0.158
7.50	0.081	0.088	0.099	0.108	0.115	0.121	0.126	0.152	0.156	0.158
7.75	0.079	0.086	0.097	0.106	0.114	0.120	0.125	0.151	0.156	0.158
8.00	0.077	0.084	0.095	0.104	0.112	0.118	0.124	0.151	0.156	0.158
8.25	0.076	0.082	0.093	0.102	0.110	0.117	0.122	0.150	0.156	0.158
8.50	0.074	0.080	0.091	0.101	0.108	0.115	0.121	0.150	0.156	0.158
8.75	0.072	0.078	0.089	0.099	0.107	0.114	0.119	0.150	0.156	0.158
9.00	0.071	0.077	0.088	0.097	0.105	0.112	0.118	0.149	0.156	0.158
9.25	0.069	0.075	0.086	0.096	0.104	0.110	0.116	0.149	0.156	0.158
9.50	0.068	0.074	0.085	0.094	0.102	0.109	0.115	0.148	0.156	0.158
9.75	0.066	0.072	0.083	0.092	0.100	0.107	0.113	0.148	0.156	0.158
10.00	0.065	0.071	0.082	0.091	0.099	0.106	0.112	0.147	0.156	0.158
20.00	0.035	0.039	0.046	0.053	0.059	0.065	0.071	0.124	0.148	0.156
50.00	0.014	0.016	0.019	0.022	0.025	0.028	0.031	0.071	0.113	0.142
100.00	0.007	0.008	0.010	0.011	0.013	0.014	0.016	0.039	0.071	0.113

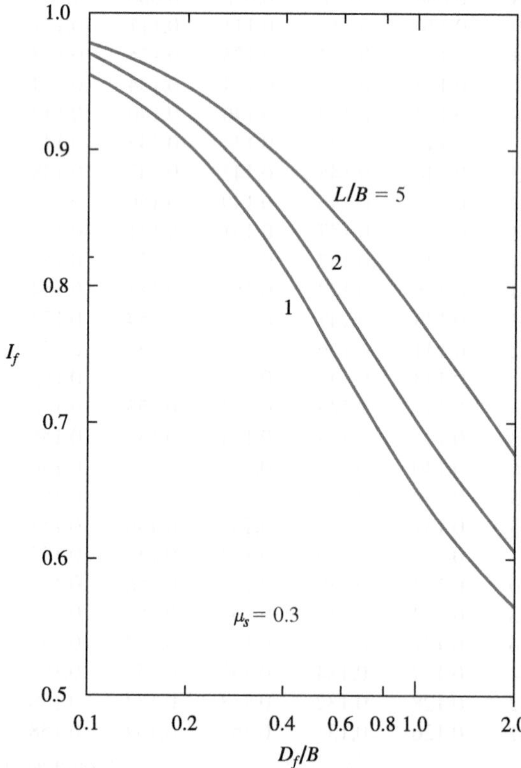

Figure 12.12 Variation of I_f with D_f/B and L/B ($\mu_s = 0.3$)

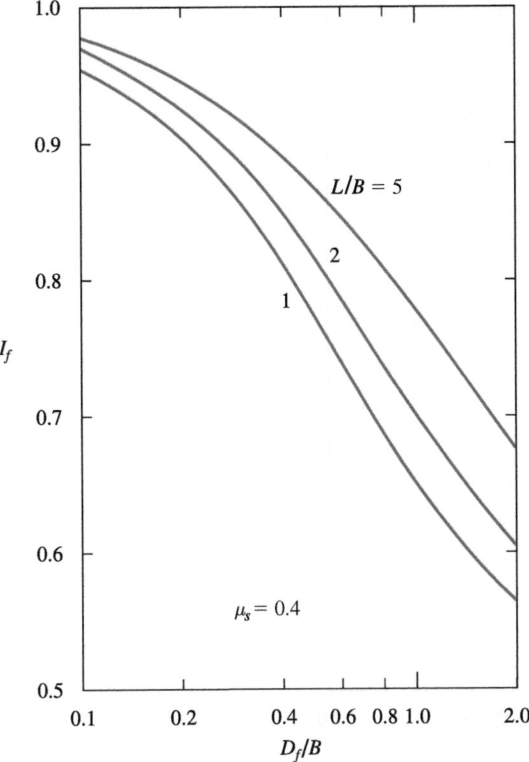

Figure 12.13 Variation of I_f with D_f/B and L/B ($\mu_s = 0.4$)

The elastic settlement of a *rigid foundation* can be estimated as

$$S_{e\,(\text{rigid})} \approx 0.93 S_{e\,(\text{flexible, center})} \tag{12.31}$$

Due to the nonhomogeneous nature of soil deposits, the magnitude of E_s may vary with depth. For that reason, Bowles (1987) recommended using a weighted average value of E_s in Eq. (12.23), or

$$E_s = \frac{\Sigma E_{s(i)} \Delta z}{\bar{z}} \tag{12.32}$$

where

$E_{s(i)}$ = soil modulus of elasticity within a depth Δz
$\bar{z}$ = H or $5B$, whichever is smaller

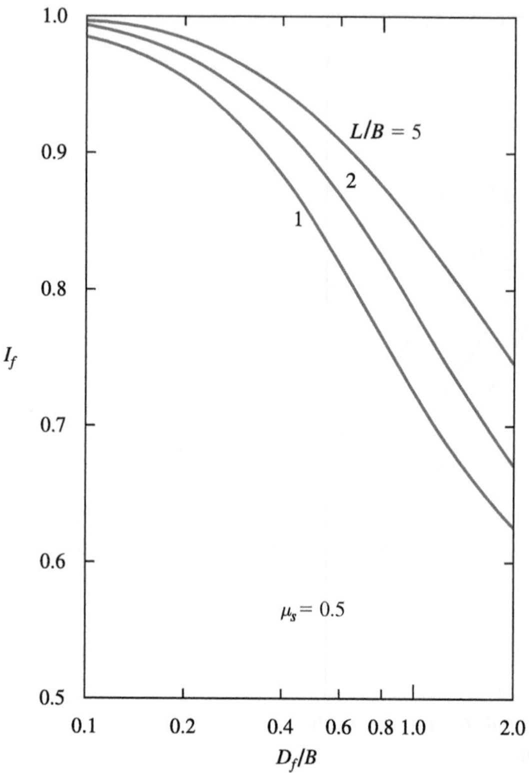

Figure 12.14 Variation of I_f with D_f/B and L/B ($\mu_s = 0.5$)

12.9 Range of Material Parameters for Computing Elastic Settlement

Section 12.8 presented the equations for calculating the elastic settlement of foundations. Those equations contain the elastic parameters, such as E_s and μ_s. If the laboratory test results for these parameters are not available, certain realistic assumptions have to be made. Table 12.5 shows the approximate ranges of the elastic parameters for various soils.

Several investigators have correlated the values of the modulus of elasticity, E_s, with the field standard penetration number, N_{60} and the cone penetration resistance, q_c. Mitchell and Gardner (1975) compiled a list of these correlations. Schmertmann (1970) indicated that the modulus of elasticity of sand may be given by

$$\frac{E_s}{p_a} = 8N_{60} \tag{12.33}$$

where

N_{60} = standard penetration resistance
p_a = atmospheric pressure $\approx 2000\ \text{lb/ft}^2$ ($\approx 100\ \text{kN/m}^2$)

Table 12.5 Elastic Parameters of Various Soils

Type of soil	Modulus of elasticity, E_s lb/in^2	Modulus of elasticity, E_s MN/m^2	Poisson's ratio, μ_s
Loose sand	1500–3500	10–24	0.20–0.40
Medium dense sand	2500–4000	17–28	0.25–0.40
Dense sand	5000–8000	34–55	0.30–0.45
Silty sand	1500–2500	10–17	0.20–0.40
Sand and gravel	10,000–25,000	69–172	0.15–0.35
Soft clay	600–3000	4–21	
Medium clay	3000–6000	21–42	0.20–0.50
Stiff clay	6000–14,000	42–96	

Similarly,

$$E_s = 2q_c \tag{12.34}$$

where q_c = static cone penetration resistance.

The modulus of elasticity of normally consolidated clays may be estimated as

$$E_s = 250c_u \text{ to } 500c_u \tag{12.35}$$

and for overconsolidated clays as

$$E_s = 750c_u \text{ to } 1000c_u \tag{12.36}$$

where c_u = undrained cohesion of clay soil.

Example 12.7

A rigid shallow foundation 3 ft × 6 ft is shown in Figure 12.15. Calculate the elastic settlement at the center of the foundation.

Solution

Given $B = 3$ ft and $L = 6$ ft. Note that $\bar{z} = 15$ ft $= 5B$. From Eq. (12.32)

$$E_s = \frac{\Sigma E_{s(i)}\Delta z}{\bar{z}}$$

$$= \frac{(1450)(6) + (1160)(3) + (1750)(6)}{15} = 1512 \text{ lb/in}^2 = 217{,}728 \text{ lb/ft}^2$$

For the *center of the foundation*,

$$\alpha = 4$$

$$m' = \frac{L}{B} = \frac{6}{3} = 2$$

$$n' = \frac{H}{\left(\frac{B}{2}\right)} = \frac{15}{\left(\frac{3}{2}\right)} = 10$$

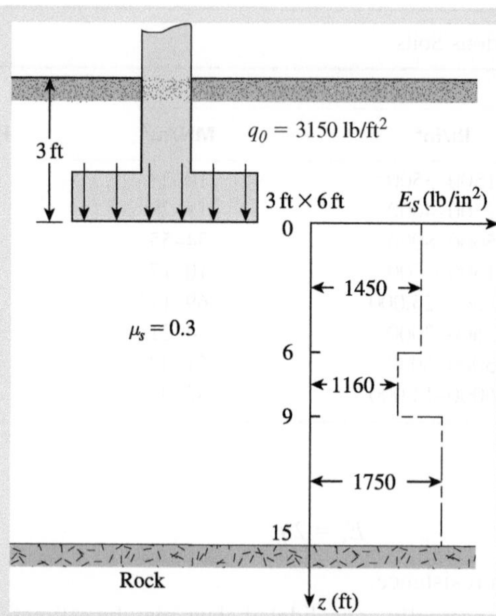

Figure 12.15

From Tables 12.3 and 12.4, $F_1 = 0.641$ and $F_2 = 0.031$. From Eq. (12.24),

$$I_s = F_1 + \frac{2 - \mu_s}{1 - \mu_s} F_2$$

$$= 0.641 + \frac{2 - 0.3}{1 - 0.3}(0.031) = 0.716$$

Again $\frac{D_f}{B} = \frac{3}{3} = 1$, $\frac{L}{B} = 2$, $\mu_s = 0.3$. From Figure 12.12, $I_f = 0.71$. Hence,

$$S_{e\,(\text{flexible})} = q_o(\alpha B')\frac{1 - \mu_s^2}{E_s} I_s I_f$$

$$= (3150)\left(4 \times \frac{3}{2}\right)\left(\frac{1 - 0.3^2}{217{,}728}\right)(0.716)(0.71) = 0.0402 \text{ ft} = 0.48 \text{ in.}$$

Since the foundation is rigid, from Eq. (12.31),

$$S_{e\,(\text{rigid})} = (0.93)(0.48) \approx \mathbf{0.45\ in.}$$

■

12.10 Plate Load Test

The ultimate load-bearing capacity of a foundation, as well as the allowable bearing capacity based on tolerable settlement considerations, can be effectively determined from the field load test, generally referred to as the *plate load test*. The plates that are used for tests in the field are usually made of steel and are 1 in. (25.4 mm) thick and 6 in. (150 mm) to 30 in. (750 mm) in diameter. Occasionally, square plates that are 12 in. × 12 in. (300 mm × 300 mm) are also used.

To conduct a plate load test, a hole is excavated with a minimum diameter of $4B$ (B is the diameter of the test plate) to a depth of D_f, the depth of the proposed foundation. The plate is placed at the center of the hole, and a load that is about one-fourth to one-fifth of the estimated ultimate load is applied to the plate in steps by means of a jack. A schematic diagram of the test arrangement is shown in Figure 12.16a. During each step of the application of the load, the settlement of the plate is observed on dial gauges. At least one hour is allowed to elapse between each application. The test should be conducted until failure or at least until the plate has gone through 1 in. (≈ 25 mm) of settlement. Figure 12.16b shows the nature of the load–settlement curve obtained from such tests, from which the ultimate load per unit area can be determined. Figure 12.17 shows a plate load test conducted in the field.

For tests in clay subjected to undrained loading,

$$q_{u(F)} = q_{u(P)} \tag{12.37}$$

where

$q_{u(F)}$ = ultimate bearing capacity of the proposed foundation
$q_{u(P)}$ = ultimate bearing capacity of the test plate

Equation (12.37) implies that the ultimate bearing capacity in clay is virtually independent of the size of the plate.

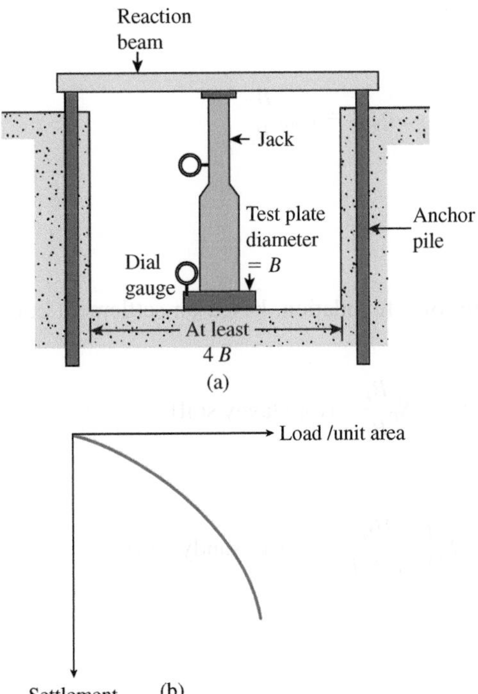

Figure 12.16 Plate load test: (a) test arrangement; (b) nature of load–settlement curve

Figure 12.17 Plate load test in the field (Photo courtesy of Braja Das)

For tests in sandy soils,

$$q_{u(F)} = q_{u(P)}\frac{B_F}{B_P} \tag{12.38}$$

where

B_F = width of the foundation
B_P = width of the test plate

The allowable bearing capacity of a foundation, based on settlement considerations and for a given intensity of load, q_o, is

$$S_F = S_P\frac{B_F}{B_P} \quad \text{(for clayey soil)} \tag{12.39}$$

and

$$S_F = S_P\left(\frac{2B_F}{B_F + B_P}\right)^2 \quad \text{(for sandy soil)} \tag{12.40}$$

where

S_F = settlement of the foundation
S_P = settlement of the test plate

The preceding relationship is based on the work of Terzaghi and Peck (1967).

12.11 Elastic Settlement Based on Standard Penetration Resistance in Sand

The net ultimate bearing capacity was defined in Eq. (12.8). In a similar manner, the net applied load per unit area (q_o) of a foundation can be given as

$$q_o = q' - \gamma D_f \tag{12.41}$$

where q' = load per unit area at the level of the foundation.

Meyerhof (1965) proposed correlations between q_o, the standard penetration resistance N_{60}, and the elastic settlement of shallow foundations in granular soils. In a slightly modified form, Bowles (1977) expressed the correlations as follows

$$q_o = \frac{N_{60}}{2.5} F_d S_e \qquad (\text{for } B \leq 4 \text{ ft}) \tag{12.42a}$$

and

$$q_o = \frac{N_{60}}{4}\left(\frac{B+1}{B}\right)^2 F_d S_e \qquad (\text{for } B > 4 \text{ ft}) \tag{12.42b}$$

where

q_o is in kip/ft^2
B is in ft
S_e = elastic settlement, in in.
F_d = depth factor $= 1 + 0.33\dfrac{D_f}{B} \leq 1.33$ (12.43)

In SI, units, Eqs. (12.42a) and (12.42b) can be written as

$$q_o(\text{kN/m}^2) = \frac{N_{60}}{0.05} F_d\left(\frac{S_e}{25}\right) \qquad (\text{for } B \leqslant 1.22 \text{ m}) \tag{12.44a}$$

and

$$q_o(\text{kN/m}^2) = \frac{N_{60}}{0.08}\left(\frac{B+0.3}{B}\right)^2 F_d\left(\frac{S_e}{25}\right) \qquad (\text{for } B > 1.22 \text{ m}) \tag{12.44b}$$

where B is in meters and S_e is in mm.
Hence, from Eqs. (12.42a) and (12.42b),

$$S_e = \frac{2.5 q_o}{F_d N_{60}} \qquad (\text{for } B \leq 4 \text{ ft}) \tag{12.45a}$$

and

$$S_e = \frac{4 q_o}{F_d N_{60}}\left(\frac{B}{B+1}\right)^2 \qquad (\text{for } B > 4 \text{ ft}) \tag{12.45b}$$

Similarly in SI units from Eqs. (12.44a) and (12.44b),

$$S_e(\text{mm}) = \frac{1.25q_o\,(\text{kN/m}^2)}{N_{60}F_d} \qquad (\text{for } B \leqslant 1.22 \text{ m}) \tag{12.46a}$$

and

$$S_e(\text{mm}) = \frac{2q_o\,(\text{kN/m}^2)}{N_{60}F_d}\left(\frac{B}{B+0.3}\right)^2 \qquad (\text{for } B > 1.22 \text{ m}) \tag{12.46b}$$

The value of N_{60} is the average value of the standard penetration resistance up to a depth of about $4B$ measured from the bottom of the foundation.

Example 12.8

For a shallow foundation on a sand deposit, given:

- Foundation: Size = 4.5 ft × 4.5 ft ($B \times B$)
  $D_f = 3$ ft
  $q_o = 4500$ lb/ft^2
- Soil: $N_{60} = 18$ (average between the bottom of the foundation up to a depth of $4B$)
  Estimate the elastic settlement of the foundation.

Solution

From Eq. (12.45b),

$$S_e = \frac{4q_o}{F_d N_{60}}\left(\frac{B}{B+1}\right)^2$$

$$F_d = 1 + 0.33\frac{D_f}{B} = 1 + 0.33\left(\frac{3}{4.5}\right) = 1.22$$

$$S_e = \frac{(4)(4.5 \text{ kip/ft}^2)}{(1.22)(18)}\left(\frac{4.5}{4.5+1}\right)^2 = \mathbf{0.55 \text{ in.}}$$

■

Example 12.9

A plate loading test carried out on a homogeneous soil using a 300 mm × 300 mm square plate settled 5.5 mm under a uniform pressure of 150 kN/m^2. What would be the settlement of a 2.0 m square foundation in the same soil:

a. if the soil is sand?
b. if the soil is clay?

Solution

$S_P = 5.5$ mm, $B_P = 0.3$ m, $B_F = 2.0$ m

Part a

If the soil is sand, from Eq. (12.40),

$$S_F = S_P\left(\frac{2B_F}{B_F + B_P}\right)^2 = 5.5 \times \left(\frac{2 \times 2.0}{2.0 + 0.3}\right)^2 = \mathbf{16.6\ mm}$$

Part b

If the soil is clay, from Eq. (12.39),

$$S_F = S_P\frac{B_F}{B_P} = 5.5 \times \frac{2.0}{0.3} = \mathbf{36.7\ mm}$$

■

Example 12.10

In a sand where $N_{60} = 22$, a 1.5 m square pad foundation is placed at a depth of 1.0 m. If the maximum allowable settlement is 25 mm, what is the maximum net applied pressure allowed on this foundation?

Solution

From Eq. (12.43), the depth factor is $F_d = 1 + 0.33\,(1/1.5) = 1.22$.

From Eq. (12.44b),

$$q_o = \frac{N_{60}}{0.08}\left(\frac{B + 0.3}{B}\right)^2 F_d\left(\frac{S_e}{25}\right) = \frac{22}{0.08}\left(\frac{1.5 + 0.3}{1.5}\right)^2(1.22)\left(\frac{25}{25}\right) = \mathbf{483\ kN/m^2}$$

■

12.12 Primary Consolidation Settlement

As mentioned before, consolidation settlement occurs over time in saturated clayey soils subjected to an increased load caused by construction of the foundation. (See Figure 12.18.) On the basis of the one-dimensional consolidation settlement equations given in Chapter 8, we write

$$S_{c(p)} = \frac{C_cH_c}{1 + e_o}\log\frac{\sigma_o' + \Delta\sigma_{av}'}{\sigma_o'} \qquad \text{(for normally consolidated clays)} \qquad (12.47)$$

$$S_{c(p)} = \frac{C_sH_c}{1 + e_o}\log\frac{\sigma_o' + \Delta\sigma_{av}'}{\sigma_o'} \qquad \text{(for overconsolidated clays with } \sigma_o' + \Delta\sigma_{av}' < \sigma_c'\text{)} \qquad (12.48)$$

$$S_{c(p)} = \frac{C_sH_c}{1 + e_o}\log\frac{\sigma_c'}{\sigma_o'} + \frac{C_cH_c}{1 + e_o}\log\frac{\sigma_o' + \Delta\sigma_{av}'}{\sigma_c'} \qquad \text{(for overconsolidated clays with } \sigma_o' < \sigma_c' < \sigma_o' + \Delta\sigma_{av}'\text{)} \qquad (12.49)$$

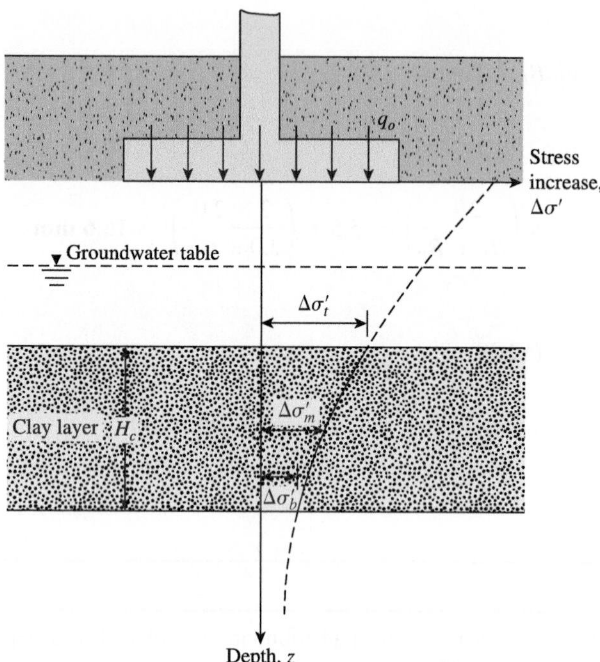

Figure 12.18 Consolidation settlement calculation

where

σ'_o = average effective pressure on the clay layer before the construction of the foundation
$\Delta\sigma'_{av}$ = average increase in effective pressure on the clay layer caused by the construction of the foundation
σ'_c = preconsolidation pressure
e_o = initial void ratio of the clay layer
C_c = compression index
C_s = swelling index
H_c = thickness of the clay layer

The procedures for determining the compression and swelling indexes were discussed in Chapter 8.

Note that the increase in effective pressure, $\Delta\sigma'$, on the clay layer is not constant with depth: The magnitude of $\Delta\sigma'$ will decrease with the increase in depth measured from the bottom of the foundation. However, the average increase in pressure may be approximated by

$$\Delta\sigma'_{av} = \frac{1}{6}(\Delta\sigma'_t + 4\Delta\sigma'_m + \Delta\sigma'_b) \tag{12.50}$$

where $\Delta\sigma'_t$, $\Delta\sigma'_m$, and $\Delta\sigma'_b$ are, respectively, the effective pressure increases at the *top, middle*, and *bottom* of the clay layer that are caused by the construction of the foundation.

The method of determining the pressure increase caused by various types of foundation load is discussed in Sections 7.5 and 7.6. The value of σ'_o can be taken as that at the middle the clay layer.

Example 12.11

A plan of a foundation 3 ft × 6 ft is shown in Figure 12.19. Estimate the consolidation settlement of the foundation.

Solution

The clay is normally consolidated. Thus,

$$S_{c(p)} = \frac{C_c H_c}{1 + e_o} \log \frac{\sigma'_o + \Delta\sigma'_{av}}{\sigma'_o}$$

so

$$\sigma'_o = (7.5)(105) + (1.5)(111 - 62.4) + (3.75)(102 - 62.4)$$

$$= 787.5 + 72.9 + 148.5 = 1008.9 \text{ lb/ft}^2$$

From Eq. (12.50),

$$\Delta\sigma'_{av} = \tfrac{1}{6}(\Delta\sigma'_t + 4\Delta\sigma'_m + \Delta\sigma'_b)$$

Now the following table can be prepared (*Note:* L = 6 ft; B = 3 ft):

$m_1 = L/B$	z(ft)	$z/(B/2) = n_1$	I_3^a	$\Delta\sigma' = q_o I_3^b$
2	6	4	0.190	$608 = \Delta\sigma'_t$
2	6 + 7.5/2 = 9.75	6.5	≈0.085	$272 = \Delta\sigma'_m$
2	6 + 7.5 = 13.5	9	0.045	$144 = \Delta\sigma'_h$

[a]Table 7.4
[b]Eq. (7.18)

Now,

$$\Delta\sigma'_{av} = \tfrac{1}{6}(608 + 4 \times 272 + 144) = 306.7 \text{ lb/ft}^2$$

so

$$S_{c(p)} = \frac{(0.32)(7.5 \times 12)}{1 + 0.8} \log\left(\frac{1008.9 + 306.7}{1008.9}\right) = \mathbf{1.84\ in.}$$

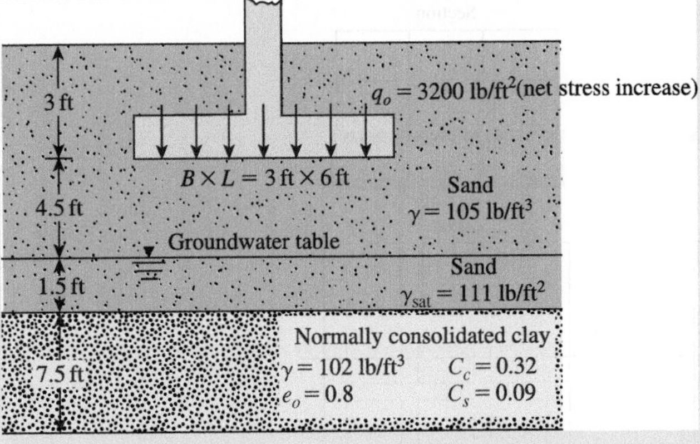

Figure 12.19 Calculation of primary consolidation settlement for a foundation ■

Mat Foundations

12.13 Common Types of Mat Foundations

Mat foundations are shallow foundations. This type of foundation, which is sometimes referred to as a *raft foundation*, is a combined footing that may cover the entire area under a structure supporting several columns and walls. Mat foundations are sometimes preferred for soils that have low load-bearing capacities but that will have to support high column and/or wall loads. Under some conditions, spread footings would have to cover more than half the building area, and mat foundations might be more economical. Several types of mat foundations are currently used. Some of the common types are shown schematically in Figure 12.20 and include the following:

1. Flat plate (Figure 12.20a). The mat is of uniform thickness.
2. Flat plate thickened under columns (Figure 12.20b).
3. Beams and slab (Figure 12.20c). The beams run both ways, and the columns are located at the intersection of the beams.

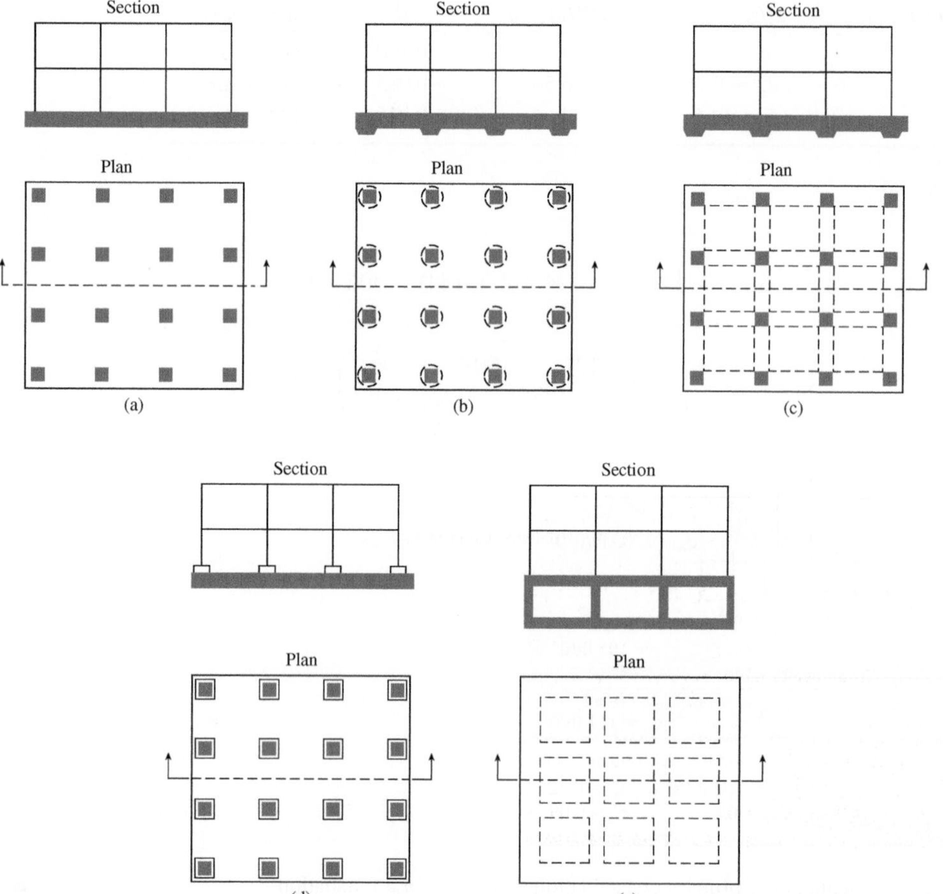

Figure 12.20 Common types of mat foundations

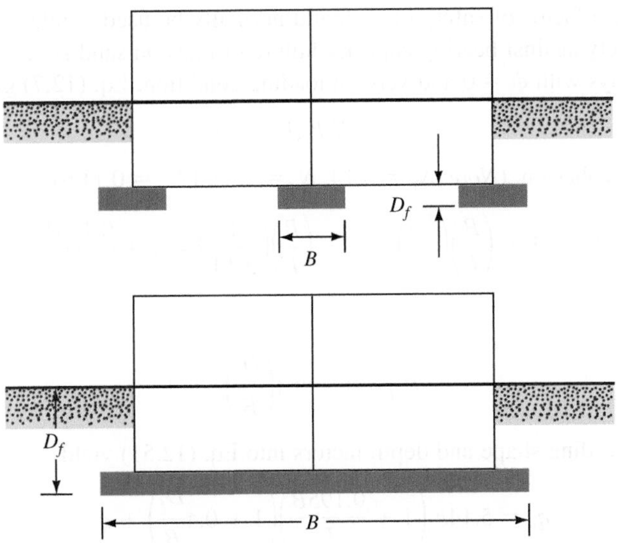

Figure 12.21 Isolated foundation and mat foundation comparison (B = width. D_f = depth)

4. Flat plates with pedestals (Figure 12.20d).
5. Slab with basement walls as a part of the mat (Figure 12.20e). The walls act as stiffeners for the mat.

Mats may be supported by piles. The piles help in reducing the settlement of a structure built over highly compressible soil. Where the water table is high, mats are often placed over piles to control buoyancy.

Figure 12.21 shows the difference between the depth D_f and the width B of isolated foundations and mat foundation.

12.14 Bearing Capacity of Mat Foundations

The *gross ultimate bearing capacity* of a mat foundation can be determined by the same equation used for shallow foundations, or

$$q_u = c'N_c F_{cs} F_{cd} F_{ci} + qN_q F_{qs} F_{qd} F_{qi} + \frac{1}{2}\gamma BN_\gamma F_{\gamma s} F_{\gamma d} F_{\gamma i} \tag{12.7}$$

Tables 12.1 and 12.2 give the proper values of the bearing capacity factors and the shape, depth, and load inclination factors. The term B in Eq. (12.7) is the smallest dimension of the mat.

The *net ultimate bearing capacity* is

$$q_{\text{net}(u)} = q_u - q \tag{12.8}$$

A suitable factor of safety should be used to calculate the net *allowable* bearing capacity. For mats on clay, the factor of safety should not be less than 3 under dead load and maximum live load. However, under the most extreme conditions, the factor of safety should be at least 1.75 to 2. For mats constructed over sand, a factor of safety of 3 should normally be used. Under most working conditions, the factor of safety against bearing capacity failure of mats on sand is very large.

For saturated clays with $\phi = 0$ and vertical loading condition. Eq. (12.7) gives

$$q_u = c_u N_c F_{cs} F_{cd} + q \tag{12.51}$$

where c_u = undrained cohesion. (*Note*: $N_c = 5.14$, $N_q = 1$, and $N_\gamma = 0$.) From Table 12.2, for $\phi = 0$,

$$F_{cs} = 1 + \left(\frac{B}{L}\right)\left(\frac{N_q}{N_c}\right) = 1 + \left(\frac{B}{L}\right)\left(\frac{1}{5.14}\right) = 1 + \frac{0.195B}{L}$$

and

$$F_{cd} = 1 + 0.4\left(\frac{D_f}{B}\right)$$

Substitution of the preceding shape and depth factors into Eq. (12.51) yields

$$q_u = 5.14c_u\left(1 + \frac{0.195B}{L}\right)\left(1 + 0.4\frac{D_f}{B}\right) + q \tag{12.52}$$

Hence, the net ultimate bearing capacity is

$$q_{\text{net}(u)} = q_u - q = 5.14c_u\left(1 + \frac{0.195B}{L}\right)\left(1 + 0.4\frac{D_f}{B}\right) \tag{12.53}$$

For FS = 3, the net allowable soil bearing capacity becomes

$$q_{\text{all(net)}} = \frac{q_{\text{net}(u)}}{\text{FS}} = 1.713c_u\left(1 + \frac{0.195B}{L}\right)\left(1 + 0.4\frac{D_f}{B}\right) \tag{12.54}$$

The net allowable bearing capacity for mats constructed over granular soil deposits can be adequately determined from standard penetration resistance numbers. In most cases of design of a shallow foundation (isolated spread footing), it is assumed that the net allowable bearing capacity corresponds to 1 in. (25.4 mm) of settlement with a differential settlement of $\frac{3}{4}$ in. (19 mm). However, the width of the mat foundation (B) is much larger than an isolated spread footing. Hence, for a mat foundation, the depth of the zone of influence of stress is much larger. Thus, the loose soil pockets under a mat may be more evenly distributed, resulting in a smaller differential settlement. On that assumption, for a maximum mat settlement of 2 in. ($\approx$50 mm), the differential settlement would be about $\frac{3}{4}$ in. ($\approx$19 mm). Using the logic and conservatively assuming $F_d = 1$ and $S_e = 2$ in. ($\approx$50 mm), Eq. (12.42b) gives

$$q_o(\text{kip/ft}^2) = \frac{N_{60}}{4}\left(\frac{B+1}{B}\right)^2(1)(2) = 0.5N_{60}\left(\frac{B+1}{B}\right)^2$$

However, if B is large, $(B + 1)$ ft $\approx B$ ft. Hence,

$$q_o(\text{kip/ft}^2) \approx 0.5N_{60} \tag{12.55a}$$

In SI units, substituting $F_d = 1$ and $S_e = 50$ mm in Eq. 12.44b,

$$q_o(\text{kN/m}^2) = \frac{N_{60}}{0.08}\left(\frac{B+0.3}{B}\right)^2(1)\left(\frac{50}{25}\right) = 25N_{60}\left(\frac{B+0.3}{B}\right)^2$$

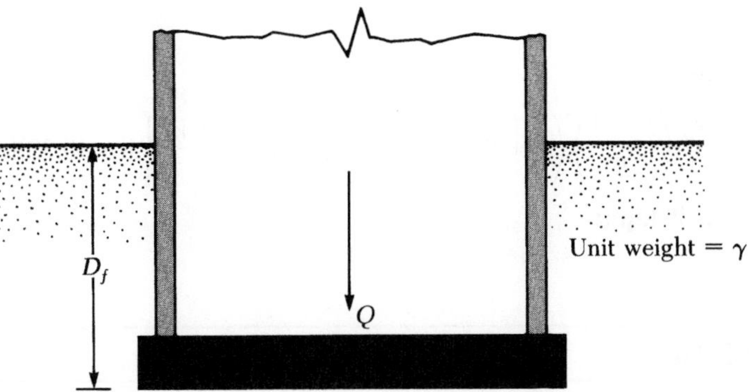

Figure 12.22 Definition of net pressure on soil caused by a mat foundation

For large B,

$$q_o(\text{kN/m}^2) \approx 25N_{60} \tag{12.55b}$$

The net pressure applied on a foundation (see Figure 12.22) may be expressed as

$$q_o = \frac{Q}{A} - \gamma D_f \tag{12.56}$$

where

Q = dead weight of the structure and the live load
A = area of the mat

Example 12.12

Determine the net ultimate bearing capacity of a mat foundation measuring 39 ft × 27 ft on a saturated clay with c_u = 1960 lb/ft², $\phi = 0$, and $D_f = 6$ ft.

Solution

From Eq. (12.53), we have

$$q_{\text{net}(u)} = 5.14c_u\left[1 + \left(\frac{0.195B}{L}\right)\right]\left[1 + 0.4\left(\frac{D_f}{B}\right)\right]$$

$$= (5.14)(1960)\left[1 + \left(\frac{0.195 \times 27}{39}\right)\right]\left[1 + 0.4\left(\frac{6}{27}\right)\right]$$

$$= 12{,}450 \text{ lb/ft}^2 = \mathbf{12.45\ kip/ft^2}$$

■

Example 12.13

What will be the net allowable bearing capacity of a mat foundation with dimensions of 39 ft × 27 ft constructed over a sand deposit? Here, $D_f = 6$ ft, allowable settlement = 1 in., and average penetration number $N_{60} = 10$.

Solution

From Eq. (12.42b):

$$q_o = \frac{N_{60}}{4}\left(\frac{B+1}{B}\right)^2\left(1+0.33\frac{D_f}{B}\right)S_e = \left(\frac{10}{4}\right)\left(\frac{27+1}{27}\right)^2\left[1+(0.33)\left(\frac{6}{27}\right)\right](1)$$

$$= \mathbf{2.89\ kip/ft^2}$$

■

Example 12.14

Determine the net allowable bearing capacity of a 10 m × 15 m raft foundation resting on a clay where c_u = 80 kN/m² and D_f = 3 m. Allow a factor of safety of 3.

Solution

From Eq. (12.54),

$$q_{\text{all(net)}} = 1.713c_u\left(1+\frac{0.195B}{L}\right)\left(1+0.4\frac{D_f}{B}\right) = 1.713(80)\left(1+\frac{0.195\times 10}{15}\right)\left(1+0.4\times\frac{3}{10}\right)$$

$$= \mathbf{173.4\ kN/m^2}$$

■

Example 12.15

Determine the allowable bearing capacity of a 10 m × 15 m mat foundation, placed at a depth of 3 m in sand (N_{60} = 27), which can accept 40 mm settlement.

Solution

From Eq. (12.43),

$$F_d = 1 + 0.33(D_f/B) = 1 + 0.33(3/10) = 1.10$$

From Eq. (12.44b),

$$q_{\text{net}} = \frac{N_{60}}{0.08}\left(\frac{B+0.3}{B}\right)^2 F_d\left(\frac{S_e}{25}\right) = \frac{27}{0.08}\left(\frac{10+0.3}{10}\right)^2(1.10)\left(\frac{40}{25}\right) = \mathbf{630\ kN/m^2}$$

■

12.15 Compensated Foundations

The settlement of a mat foundation can be reduced by decreasing the net pressure increase on soil and by increasing the depth of embedment, D_f. This increase is particularly important for mats on soft clays, where large consolidation settlements are expected. From Eq. (12.56), the net average applied pressure on soil is

$$q_o = \frac{Q}{A} - \gamma D_f$$

For no increase of the net soil pressure on soil below a mat foundation, q_o should be 0. Thus,

$$D_f = \frac{Q}{A\gamma} \tag{12.56a}$$

This relation for D_f is usually referred to as the depth of embedment of a *fully compensated foundation*.

The factor of safety against bearing capacity failure for partially compensated foundations (that is, $D_f < Q/A\gamma$) may be given as

$$\text{FS} = \frac{q_{\text{net}(u)}}{q_o} = \frac{q_{\text{net}(u)}}{\dfrac{Q}{A} - \gamma D_f} \tag{12.57}$$

For saturated clays, the factor of safety against bearing capacity failure can thus be obtained by substituting Eq. (12.53) into Eq. (12.57):

$$\text{FS} = \frac{5.14c_u\left(1 + \dfrac{0.195\,B}{L}\right)\left(1 + 0.4\dfrac{D_f}{B}\right)}{\dfrac{Q}{A} - \gamma D_f} \tag{12.58}$$

Example 12.16

Refer to Figure 12.22. The mat has dimensions of 90 ft × 120 ft, and the live load and dead load on the mat are 45×10^3 kip. The mat is placed over a layer of soft clay that has a unit weight of 120 lb/ft^3. Find D_f for a fully compensated foundation.

Solution

From Eq. (12.56a), we have

$$D_f = \frac{Q}{A\gamma} = \frac{45 \times 10^6 \text{ lb}}{(90 \times 120)(120)} = \mathbf{34.7\ ft}$$

■

Example 12.17

Refer to Example 12.16. For the clay, $c_u = 270\ \text{lb/ft}^2$. If the required factor of safety against bearing capacity failure is 3, determine the depth of the foundation.

Solution

From Eq. (12.58), we have

$$\text{FS} = \frac{5.14c_u\left(1 + \dfrac{0.195\,B}{L}\right)\left(1 + 0.4\dfrac{D_f}{B}\right)}{\dfrac{Q}{A} - \gamma D_f}$$

Here, FS = 3, $c_u = 270\ \text{lb/ft}^2$, $B/L = 90/120 = 0.75$, and $Q/A = (45 \times 10^6)/(90 \times 120) = 4166.7\ \text{lb/ft}^2$. Substituting these values into Eq. (12.58) yields

$$3 = \frac{(5.14)(270)[1 + (0.195)(0.75)]\left[1 + 0.4\left(\frac{D_f}{90}\right)\right]}{4166.7 - (120)D_f}$$

$$12{,}500.1 - 360\,D_f = 1590.77 + 7.07\,D_f$$

$$10{,}909.33 = 367.07D_f$$

or

$$D_f \approx \mathbf{29.72\ ft}$$

■

12.16 Summary

Shallow foundations include pad, strip, and mat foundations. Their depth below the ground level is not very much larger than the breadth. They are generally designed to satisfy two conditions: (a) there must be no failure within the underlying soil and (b) the settlement must be within tolerable limits.

Eccentricity and inclination in the applied load reduce the bearing capacity of a shallow foundation. The general bearing capacity equation with shape, depth and inclination factors can be used for computing the bearing capacity of shallow foundations in all soils.

As discussed in the beginning of Chapter 8, the settlement of soils consists of three components: elastic settlement, primary consolidation settlement, and secondary consolidation settlement. Methods for computing elastic settlements of flexible and rigid footings are discussed in this chapter. Flexible foundations apply uniform pressure to the underlying soil and will settle unevenly; and rigid foundations undergo uniform settlements and apply non-uniform pressures to the underlying soil.

Mat foundations can tolerate larger settlements than isolated spread footings. In sands, the bearing capacity is rarely a problem for mat foundations. By placing the mat deeper, the net applied pressure can be significantly reduced, which increases the factor of safety. These are called compensated foundations, where part of the building load is compensated by the soil overburden that was removed.

Problems

12.1 State whether the following are true or false.

a. General shear failure takes place more in loose sand than dense sand.

b. The higher the friction angle of a soil, the higher is the allowable bearing capacity of a foundation resting on it.

c. In sands, the larger the N_{60}, the larger is the settlement.

d. The factor of safety for a fully compensated foundation is infinity.

e. Larger foundations settle more.

12.2 For a continuous foundation (i.e. $B/L = 0$), the following are given:
$B = 4$ ft, $D_f = 3$ ft, $\gamma = 110\ \text{lb/ft}^3$, $\phi' = 25°$, and $c' = 600\ \text{lb/ft}^2$
Use a factor of safety of 4 to determine the gross allowable vertical load-bearing capacity.

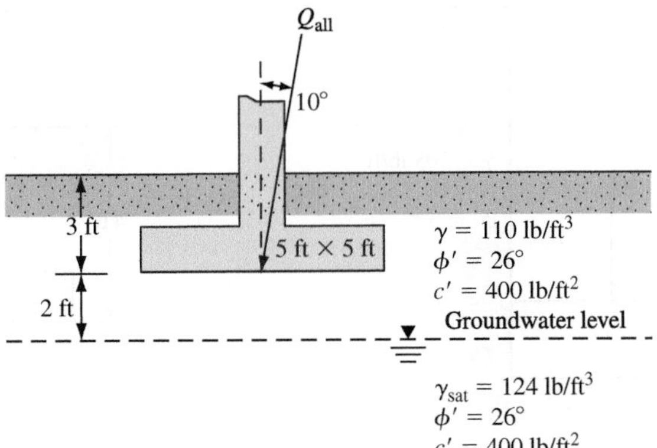

Figure 12.23

12.3 Determine the maximum column load that be applied on a 1.5 m × 1.5 m square foundation placed at a depth of 1.0 m within a soil, where $\gamma = 19.0\ \text{kN/m}^3$, $c' = 10\ \text{kN/m}^2$ and $\phi' = 24°$. Allow a factor of safety of 3.0.

12.4 A square column foundation is 6 ft × 6 ft in plan. Let $D_f = 4.5$ ft, $\gamma = 105\ \text{lb/ft}^3$, $\phi' = 36°$, and $c' = 0$. Use a factor of safety of 3 to determine the gross allowable vertical load the column could carry.

12.5 For the column foundation shown in Figure 12.23, determine the gross allowable inclined load Q_{all}. Use FS = 4.

12.6 A 2.0-m-wide strip foundation is placed in sand at 1.0 m depth. The properties of the sand are $\gamma = 19.5\ \text{kN/m}^3$, $c' = 0$, and $\phi' = 34°$. Determine the maximum wall load that the foundation can carry with a factor of safety of 3.0.

12.7 Redo problem 12.6 using Terzaghi's original bearing capacity equation [Eq. (12.3)].

12.8 A column foundation is shown in Figure 12.24. Determine the net allowable load the foundation can carry. Use FS = 3.

12.9 An eccentrically loaded foundation is shown in Figure 12.25. Determine the ultimate load Q_u that the foundation can carry.

12.10 A 1.5 m × 2.0 m foundation is placed at 1.0 m depth in a clay soil, where $\gamma = 19.0\ \text{kN/m}^3$, $c_u = 40\ \text{kN/m}^2$, and $\phi = 0$. What is the maximum column load that the foundation can carry with factor of safety of 3?

12.11 An eccentrically loaded foundation is shown in Figure 12.26. Use FS = 4, and determine the allowable load that the foundation can carry.

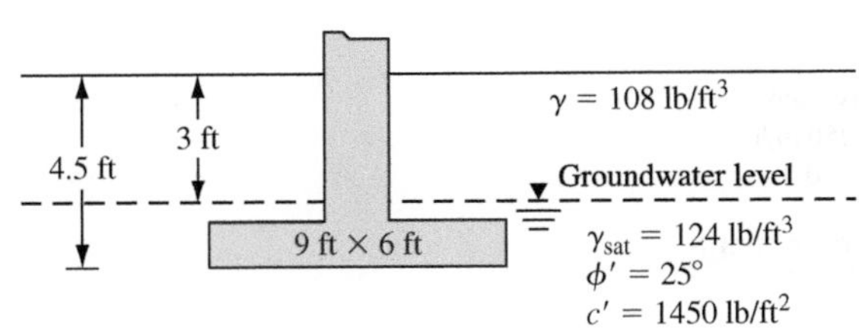

Figure 12.24

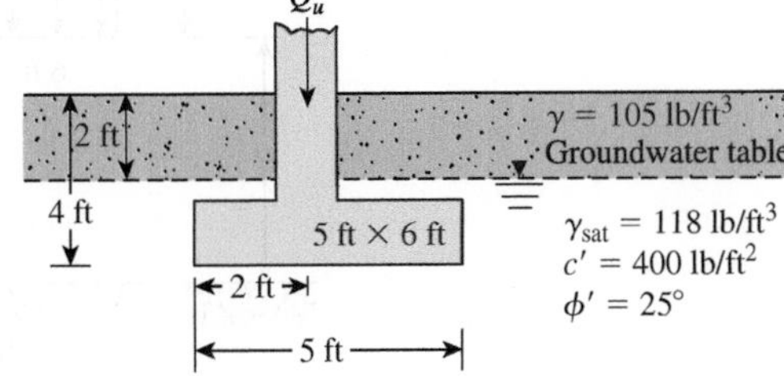

Figure 12.25

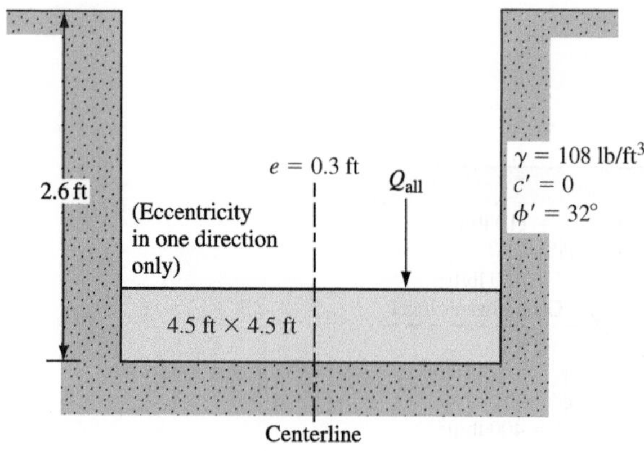

Figure 12.26

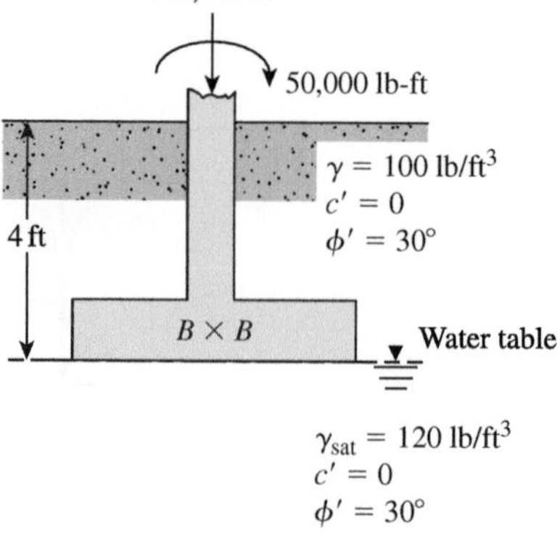

Figure 12.27

12.12 A square foundation is shown in Figure 12.27. Use FS = 6, and determine the size of the foundation.

12.13 A planned flexible load area (see Figure 12.28) is to be 6 ft × 10 ft and carries a uniformly distributed load of 4000 lb/ft^2. Estimate the elastic settlement below the center of the loaded area. Assume that $D_f = 5$ ft and $H = \infty$.

12.14 Determine the eccentricity and the inclination of the foundation load in the two cases shown in Figure 12.29.

12.15 A 5 m × 10 m flexible foundation (raft) applies a net uniform pressure of 300 kN/m^2 to the underlying soil. The foundation is placed 5 m below the ground level, and there is bedrock at a depth of 5 m below the bottom of the raft. The modulus of elasticity and Poisson's ratio of the soil are 40 MN/m^2 and 0.3, respectively. Determine the elastic settlement at the center and corner of the foundation. What would be the settlement if the foundation is made rigid?

12.16 Figure 12.11 shows a foundation 10 ft × 6.25 ft resting on a sand deposit. The net load per unit area at the level of the foundation, q_o, is 3000 lb/ft^2. For the sand, $\mu_s = 0.3$, $E_s = 3200$ lb/in.^{2}, $D_f = 2.5$ ft, and $H = 32$ ft. Assume that the foundation is rigid, and determine the elastic settlement that the foundation would undergo.

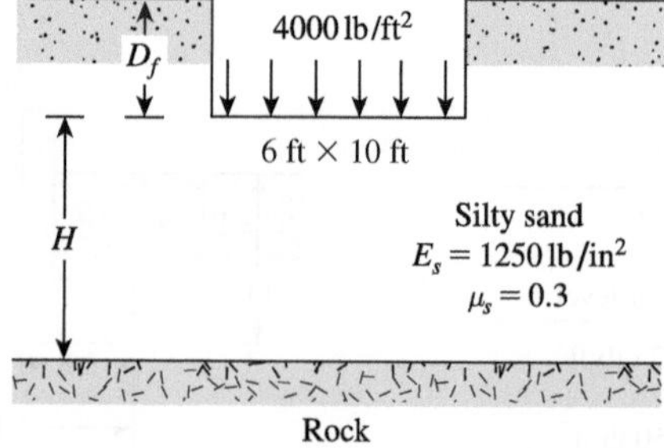

Figure 12.28

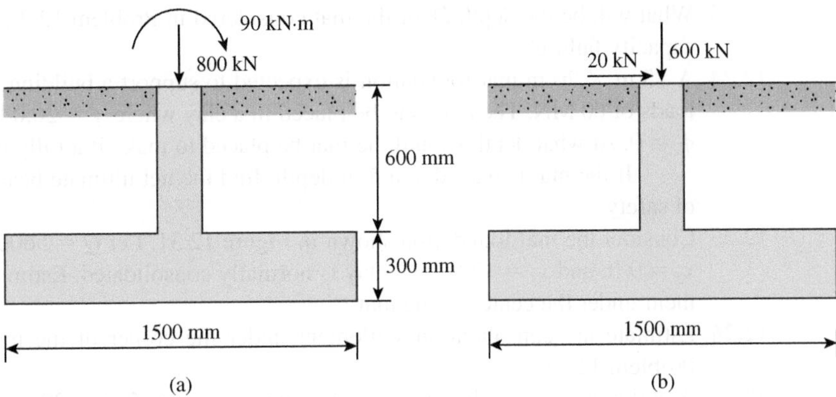

Figure 12.29

12.17 A square column foundation is shown in Figure 12.30. Estimate the consolidation of the clay layer.

12.18 Determine the net ultimate bearing capacity of a mat foundation measuring 45 ft × 30 ft on a saturated clay with c_u = 1950 lb/ft^2, ϕ = 0, and D_f = 6.5 ft.

12.19 Determine the net ultimate bearing of a 10 m × 20 m raft placed at a depth of 6 m below the ground level in a clay where c_u = 40 kN/m^2 and ϕ = 0.

12.20 What will be the net allowable bearing capacity of a mat foundation with dimensions of 45 ft × 30 ft constructed over a sand deposit? Let D_f = 6 ft, allowable settlement = 1.5 in., and average penetration number N_{60} = 10.

12.21 Repeat Problem 12.20 for an allowable settlement of 2 in.

12.22 Consider a mat foundation with dimensions of 60 ft × 39 ft. The dead and live load on the mat is 9000 kip. The mat is to be placed on a clay with c_u = 850 lb/ft^2. The unit weight of the clay is 111 lb/ft^3. Find the depth D_f of the mat for a fully compensated foundation.

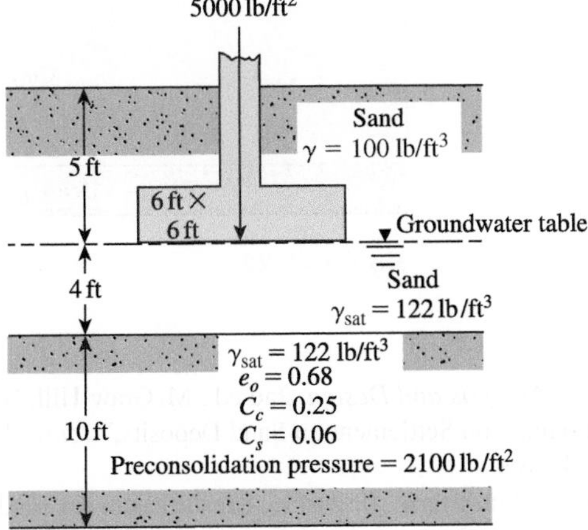

Figure 12.30

12.23 What will be the depth D_f of the mat considered in Problem 12.22 for FS = 3 against bearing capacity failure?

12.24 A 15 m × 20 m mat foundation is expected to support a building that includes dead and live loads of 60 MN. The mat is to be placed in a clay where $\gamma = 20.0$ kN/m^3, $c_u = 50$ kN/m^2, and $\phi = 0$. At what depth should the mat be placed to make it a fully compensated foundation?

If the mat is placed at a 5 m depth, find the net ultimate bearing capacity and the factor of safety.

12.25 Consider the mat foundation shown in Figure 12.31. Let $Q = 5600$ kip, $D_f = 4.5$ ft, $x_1 = 6$ ft, $x_2 = 9$ ft, and $x_3 = 12$ ft. The clay is normally consolidated. Estimate the consolidation settlement under the center of the mat.

12.26 Estimate the consolidation settlement under the corner of the mat foundation described in Problem 12.25.

12.27 A building shown in Figure 12.32 is supported by a 15 m × 20 m mat foundation that applies a net pressure of 60 kN/m^2 to the soil below. The soil properties are as follows:

Sand: $\gamma = 19.0$ kN/m^3

Clay: $\gamma_{sat} = 19.5$ kN/m^3, $e_o = 0.85$, $C_c = 0.43$, $C_s = 0.07$, $OCR = 2.5$

Determine the consolidation settlement under the center of the mat.

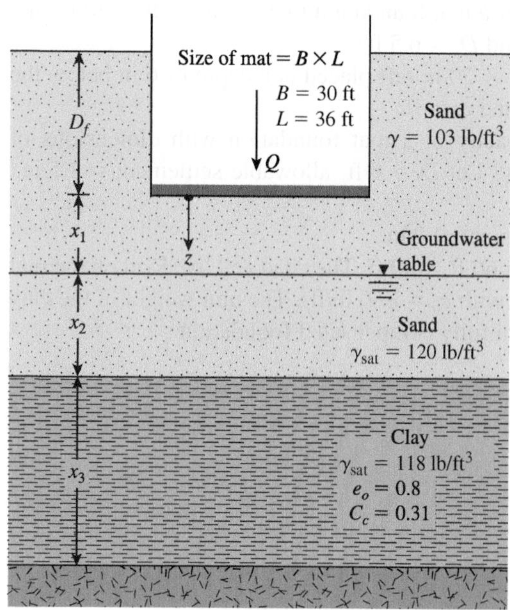

Figure 12.31

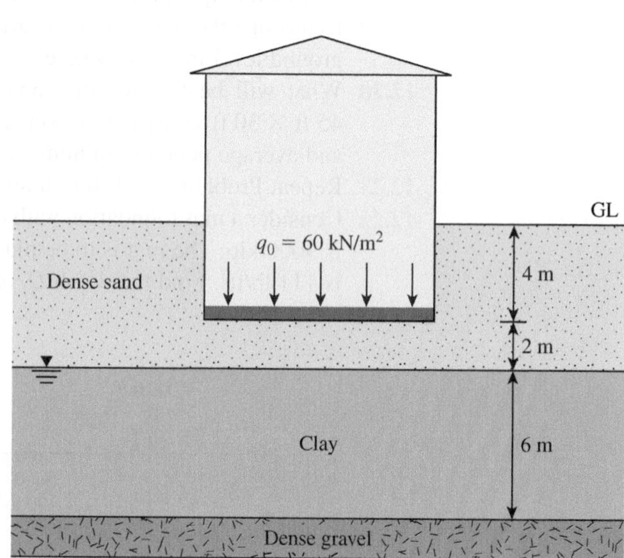

Figure 12.32

References

Bowles, J. E. (1977). *Foundation Analysis and Design*, 2nd ed., McGraw-Hill, New York.

Bowles, J. E. (1987). "Elastic Foundation Settlement on Sand Deposits," *Journal of Geotechnical Engineering*, ASCE, Vol. 113, No. 8, 846–860.

Caquot, A. and Kerisel, J. (1953). "Sur le terme de surface dans le calcul des fondations en milieu pulverulent," *Proceedings*, Third International Conference on Soil Mechanics and Foundation Engineering, Zürich, Vol. I, 336–337.

De Beer, E. E. (1970). "Experimental Determination of the Shape Factors and Bearing Capacity Factors of Sand," *Geotechnique*, Vol. 20, No. 4, 387–411.

Fox, E. N. (1948). "The Mean Elastic Settlement of a Uniformaly Loaded Area at a Depth Below the Ground Surface," *Proceedings*, 2nd International Conference on Soil Mechanics and Foundation Engineering, Rotterdam, Vol. 1, pp. 129–132.

Hanna, A. M. and Meyerhof, G. G. (1981). "Experimental Evaluation of Bearing Capacity of Footings Subjected to Inclined Loads," *Canadian Geotechnical Journal*, Vol. 18, No. 4, 599–603.

Hansen, J. B. (1970). "A Revised and Extended Formula for Bearing Capacity," Danish Geotechnical Institute, *Bulletin 28*, Copenhagen.

Meyerhof, G. G. (1953). "The Bearing Capacity of Foundations Under Eccentric and Inclined Loads," *Proceedings*, Third International Conference on Soil Mechanics and Foundation Engineering, Zürich, Vol. 1, 440–445.

Meyerhof, G. G. (1963). "Some Recent Research on the Bearing Capacity of Foundations," *Canadian Geotechnical Journal*, Vol. 1, No. 1, 16–26.

Meyerhof, G. G. (1965). "Shallow Foundations," *Journal of the Soil Mechanics and Foundations Division*, ASCE, Vol. 91, No. SM2, 21–31.

Mitchell, J. K., and Gardner, W. S. (1975). "*In Situ* Measurement of Volume Change Characteristics," *Proceedings, Specialty Conference*, American Society of Civil Engineers, Vol. 2, pp. 279–345.

Prandtl, L. (1921). "über die Eindringungsfestigkeit (Härte) plastischer Baustoffe und die Festigkeit von Schneiden," *Zeitschrift für angewandte Mathematik und Mechanik*, Vol. 1, No. 1, 15–20.

Reissner, H. (1924). "Zum Erddruckproblem," *Proceedings*, First International Congress of Applied Mechanics, Delft, 295–311.

Schmertmann, J. H. (1970). "Static Cone to Compute Settlement Over Sand," *Journal of the Soil Mechanics and Foundations Division*, American Society of Civil Engineers, Vol. 96, No. SM3, pp. 1011–1043.

Steinbrenner, W. (1934). "Tafeln zur Setzungsberechnung," *Die Strasse*, Vol. 1, pp. 121–124.

Terzaghi, K. (1943). *Theoretical Soil Mechanics*, Wiley, New York.

Terzaghi, K. and Peck, R. B. (1967). *Soil Mechanics in Engineering Practice*, 2nd ed., Wiley, New York.

Vesic, A. S. (1963). "Bearing Capacity of Deep Foundations in Sand," *Highway Research Record No. 39*, National Academy of Sciences, 112–153.

Vesic, A. S. (1973). "Analysis of Ultimate Loads of Shallow Foundations," *Journal of the Soil Mechanics and Foundations Division*, American Society of Civil Engineers, Vol. 99, No. SM1, 45–73.

13 Deep Foundations

13.1 Introduction

In Chapter 12, we briefly discussed pile and drilled shaft foundations. They were identified as deep foundations. In this chapter, we will discuss the pile foundations and drilled shafts in more details. In this chapter, you will learn the following.

- Different types of piles.
- Engineering applications of pile foundations.
- Calculation of ultimate bearing capacity of piles.
- Single piles versus pile groups.
- Pile driving formulae.
- Pile load tests.
- Bearing capacity of a pile group.
- Settlement of a pile group.
- Drilled shafts.

Pile Foundations

13.2 Pile Foundations—General

Piles are structural members that are made of steel, concrete, or timber. They are used to build pile foundations, which are deep and which cost more than shallow foundations. Despite the cost, the use of piles often is necessary to ensure structural safety. The following list identifies some of the conditions that require pile foundations (Vesic, 1977):

1. When one or more upper soil layers are highly compressible and too weak to support the load transmitted by the superstructure, piles are used to transmit the load to underlying bedrock or a stronger soil layer, as shown in Figure 13.1a. When bedrock is not encountered at a reasonable depth below the ground surface, piles are used to transmit the structural load to the soil

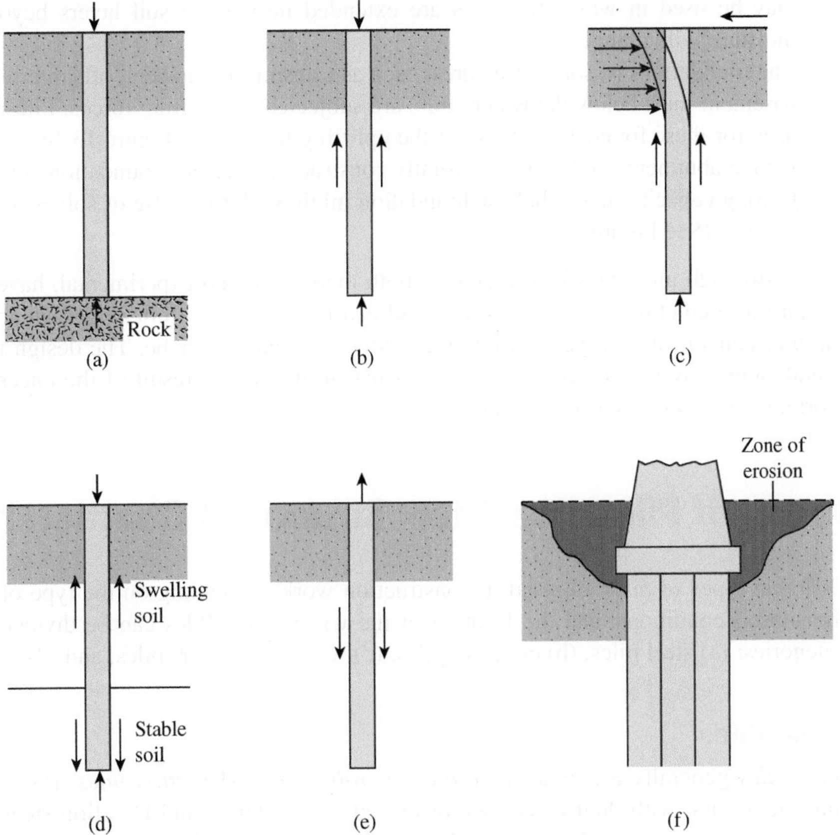

Figure 13.1 Conditions that require the use of pile foundations

gradually. The resistance to the applied structural load is derived mainly from the frictional resistance developed at the soil–pile interface. (See Figure 13.1b.)

2. When subjected to horizontal forces (see Figure 13.1c), pile foundations resist by bending, while still supporting the vertical load transmitted by the superstructure. This type of situation is generally encountered in the design and construction of earth-retaining structures and foundations of tall structures such as chimneys and transmission towers. That are subjected to high wind or to earthquake forces.
3. In many cases, expansive and collapsible soils may be present at the site of a proposed structure. These soils may extend to a great depth below the ground surface. Expansive soils swell and shrink as their moisture content increases and decreases, and the pressure of the swelling can be considerable. If shallow foundations are used in such circumstances, the structure may suffer considerable damage. However, pile foundations may be considered as an alternative when piles are extended beyond the active zone, which is where swelling and shrinking occur. (See Figure 13.1d)

   Soils such as loess are collapsible in nature. When the moisture content of these soils increases, their structures may break down. A sudden decrease in the void ratio of soil induces large settlements of structures supported by shallow foundations. In such cases, pile foundations may be used in which the piles are extended into stable soil layers beyond the zone where moisture will change.

4. The foundations of some structures, such as transmission towers, offshore platforms, and basement mats below the water table, are subjected to uplifting forces. Piles are sometimes used for these foundations to resist the uplifting force. (See Figure 13.1e.)
5. Bridge abutments and piers are usually constructed over pile foundations to avoid the loss of bearing capacity that a shallow foundation might suffer because of soil erosion at the ground surface. (See Figure 13.1f.)

Although numerous investigations, both theoretical and experimental, have been conducted in the past to predict the behavior and the load-bearing capacity of piles in granular and cohesive soils, the mechanisms are not yet entirely understood and may never be. The design and analysis of pile foundations may thus be considered somewhat of an art as a result of the uncertainties involved in working with some subsoil conditions.

13.3 Types of Piles and Their Structural Characteristics

Different types of piles are used in construction work, depending on the type of load to be carried, the subsoil conditions, and the location of the water table. Piles can be divided into the following categories: (a) steel piles, (b) concrete piles, (c) wooden (timber) piles, and (d) composite piles.

Steel Piles

Steel piles generally are either *pipe piles* or *rolled steel* H-*section piles*. Pipe piles can be driven into the ground with their ends open or closed. Wide-flange and I-section steel beams can also be used as piles. However, H-section piles are usually preferred because their web and flange thicknesses are equal. (In wide-flange and I-section beams, the web thicknesses are smaller than the thicknesses of the flange.) Tables 13.1a and b gives the dimensions of some standard H-section steel

Table 13.1a Common H-Pile Sections used in the United States (English Units)

Designation size (in.) × weight (lb/ft)	Depth d_1 (in.)	Section area (in^2)	Flange and web thickness w (in.)	Flange width d_2 (in.)	Moment of inertia (in^4) I_{XX}	Moment of inertia (in^4) I_{YY}
HP 8 × 36	8.02	10.6	0.445	8.155	119	40.3
HP 10 × 57	9.99	16.8	0.565	10.225	294	101
× 42	9.70	12.4	0.420	10.075	210	71.7
HP 12 × 84	12.28	24.6	0.685	12.295	650	213
× 74	12.13	21.8	0.610	12.215	570	186
× 63	11.94	18.4	0.515	12.125	472	153
× 53	11.78	15.5	0.435	12.045	394	127
HP 13 × 100	13.15	29.4	0.766	13.21	886	294
× 87	12.95	25.5	0.665	13.11	755	250
× 73	12.74	21.6	0.565	13.01	630	207
× 60	12.54	17.5	0.460	12.90	503	165

(continued)

Designation size (in.) × weight (lb/ft)	Depth d_1 (in.)	Section area (in^2)	Flange and web thickness w (in.)	Flange width d_2 (in.)	Moment of inertia (in^4)	
					I_{XX}	I_{YY}
HP 14 × 117	14.21	34.4	0.805	14.89	1220	443
× 102	14.01	30.0	0.705	14.78	1050	380
× 89	13.84	26.1	0.615	14.70	904	326
× 73	13.61	21.4	0.505	14.59	729	262

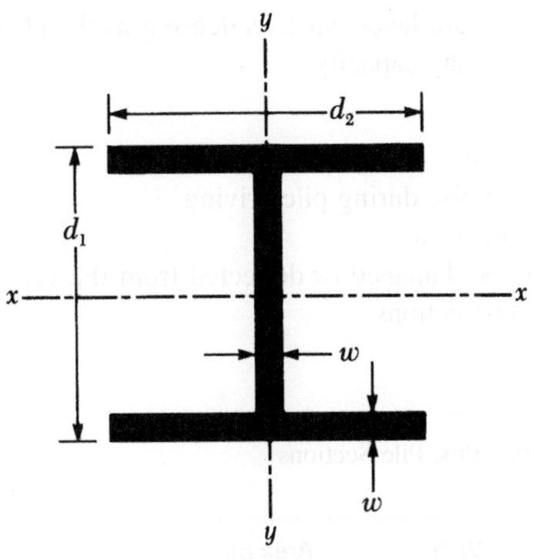

Table 13.1b Common H-Pile Sections used in the United States (SI Units)

Designation size (mm) × weight (kg/m)	Depth d_1 (mm)	Section area (m^2 × 10^{-3})	Flange and web thickness w (mm)	Flange width d_2 (mm)	Moment of inertia (m^4 × 10^{-5})	
					I_{XX}	I_{YY}
HP 200 × 53	204	6.84	11.3	207	49.4	16.8
HP 250 × 85	254	10.8	14.4	260	123	42
× 62	246	8.0	10.6	256	87.5	24
HP 310 × 125	312	15.9	17.5	312	271	89
× 110	308	14.1	15.49	310	237	77.5
× 93	303	11.9	13.1	308	197	63.7
× 79	299	10.0	11.05	306	164	62.9
HP 330 × 149	334	19.0	19.45	335	370	123
× 129	329	16.5	16.9	333	314	104
× 109	324	13.9	14.5	330	263	86
× 89	319	11.3	11.7	328	210	69
HP 360 × 174	361	22.2	22.45	378	508	184
× 152	356	19.4	17.91	376	437	158
× 132	351	16.8	15.62	373	374	136
× 108	346	13.8	12.82	371	303	109

piles used in the United States. Tables 13.2a and b show selected pipe sections frequently used for piling purposes.

The following are some general facts about steel piles.

- Usual length: 50 ft to 200 ft (≈ 15 m to 60 m)
- Usual load: 67 kip to 265 kip (≈ 300 kN to 1200 kN)
- Advantages:
  a. Easy to handle with respect to cutoff and extension to the desired length
  b. Can stand high driving stresses
  c. Can penetrate hard layers such as dense gravel and soft rock
  d. High load-carrying capacity
- Disadvantages:
  a. Relatively costly
  b. High level of noise during pile driving
  c. Subject to corrosion
  d. H-piles may be damaged or deflected from the vertical during driving through hard layers or past major obstructions

Table 13.2a Selected Pipe Pile Sections (English Units)

Outside diameter (in.)	Wall thickness (in.)	Area of steel (in^2)
$8\frac{5}{8}$	0.125	3.34
	0.188	4.98
	0.219	5.78
	0.312	8.17
10	0.188	5.81
	0.219	6.75
	0.250	7.66
12	0.188	6.96
	0.219	8.11
	0.250	9.25
16	0.188	9.34
	0.219	10.86
	0.250	12.37
18	0.219	12.23
	0.250	13.94
	0.312	17.34
20	0.219	13.62
	0.250	15.51
	0.312	19.30
24	0.250	18.7
	0.312	23.2
	0.375	27.8
	0.500	36.9

Table 13.2b Selected Pipe Pile Sections (SI Units)

Outside diameter (mm)	Wall thickness (mm)	Area of steel (cm^2)
219	3.17	21.5
	4.78	32.1
	5.56	37.3
	7.92	52.7
254	4.78	37.5
	5.56	43.6
	6.35	49.4
305	4.78	44.9
	5.56	52.3
	6.35	59.7
406	4.78	60.3
	5.56	70.1
	6.35	79.8
457	5.56	80
	6.35	90
	7.92	112
508	5.56	88
	6.35	100
	7.92	125
610	6.35	121
	7.92	150
	9.53	179
	12.70	238

Concrete Piles

Concrete piles may be divided into two basic categories: (a) precast piles and (b) cast-*in-situ* piles. *Precast piles* can be prepared by using ordinary reinforcement, and they can be square or octagonal in cross section. (See Figure 13.2.) Reinforcement is provided to enable the pile to resist the bending moment developed during pickup and transportation, the vertical load, and the bending moment caused by a lateral load. The piles are cast to desired lengths and cured before being transported to the work sites.

Some general facts about concrete piles are as follows:

- Usual length: 30 ft to 50 ft (≈ 10 m to 15 m)
- Usual load: 67 kip to 675 kip (≈ 300 kN to 3000 kN)
- Advantages:
  a. Can be subjected to hard driving
  b. Corrosion resistant
  c. Can be easily combined with a concrete superstructure
- Disadvantages:
  a. Difficult to achieve proper cutoff
  b. Difficult to transport

Precast piles can also be prestressed by the use of high-strength steel prestressing cables. The ultimate strength of these cables is about 260 ksi (≈ 1800 MN/m^2). During casting of the piles, the cables are pretensioned to about 130 to 190 ksi (≈ 900 to 1300 MN/m^2), and concrete is poured around them. After curing, the cables are cut, producing a compressive force on the pile section. Tables 13.3a and b give additional information about prestressed concrete piles with square and octagonal cross sections.

Some general facts about precast prestressed piles are as follows:

- Usual length: 30 ft to 150 ft (≈ 10 m to 45 m)
- Maximum length: 200 ft (≈ 60 m)
- Maximum load: 1700 kip to 1900 kip (≈ 7500 kN to 8500 kN)

The advantages and disadvantages are the same as those of precast piles.

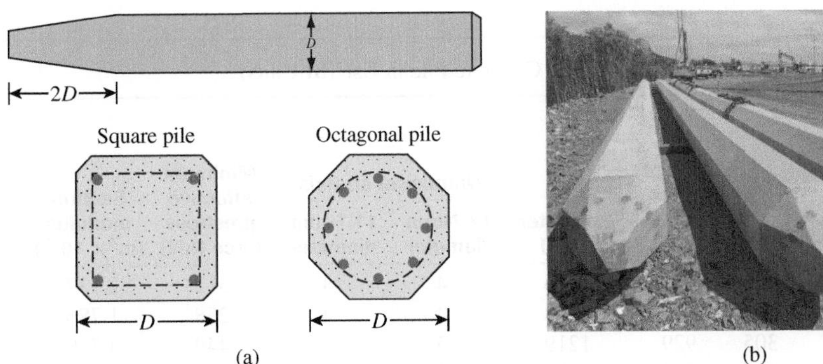

Figure 13.2 Precast piles with ordinary reinforcement: (a) schematic diagram; (b) photograph of octagonal precast piles ready for driving (*Courtesy of N. Sivakugan, James Cook University, Australia*)

Table 13.3a Typical Prestressed Concrete Pile in Use (English Units)

Pile shape[a]	D (in.)	Area of cross section (in²)	Perimeter (in.)	Number of strands: $\frac{1}{2}$-in diameter	Number of strands: $\frac{7}{16}$-in diameter	Minimum effective prestress force (kip)	Section modulus (in³)	Design bearing capacity (kip), Strength of concrete: 5000 psi	Design bearing capacity (kip), Strength of concrete: 6000 psi
S	10	100	40	4	4	70	167	125	175
O	10	83	33	4	4	58	109	104	125
S	12	144	48	5	6	101	288	180	216
O	12	119	40	4	5	83	189	149	178
S	14	196	56	6	8	137	457	245	295
O	14	162	46	5	7	113	300	203	243
S	16	256	64	8	11	179	683	320	385
O	16	212	53	7	9	148	448	265	318
S	18	324	72	10	13	227	972	405	486
O	18	268	60	8	11	188	638	336	402
S	20	400	80	12	16	280	1333	500	600
O	20	331	66	10	14	234	876	414	503
S	22	484	88	15	20	339	1775	605	727
O	22	401	73	12	16	281	1166	502	602
S	24	576	96	18	23	403	2304	710	851
O	24	477	80	15	19	334	2123	596	716

[a]S = square section; O = octagonal section

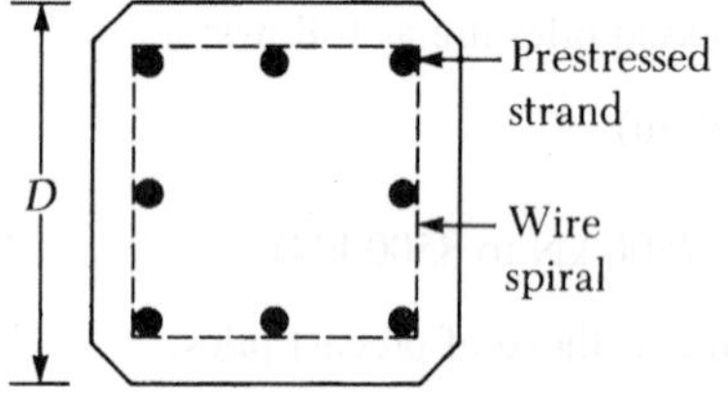

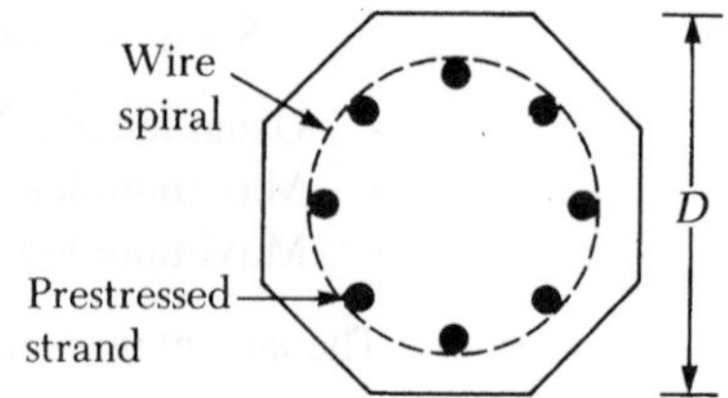

Table 13.3b Typical Prestressed Concrete Pile in Use (SI Units)

Pile shape[a]	D (mm)	Area of cross section (cm²)	Perimeter (mm)	Number of strands: 12.7-mm diameter	Number of strands: 11.1-mm diameter	Minimum effective prestress force (kN)	Section modulus ($m^3 \times 10^{-3}$)	Design bearing capacity (kN), Strength of concrete (MN/m²): 34.5	Design bearing capacity (kN), Strength of concrete (MN/m²): 41.4
S	254	645	1016	4	4	312	2.737	556	778
O	254	536	838	4	4	258	1.786	462	555
S	305	929	1219	5	6	449	4.719	801	962
O	305	768	1016	4	5	369	3.097	662	795
S	356	1265	1422	6	8	610	7.489	1091	1310
O	356	1045	1168	5	7	503	4.916	901	1082

(*continued*)

Pile shape[a]	D (mm)	Area of cross section (cm^2)	Perimeter (mm)	Number of strands: 12.7-mm diameter	Number of strands: 11.1-mm diameter	Minimum effective prestress force (kN)	Section modulus ($m^3 \times 10^{-3}$)	Design bearing capacity (kN), Strength of concrete (MN/m^2): 34.5	Design bearing capacity (kN), Strength of concrete (MN/m^2): 41.4
S	406	1652	1626	8	11	796	11.192	1425	1710
O	406	1368	1346	7	9	658	7.341	1180	1416
S	457	2090	1829	10	13	1010	15.928	1803	2163
O	457	1729	1524	8	11	836	10.455	1491	1790
S	508	2581	2032	12	16	1245	21.844	2226	2672
O	508	2136	1677	10	14	1032	14.355	1842	2239
S	559	3123	2235	15	20	1508	29.087	2694	3232
O	559	2587	1854	12	16	1250	19.107	2231	2678
S	610	3658	2438	18	23	1793	37.756	3155	3786
O	610	3078	2032	15	19	1486	34.794	2655	3186

[a]S = square section; O = octagonal section

Cast-in-situ, or *cast-in-place piles* are built by making a hole in the ground and then filling it with concrete. Various types of cast-in-place concrete piles are currently used in construction, and most of them have been patented by their manufacturers. These piles may be divided into two broad categories: (a) cased and (b) uncased. Both types may have a pedestal at the bottom.

Cased piles are made by driving a steel casing into the ground with the help of a mandrel placed inside the casing. When the pile reaches the proper depth the mandrel is withdrawn and the casing is filled with concrete. Figures 13.3a, 13.3b, 13.3c, and 13.3d show some examples of cased piles without a pedestal. Figure 13.3e shows a cased pile with a pedestal. The pedestal is an expanded concrete bulb that is formed by dropping a hammer on fresh concrete.

Some general facts about cased cast-in-place piles are as follows.

- Usual length: 15 ft to 50 ft ($\approx$ 5 m to 15 m)
- Maximum length: 100 ft to 130 ft ($\approx$ 30 m to 40 m)
- Usual load: 45 kip to 115 kip ($\approx$ 200 kN to 500 kN)
- Approximate maximum load: 180 kip (800 kN)
- Advantages:
  a. Relatively cheap
  b. Allow for inspection before pouring concrete
  c. Easy to extend
- Disadvantages:
  a. Difficult to splice after concreting
  b. Thin casings may be damaged during driving

Figures 13.3f and 13.3g are two types of uncased pile, one with a pedestal and the other without. The uncased piles are made by first driving the casing to the desired depth and then filling it with fresh concrete. The casing is then gradually withdrawn.

Following are some general facts about uncased cast-in-place concrete piles:

- Usual length: 15 ft to 50 ft ($\approx$ 5 m to 15 m)
- Maximum length: 100 ft to 130 ft ($\approx$ 30 m to 40 m)

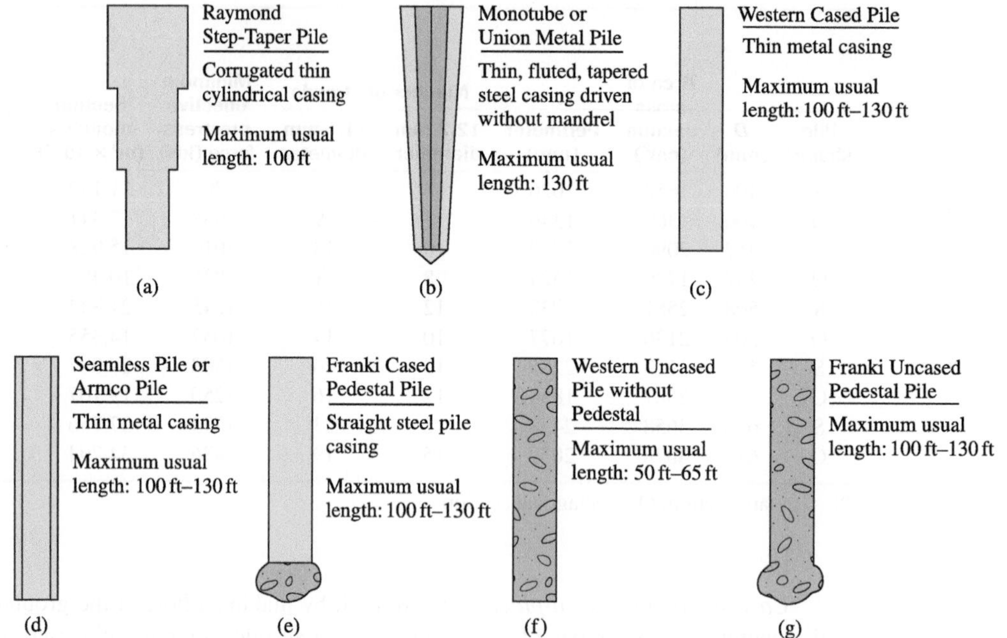

Figure 13.3 Cast-in-place concrete piles

- Usual load: 67 kip to 115 kip (≈ 300 kN to 500 kN)
- Approximate maximum load: 160 kip (≈ 700 kN)
- Advantages:
  a. Initially economical
  b. Can be finished at any elevation
- Disadvantages:
  a. Voids may be created if concrete is placed rapidly
  b. Difficult to splice after concreting
  c. In soft soils, the sides of the hole may cave in, squeezing the concrete

Timber Piles

Timber piles are tree trunks that have had their branches and bark carefully trimmed off. The maximum length of most timber piles is 30 to 65 ft (≈ 9 m to 20 m). To qualify for use as a pile, the timber should be straight, sound, and without any defects. The American Society of Civil Engineers' *Manual of Practice*, No. 17 (1959) divided timber piles into three classes:

1. *Class A piles* carry heavy loads. The minimum diameter of the butt should be 14 in. (356 mm).
2. *Class B piles* are used to carry medium loads. The minimum butt diameter should be 12 to 13 in. (305 to 330 mm).
3. *Class C piles* are used in temporary construction work. They can be used permanently for structures when the entire pile is below the water table. The minimum butt diameter should be 12 in. (305 mm).

In any case, a pile tip should not have a diameter less than 6 in. (152 mm).

Timber piles cannot withstand hard driving stress; therefore, the pile capacity is generally limited. Steel shoes may be used to avoid damage at the pile tip (bottom). The tops of timber piles may also be damaged during the driving operation. The crushing of the wooden fibers caused by the impact of the hammer is referred to as *brooming*. To avoid damage to the top of the pile, a metal band or a cap may be used.

The usual length of wooden piles is 15 ft to 50 ft (4.5 to 15 m). The maximum length is about 100 ft to 130 ft (30.5 m to 40 m). The usual load carried by wooden piles is 65 kip to 115 kip (290 kN to 510 kN).

Composite Piles

The upper and lower portions of *composite piles* are made of different materials. For example, composite piles may be made of steel and concrete or timber and concrete. Steel-and-concrete piles consist of a lower portion of steel and an upper portion of cast-in-place concrete. This type of pile is used when the length of the pile required for adequate bearing exceeds the capacity of simple cast-in-place concrete piles. Timber-and-concrete piles usually consist of a lower portion of timber pile below the permanent water table and an upper portion of concrete. In any case, forming proper joints between two dissimilar materials is difficult, and for that reason, composite piles are not widely used. Presently, fiber-reinforced concrete and fiber-reinforced polymer (FRP) are also being used from time to time.

13.4 Estimating Pile Length

Selecting the type of pile to be used and estimating its necessary length are fairly difficult tasks that require good judgment. In addition to being broken down into the classification given in Section 13.3, piles can be divided into three major categories, depending on their lengths and the mechanisms of load transfer to the soil (a) point bearing piles, (b) friction piles, and (c) compaction piles.

Point Bearing Piles

If soil-boring records establish the presence of bedrock or rocklike material at a site within a reasonable depth, piles can be extended to the rock surface. (See Figure 13.4a.) In this case, the ultimate capacity of the piles depends entirely on the load-bearing capacity of the underlying material; thus, the piles are called *point bearing piles*. In most of these cases, the necessary length of the pile can be fairly well established.

If, instead of bedrock, a fairly compact and hard stratum of soil is encountered at a reasonable depth, piles can be extended a few feet into the hard stratum. (See Figure 13.4b.) Piles with pedestals can be constructed on the bed of the hard stratum, and the ultimate pile load may be expressed as

$$Q_u = Q_p + Q_s \tag{13.1}$$

where

Q_p = load carried at the pile point
Q_s = load carried by skin friction developed at the side of the pile (caused by shearing resistance between the soil and the pile)

If Q_s is very small,

$$Q_s \approx Q_p \tag{13.2}$$

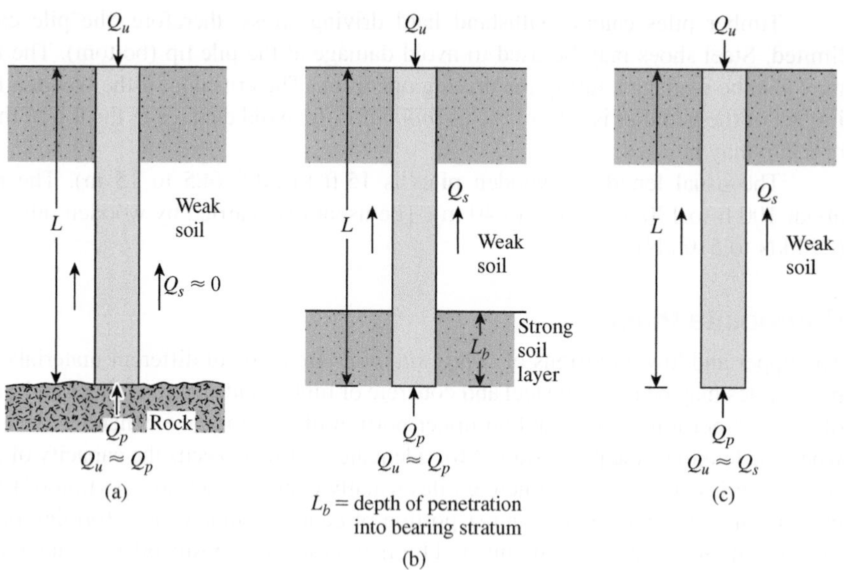

Figure 13.4 (a) and (b) Point bearing piles; (c) friction piles

In this case, the required pile length may be estimated accurately if proper subsoil exploration records are available.

Friction Piles

When no layer of rock or rocklike material is present at a reasonable depth at a site, point bearing piles become very long and uneconomical. In this type of subsoil, piles are driven through the softer material to specified depths. (See Figure 13.4c.) The ultimate load of the piles may be expressed by Eq. (13.1). However, if the value of Q_p is relatively small, then

$$Q_u \approx Q_s \tag{13.3}$$

These piles are called *friction piles*, because most of their resistance is derived from skin friction. However, the term *friction pile*, although used often in the literature, is a misnomer: In clayey soils, the resistance to applied load is also caused by *adhesion*.

The lengths of friction piles depend on the shear strength of the soil, the applied load, and the pile size. To determine the necessary lengths of these piles, an engineer needs a good understanding of soil–pile interaction, good judgment, and experience. Theoretical procedures for calculating the load-bearing capacity of piles are presented later in the chapter.

Compaction Piles

Under certain circumstances, piles are driven in granular soils to achieve proper compaction of soil close to the ground surface. These piles are called *compaction piles*. The lengths of compaction piles depend on factors such as (a) the relative density of the soil before compaction, (b) the desired relative density of the soil after compaction, and (c) the required depth of compaction. These piles are generally short; however, some field tests are necessary to determine a reasonable length.

13.5 Installation of Piles

Most piles are driven into the ground by means of *hammers* or *vibratory drivers*. In special circumstances, piles can also be inserted by *jetting* or *partial augering*. The types of hammer used for pile driving include (a) the drop hammer, (b) the single-acting air or steam hammer, (c) the double-acting and differential air or steam hammer, and (d) the diesel hammer. In the driving operation, a cap is attached to the top of the pile. A cushion may be used between the pile and the cap. The cushion has the effect of reducing the impact force and spreading it over a longer time; however, the use of the cushion is optional. A hammer cushion is placed on the pile cap. The hammer drops on the cushion.

Figure 13.5 illustrates various hammers. A drop hammer (see Figure 13.5a) is raised by a winch and allowed to drop from a certain height H. It is the oldest type of hammer used for pile driving. The main disadvantage of the drop hammer is its slow rate of blows. The principle of the single-acting air or steam hammer is shown in Figure 13.5b. The striking part, or ram, is raised by air or steam pressure and then drops by gravity. Figure 13.5c shows the operation of the double-acting and differential air or steam hammer. Air or steam is used both to raise the ram and to push it downward, thereby increasing the impact velocity of the ram. The diesel hammer (see Figure 13.5d) consists essentially of a ram, an anvil block, and a fuel-injection system. First the ram is raised and fuel is injected near the anvil. Then the ram is released. When the ram drops, it compresses the air–fuel mixture, which ignites. This action, in effect, pushes the pile downward and raises the ram. Diesel hammers work well under hard driving conditions. In soft soils, the downward movement of the pile is rather large, and the upward

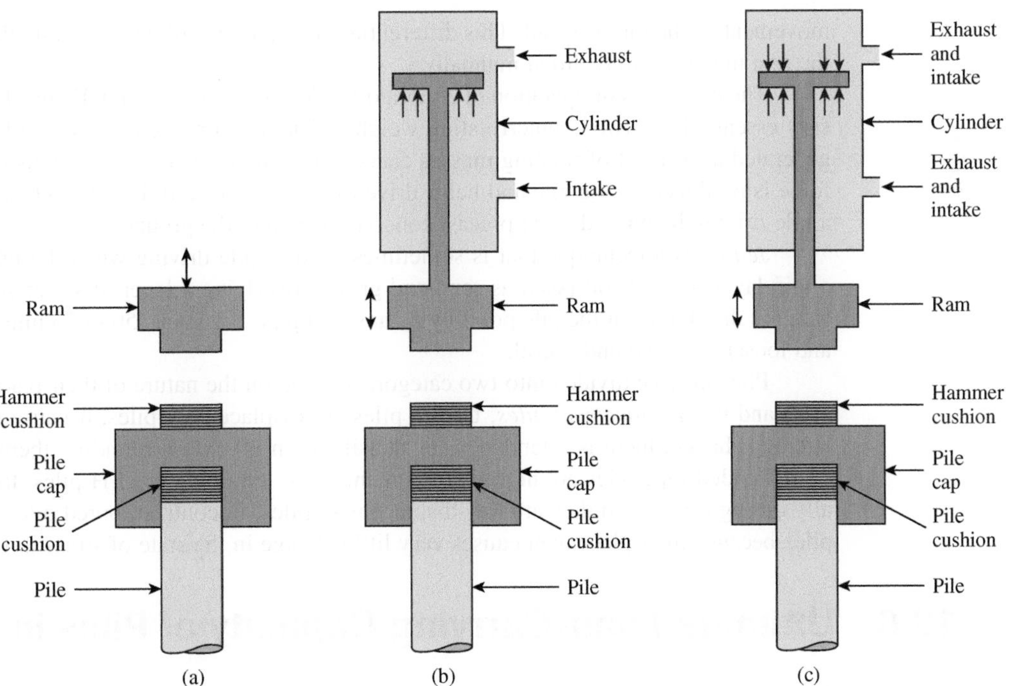

Figure 13.5 Pile-driving equipment: (a) drop hammer; (b) single-acting air or steam hammer; (c) double-acting and differential air or steam hammer

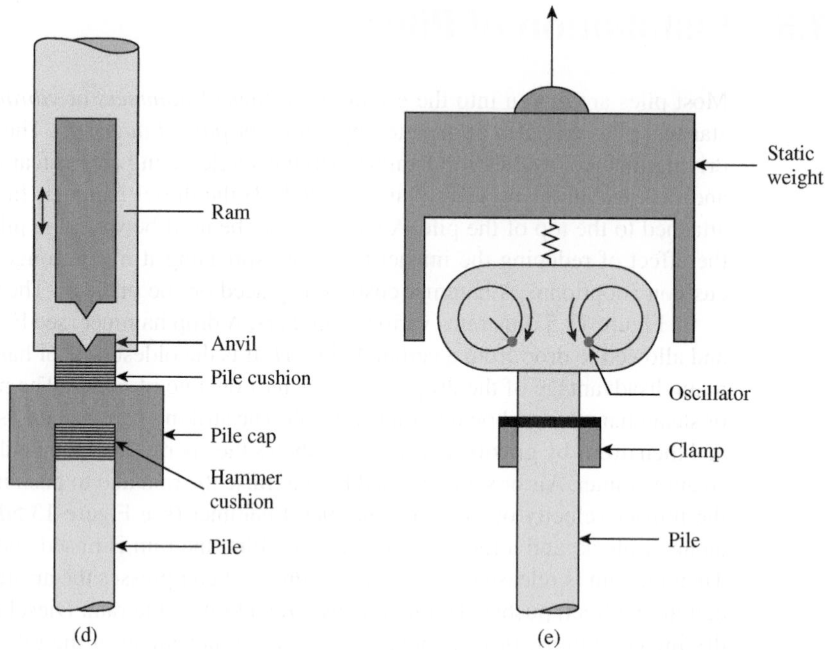

Figure 13.5 (*continued*) (d) diesel hammer; (e) vibratory pile driver

movement of the ram is small. This differential may not be sufficient to ignite the air–fuel system, so the ram may have to be lifted manually.

The principles of operation of a vibratory pile driver are shown in Figure 13.5e. This driver consists essentially of two counterrotating weights. The horizontal components of the centrifugal force generated as a result of rotating masses cancel each other. As a result, a sinusoidal dynamic vertical force is produced on the pile and helps drive the pile downward. Figure 13.6 shows a photograph of a pile-driving hammer driving precast concrete piles into the ground.

Jetting is a technique that is sometimes used in pile driving when the pile needs to penetrate a thin layer of hard soil (such as sand and gravel) overlying a layer of softer soil. In this technique, water is discharged at the pile point by means of a pipe 2 to 3 in. (50 to 75 mm) in diameter to wash and loosen the sand and gravel.

Piles may be divided into two categories based on the nature of their placement: *displacement piles* and *nondisplacement piles*. Driven piles are displacement piles, because they move some soil laterally; hence, there is a tendency for densification of soil surrounding them. Concrete piles and closed-ended pipe piles are high-displacement piles. However, steel H-piles displace less soil laterally during driving, so they are low-displacement piles. In contrast, bored piles are nondisplacement piles because their placement causes very little change in the state of stress in the soil.

13.6 Ultimate Load-Carrying Capacity of Piles in Sand

The ultimate load-carrying capacity, Q_u, has been expressed in Eq. (13.1) as

$$Q_u = Q_p + Q_s$$

Figure 13.6 Installing octagonal precast concrete piles using a pile driving hammer (*Courtesy of N. Sivakugan, James Cook University, Australia*)

where

Q_p = load-carrying capacity of the pile point
Q_s = frictional resistance (skin friction) derived from the soil–pile interface

In sand, the point load Q_p can be expressed as

$$Q_p = A_p q_p = A_p q' N_q^* \tag{13.4}$$

where

A_p = area of the pile tip
q_p = unit point resistance
q' = effective vertical stress at the level of the pile tip
N_q^* = bearing capacity factor

The variation of N_q^* with soil friction angle (ϕ') is shown in Figure 13.7 (Meyerhof, 1976). However, research has shown that the unit point resistance increases with L/D (D = width or diameter of the pile) and reaches a limiting value of q_l as shown in Figure 13.8b. So, Eq. (13.4) can be modified as

$$Q_p = A_p q' N_q^* \leq A_p q_l \tag{13.5}$$

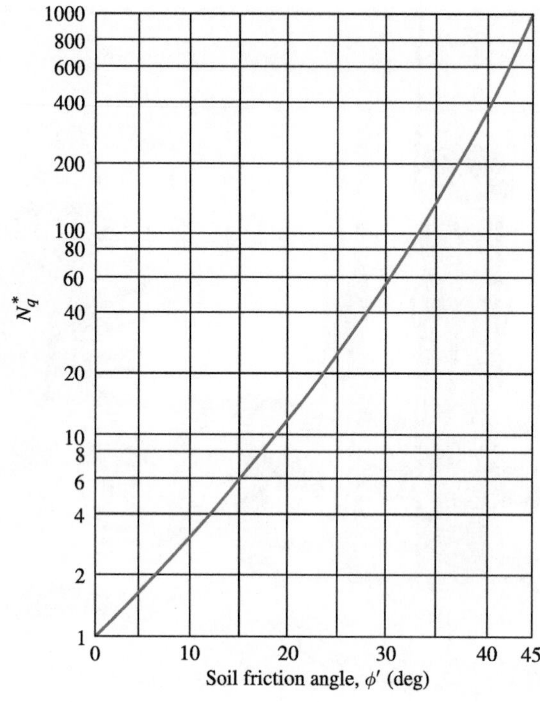

Figure 13.7 Meyerhof's bearing capacity factor, N_q^*

Meyerhof (1976) suggested that

$$q_l(\text{lb/ft}^2) = 1000 N_q^* \tan\phi' \tag{13.6a}$$

In SI units

$$q_l(\text{kN/m}^2) = 48\, N_q^* \tan\phi' \tag{13.6b}$$

The frictional resistance, Q_s, can be estimated as

$$Q_s = \Sigma p \Delta L f \tag{13.7}$$

where

p = perimeter of the pile cross section
f = unit frictional resistance

The magnitude of f can be expressed as

$$f = K\sigma_o' \tan\delta' \tag{13.8}$$

where

σ_o' = effective vertical stress at any depth measured from the ground surface
δ' = soil–pile friction angle (ranges from 0.5 ϕ' to 0.8 ϕ')

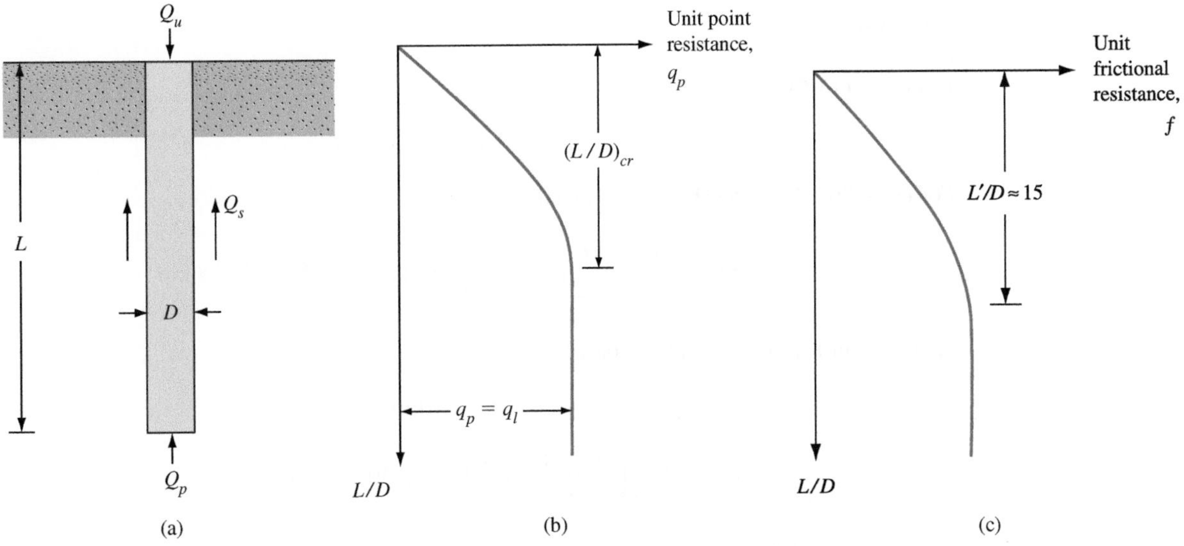

Figure 13.8 Variation of q_p and f with L/D

Based on results presently available, the following average values of K are recommended for use in Eq. (13.8):

Pile type	K
Bored or jetted	$\approx K_o = 1 - \sin\phi'$
Low-displacement driven	$\approx K_o = 1 - \sin\phi'$ to $1.4K_o = 1.4(1 - \sin\phi')$
High-displacement driven	$\approx K_o = 1 - \sin\phi'$ to $1.8K_o = 1.8(1 - \sin\phi')$

The effective vertical stress, σ'_o, for use in Eq. (13.8) increases with pile depth to a maximum limit at a depth of 15 to 20 pile diameters and remains constant thereafter, as shown in Figure 13.8c. This critical depth, L', depends on several factors, such as the soil friction angle and compressibility and relative density. A conservative estimate is to assume that

$$L' = 15D \tag{13.9}$$

Example 13.1

A concrete pile is 50 ft (L) long and 16 in. × 16 in. in cross section. The pile is fully embedded in sand for which $\gamma = 110 \text{ lb/ft}^3$ and $\phi' = 30°$. Determine

a. Ultimate load, Q_p
b. Ultimate load, Q_s. Use $K = 1.3$ and $\delta' = 0.8\phi'$
c. Allowable total load-carrying capacity of the pile. Use FS = 4.

Solution

Part a

From Eq. (13.5),

$$Q_p = A_p q' N_q^* = A_p \gamma L N_q^*$$

For $\phi' = 30°$, $N_q^* \approx 55$ (Figure 13.7), so

$$Q_p = \left(\frac{16 \times 16}{12 \times 12}\,\text{ft}^2\right)\left(\frac{110 \times 50}{1000}\,\text{kip/ft}^2\right)(55) = 537.8\ \text{kip}$$

Again, from Eqs. (13.5) and (13.6a),

$$Q_p = A_p q_l = A_p N_q^* \tan\phi' (\text{kip/ft}^2)$$

$$= \left(\frac{16 \times 16}{12 \times 12}\,\text{ft}^2\right)(55)\tan 30 = 56.45\ \text{kip}$$

Hence, $Q_p =$ **56.45 kip**.

Part b

From Eq. (13.9),

$$L' \approx 15D = 15\left(\frac{16}{12}\,\text{ft}\right) = 20\ \text{ft}$$

From Eq. (13.8), at $z = 0$, $\sigma_o' = 0$, so $f = 0$. Again, at $z = L' = 20$ ft,

$$\sigma_o' = \gamma L' = \frac{(110)(20)}{1000} = 2.2\ \text{kip/ft}^2$$

So

$$f = K\sigma_o' \tan\delta' = (1.3)(2.2)[\tan(0.8 \times 30)] = 1.273\ \text{kip/ft}^2$$

Thus,

$$Q_s = \left(\frac{f_{z=0} + f_{z=20\,\text{ft}}}{2}\right)pL' + f_{z=20\,\text{ft}}\,p(L - L')$$

$$= \left(\frac{0 + 1.273}{2}\right)\left(4 \times \frac{16}{12}\right)(20) + (1.273)\left(4 \times \frac{16}{12}\right)(50 - 20)$$

$$= 67.9 + 203.7 = \mathbf{271.6\ kip}$$

Part c

$$Q_{\text{all}} = \frac{Q_p + Q_s}{\text{FS}} = \frac{56.45 + 271.6}{4} \approx \mathbf{82\ kip}$$

■

Example 13.2

A 500-mm-diameter and 20-m-long concrete pile is driven into a sand where $\gamma = 18.5$ kN/m^3 and $\phi' = 32°$. Assuming $\delta' = 0.7\ \phi'$ and $K = 1.5\ K_o$, determine the load carrying capacity of the pile with a factor of safety of 3.

Solution

$$K_o = 1 - \sin 32 = 0.470;\ N_q^* = 78;\ \text{and}\ \delta' = 22.4°$$

Therefore, $K = 1.5 \times 0.470 = 0.705$

$$A_p = \frac{\pi}{4}D^2 = \frac{\pi}{4} \times 0.5^2 = 0.196\ \text{m}^2 \text{ and perimeter } p = \pi D = 1.571\ \text{m}$$

Point Load Q_P

$$q'N_q^* = (20 \times 18.5)(78) = 28{,}860\ \text{kN/m}^2$$

$$q_l = 48N_q^* \tan\phi' = (48)(78)(\tan 32) = 2340\ \text{kN/m}^2$$

Using the smaller of the above two values, $Q_p = 0.196 \times 2340 =$ **459 kN**

Shaft Load Q_s
The critical depth is $L' = 15D = 7.5$ m. Let's apply Eq. (13.8) for computing the skin friction as $f = K\sigma_o' \tan\delta'$:

At $z = 0$, $f = 0$
At $z = 7.5$ m, $f = (0.705)(7.5 \times 18.5)(\tan 22.4) = 40.3$ kN/m^2
At $z = 20$ m, $f = 40.3$ kN/m^2

$$Q_s = \left(\frac{0 + 40.3}{2}\right)(7.5)(1.571) + (40.3)(20 - 7.5)(1.571) = \mathbf{1029\ kN}$$

The ultimate load $Q_u = 459 + 1029 = 1488$ kN.
The allowable load $Q_{all} = 1488/3 =$ **496 kN.** ■

13.7 Ultimate Load-Carrying Capacity of Piles in Clay ($\phi = 0$ Condition)

The ultimate load-carrying capacity of piles in saturated clay also can be expressed by Eq. (13.1) as

$$Q_u = Q_p + Q_s$$

The ultimate point load is given by the expression

$$Q_p = A_p q_p = A_p c_u N_c^* \tag{13.10}$$

where

c_u = undrained cohesion at the pile tip
N_c^* = bearing capacity factor = 9.0

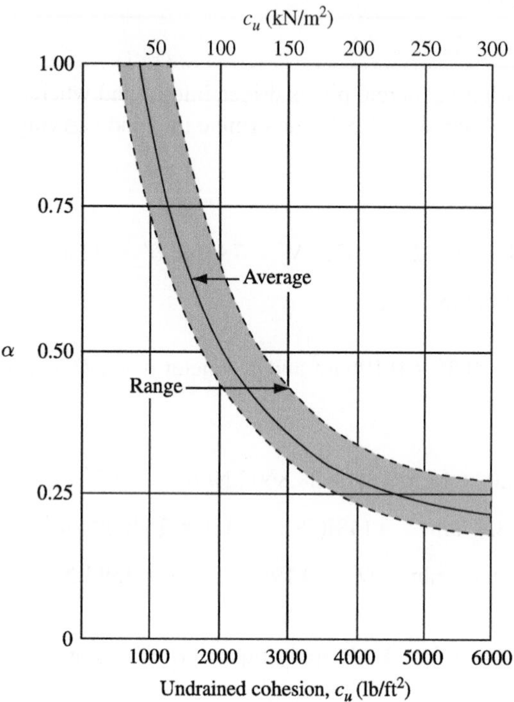

Figure 13.9 Variation of α with undrained cohesion of clay

Thus,

$$Q_p = 9A_p c_u \tag{13.11}$$

The skin resistance, Q_s, can be given by the expression

$$Q_s = \Sigma p \Delta L c_a = \Sigma p \Delta L \alpha c_u \tag{13.12}$$

where

c_a = adhesion at the soil–pile interface
α = a nondimensional quantity which is a function of c_u (Figure 13.9)

Example 13.3

A concrete pile 16 in. × 16 in. in cross section is shown in Figure 13.10. Determine the allowable load that the pile can carry (FS = 4). Use the α method for determination of the skin resistance.

Solution

From Eq. (13.10),

$$Q_p = A_p q_p = A_p c_u N_c^* = \frac{\left(\frac{16}{12}\right)^2 (1800)(9)}{1000} = 28.8 \text{ kips}$$

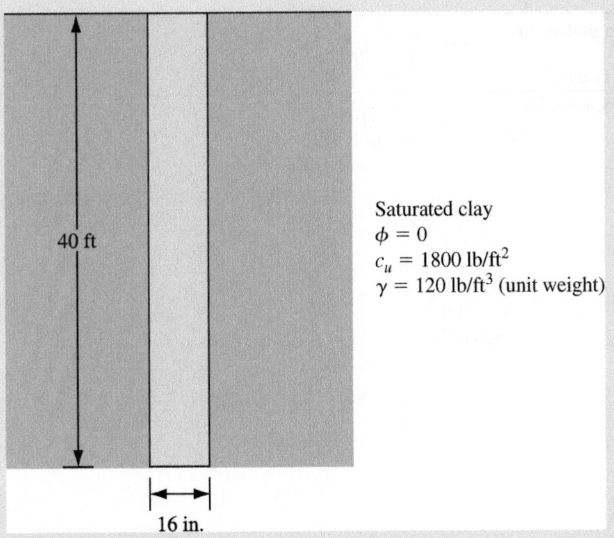

Figure 13.10

Again, from Eq. (13.12),

$$Q_s = \alpha c_u pL$$

From the average plot of Figure 13.9, for $c_u = 1800$ lb/ft^2, $\alpha = 0.55$. So

$$Q_s = \frac{(0.55)(1800)\left(4 \times \frac{16}{12}\right)(40)}{1000} = 211.2 \text{ kips}$$

$$Q_{all} = \frac{Q_p + Q_s}{\text{FS}} = \frac{28.8 + 211.2}{4} = \mathbf{60\ kips}$$

■

Example 13.4

A driven pipe pile in clay is shown in Figure 13.11. The pipe has an outside diameter of 16 in., and the wall thickness is 0.25 in.

a. Calculate the net point bearing capacity.
b. Calculate the skin resistance.
c. Estimate the allowable pile capacity. Use FS = 4.

Solution

Area of cross section of pile including the soil inside the pile =

$$A_p = \frac{\pi}{4}D^2 = \frac{\pi}{4}(16/12)^2 = 1.396 \text{ ft}^2$$

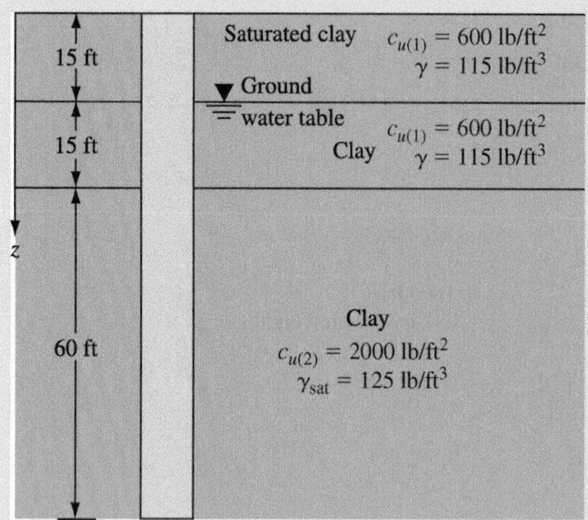

Figure 13.11

Part a

Calculation of point-bearing capacity from Eq. (13.10),

$$Q_p = A_p q_p = A_p N_c^* c_{u(2)} = (1.396)(9)(2000) = 25{,}128 \text{ lb}$$

Part b

Calculation of skin resistance

$$Q_s = \Sigma \alpha c_u p \Delta L$$

For the top soil layer, $c_u(1) = 600$ lb/ft^2. According to the average plot of Figure 13.9, $\alpha_1 = 1.0$. Similarly, for the bottom soil layer, $c_{u(2)} = 2000$ lb/ft^2; $\alpha_2 = 0.5$. Thus

$$\begin{aligned} Q_s &= \alpha_1 c_{u(1)}[(\pi)(16/12)]30 + \alpha_2 c_{u(2)}[(\pi)(16/12)]60 \\ &= (1)(600)[(\pi)(16/12)]30 + (0.5)(2000)[(\pi)(16/12)]60 \\ &= 75{,}398 + 251{,}327 = 326{,}725 \text{ lb} \end{aligned}$$

Part c

Allowable pile capacity,

$$Q_{all} = \frac{Q_p + Q_s}{\text{FS}} = \frac{25{,}128 + 326{,}725}{4} = 87{,}963 \text{ lb} \approx \mathbf{88\ kip}$$

■

Example 13.5

Determine the maximum load that can be allowed on a 500 mm diameter and 18 m long pile driven into a clay where $\gamma = 20.0$ kN/m^3 and $c_u = 60$ kN/m^2. Allow factor of safety of three.

What percentage of the ultimate load is being carried by the pile shaft?

Solution

Point Load Q_p

From Eq (13.11),

$$Q_p = 9A_p c_u = (9)\left(\frac{\pi}{4} \times 0.5^2\right)(60) = 106 \text{ kN}$$

Shaft Load Q_s

For $c_u = 60$ kN/m², from Figure 13.9, $\alpha = 0.78$. Hence,

$$Q_s = (\pi \times 0.5 \times 18.0)(0.78)(60) = 1323 \text{ kN}$$

$$Q_u = 106 + 1323 = 1429 \text{ kN}$$

$$Q_{\text{all}} = 1429/3 = \mathbf{476\ kN}$$

Percentage of the ultimate load carried by the pile shaft $= \dfrac{1323}{1429} \times 100 = \mathbf{93\%}$.

This is a friction pile, where most of the load is carried by the shaft. ■

13.8 Pile-Driving Formulas

To develop the desired load-carrying capacity, a point bearing pile must sufficiently penetrate the dense soil layer or have sufficient contact with a layer of rock. This requirement cannot always be satisfied by driving a pile to a predetermined depth, because soil profiles vary. For that reason, several equations have been developed to calculate the ultimate capacity of a pile during driving. These dynamic equations are widely used in the field to determine if the pile has reached satisfactory bearing value at the predetermined depth. One of the earliest of these dynamic equations—commonly referred to as the *Engineering News Record* (*ENR*) *formula*—is derived on the basis of the work-energy theory. This means that

Energy imparted by the hammer per blow =
(pile resistance) × (penetration per hammer blow)

According to the ENR formula, the pile resistance is the ultimate load Q_u and can be expressed as

$$Q_u = \frac{W_R h}{S + C} \tag{13.13}$$

where

W_R = weight of the ram (see Figure 13.5)
h = height of fall of the ram
S = penetration of pile per hammer blow, known as set
C = a constant

The pile penetration, S, is usually based on the average value obtained from the last few driving blows. In the equation's original form, the following values of C were recommended.

For drop hammers:

$$C = \begin{cases} 1 \text{ in. (if the units of } S \text{ and } h \text{ are in inches)} \\ 25.4 \text{ mm (if the units of } S \text{ and } h \text{ are in mm)} \end{cases}$$

For steam hammers:

$$C = \begin{cases} 0.1 \text{ in. (if the units of } S \text{ and } h \text{ are in inches)} \\ 2.54 \text{ mm (if the units of } S \text{ and } h \text{ are in mm)} \end{cases}$$

Also, a factor of safety, FS = 6, was recommended to estimate the allowable pile capacity. Note that, for single- and double-acting hammers, the term $W_R h$ can be replaced by EH_E (where E = hammer efficiency and H_E = rated energy of hammer). Thus,

$$Q_u = \frac{EH_E}{S + C} \tag{13.14}$$

The ENR pile-driving formula has gone through several revisions over the years. A recent form—the *modified ENR formula*—can be given as

$$Q_u = \frac{EW_R h}{S + C} \cdot \frac{W_R + n^2 W_p}{W_R + W_p} \tag{13.15}$$

where

E = hammer efficiency
C = 0.1 in. if the units of S and h are in inches (or 2.54 mm if the units of S and h are in mm)
W_p = weight of the pile
n = coefficient of restitution between the ram and the pile cap

The efficiencies of various pile-driving hammers, E, are in the following ranges:

Hammer type	Efficiency, E
Single- and double-acting hammers	0.7 to 0.85
Diesel hammers	0.8 to 0.9
Drop hammers	0.7 to 0.9

Representative values of the coefficient of restitution, n, are given in the following table.

Pile material	Coefficient of restitution, n
Cast iron hammer and concrete piles (without cap)	0.4 to 0.5
Wood cushion on steel piles	0.3 to 0.4
Wooden piles	0.25 to 0.3

A factor of safety (FS) of 4 to 6 may be used in Eq. (13.15) to obtain the allowable load-bearing capacity of a pile.

Example 13.6

A precast concrete pile 12 in. × 12 in. in cross section is driven by a hammer. Given:

- Maximum rated hammer energy = 26 kips · ft
- Weight of ram = 8 kips
- Total length of pile = 65 ft
- Hammer efficiency = 0.8
- Coefficient of restitution = 0.45
- Weight of pile cap = 0.72 kip
- Number of blows for last 1 in. of penetration = 5

Estimate the allowable pile capacity by using the following:

a. Equation (13.14) (use FS = 6)
b. Equation (13.15) (use FS = 4)

Solution

Part a

Use of Eq. (13.14):

$$Q_u = \frac{EH_E}{S + C}$$

Given:

$$E = 0.8, H_E = 26 \text{ kips}\cdot\text{ft}$$

$$S = \frac{1}{5} = 0.2 \text{ in.}$$

So

$$Q_u = \frac{\overbrace{(0.8)(26)(12)}^{\substack{\text{kips . in.}\\ \downarrow}}}{0.2 + 0.1} = 832 \text{ kips}$$

Hence,

$$Q_{\text{all}} = \frac{Q_u}{\text{FS}} = \frac{832}{6} = \mathbf{138.7\ kips}$$

Part b

Use of Eq. (13.15):

$$Q_u = \frac{EW_R h}{S + C} = \frac{W_R + n^2 W_p}{W_R + W_p}$$

$$\text{Weight of piles} = WA_p\gamma_c = (65 \text{ ft})(1 \text{ ft} \times 1 \text{ ft})(150 \text{ lb/ft}^3)$$
$$= 9750 \text{ lb} = 9.75 \text{ kips}$$

$$W_p = \text{weight of pile} + \text{weight of cap}$$

$$= 9.75 + 0.72 = 10.47 \text{ kips}$$

So

$$Q_u = \left[\frac{(0.8)(26)(12)}{0.2 + 0.1}\right]\left[\frac{8 + (0.45)^2(10.47)}{8 + 10.47}\right]$$

$$= (832)(0.548) = 455.9 \text{ kips}$$

$$Q_{\text{all}} = \frac{Q_u}{\text{FS}} = \frac{455.9}{4} = \mathbf{114\ kips}$$

■

Example 13.7

A single-acting stream hammer (Vulcan Model 08) is used to drive a 400-mm-diameter and 20-m-long precast concrete pile into the ground. The following data are recorded.

- Weight of ram = 35.6 kN
- Stroke (height of fall of the ram) = 1.0 m
- Maximum rated hammer energy = 35.6 kN · m
- Hammer efficiency = 0.8
- Coefficient of restitution = 0.4
- Weight of pile cap = 5 kN
- Number of blows recorded for the last 25 mm penetration = 6
- Unit weight of concrete = 24 kN/m^3

Determine the allowable load for the pile with a factor of safety of 4 and using the modified ENR formula (Eq. 13.15).

Solution

$$W_p = W_{\text{pile}} + W_{\text{cap}} = \frac{\pi}{4} \times 0.4^2 \times 20 \times 24 + 5.0 = 65.3 \text{ kN}$$

$$S = 6/25 = 0.24 \text{ blows/mm}$$

$$Q_u = \frac{EW_R h}{S + C} \times \frac{W_R + n^2 W_p}{W_R + W_p} = \frac{(0.8)(35.6)(1000)}{0.24 + 2.54} \times \frac{35.6 + (0.4^2)(65.3)}{35.6 + 65.3} = 4675.4 \text{ kN}$$

$$Q_{\text{all}} = 4675.4/4 = \mathbf{1169\ kN}$$

■

13.9 Pile Load Tests

In most large projects, a specific number of load tests must be conducted on piles. This is primarily because of the unreliability of prediction methods. Vertical and lateral load-bearing capacity of a given pile can be tested in the field. Figure 13.12a shows a schematic diagram of the pile load test

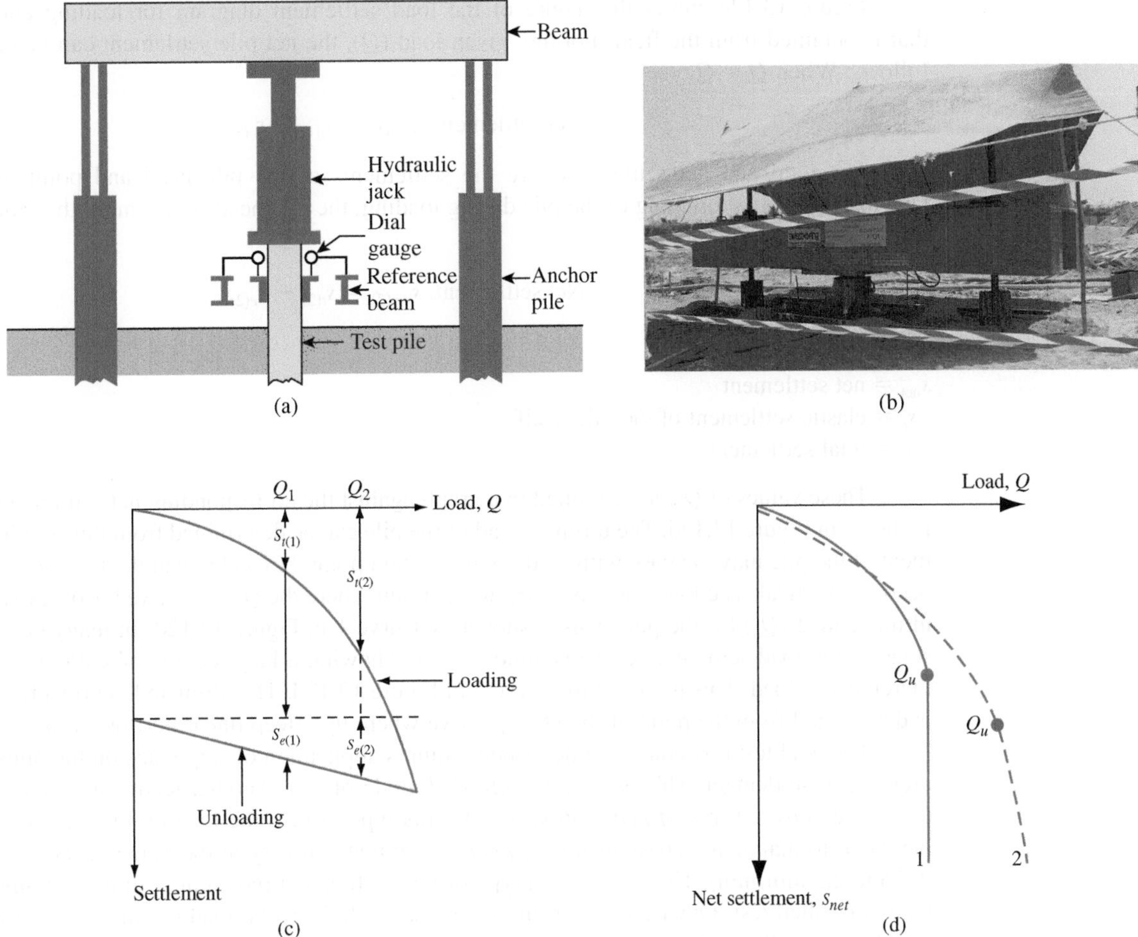

Figure 13.12 (a) Schematic diagram of pile load test arrangement; (b) photograph of a pile load test; (c) plot of load against total settlement; (d) plot of load against net settlement

arrangement in the field. This arrangement is for testing the pile in *axial compression*. Figure 13.12b shows a photograph of pile load test in programs. The load to the pile is applied by a hydraulic jack. Step loads are applied to the pile, and sufficient time is allowed to elapse after each step load so that the settlement rate reaches a small value. The settlement of the pile is recorded by means of dial gauges. The amount of load to be applied for each step will vary depending on local building codes. Most building codes require that each step load be about one-fourth of the proposed working load and should be carried out to at least a total load of two times the proposed working load. After reaching the desired pile load, the pile is gradually unloaded.

Load tests on piles in sand can be carried out immediately after the piles are driven. However, care should be taken in deciding the time lapse between driving and starting the load test when piles are embedded in clay. This time lapse can range from 30 to 60 days or more, because the soil requires some time to gain its *thixotropic strength*.

Figure 13.12c shows the nature of the load settlement diagram for loading and unloading that is obtained from the field. For any given load (Q), the net pile settlement can be calculated as follows: When $Q = Q_1$

$$\text{Net settlement, } s_{\text{net}(1)} = s_{t(1)} - s_{e(1)}$$

The total and net settlements are the settlements of the pile head and point, respectively. Due to the elastic shortening of the pile during loading, the pile head settles more than the pile point. When $Q = Q_2$

$$\text{Net settlement, } s_{\text{net}(2)} = s_{t(2)} - s_{e(2)}$$

where

s_{net} = net settlement
s_e = elastic settlement of the pile itself
s_t = total settlement

These values of Q can be plotted in a graph against the corresponding net settlement (s_{net}). This is shown in Figure 13.12d. The ultimate load of the pile can be determined from this graph. The settlement of the pile may increase with load up to a certain point, beyond which the load-settlement curve becomes vertical. The load corresponding to the point where the $Q-s_{\text{net}}$ curve becomes vertical is the ultimate load (Q_u) for the pile. This is shown by Curve 1 in Figure 13.12d. In many cases, the latter stage of the load-settlement curve is almost linear, showing a large degree of settlement for a small increment of load. This is shown by Curve 2 in Figure 13.12d. The ultimate load (Q_u) for such a case is determined from the point of the $Q-s_{\text{net}}$ curve where this steep linear portion starts.

The load test procedure just described requires application of step loads on the piles and measurement of settlement. This is a *load-controlled* mode of test. Another technique used for pile load test is the *constant-rate-of-penetration* test. In this type of test, the load on the pile is continuously increased to maintain a constant *rate of penetration* that can vary in the range of 0.01 to 0.1 in./min (0.25 to 2.5 mm/min). This test gives a type of load-settlement plot similar to that obtained from the load-controlled test. Other modes of pile load tests include cyclic loading, in which an incremental load is repeatedly applied and removed.

These days, it is very common to see *pile-driving analyzers* (*PDA*) used for monitoring pile driving. It can be used to determine the load carrying capacity of a driven pile, and is seen as an alternative to a pile load test. Figure 13.13 shows a pile driving analyser connected to a square concrete pile that is being driven into the ground.

13.10 Group Piles—Efficiency

In most cases, piles are used in groups, as shown in Figure 13.14, to transmit the structural load to the soil. A reinforced concrete *pile cap* is constructed over *group piles*. Figure 13.14a shows a 2 × 5 pile group consisting of ten octagonal precast concrete piles that have been driven into the ground. Figure 13.14b shows the reinforcement cage for pile cap on top of four concrete piles. The pile cap can be in contact with the ground, as in most of the cases (Figure 13.15a), or it may be well above the ground, as in the case of construction of offshore platforms (Figure 13.15b).

The preceding sections have discussed the load-bearing capacity of single piles. Determination of the load-bearing capacity of group piles is an extremely complicated problem and has not yet

(a) (b)

Figure 13.13 Using pile driving analyser: (a) instrumented square pile; (b) data logger connected to the pile while driving. (Photo courtesy of N. Sivakugan, james Cook University, Australia)

been fully resolved. When the piles are placed close to each other, it is reasonable to assume that the stresses transmitted by the piles to the soil medium will overlap (Figure 13.15c), and this may reduce the load-bearing capacity of the piles. Ideally, the piles in a group should be spaced in such a way that the load-bearing capacity of the group should not be less than the sum of the bearing capacity of the individual piles. In practice, the center-to-center pile spacings (d) are kept to a minimum of 2.5D. However, in ordinary situations, they are kept at about 3–3.5D.

The efficiency of the load-bearing capacity of a group pile may be defined as

$$\eta = \frac{Q_{g(u)}}{\Sigma Q_u} \tag{13.16}$$

(a) (b)

Figure 13.14 Pile group: (a) 2 × 5 pile group; (b) reinforcement cage for the pile cap over a 2 × 2 pile group

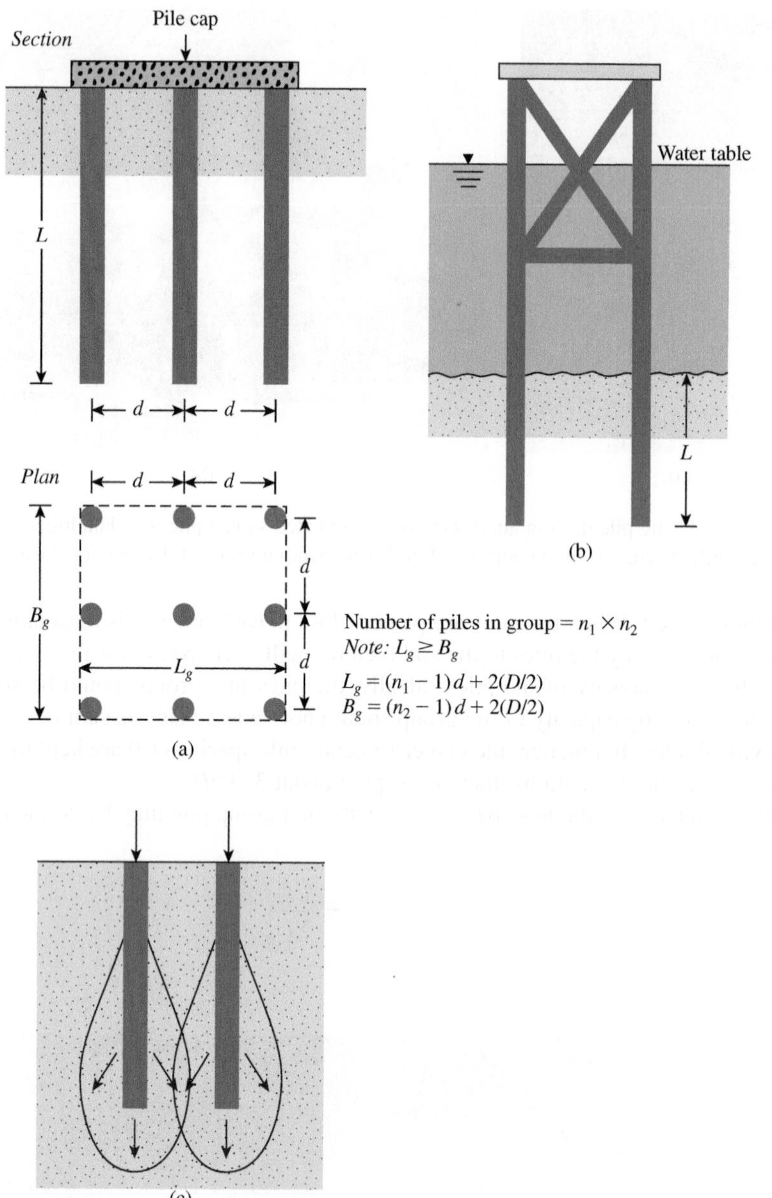

Figure 13.15 Pile groups

where

η = group efficiency

$Q_{g(u)}$ = ultimate load-bearing capacity of the group pile

Q_u = ultimate load-bearing capacity of each pile without the group effect

Piles in Sand

Many structural engineers use a simplified analysis to obtain the group efficiency for *friction* piles in sand. This can be explained with the aid of Figure 13.15a. Depending on their spacing within the group, the piles may act in one of the two following ways: (1) as a *block* with dimensions $L_g \times B_g \times L$, or (2) as *individual piles*. If the piles act as a block, the frictional capacity can be given as $f_{av} p_g L \approx Q_{g(u)}$. [*Note:* p_g = perimeter of the cross section of block = $2(n_1 + n_2 - 2)d + 4D$, and f_{av} = average unit frictional resistance.] Similarly, for each pile acting individually, $Q_u \approx pLf_{av}$. (*Note:* p = perimeter of the cross section of each pile.) Thus,

$$\eta = \frac{Q_{g(u)}}{\Sigma Q_u} = \frac{f_{av}[2(n_1 + n_2 - 2)d + 4D]L}{n_1 n_2 p L f_{av}}$$

$$= \frac{2(n_1 + n_2 - 2)d + 4D}{p n_1 n_2} \tag{13.17}$$

Hence,

$$Q_{g(u)} = \left[\frac{2(n_1 + n_2 - 2)d + 4D}{p n_1 n_2}\right]\Sigma Q_u \tag{13.18}$$

From Eq. (13.18), if the center-to-center spacings (d) are large, one may obtain $\eta > 1$. In that case, the piles will behave as individual piles. Thus, in practice, if $\eta < 1$,

$$Q_{g(u)} = \eta \Sigma Q_u$$

and, if $\eta \geq 1$,

$$Q_{g(u)} = \Sigma Q_u$$

Another equation that is quoted often among design engineers is the *Converse-Labarre equation*, which can be stated as

$$\eta = 1 - \left[\frac{(n_1 - 1)n_2 + (n_2 - 1)n_1}{90 n_1 n_2}\right]\theta \tag{13.19}$$

$$\text{where } \theta \text{ (deg)} = \tan^{-1}(D/d). \tag{13.20}$$

Example 13.8

A 3 × 4 pile group consists of 12 driven precast concrete piles that are 400 mm in diameter and 15 m in length. The center-to-center spacing of the piles is 1200 mm. Determine the efficiency of the pile group using Eqs. (13.17) and (13.19).

Solution

Given: $n_1 = 3$, $n_2 = 4$, $d = 1200$ mm, and $D = 400$ mm. The perimeter of a pile is $p = \pi D =$ 1256.6 mm.

From Eq. (13.17),

$$\eta = \frac{2(n_1 + n_2 - 2)d + 4D}{pn_1n_2} = \frac{2(3 + 4 - 2)(1200) + 4(400)}{1256.6 \times 3 \times 4} = \mathbf{0.90 \text{ or } 90\%}$$

From Eq. (13.19),

$$\eta = 1 - \left[\frac{(n_1 - 1)n_2 + (n_2 - 1)n_1}{90n_1n_2}\right]\theta \text{ [where } \theta(\text{deg}) = \tan^{-1}(400/1200) = 18.43]$$

$$\eta = 1 - \left[\frac{(3 - 1)4 + (4 - 1)3}{90(3)(4)}\right](18.43) = \mathbf{0.71 \text{ or } 71\%}$$ ■

Piles in Clay

The ultimate load-bearing capacity of group piles in clay may be estimated with the following procedure:

1. Determine $\Sigma Q_u = n_1n_2(Q_p + Q_s)$. From Eq. (13.11),

$$Q_p = A_p[9c_{u(p)}]$$

where $c_{u(p)}$ = undrained cohesion of the clay at the pile tip. Also, from Eq. (13.12),

$$Q_s = \Sigma \alpha p c_u \Delta L$$

So

$$\Sigma Q_u = n_1n_2[9A_pc_{u(p)} + \Sigma \alpha p c_u \Delta L] \tag{13.21}$$

2. Determine the ultimate capacity by assuming that the piles in the group act as a block with dimensions of $L_g \times B_g \times L$. The skin resistance of the block is

$$\Sigma p_g c_u \Delta L = \Sigma 2(L_g + B_g)c_u \Delta L$$

Calculate the point bearing capacity from

$$A_p q_p = A_p c_{u(p)} N_c^* = (L_g B_g) c_{u(p)} N_c^*$$

The variation of N_c^* with L/B_g and L_g/B_g is illustrated in Figure 13.16. Thus, the ultimate load is

$$\Sigma Q_u = L_g B_g c_{u(p)} N_c^* + \Sigma 2(L_g + B_g)c_u \Delta L \tag{13.22}$$

3. Compare the values obtained from Eqs. (13.21) and (13.22). The *lower* of the two values is $Q_{g(u)}$.

Piles in Rock

For point bearing piles resting on rock, most building codes specify that $Q_{g(u)} = \Sigma Q_u$, provided that the minimum center-to-center spacing of piles is D + 12 in. (300 mm). For H-piles and piles with square cross sections, the magnitude of D is equal to the diagonal dimension of the pile cross section.

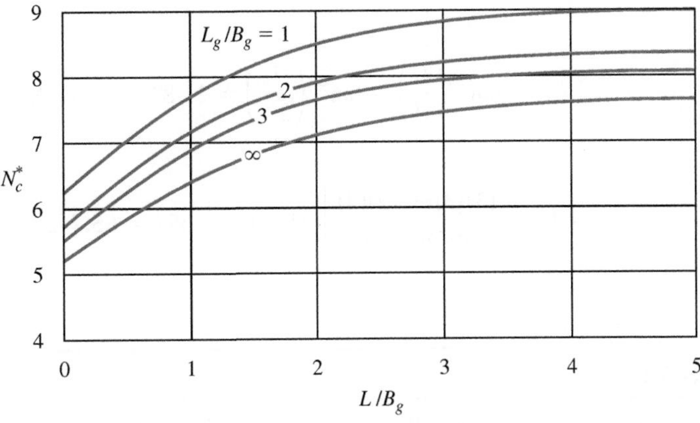

Figure 13.16 Variation of N_c^* with L_g/B_g and L/B_g

Example 13.9

Refer to Figure 13.15a. Given: $n_1 = 4$, $n_2 = 3$, $D = 12$ in., and $d = 2.5D$. The piles are square in cross section and are embedded in sand. Use Eq. (13.17) to obtain the group efficiency.

Solution

From Eq. (13.17),

$$\eta = \frac{2(n_1 + n_2 - 2)d + 4D}{pn_1n_2}$$

$$d = 2.5D = (2.5)(12) = 30 \text{ in.}$$

$$p = 4D = (4)(12) = 48 \text{ in.}$$

So

$$\eta = \frac{2(4 + 3 - 2)30 + 48}{(48)(4)(3)} = 0.604 = \mathbf{60.4\%}$$

■

Example 13.10

Redo Example 13.9 using Eq. (13.19).

Solution

According to Eq. (13.19)

$$\eta = 1 - \left[\frac{(n_1 - 1)n_2 + (n_2 - 1)n_1}{90n_1n_2}\right]\tan^{-1}\left(\frac{D}{d}\right)$$

However,

$$\tan^{-1}\left(\frac{D}{d}\right) = \tan^{-1}\left(\frac{1}{2.5}\right) = 21.8°$$

So

$$\eta = 1 - \left[\frac{(3)(3) + (2)(4)}{(90)(3)(4)}\right](21.8°) = 0.657 = \mathbf{65.7\%}$$

■

Example 13.11

A 2 × 4 pile group consists of four 450 mm diameter and 8 m long piles driven into a clay at 1200 mm center-to-center spacing. The undrained cohesion of the clay is 75 kN/m^2. Determine the maximum load that can be allowed on the pile group, with a factor of safety of 3.

Solution

Given: $L = 8.0$ m, $A_p = (\pi/4)(0.45^2) = 0.159$ m^2, and $p = \pi D = 1.414$ m.

From Figure 13.9, for $c_u = 75$ kN/m^2, $\alpha = 0.63$.

From Eq. (13.21),

$$\Sigma Q_u = n_1 n_2[9A_p c_u + (\alpha c_u)(pL)] = (2 \times 4)[(9)(0.159)(75) + (0.63 \times 75)(1.414 \times 8.0)]$$
$$= \mathbf{5134.5\ kN}$$

$$B_g = 1200 + 450 = 1650 \text{ mm}$$

$$L_g = 3 \times 1200 + 450 = 4050 \text{ mm}$$

For $L/B_g = 8/1.65 = 4.84$ and $L_g/B_g = 4.05/1.65 = 2.45$, from Figure 13.16, $N_c^* = 8.2$.

From Eq. (13.22),

$$\Sigma Q_u = L_g B_g c_u N_c^* + 2(L_g + B_g)(L)(c_u) = (4.05)(1.65)(75)(8.2) + (2)(4.05 + 1.65)(8.0)(75)$$
$$= \mathbf{10{,}949.7\ kN}$$

Using the lesser of the two values, the allowable load on the pile group is given as 5134.5/3 = 1712 kN.

■

Example 13.12

The section of a 3 × 4 group pile in a layered saturated clay is shown in Figure 13.17. The piles are square in cross section (14 in. × 14 in.). The center-to-center spacing, d, of the piles is 35 in. Determine the allowable load-bearing capacity of the pile group. Use FS = 4.

Solution

From Eq. (13.21),

$$\Sigma Q_u = n_1 n_2 [9A_p c_{u(p)} + \alpha_1 p c_{u(1)} L_1 + \alpha_2 p c_{u(2)} L_2]$$

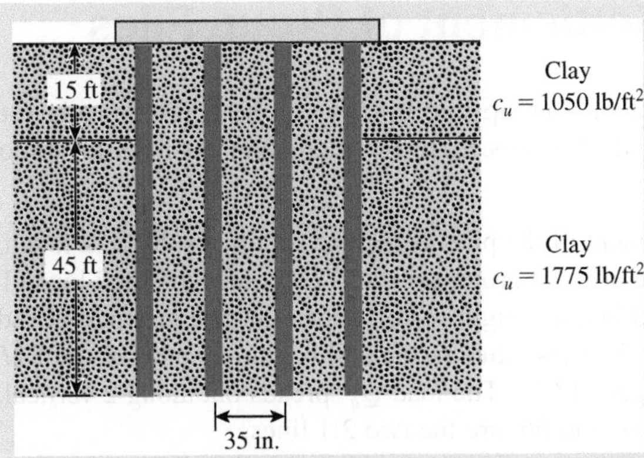

Figure 13.17

From Figure 13.9, $c_{u(1)} = 1050\ \text{lb/ft}^2$; $\alpha_1 = 0.86$ and $c_{u(2)} = 1775\ \text{lb/ft}^2$; $\alpha_2 = 0.6$.

$$\Sigma Q_u = \frac{(3)(4)}{1000}\left[\begin{array}{c}(9)\left(\dfrac{14}{12}\right)^2(1775) + (0.86)\left(4 \times \dfrac{14}{12}\right)(1050)(15) \\ + (0.6)\left(4 \times \dfrac{14}{12}\right)(1775)(45)\end{array}\right] = 3703\ \text{kip}$$

For piles acting as a group,

$$L_g = (3)(35) + 14 = 119\ \text{in.} = 9.92\ \text{ft}$$

$$B_g = (2)(35) + 14 = 84\ \text{in.} = 7\ \text{ft}$$

$$\frac{L_g}{B_g} = \frac{9.92}{7} = 1.42$$

$$\frac{L}{B_g} = \frac{60}{7} = 8.57$$

From Figure 13.16, $N_c^* = 8.75$. From Eq. (13.22)

$$\begin{aligned}\Sigma\, Q_u &= L_g B_g c_{u(p)} N_c^* + \Sigma\, 2(L_g + B_g)c_u\, \Delta L \\ &= (9.92)(7)(1775)(8.75) + (2)(9.92 + 7)[(1050)(15) + (1775)(45)] \\ &= 4313\ \text{kip}\end{aligned}$$

Hence, $\Sigma\, Q_u = 3703$ kip.

$$\Sigma\, Q_{\text{all}} = \frac{3703}{\text{FS}} = \frac{3703}{4} \approx \mathbf{926\ kip}$$

■

13.11 Consolidation Settlement of Group Piles

The consolidation settlement of a group pile in clay can be approximately estimated by using the 2:1 stress distribution method. The procedure of calculation involves the following steps (refer to Figure 13.18):

1. Let the depth of embedment of the piles be L. The group is subjected to a total load of Q_g. If the pile cap is below the original ground surface, Q_g equals the total load of the superstructure on the piles minus the effective weight of soil above the pile group removed by excavation.
2. Assume that the load Q_g is transmitted to the soil beginning at a depth of $2L/3$ from the top of the pile, as shown in Figure 13.18. The load Q_g spreads out along 2 vertical: 1 horizontal lines from this depth. Lines aa' and bb' are the two 2:1 lines.
3. Calculate the stress increase caused at the middle of each soil layer by the load Q_g:

$$\Delta\sigma_i' = \frac{Q_g}{(B_g + z_i)(L_g + z_i)} \tag{13.23}$$

where

$\Delta\sigma_i'$ = stress increase at the middle of layer i
L_g, B_g = length and width of the plan of pile group, respectively
z_i = distance from $z = 0$ to the middle of the clay layer, i

For example, in Figure 13.18 for layer 2, $z_i = L_1/2$; for layer 3, $z_i = L_1 + L_2/2$; and for layer 4, $z_i = L_1 + L_2 + L_3/2$. Note, however, that there will be no stress increase in clay layer 1 because it is above the horizontal plane ($z = 0$) from which the stress distribution to the soil starts.

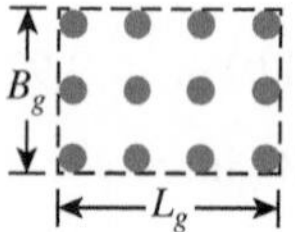

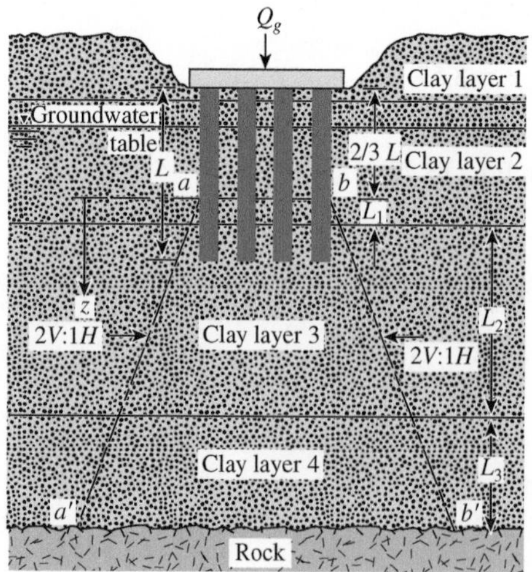

Figure 13.18 Consolidation settlement of group piles

4. Calculate the settlement of each layer caused by the increased stress:

$$\Delta s_i = \left[\frac{\Delta e_{(i)}}{1 + e_{o(i)}} \right] H_i \qquad (13.24)$$

where

Δs_i = consolidation settlement of layer i
$\Delta e_{(i)}$ = change of void ratio caused by the stress increase in layer i
e_o = initial void ratio of layer i (before construction)
H_i = thickness of layer i (*Note:* In Figure 13.18, for layer 2 $H_i = L_1$; for layer 3, $H_i = L_2$; and for layer 4, $H_i = L_3$.)

Relations for $\Delta e_{(i)}$ are given in Chapter 8.

5. Total consolidation settlement of the pile group is then

$$\Delta s_g = \Sigma \Delta s_i \qquad (13.25)$$

Note that consolidation settlement of piles may be initiated by fills placed nearby, adjacent floor loads, and lowering of water tables.

Example 13.13

A group pile in clay is shown in Figure 13.19. Determine the consolidation settlement of the pile groups. All clays are normally consolidated.

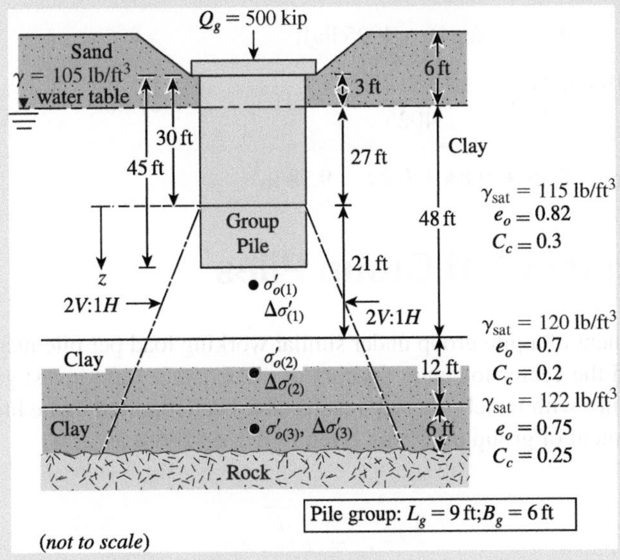

Figure 13.19

Solution

The stress distribution pattern is shown in Figure 13.19. Hence

$$\Delta\sigma'_{(1)} = \frac{Q_g}{(L_g + z_1)(B_g + z_1)} = \frac{(500)(1000)}{\left(9 + \frac{21}{2}\right)\left(6 + \frac{21}{2}\right)} = 1554 \text{ lb/ft}^2$$

$$\Delta\sigma'_{(2)} = \frac{(500)(1000)}{(9 + 27)(6 + 27)} = 421 \text{ lb/ft}^2$$

$$\Delta\sigma'_{(3)} = \frac{(500)(1000)}{(9 + 36)(6 + 36)} = 265 \text{ lb/ft}^2$$

$$\Delta s_1 = \frac{C_{c(1)}H_1}{1 + e_{o(1)}} \log\left[\frac{\sigma'_{o(1)} + \Delta\sigma'_{(1)}}{\sigma'_{o(1)}}\right]$$

$$\sigma'_{o(1)} = (6)(105) + \left(27 + \frac{21}{2}\right)(115 - 62.4) = 2603 \text{ lb/ft}^2$$

$$\Delta s_1 = \frac{(0.3)(21)}{1 + 0.82} \log\left(\frac{2603 + 1554}{2603}\right) = 0.703 \text{ ft} = 8.45 \text{ in.}$$

$$\Delta s_2 = \frac{C_{c(2)}H_2}{1 + e_{o(2)}} \log\left[\frac{\sigma'_{o(2)} + \Delta\sigma'_{(2)}}{\sigma'_{o(2)}}\right]$$

$$\sigma'_{o(2)} = (6)(105) + (27 + 21)(115 - 62.4) + (6)(120 - 62.4) = 3500 \text{ lb/ft}^2$$

$$\Delta s_2 = \frac{(0.2)(12)}{1 + 0.7} \log\left(\frac{3500 + 421}{3500}\right) = 0.07 \text{ ft} = 0.84 \text{ in.}$$

$$\sigma'_{o(3)} = (6)(105) + (48)(115 - 62.4) + (12)(120 - 62.4) + (3)(122 - 62.4) = 4025 \text{ lb/ft}^2$$

$$\Delta s_2 = \frac{(0.25)(6)}{1 + 0.75} \log\left(\frac{4025 + 265}{4025}\right) = 0.024 \text{ ft} \approx 0.29 \text{ in.}$$

Total settlement, $\Delta s_g = 8.45 + 0.84 + 0.29 =$ **9.58 in.** ■

13.12 Elastic Settlement of Group Piles

In general, the settlement of a pile group under similar working load per pile increases with the width of the group (B_g) and the center-to-center spacing of piles (d). Several investigations relating to the settlement of group piles with widely varying results have been reported in the literature. The simplest relation for the settlement of group piles was given by Vesic (1969) as

$$s_{g(e)} = \sqrt{\frac{B_g}{D}}\, s_e \tag{13.26}$$

where

$s_{g(e)}$ = elastic settlement of group piles
B_g = width of pile group section
D = width or diameter of each pile in the group
s_e = elastic settlement of each pile at comparable working load

For pile groups in sand and gravel, Meyerhof (1976) suggested the following empirical relation for elastic settlement:

$$s_{g(e)}\,(\text{in.}) = \frac{2q\sqrt{B_g}\,I}{N_{60}} \tag{13.27a}$$

In SI units,

$$s_{g(e)}\,(\text{mm}) = \frac{0.96\,q\,\sqrt{B_g}\,I}{N_{60}} \tag{13.27b}$$

where

$q = Q_g/(L_g B_g)$ [in ton/ft^2 in Eq. (13.27a); and kN/m^2 in Eq. (13.27b)] (13.28)
L_g and B_g = length and width of the pile group section respectively [ft in Eq. (13.27a); and m in Eq. (13.27b)]
N_{60} = average standard penetration number within seat of settlement ($\approx B_g$ deep below the tip of the piles)
I = influence factor $= 1 - L/8B_g \geq 0.5$ (13.29)
L = length of embedment of piles [ft in Eq. (13.27a); and m in Eq. (13.27b)]

Drilled Shafts

13.13 Drilled Shafts–General

Drilled shafts are cast-in-place piles that generally have a diameter of about 30 in. (762 mm) or more. The use of drilled-shaft foundations has many advantages:

1. A single drilled shaft may be used instead of a group of piles and the pile cap.
2. Constructing drilled shafts in deposits of dense sand and gravel is easier than driving piles.
3. Drilled shafts may be constructed before grading operations are completed.
4. When piles are driven by a hammer, the ground vibration may cause damage to nearby structures, which the use of drilled shafts avoids.
5. Piles driven into clay soils may produce ground heaving and cause previously driven piles to move laterally. This does not occur during construction of drilled shafts.
6. There is no hammer noise during the construction of drilled shafts, as there is during pile driving.
7. Because the base of a drilled shaft can be enlarged, it provides great resistance to the uplifting load.

8. The surface over which the base of the drilled shaft is constructed can be visually inspected.
9. Construction of drilled shafts generally utilizes mobile equipment, which, under proper soil conditions, may prove to be more economical than methods of constructing pile foundations.
10. Drilled shafts have high resistance to lateral loads.

There are also several drawbacks to the use of drilled-shaft construction. The concreting operation may be delayed by bad weather and always needs close supervision. Also, as in the case of braced cuts, deep excavations for drilled shafts may cause substantial ground loss and damage to nearby structures.

13.14 Types of Drilled Shafts

Drilled shafts are classified according to the ways in which they are designed to transfer the structural load to the substratum. Figure 13.20a shows a drilled shaft that has a *straight shaft*. It extends through the upper layer(s) of poor soil, and its tip rests on a strong load-bearing soil layer or rock. The shaft can be cased with steel shell or pipe when required (as in the case of cased, cast-in-place concrete piles). For such shafts, the resistance to the applied load may develop from end bearing and also from side friction at the shaft perimeter and soil interface.

A *drilled shaft with bell* (Figures 13.20b and c) consists of a straight shaft with a bell at the bottom, which rests on good bearing soil. The bell can be constructed in the shape of a dome (Figure 13.20b), or it can be angled (Figure 13.20c). For angled bells, the underreaming tools commercially available can make 30° to 45° angles with the vertical.

Straight shafts also can be extended into an underlying rock layer (Figure 13.20d). In calculating the load-bearing capacity of such drilled shafts, engineers take into account the end bearing and the shear stress developed along the shaft perimeter and rock interface.

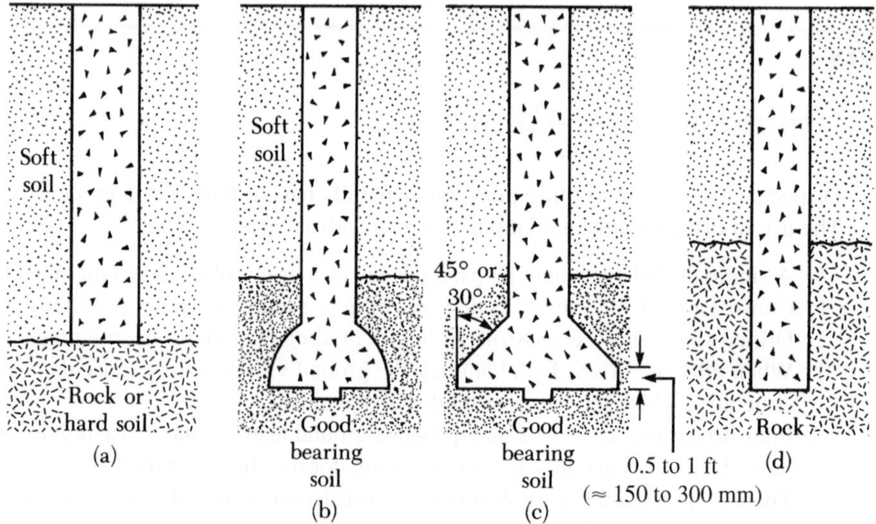

Figure 13.20 Types of drilled shaft: (a) straight shaft; (b) and (c) shaft with bell; (d) straight shafts socketed into rock

13.15 Construction Procedures

The most common construction procedure used in the United States involves rotary drilling. There are three major types of construction methods, and they may be classified as (a) dry method, (b) casing method, and (c) wet method. A brief description of each method follows.

Dry Method of Construction

This method is employed in soils and rocks that are above the water table and will not cave in when the hole is drilled to full depth. The sequence of construction, as shown in Figure 13.21 is as follows:

1. The excavation is completed (and belled if desired) using proper drilling tools, and the spoils from the hole are deposited nearby (Figure 13.21a).
2. Concrete is then poured into the cylindrical hole (Figure 13.21b).
3. If desired a rebar cage is placed only in the upper portion of the shaft (Figure 13.21c).
4. The concreting is then completed and the drilled shaft will be as shown in Figure 13.21d.

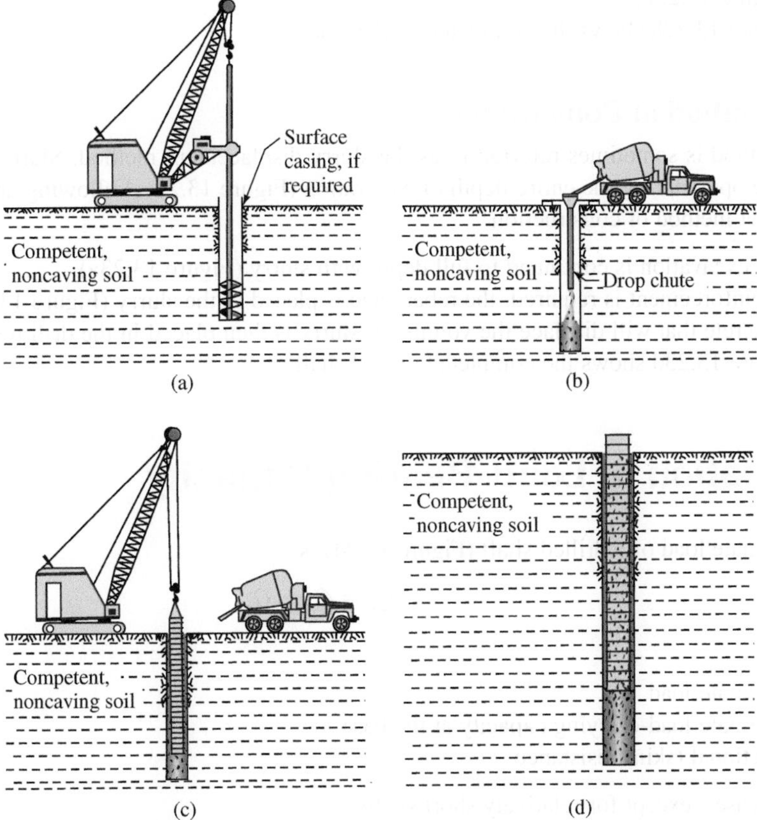

Figure 13.21 Dry method of construction: (a) initiating drilling, (b) starting concrete pour, (c) placing rebar cage, (d) completed shaft (*Courtesy of O'Neill and Reese, 1999*)

Casing Method of Construction

This method is used in soils or rocks where caving or excessive deformation is likely to occur when the borehole is excavated. The sequence of construction is shown in Figure 13.22 and may be explained as follows:

1. The excavation procedure is initiated as in the case of the dry method of construction described earlier (Figure 13.22a).
2. When the caving soil is encountered, bentonite slurry is introduced into the borehole (Figure 13.22b). Drilling is continued until the excavation goes past the caving soil and a layer of impermeable soil or rock is encountered.
3. A casing is then introduced into the hole (Figure 13.22c).
4. The slurry is bailed out of the casing using a submersible pump (Figure 13.22d).
5. A smaller drill that can pass through the casing is introduced into the hole and the excavation is continued (Figure 13.22e).
6. If needed, the base of the excavated hole can then be enlarged using an under-reamer (Figure 13.22f).
7. If reinforcing steel is needed, the rebar cage needs to extend the full length of the excavation. Concrete is then poured into the excavation and the casing is gradually pulled out (Figure 13.22g).
8. Figure 13.22h shows the completed drilled shaft.

Wet Method of Construction

This method is sometimes referred to as the slurry displacement method. Slurry is used to keep the borehole open during the entire depth of excavation (Figure 13.23). Following are the steps involved in the wet method of construction.

1. The excavation is continued to full depth with slurry (Figure 13.23a).
2. If reinforcement is required, the rebar cage is placed in the slurry (Figure 13.23b).
3. Concrete that will displace the volume of slurry is then placed in the drill hole (Figure 13.23c).
4. Figure 13.23d shows the completed drilled shaft.

13.16 Estimation of Load-Bearing Capacity

The ultimate load of a drilled shaft (Figure 13.24) is

$$Q_u = Q_p + Q_s \tag{13.30}$$

where

Q_u = ultimate load
Q_p = ultimate load-carrying capacity at the base
Q_s = frictional (skin) resistance

In most cases, except for relatively short shafts

$$Q_p = A_p(c'N_c^* + q'N_q^*) \tag{13.31}$$

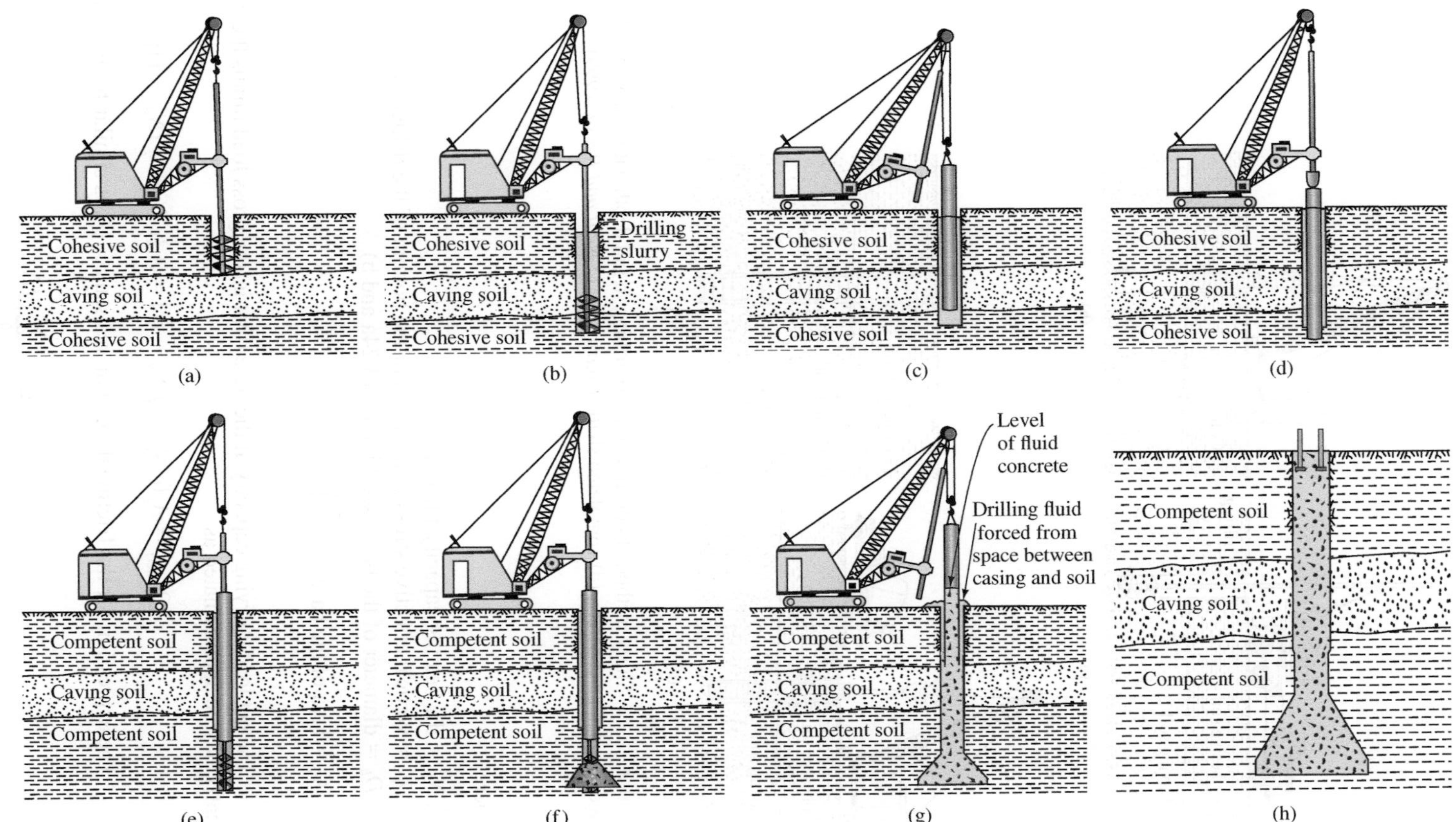

Figure 13.22 Casing method of construction: (a) initiating drilling, (b) drilling with slurry, (c) introducing casing, (d) casing is sealed and slurry is being removed from interior of casing, (e) drilling below casing, (f) underreaming, (g) removing casing, and (h) completed shaft (*Courtesy of O'Neill and Reese, 1999*)

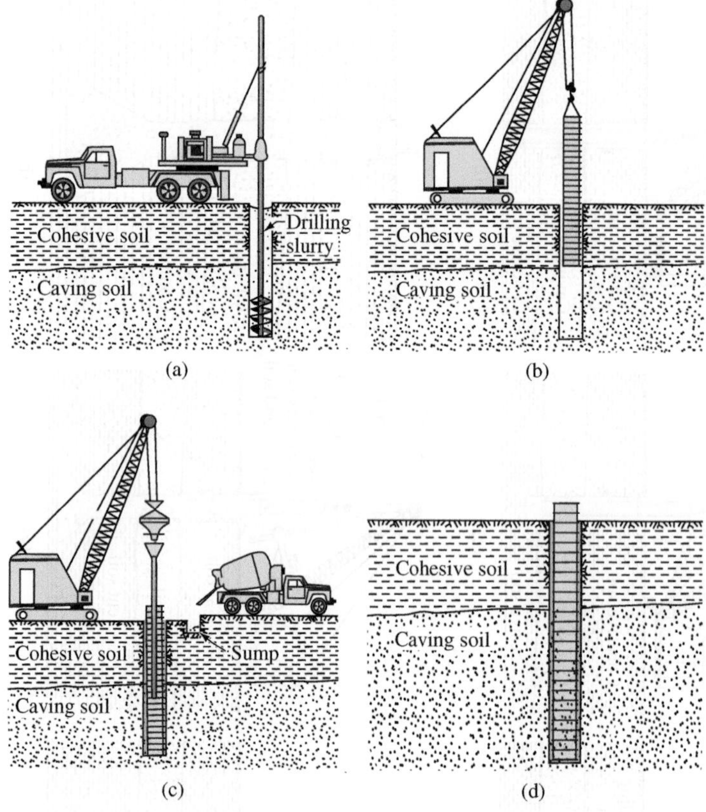

Figure 13.23 Slurry method of construction: (a) drilling to full depth with slurry, (b) placing rebar cage, (c) placing concrete, (d) completed shaft (*Courtesy of O'Neill and Reese, 1999*)

where

N_c^*, N_q^* = the bearing capacity factors
q' = vertical effective stress at the level of the bottom of the drilled shaft
D_b = diameter of the base (see Figures 13.24a and b)
A_p = area of the base = $\left(\frac{\pi}{4}\right)(D_b)^2$

The net load-carrying capacity at the base (that is, the gross load minus the weight of the drilled shaft) may be approximated as

$$Q_{p(\text{net})} = A_p(c'N_c^* + q'N_q^* - q') = A_p[c'N_c^* + q'(N_q^* - 1)] \tag{13.32}$$

The expression for the frictional, or skin, resistance, Q_s, is similar to that for piles:

$$Q_s = \sum_0^{L_1} pf\Delta L \tag{13.33}$$

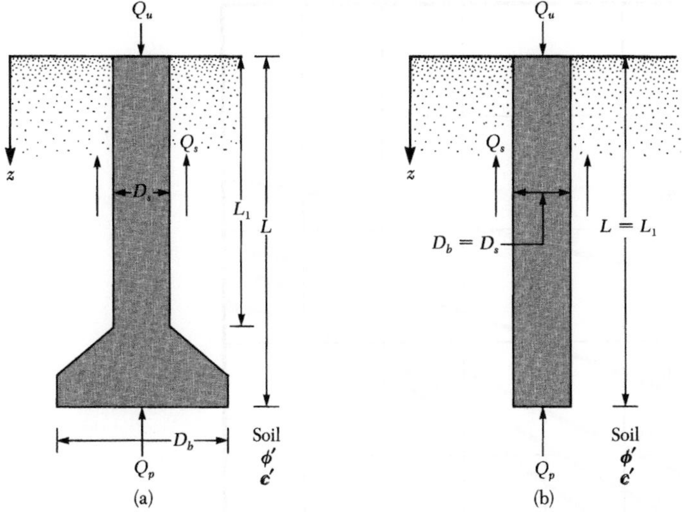

Figure 13.24 Ultimate bearing capacity of drilled shafts: (a) with bell; (b) straight shaft

where

p = shaft perimeter = πD_s
f = unit frictional (or skin) resistance

Drilled Shafts in Sand

For shafts in sand, $c' = 0$ and hence Eq. (13.32) simplifies to

$$Q_{p(\text{net})} = A_p q'(N_q^* - 1) \tag{13.34}$$

The magnitude of $Q_{p(\text{net})}$ can be reasonably estimated from a relationship based on the analysis of Berezantzev et al. (1961), which is a slight modification of Eq. (13.34), or

$$Q_{p(\text{net})} = A_p q'(\omega N_q^* - 1) \tag{13.35}$$

where

$$N_q^* = 0.21 e^{0.17\phi'} \quad (\textit{Note: } \phi' \text{ is in degrees}) \tag{13.36}$$

$$\omega = \text{a correction factor} = f\left(\frac{L}{D_b}, \phi'\right) \text{ see Figure 13.25}$$

The frictional resistance at ultimate load, Q_s, developed in a drilled shaft may be calculated from the relation given in Eq. (13.33), in which

$$p = \text{shaft perimeter} = \pi D_s$$

$$f = \text{unit frictional (or skin) resistance} = K\sigma_o' \tan \delta' \tag{13.37}$$

where

K = earth pressure coefficient $\approx K_o = 1 - \sin \phi'$
σ_o' = effective vertical stress at any depth z

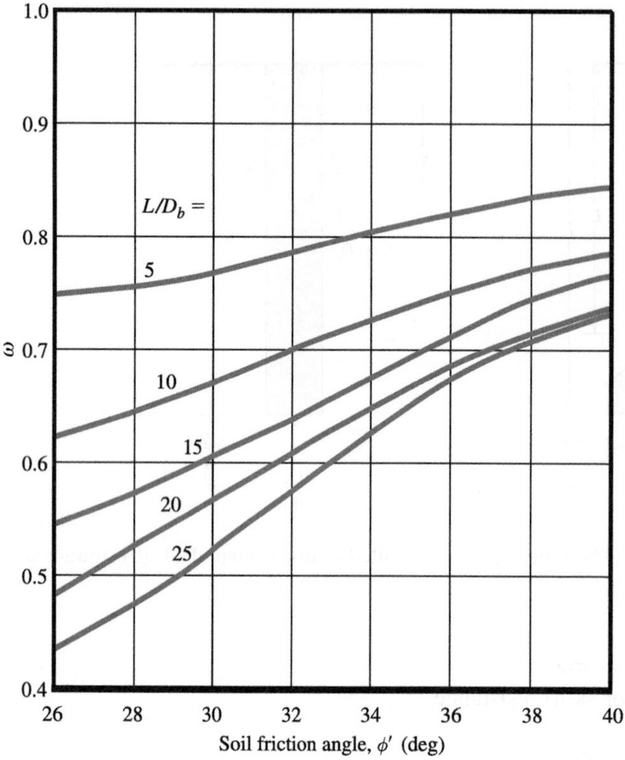

Figure 13.25 Variation of ω with ϕ' and L/D_b

The value of σ_o' will increase to a depth of about $15D_s$ and will remain constant there-after, as shown in Figure 13.7c.

An appropriate factor of safety should be applied to the ultimate load to obtain the net allowable load, or

$$Q_{u(\text{net})} = \frac{Q_{p(\text{net})} + Q_s}{\text{FS}} \tag{13.38}$$

Drilled Shafts in Clay

From Eq. (13.32), for saturated clays with $\phi = 0$, $N_q^* = 1$; hence, the net base resistance becomes

$$Q_{p(\text{net})} = A_p c_u N_c^* \tag{13.39}$$

where

c_u = undrained cohesion
N_c^* = bearing capacity factor = $1.33[(\ln I_r) + 1]$ (for $L \geq 3D_b$)] (13.40)
I_r = soil rigidity index

For $\phi = 0$ condition, I_r can be defined as

$$I_r = \frac{E_s}{3c_u}$$

where E_s = modulus of elasticity of soil.

O'Neill and Reese (1999) provided an approximate relationship between c_u and $E_s/3c_u$. This is shown in Figure 13.26 along with the corresponding N_c^* values. For all practical purposes, if c_u is equal to or greater than 2000 lb/ft², the magnitude of N_c^* is 9.

The expression for the skin resistance of drilled shafts in clay is similar to Eq. (13.12), or

$$Q_s = \sum_{L=0}^{L=L_1} \alpha^* c_u p \,\Delta L \tag{13.41}$$

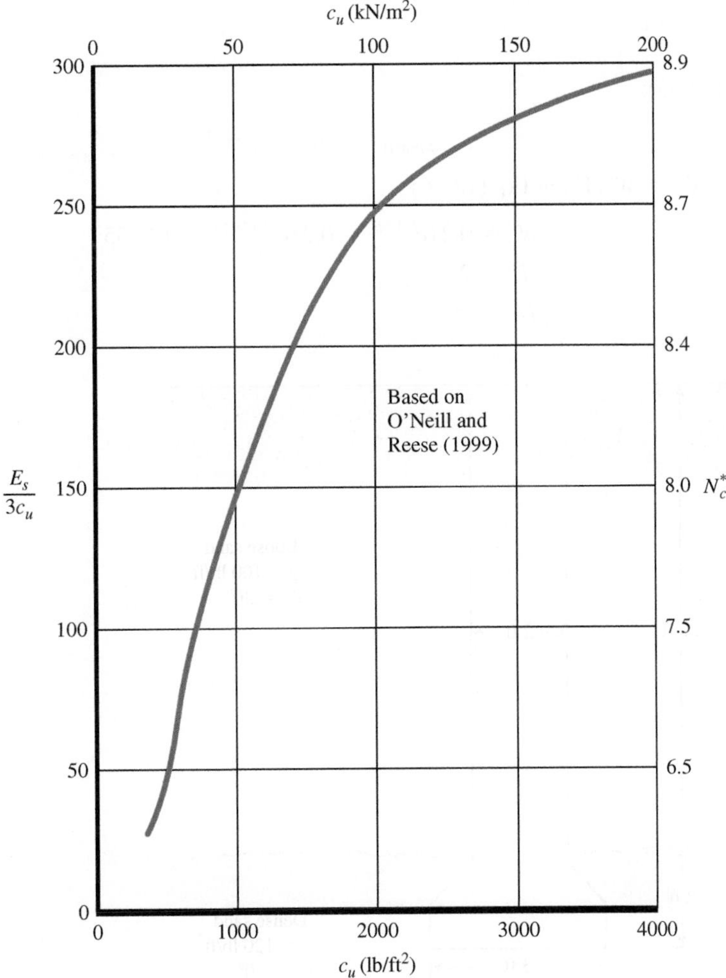

Figure 13.26 Approximate variation of $\frac{E_s}{3c_u}$ with c_u

where p = perimeter of the shaft cross section. The value of α^* that can be used in Eq. (13.41) has not been fully established. However, the field test results available at this time indicate that α^* may vary between 1.0 and 0.3. Conservatively we may assume that

$$\alpha^* = 0.4 \tag{13.42}$$

Example 13.14

A drilled shaft is shown in Figure 13.27. Determine:

a. $Q_{p(\text{net})}$
b. Q_s. Use $\delta'/\phi' = 0.5$.
c. Allowable load, Q_{all}, for a factor of safety (FS) of 4

Solution

Part a

From Eq. (13.35),

$$Q_{p(\text{net})} = A_p q'(\omega N_q^* - 1)$$

At the base, $\phi' = 40°$. From Eq. (13.36),

$$N_q^* = 0.21e^{0.17\phi'} = 0.21e^{(0.17)(40)} = 188.55$$

$$\frac{L}{D_b} = \frac{22}{3} = 7.33$$

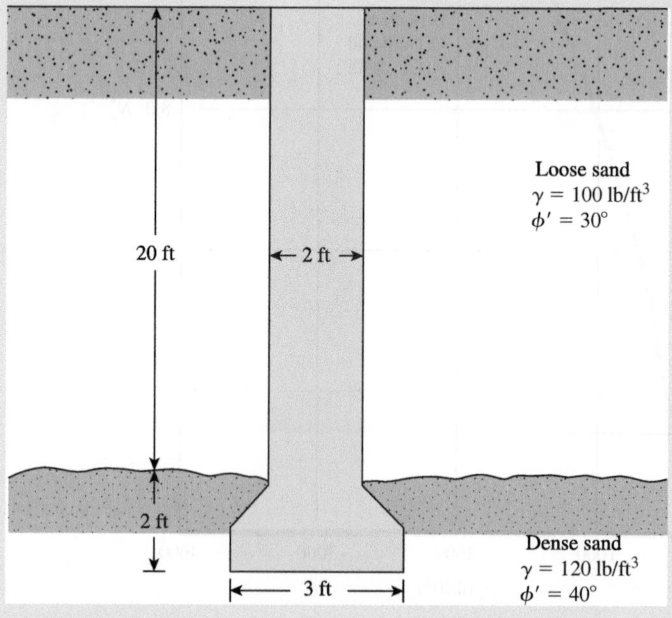

Figure 13.27 Drilled shaft supported by a dense layer of sand

From Figure 13.25, for $L/D_b = 7.33$ and $\phi' = 40°$, the magnitude of ω is about 0.83.

$$q' = (20)(100) + (2)(120) = 2240 \text{ lb/ft}^2 = 2.24 \text{ kip/ft}^2$$

$$A_p = \frac{\pi}{4}(D_b)^2 = \frac{\pi}{4}(3)^2 = 7.07 \text{ ft}^2$$

$$Q_{p(\text{net})} = (7.07)(2.24)[(0.83)(188.55) - 1] = \mathbf{2462.6 \text{ kip}}$$

Part b

For this case,

$$\frac{L_1}{D_s} = \frac{20}{2} = 10 < 15$$

From Eq. (13.33),

$$Q_s = \Sigma pf\,\Delta L$$

$$P = \pi D_s = (\pi)(2) = 6.28 \text{ ft}$$

$$f = K\sigma_o' \tan \delta'$$

$$\frac{\delta'}{\phi'} = 0.5;\ \delta' = (0.5)(30) = 15°$$

$$K = 1 - \sin \phi' = 1 - \sin 30 = 0.5$$

$$Q_s = (p)\left[\frac{\sigma_o'(\text{at } z = 0 \text{ ft}) + \sigma_o'(\text{at } z = 20 \text{ ft})}{2}\right](K)(\tan 15)(L)$$

$$\sigma_o' = (\text{at z} = 0 \text{ ft}) = 0$$

$$\sigma_o'(\text{at } z = 20 \text{ ft}) = (100)(20) = 2000 \text{ lb/ft}^2 = 2 \text{ kip/ft}^2$$

$$Q_s = (6.28)\left(\frac{0 + 2}{2}\right)(0.5)(\tan 15)(20) = \mathbf{16.8 \text{ kip}}$$

Part c

$$Q_{\text{all}} = \frac{Q_{p(\text{net})} + Q_s}{\text{FS}} = \frac{2462.6 + 16.8}{4} = \mathbf{619.9 \text{ kip}}$$

■

Example 13.15

Figure 13.28 shows a drilled shaft in sands. Determine the maximum load that can be allowed on it with a factor of safety of 4. Assume $\delta' = 0.6\,\phi'$.

Solution

Point Load $Q_{p(\text{net})}$

From Eq.(13.35),

$$Q_{p(\text{net})} = A_p q'(\omega N_q^* - 1)$$

$$A_p = \frac{\pi}{4} \times 1.6^2 = 2.01 \text{ m}^2$$

$$q' = 8 \times 17.5 + 1 \times 18.5 = 158.5 \text{ kN/m}^2$$

$$L = 9.0 \text{ m},\ D_b = 1.6 \text{ m},\ L_1 = 8.0 \text{ m},\ D_s = 0.8 \text{ m}$$

$$L/D_b = 9.0/1.6 = 5.63$$

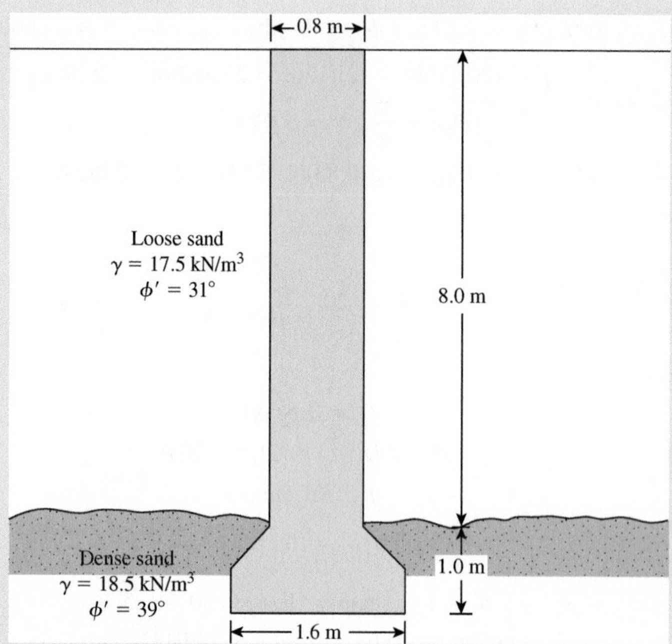

Figure 13.28

From Figure 13.25, $\omega = 0.83$.
From Eq. (13.36),

$$N_q^* = 0.21e^{0.17\phi'} = 0.21 \times e^{0.17\times 39} = 159.1$$

$$Q_{p(\text{net})} = (2.01)(158.5)(0.83 \times 159.1 - 1) = 41{,}751.5 \text{ kN}$$

Shaft Load Q_s

$$\phi' = 31°; \text{ and } \delta' = 18.6°$$

$$K_o = 1 - \sin\phi' = 0.485$$

$$p = \pi D_s = 2.51 \text{ m}$$

The critical depth is $15D_s = 12.0$ m, which is more than L_1. Therefore, the frictional resitance $f(= K\,\sigma_o' \tan\delta')$ along the pile shaft will increase over the entire length of the shaft.

$$\text{At } z = 0, f = 0$$

$$\text{At } z = 8.0 \text{ m}, f = (0.485)(8 \times 17.5)(\tan 18.6) = 22.9 \text{ kN/m}^2$$

$$Q_s = \left(\frac{0 + 22.9}{2}\right)(2.51 \times 8.0) = 229.9 \text{ kN}$$

$$Q_{\text{all}} = \frac{41751.5 + 229.9}{4} = \mathbf{10{,}495.4\ kN}$$

■

Example 13.16

Figure 13.29 shows a drilled shaft without a bell. Here, $L_1 = 27$ ft, $L_2 = 8.5$ ft, $D_s = 3.3$ ft, $c_{u(1)} = 1000$ lb/ft^2, and $c_{u(2)} = 2175$ lb/ft^2. Determine

a. The net ultimate point bearing capacity
b. The ultimate skin resistance
c. The working load, $Q_{\text{all}}(\text{FS} = 3)$

Solution

Part a

From Eq. (13.39),

$$Q_{p(\text{net})} = A_p c_u N_c^* = A_p c_{u(2)} N_c^* = \left[\left(\frac{\pi}{4}\right)(3.3)^2\right](2175)(9)$$

$$= 167{,}425 \text{ lb} \approx \mathbf{167.4 \text{ kip}}$$

(*Note:* Since $c_{u(2)} > 2000$ lb/ft^2, $N_c^* \approx 9$.)

Part b

From Eq. (13.41),

$$Q_s = \Sigma \alpha^* c_u p \Delta L$$

From Eq. (13.42),

$$\alpha^* = 0.4$$
$$p = \pi D_s = (3.14)(3.3) = 10.37 \text{ ft}$$

and

$$Q_s = (0.4)(10.37)[(1000 \times 27) + (2175 \times 8.5)]$$
$$= 188{,}682 \text{ lb} \approx \mathbf{188.7 \text{ kip}}$$

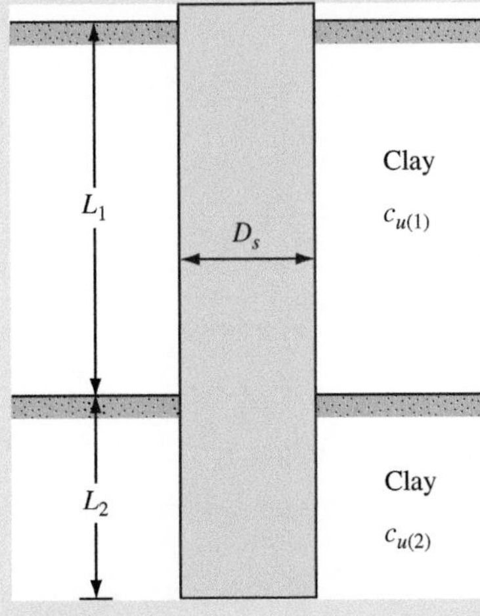

Figure 13.29 A drill shaft without a bell

Part c

$$Q_{\text{all}} = \frac{Q_{p(\text{net})} + Q_s}{\text{FS}} = \frac{167.4 + 188.7}{3} = \mathbf{118.7\ kip}$$

■

Example 13.17

Determine the maximum load that can be placed on a drilled shaft in clay as shown in Figure 13.30. Allow a factor of safety of 3.

Solution

Point Load $Q_{P(\text{net})}$
From Eq. (13.39),

$$Q_{p(\text{net})} = A_p c_u N_c^*$$

From Figure 13.26, for $c_u = 120\ \text{kN/m}^2$, $N_c^* = 8.75$ (or it can be taken as approximately 9). Hence,

$$Q_{p(\text{net})} = \left(\frac{\pi}{4} \times 1.5^2\right)(120)(8.75) = 1855.5\ \text{kN}$$

Shaft Load Q_s
From Eq. (13.41),

$$Q_s = \Sigma \alpha^* c_u p \Delta L$$

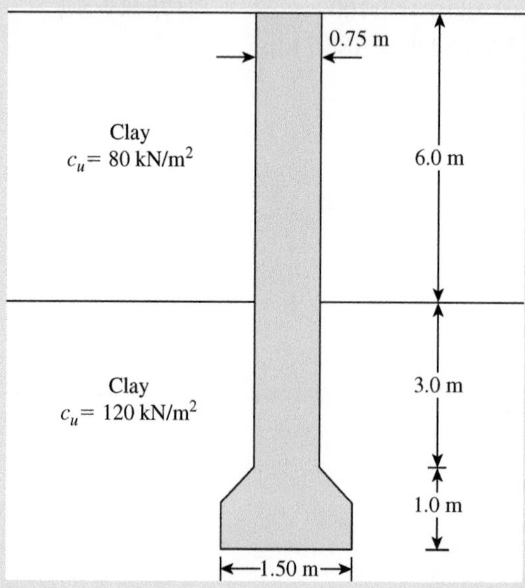

Figure 13.30

From Eq. (13.42), $\alpha^* = 0.4$. Thus,

$$p = \pi \times 0.75 = 2.36 \text{ m}$$

$$Q_s = (0.4)(2.36)[6 \times 80 + 3 \times 120] = 793.0 \text{ kN}$$

$$Q_{\text{all}} = \frac{1855.5 + 793.0}{3} = \mathbf{882.8\ kN}$$

■

13.17 Allowable Load-Bearing Capacity Based on Settlement

In most circumstances, the load-bearing capacity of drilled shafts will be controlled by the elastic settlements they are likely to undergo. Based on field tests, Reese and O'Neill (1989) developed relationships between settlement and allowable loads in sand and clay. These relationships are summarized in Tables 13.4, 13.5, 13.6, and 13.7. They are based on the average plots obtained from field tests.

In granular soils, the ultimate load-carrying capacity of the drilled shaft is mainly due to the point resistance, which is very much greater than that of the shaft load. This can be seen from Examples 13.14 and 13.15. This is true for straight shafts as well.

Table 13.4 Normalized Base Load Transfer with Settlement for Cohesionless Soils (Based on Average Curve)

$\frac{\text{Settlement}}{D_b}$ (%)	$\frac{Q_{p(\text{all})}}{Q_{p(\text{net})}}$	$\frac{\text{Settlement}}{D_b}$ (%)	$\frac{Q_{p(\text{all})}}{Q_{p(\text{net})}}$
0	0	6	1.10
1	0.32	7	1.20
2	0.56	8	1.29
3	0.73	9	1.38
4	0.87	10	1.44
5	0.98		

Table 13.5 Normalized Side Load Transfer with Settlement for Cohesionless Soils (Based on Average Curve)

$\frac{\text{Settlement}}{D_s}$ (%)	$\frac{Q_{s(\text{all})}}{Q_s}$	$\frac{\text{Settlement}}{D_s}$ (%)	$\frac{Q_{s(\text{all})}}{Q_s}$
0	0	0.8	0.974
0.1	0.371	1.0	0.987
0.2	0.590	1.2	0.974
0.3	0.744	1.4	0.968
0.4	0.846	1.6	0.960
0.5	0.910	1.8	0.940
0.6	0.936	2.0	0.920

Table 13.6 Normalized Base Load Transfer with Settlement for Cohesive Soils (Based on Average Curve)

$\frac{\text{Settlement}}{D_b}$ (%)	$\frac{Q_{p(\text{all})}}{Q_{p(\text{net})}}$	$\frac{\text{Settlement}}{D_b}$ (%)	$\frac{Q_{p(\text{all})}}{Q_{p(\text{net})}}$
0	0	4.0	0.951
0.5	0.363	5.0	0.971
1.0	0.578	6.0	0.971
1.5	0.721	7.0	0.971
2.0	0.804	8.0	0.971
2.5	0.863	9.0	0.971
3.0	0.902	10.0	0.971

Table 13.7 Normalized Side Load Transfer with Settlement for Cohesive Soils (Based on Average Curve)

$\frac{\text{Settlement}}{D_s}$ (%)	$\frac{Q_{s(\text{all})}}{Q_s}$	$\frac{\text{Settlement}}{D_s}$ (%)	$\frac{Q_{s(\text{all})}}{Q_s}$
0	0	0.8	0.95
0.1	0.48	1.0	0.94
0.2	0.74	1.2	0.92
0.3	0.86	1.4	0.91
0.4	0.91	1.6	0.89
0.6	0.95	1.8	0.85
0.7	0.955	2.0	0.82

It can be seen from Tables 13.4 and 13.5 that for a straight shaft in a cohesionless soil, when the settlement is 1% of the pile diameter, only 32% of the point resistance is mobilized, while 98.7% of the shaft resistance is mobilized.

Example 13.18

Refer to Example 13.14. Estimate Q_{all} for an allowable settlement of 0.5 in.

Solution

We have

$$\frac{\text{Allowable settlement}}{D_s} = \frac{0.5}{(2)(12)} = 0.021 = 2.1\%$$

From Table 13.5,

$$\frac{Q_{s(\text{all})}}{Q_s} \approx 0.92$$

$$Q_{s(\text{all})} = (0.92)(16.8) = 15.46 \text{ kip}$$

Again

$$\frac{\text{Allowable settlement}}{D_b} = \frac{0.5}{(3)(12)} = 0.014 = 1.4\%$$

From Table 13.4,

$$\frac{Q_{p(\text{all})}}{Q_{p(\text{net})}} \approx 0.42$$

$$Q_{p(\text{all})} = (0.42)(2462.6) = 1034.3 \text{ kip}$$

$$Q_{\text{all}} = Q_{p(\text{all})} + Q_{s(\text{all})} = 1034.3 + 15.46 \approx \mathbf{1050 \text{ kip}}$$ ■

Example 13.19

Estimate the allowable load for the drilled shaft in Example 13.15, if the allowable settlement is limited to 15 mm.

Solution

Point Load $Q_{p(\text{all})}$

$$\text{Allowable settlement}/D_b = 15.0/1600 = 0.0094 \text{ or } 0.94\%$$

From Table 13.4, $Q_{p(\text{all})}/Q_{p(\text{net})} = 0.30$. Therefore, $Q_{p(\text{all})} = 0.30 \times 41751.5 = 12{,}525.4$ kN.

Shaft Load $Q_{s(\text{all})}$

$$\text{Allowable settlement}/D_s = 15.0/800 = 0.0188 \text{ or } 1.88\%$$

From Table 13.5, $Q_{s(\text{all})}/Q_s = 0.93$. Therefore, $Q_{s(\text{all})} = 0.93 \times 229.9 = 213.8$ kN.

$$Q_{\text{all}} = Q_{p(\text{all})} + Q_{s(\text{all})} = 12525.4 + 213.8 = \mathbf{12739 \text{ kN}}$$ ■

Example 13.20

Refer to Example 13.16. Estimate Q_{all} for an allowable settlement of 0.75 in.

Solution

$$\frac{\text{Allowable settlement}}{D_s} = \frac{0.75 \text{ in.}}{(3.3)(12)} = 0.019 = 1.9\%$$

From Table 13.7,

$$\frac{Q_{s(\text{all})}}{Q_s} \approx 0.84$$

$$Q_{s(\text{all})} = (0.84)(188.7) = 158.5 \text{ kip}$$

Again,

$$\frac{\text{Allowable settlement}}{D_b} = \frac{0.75 \text{ in.}}{(3.3)(12)} = 0.019 = 1.9\%$$

From Table 13.6,

$$\frac{Q_{p(\text{all})}}{Q_{p(\text{net})}} \approx 0.79$$

$$Q_{p(\text{all})} = (0.79)(167.4) = 132.2 \text{ kip}$$

$$Q_{\text{all}} = Q_{p(\text{all})} + Q_{s(\text{all})} = 132.2 + 158.5 = \mathbf{290.7\ kip}$$

■

13.18 Summary

Deep foundations include piles, pile groups, and drilled shafts. Piles are made of timber, steel, or concrete. They are generally driven into the ground by hammers or vibratory drivers. Drilled shafts are cast-in-place piles which have diameters of 30 in. (750 mm) or more.

The ultimate load-carrying capacity of piles or drilled shafts in sands and clays can be determined from their dimensions and the shear strength parameters of the soil. From these, the length necessary to carry a specific load can be determined. Pile-driving formulas work on the principle that the energy imparted to the pile head during a blow is transferred into the work done at the pile point. Due to the approximations involved in these equations, a larger factor of safety is used in arriving at the allowable bearing capacity.

Pile load tests are carried out to verify the calculated allowable loads. These days, pile-driving analyzers (PDA) are widely used for determining the load carrying capacities of piles, without actually loading them to failure.

Problems

13.1 State whether the following are true or false.

a. Steel H-piles are high displacement driven piles.

b. The adhesion factor α is greater for stiff clays than for soft clays.

c. The shaft load is always greater than the point load.

d. The larger the set S, the larger is the allowable pile load.

e. When the pile is in compression, the settlement of the pile point is always less than the settlement of the pile head.

13.2 A concrete pile is 75 ft long and 12 in. × 12 in. in cross section. The pile is fully embedded in sand, for which $\gamma = 112$ lb/ft^3 and $\phi' = 35°$. Calculate:

a. The ultimate point load, Q_p

b. The total frictional resistance for $K = 1.3$ and $\delta' = 0.8\phi'$

13.3 Redo Problem 13.2 for $\gamma = 118$ lb/ft^3 and $\phi' = 40°$.

13.4 A 400-mm-diameter and 15-m-long concrete pile is driven into a sand where $\gamma = 18.0$ kN/m^3 and $\phi' = 31°$. Assuming $\delta' = 0.65\,\phi'$ and $K = 1.4\,K_o$, determine the load-carrying capacity of the pile with a factor of safety of 2.5.

13.5 A driven closed-ended pile, circular in cross section, is shown in Figure 13.31. Calculate:

a. The ultimate point load

b. The ultimate frictional resistance Q_s. (Use $K = 1.4$ and $\delta' = 0.6\phi'$.)

c. The allowable load of the pile. (Use FS = 4)

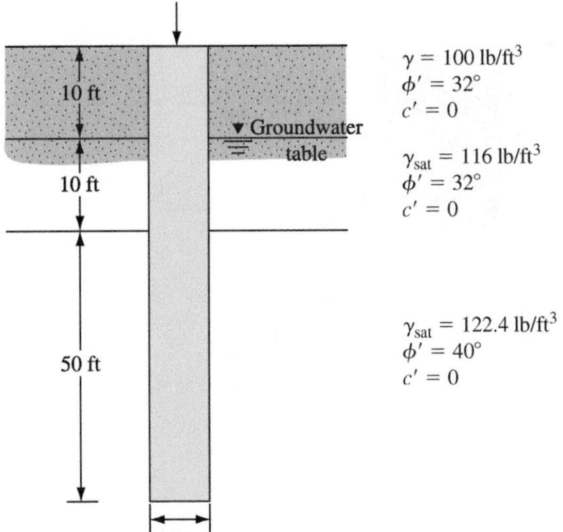

Figure 13.31

13.6 A 400 mm × 400 mm square precast concrete pile with a 15 m length is driven into sand where $\gamma = 18.0$ kN/m^3 and $\phi' = 33°$. Assuming $\delta' = 0.7\ \phi'$ and $K = 1.4\ K_o$, determine the load-carrying capacity of the pile with factor of satety of 3.0.

13.7 Determine the maximum load that can be allowed on a 450-mm-diameter driven pile shown in Figure 13.32, allowing a safety factor of 3. Use $K = 1.5\ K_o$ and $\delta' = 0.65\phi'$.

13.8 A concrete pile 60 ft long having a cross section of 15 in. × 15 in. is fully embedded in a saturated clay layer for which $\gamma_{sat} = 118$ lb/ft^3, $\phi = 0$, and $c_u = 1400$ lb/ft^2. Determine the allowable load that the pile can carry. (Let FS = 3.)

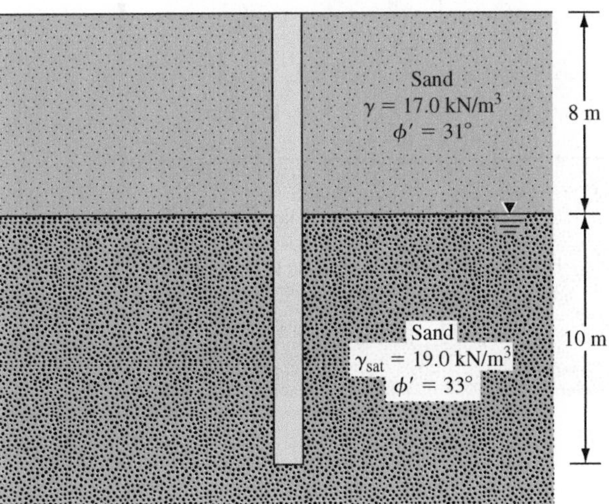

Figure 13.32

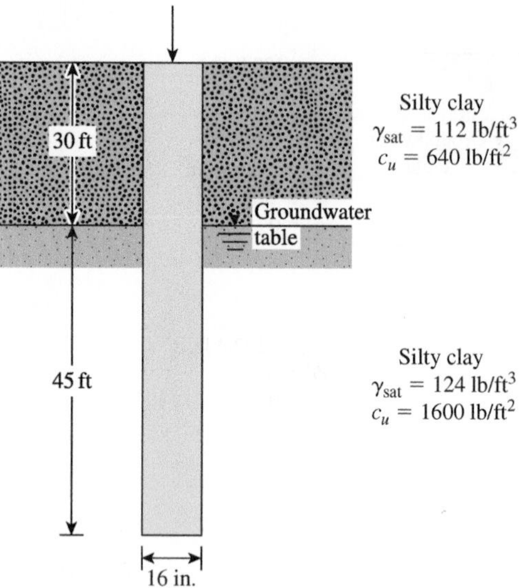

Figure 13.33

13.9 A concrete pile 16 in. × 16 in. in cross section is shown in Figure 13.33. Calculate the ultimate skin resistance.

13.10 Determine the maximum load that can be allowed on the 450 mm diameter pile shown in Figure 13.34, with safety factor of 3.

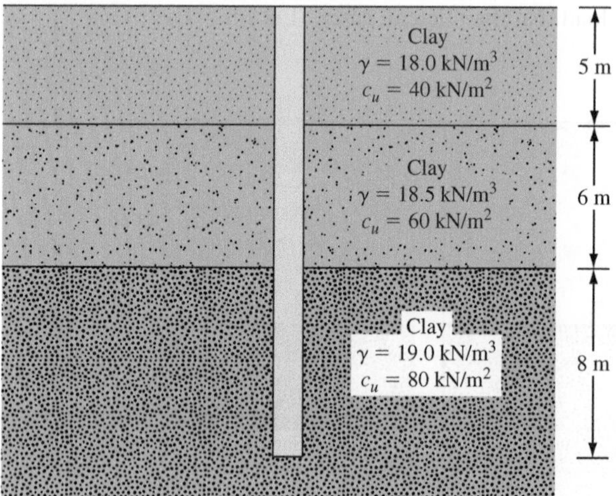

Figure 13.34

13.11 A steel H pile (Section HP 13 × 100) is driven by a steam hammer. The maximum rated hammer energy is 36 kip-ft, the weight of the ram is 14 kip, and the length of the pile is 80 ft. Also, we have

$$
\begin{aligned}
\text{Coefficient of restitution} &= 0.35 \\
\text{Weight of the pile cap} &= 1.2 \text{ kip} \\
\text{Hammer efficiency} &= 0.85 \\
\text{Number of blows for the last inch of penetration} &= 10 \\
E_p &= 30 \times 10^6 \text{ lb/in.}^2
\end{aligned}
$$

Estimate the pile capacity, using Eq. (13.14). Take FS = 6.

13.12 Solve Problem 13.11, using the modified ENR formula. [See Eq. (13.15)] Take FS = 4.

13.13 A diesel hammer (Delmag D-12) with a maximum energy rating of 30.0 kN·m and a hammer efficiency of 85% is used to drive a 15-m-long concrete pile having a square cross section of 400 mm × 400 mm. The following data are given.

- Weight of the ram = 12.2 kN
- Weight of the pile cap = 4.0 kN
- Coefficient of restitution = 0.35
- Unit weight of concrete = 24.0 kN/m^3

To derive an allowable load-carrying capacity of 400 kN from the pile based on modified ENR formula and with a factor of safety of 4, what should be the set S?

13.14 Consider a group pile. (See Figure 13.15a.) If $n_1 = 4$, $n_2 = 3$, the pile diameter, $D = 16$ in., and the pile spacing, $d = 36$ in., determine the efficiency of the group pile. Use Eq. (13.17).

13.15 Solve Problem 13.14, using the Converse–Labarre equation. [Eq. (13.19).]

13.16 In Problem 13.14, if the center-to-center pile spacings are increased to 48 in., what will be the group efficiency?

13.17 Consider the group pile described in Problem 13.14. Assume that the piles are embedded in a saturated homogeneous clay having $c_u = 2040$ lb/ft^2. The length of the piles is 60 ft. Find the allowable load-carrying capacity of the group pile. Use FS = 3.

13.18 A section of a 3 × 4 group pile in a layered saturated clay is shown in Figure 13.35. The piles are square in cross section (14 in. × 14 in.). The center-to-center spacing d of the piles is 35 in. Determine the allowable load-bearing capacity of the group pile. Use FS = 4.

13.19 Figure 13.36 shows a 3 × 5 pile group consisting of 15 concrete piles with a 400 mm diameter and 12 m in length. What would be the maximum load that can be allowed on the mat with a factor of safety of 3? The piles have a center-to-center spacing of 1200 mm.

13.20 Figure 13.37 shows a group pile in clay. Determine the consolidation settlement of the group.

13.21 A drilled shaft is shown in Figure 13.38. Determine the net allowable point bearing capacity. Assume the following values:

$$
\begin{aligned}
D_b &= 6 \text{ ft} & \gamma_c &= 100 \text{ lb/ft}^3 \\
D_s &= 3.6 \text{ ft} & \gamma_s &= 115 \text{ lb/ft}^3 \\
L_1 &= 18 \text{ ft} & \phi' &= 35^\circ \\
L_2 &= 10 \text{ ft} & c_u &= 700 \text{ lb/ft}^2
\end{aligned}
$$

Factor of safety = 4

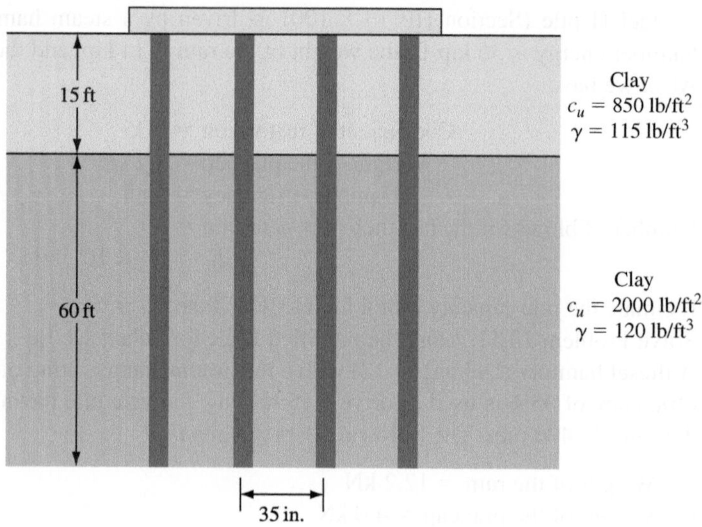

Figure 13.35

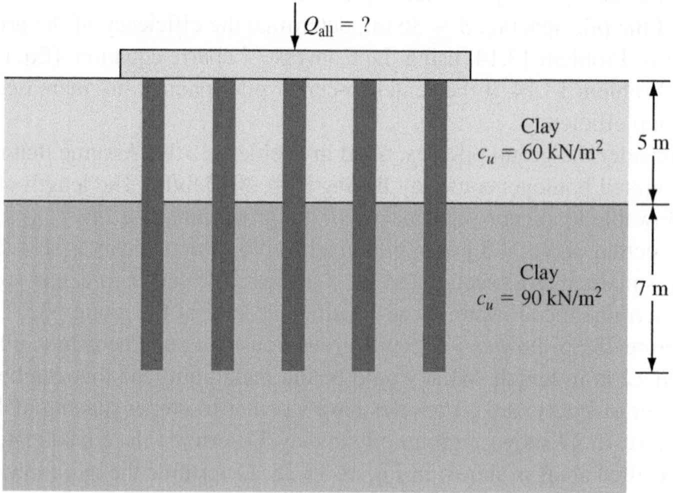

Figure 13.36

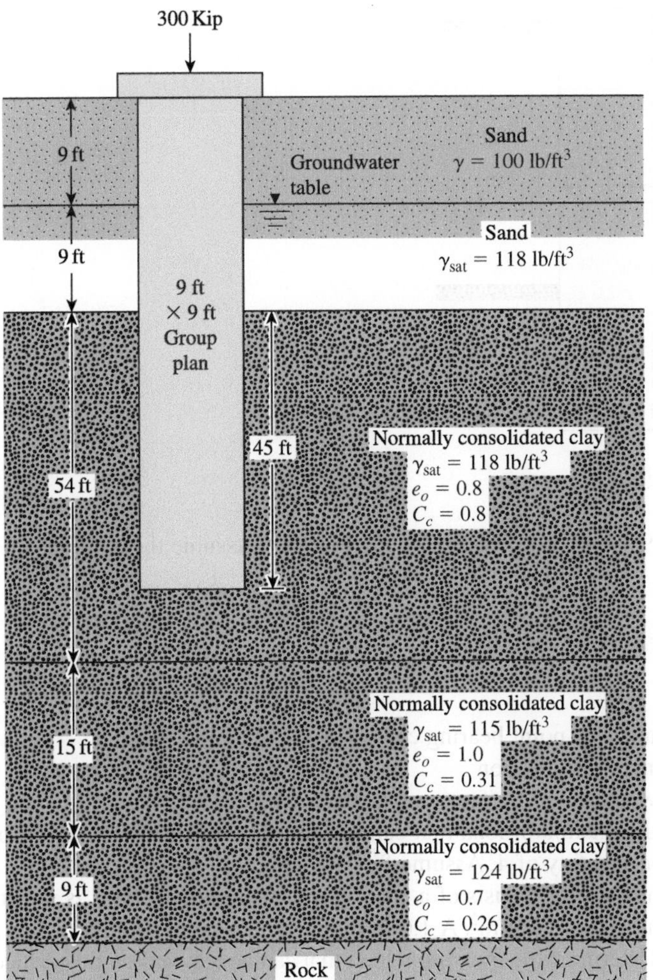

Figure 13.37

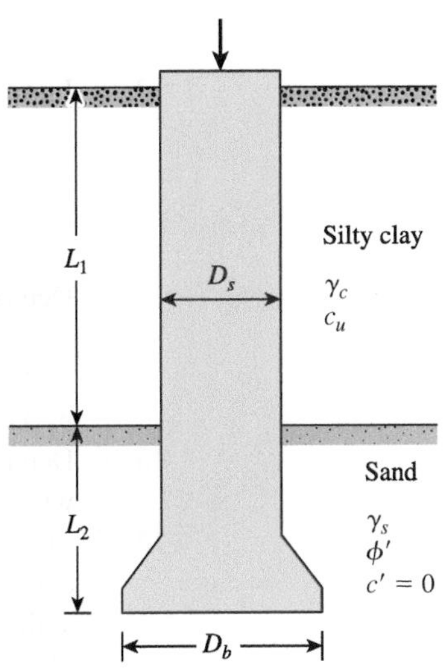

Figure 13.38

13.22 Redo Problem 13.21 with the following data:

$D_b = 5$ ft $\qquad \gamma_c = 113$ lb/ft^3
$D_s = 3$ ft $\qquad \gamma_s = 116$ lb/ft^3
$L_1 = 20.5$ ft $\qquad \phi' = 32°$
$L_2 = 7.5$ ft $\qquad c_u = 640$ lb/ft^2
Factor of safety = 4

13.23 For the drilled shaft described in Problem 13.21, what skin resistance would develop in the top 18 ft, which is in clay?

13.24 For the drilled shaft described in Problem 13.22, what skin resistance would develop in the top 20.5 ft?

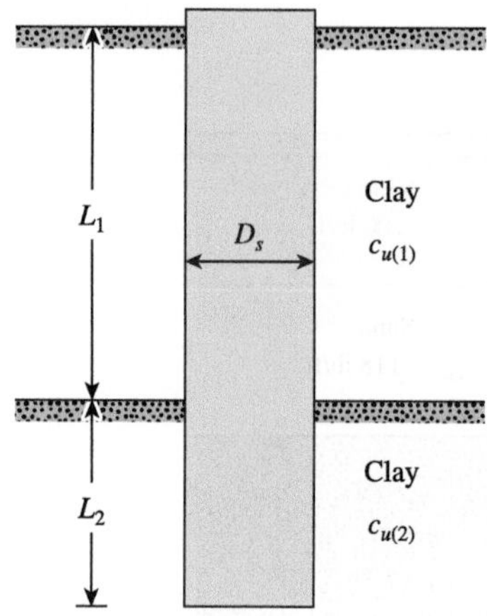

Figure 13.39

13.25 Figure 13.39 shows a drilled shaft without a bell. Assume the values:

$$L_1 = 20 \text{ ft}, c_{u(1)} = 900 \text{ lb/ft}^2$$
$$L_2 = 15 \text{ ft}, c_{u(2)} = 1500 \text{ lb/ft}^2$$
$$D_s = 5 \text{ ft}$$

Determine

a. The net ultimate point bearing capacity

b. The ultimate skin friction

c. The allowable load Q_w (factor of safety = 3)

13.26 Determine the maximum load that can be allowed on the drilled shaft shown in Figure 13.40 with a factor of safety of 4. Assume $\delta' = 0.5\,\phi'$.

13.27 The top 5 m in a site consists of a clay with $c_u = 80$ kN/m^2. This is underlain by a thick clay deposit with $c_u = 150$ kN/m^2. A 1.0-m-diameter and 15 m straight drilled shaft is constructed at this site. Determine the maximum load that can be allowed on this drilled shaft with a factor of safety of 3.

13.28 A drilled shaft in a medium sand is shown in Figure 13.41. Using the results of Reese and O'Neill, determine the following:

a. The net allowable point resistance for a base movement of 1 in.

b. The shaft frictional resistance for a base movement of 1 in.

c. The total load that can be carried by the drilled shaft for a total base movement of 1 in.

Assume the following values:

$$L = 36 \text{ ft} \qquad \gamma = 118 \text{ lb/ft}^3$$
$$L_1 = 33 \text{ ft} \qquad \phi' = 38^\circ$$
$$D_s = 3 \text{ ft}$$
$$D_b = 6 \text{ ft}$$

13.29 For the drilled shaft described in Problem 13.25, determine the load carrying capacity for a settlement of 0.5 in.

13.30 Estimate the allowable load for the drilled shaft in problem 13.26 if the allowable settlement is limited to 15 mm.

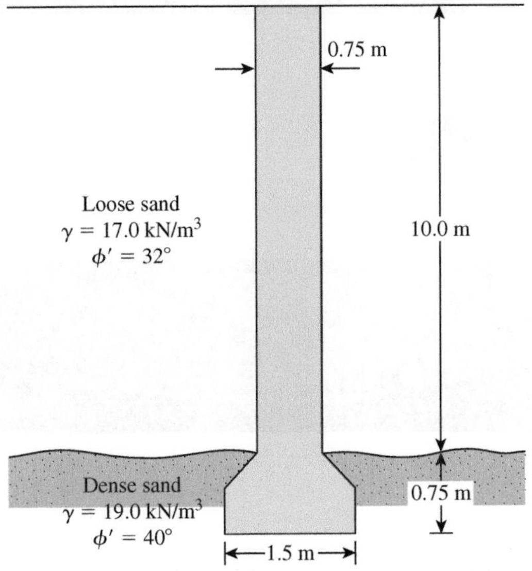

Figure 13.40

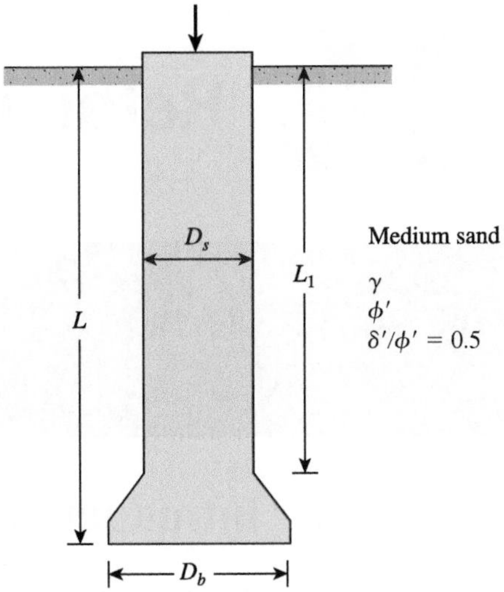

D_s
L
L_1
Medium sand
γ
ϕ'
$\delta'/\phi' = 0.5$
D_b

Figure 13.41

References

American Society of Civil Engineers (1959). "Timber Piles and Construction Timbers," *Manual of Practice*, No. 17, American Society of Civil Engineers, New York.

Berezantzev, V. G., Khristoforov, V. S., and Golubkov, V. N. (1961). "Load Bearing Capacity and Deformation of Piled Foundations," *Proceedings, Fifth International Conference on Soil Mechanics and Foundation Engineering*, Paris, Vol. 2, pp. 11–15.

Meyerhof, G. G. (1976). "Bearing Capacity and Settlement of Pile Foundations," *Journal of the Geotechnical Engineering Division*, American Society of Civil Engineers, Vol. 102, No. GT3, pp. 197–228.

O'Neill, M. W. and Reese, L. C. (1999). *Drilled Shafts: Construction Procedure and Design Methods*, FHWA Report No. IF-99-025.

Reese, L. C. and O'Neill, M. W. (1989). "New Design Method for Drilled Shafts from Common Soil and Rock Tests," *Proceedings, Foundation Engineering: Current Principles and Practices*, American Society of Civil Engineers, Vol. 2, pp. 1026–1039.

Vesic, A. S. (1969). *Experiments with Instrumented Pile Groups in Sand*, American Society for Testing and Materials; Special Technical Publication No. 444, pp. 177–222.

Vesic, A. S. (1977). *Design of Pile Foundations*, National Cooperative Highway Research Program Synthesis of Practice No. 42, Transportation Research Board, Washington, D.C.

14 Retaining Walls

14.1 Introduction

In Chapter 11, you were introduced to various theories of lateral earth pressure. Those theories will be used in this chapter to design various types of retaining walls. In this chapter, you will learn the following.

- Types of retaining walls
- Proportioning retaining walls
- Checking the stability against possible overturning
- Checking stability against possible sliding
- Checking the stability against possible bearing capacity failure

In general, retaining walls can be divided into two major categories: (a) conventional retaining walls and (b) mechanically stabilized earth walls. Conventional retaining walls can generally be classified into four varieties:

1. Gravity retaining walls
2. Semigravity retaining walls
3. Cantilever retaining walls
4. Counterfort retaining walls

Gravity retaining walls (Figure 14.1a) are constructed with plain concrete or stone masonry. They depend for stability on their own weight and any soil resting on the masonry. This type of construction is not economical for high walls.

In many cases, a small amount of steel may be used for the construction of gravity walls, thereby minimizing the size of wall sections. Such walls are generally referred to as *semigravity walls* (Figure 14.1b).

Cantilever retaining walls (Figure 14.1c) are made of reinforced concrete that consists of a thin stem and a base slab. This type of wall is economical to a height of about 25 ft ($\approx$ 7.5 m).

Counterfort retaining walls (Figure 14.1d) are similar to cantilever walls. At regular intervals, however, they have thin vertical concrete slabs known as *counterforts* that tie the wall and the base slab together. The purpose of the counterforts is to reduce the shear and the bending moments.

To design retaining walls properly, an engineer must know the basic parameters—the *unit weight, angle of friction*, and *cohesion*—of the soil retained behind the wall and the soil below the

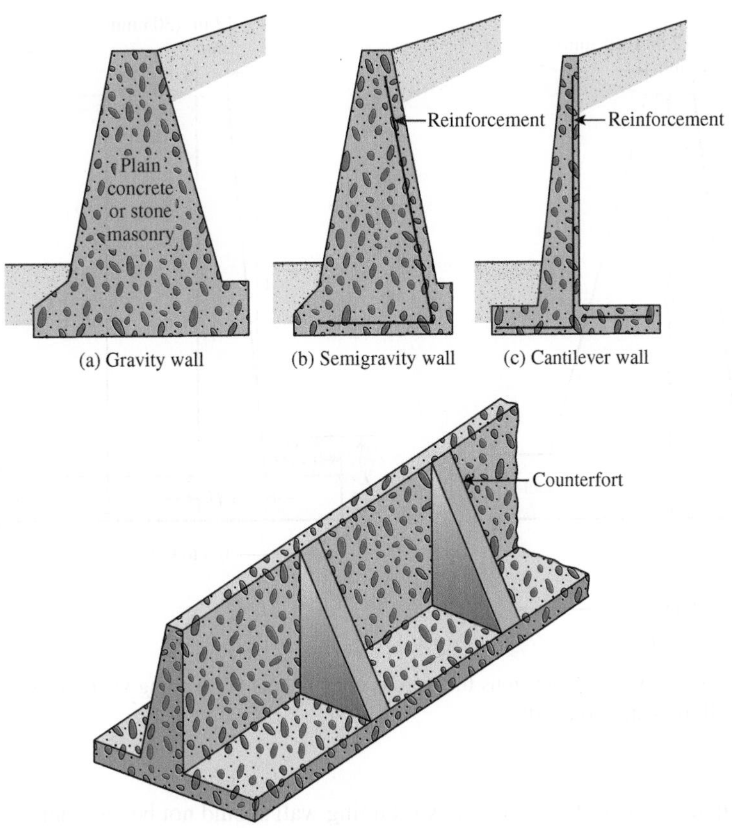

Figure 14.1 Types of retaining wall

base slab. Knowing the properties of the soil behind the wall enables the engineer to determine the lateral pressure distribution that has to be designed for.

There are two phases in the design of a conventional retaining wall. First, with the lateral earth pressure known, the structure as a whole is checked for *stability:* The structure is examined for possible *overturning, sliding,* and *bearing capacity* failures. Second, each component of the structure is checked for *strength*, and the *steel reinforcement* of each component is determined.

This chapter presents the procedures for determining the stability of the retaining wall. Checks for strength can be found in any textbook on reinforced concrete.

Gravity and Cantilever Walls

14.2 Proportioning Retaining Walls

In designing retaining walls, an engineer must assume some of their dimensions. Called *proportioning,* such assumptions allow the engineer to check trial sections of the walls for stability. If the stability checks yield undesirable results, the sections can be changed and rechecked. Figure 14.2 shows the general proportions of various retaining-wall components that can be used for initial checks.

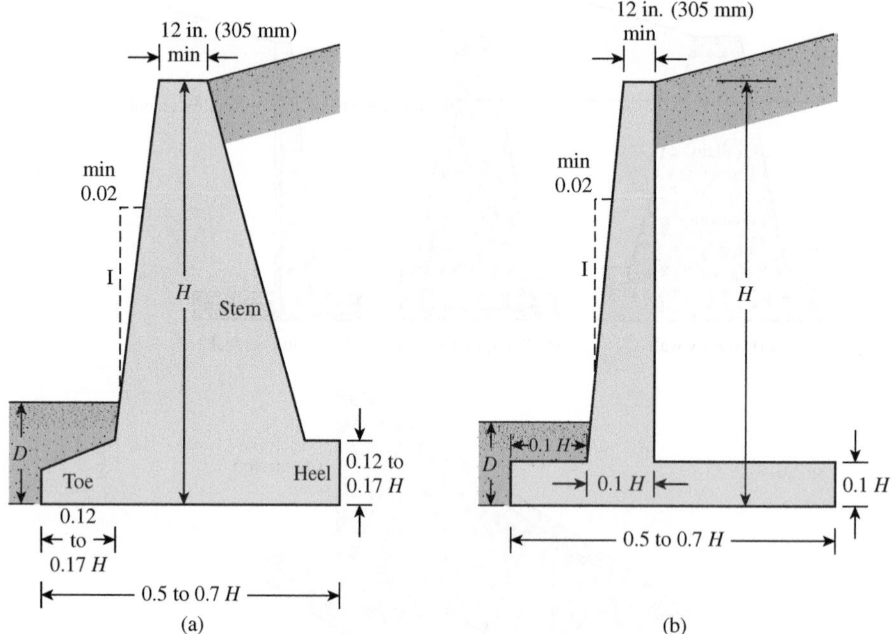

Figure 14.2 Approximate dimensions for various components of retaining wall for initial stability checks: (a) gravity wall; (b) cantilever wall

Note that the top of the stem of any retaining wall should not be less than about 12 in. (305 mm) for proper placement of concrete. The depth, D, to the bottom of the base slab should be a minimum of 2 ft ($\approx$ 0.6 mm). However, the bottom of the base slab should be positioned below the seasonal frost line.

For counterfort retaining walls, the general proportion of the stem and the base slab is the same as for cantilever walls. However, the counterfort slabs may be about 12 in. (305 mm) thick and spaced at center-to-center distances of $0.3H$ to $0.7H$.

14.3 Application of Lateral Earth Pressure Theories to Design

The fundamental theories for calculating lateral earth pressure were presented in Chapter 11. To use these theories in design, an engineer must make several simple assumptions. In the case of cantilever walls, the use of the Rankine earth pressure theory for stability checks involves drawing a vertical line AB through point A, located at the edge of the heel of the base slab in Figure 14.3a. The Rankine active condition is assumed to exist along the vertical plane AB. Rankine active earth pressure equations may then be used to calculate the lateral pressure on the face AB of the wall. In the analysis of the wall's stability, the force $P_{a(\text{Rankine})}$, the weight of soil above the heel, and the weight W_c of the concrete all should be taken into consideration. The assumption for the development of Rankine active pressure along the soil face AB is theoretically correct if the shear zone bounded by the line AC is not obstructed by the stem of the wall. The angle, η, that the line AC makes with the vertical is

$$\eta = 45 + \frac{\alpha}{2} - \frac{\phi'}{2} - \sin^{-1}\left(\frac{\sin\alpha}{\sin\phi'}\right)$$

A similar type of analysis may be used for gravity walls, as shown in Figure 14.3b. However, *Coulomb's active earth pressure theory* also may be used, as shown in Figure 14.3c. If it is used, the only forces to be considered are $P_{a(\text{Coulomb})}$ and the weight of the wall, W_c.

If Coulomb's theory is used, it will be necessary to know the range of the wall friction angle δ' with various types of backfill material. Following are some ranges of wall friction angle for masonry or mass concrete walls:

Backfill material	Range of δ' (deg)
Gravel	27–30
Coarse sand	20–28
Fine sand	15–25
Stiff clay	15–20
Silty clay	12–16

In the case of ordinary retaining walls, water table problems and hence hydrostatic pressure are not encountered. Facilities for drainage from the soils that are retained are always provided.

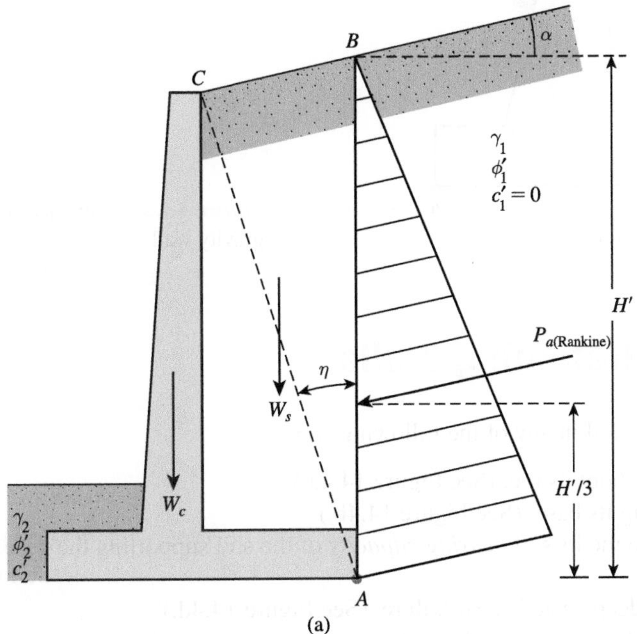

Figure 14.3 Assumption for the determination of lateral earth pressure: (a) cantilever wall

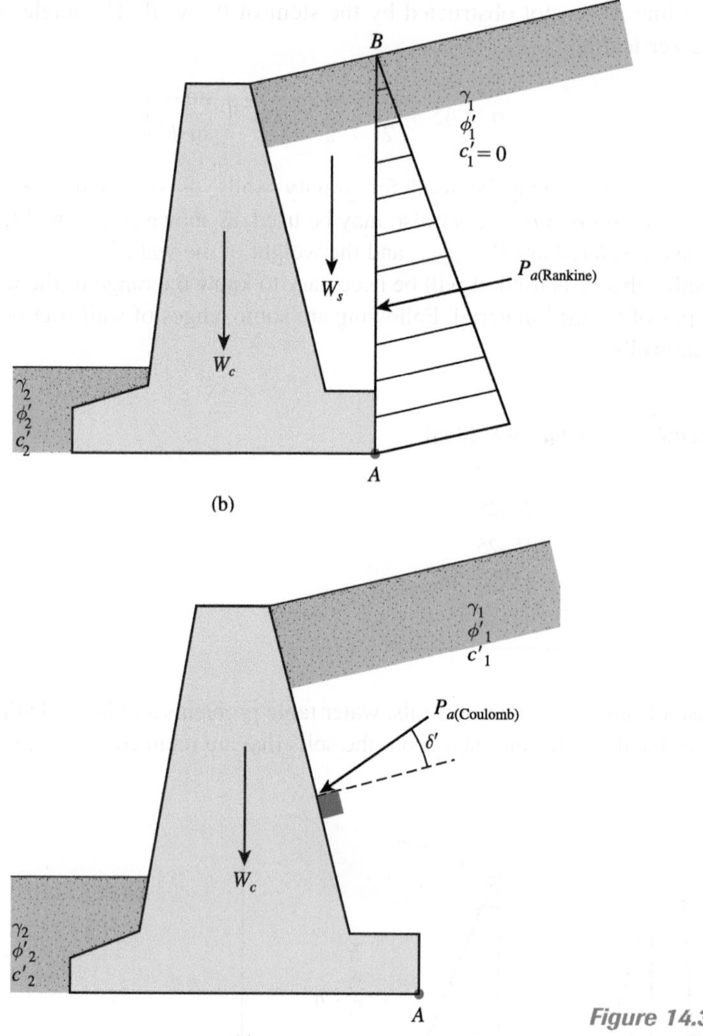

Figure 14.3 (*Continued*) (b) and (c) gravity wall

14.4 Stability of Retaining Walls

A retaining wall may fail in any of the following ways:

- It may *overturn* about its toe. (See Figure 14.4a.)
- It may *slide* along its base. (See Figure 14.4b.)
- It may fail due to the loss of *bearing capacity* of the soil supporting the base. (See Figure 14.4c.)
- It may undergo deep-seated shear failure. (See Figure 14.4d.)
- It may go through excessive settlement.

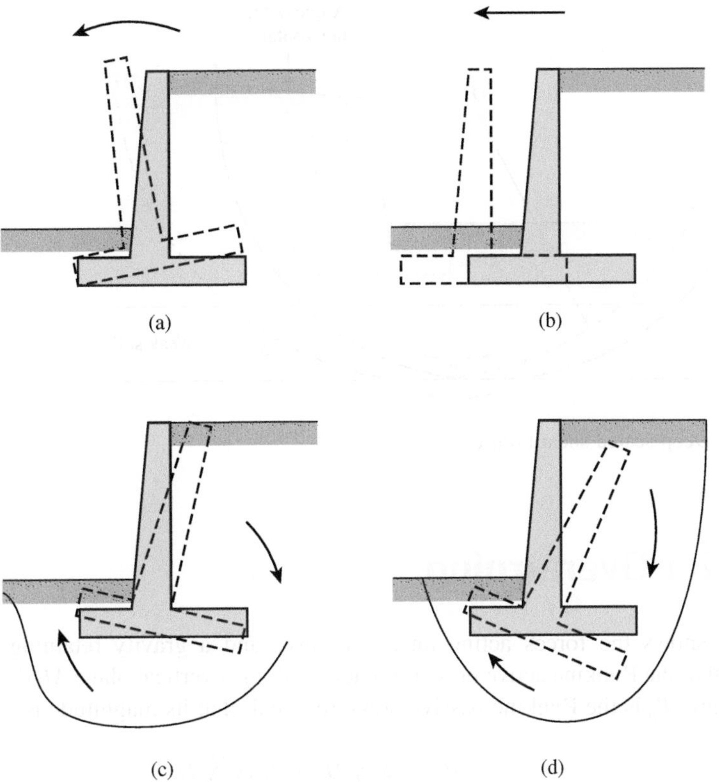

Figure 14.4 Failure of retaining wall: (a) by overturning; (b) by sliding; (c) by bearing capacity failure; (d) by deep-seated shear failure

The checks for stability against overturning, sliding, and bearing capacity failure will be described in Sections 14.5, 14.6, and 14.7. The principles used to estimate settlement were covered in Chapter 8 and will not be discussed further. When a weak soil layer is located at a shallow depth—that is, within a depth of 1.5 times the width of the base slab of the retaining wall—the possibility of excessive settlement should be considered. In some cases, the use of lightweight backfill material behind the retaining wall may solve the problem.

Deep shear failure can occur along a cylindrical surface, such as *abc* shown in Figure 14.5, as a result of the existence of a weak layer of soil underneath the wall at a depth of about 1.5 times the width of the base slab of the retaining wall. In such cases, the critical cylindrical failure surface *abc* has to be determined by trial and error, using various centers such as O. The failure surface along which the minimum factor of safety is obtained is the *critical surface of sliding*. For the backfill slope with a less than about 10°, the critical failure circle apparently passes through the edge of the heel slab (such as *def* in the figure). In this situation, the minimum factor of safety also has to be determined by trial and error by changing the center of the trial circle.

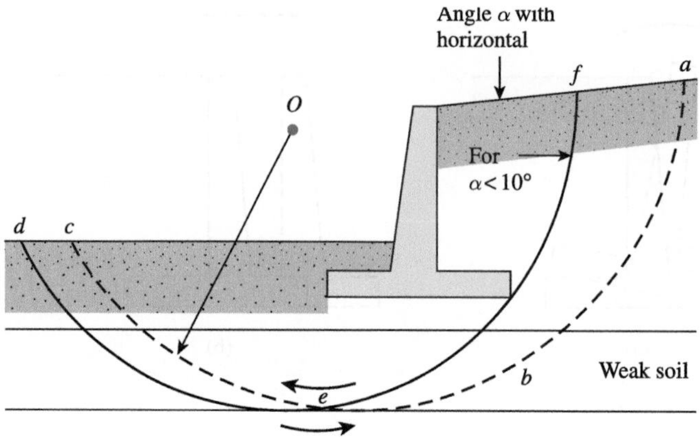

Figure 14.5 Deep-seated shear failure

14.5 Check for Overturning

Figure 14.6 shows the forces acting on a cantilever and a gravity retaining wall, based on the assumption that the Rankine active pressure is acting along a vertical plane *AB* drawn through the heel of the structure. P_p is the Rankine passive pressure; recall that its magnitude is

$$P_p = \tfrac{1}{2}K_p\gamma_2 D^2 + 2c_2'\sqrt{K_p}D$$

where

γ_2 = unit weight of soil in front of the heel and under the base slab
K_p = Rankine passive earth pressure coefficient = $\tan^2(45 + \phi_2'/2)$
c_2', ϕ_2' = cohesion and effective soil friction angle, respectively

The factor of safety against overturning about the toe—that is, about point *C* in Figure 14.6—may be expressed as

$$FS_{(\text{overturning})} = \frac{\Sigma M_R}{\Sigma M_O} \tag{14.1}$$

where

ΣM_O = sum of the moments of forces tending to overturn about point *C*
ΣM_R = sum of the moments of forces tending to resist overturning about point *C*

The overturning moment is

$$\Sigma M_O = P_h\left(\frac{H'}{3}\right) \tag{14.2}$$

where $P_h = P_a \cos\alpha$.

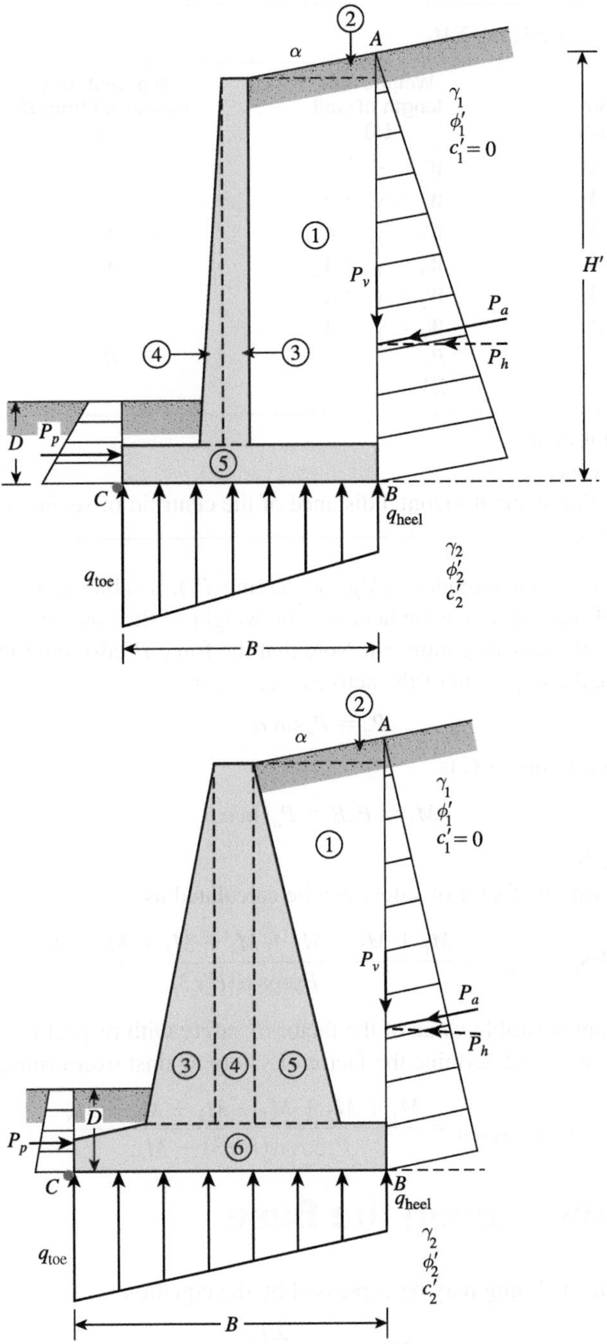

Figure 14.6 Check for overturning, assuming that the Rankine pressure is valid

Table 14.1 Procedure for Calculating ΣM_R

Section (1)	Area (2)	Weight/unit length of wall (3)	Moment arm measured from *C* (4)	Moment about *C* (5)
1	A_1	$W_1 = \gamma_1 \times A_1$	X_1	M_1
2	A_2	$W_2 = \gamma_1 \times A_2$	X_2	M_2
3	A_3	$W_3 = \gamma_c \times A_3$	X_3	M_3
4	A_4	$W_4 = \gamma_c \times A_4$	X_4	M_4
5	A_5	$W_5 = \gamma_c \times A_5$	X_5	M_5
6	A_6	$W_6 = \gamma_c \times A_6$	X_6	M_6
		P_v	B	M_v
		ΣV		ΣM_R

Note: γ_1 = unit weight of backfill
γ_c = unit weight of concrete
X_i = moment arm that is the horizontal distance of the centroid of section i from C

To calculate the resisting moment, ΣM_R (neglecting P_p), a table such as Table 14.1 can be prepared. The weight of the soil above the heel and the weight of the concrete (or masonry) are both forces that contribute to the resisting moment. Note that the force P_v also contributes to the resisting moment. P_v is the vertical component of the active force P_a, or

$$P_v = P_a \sin \alpha$$

The moment of the force P_v about C is

$$M_v = P_v B = P_a \sin \alpha B \tag{14.3}$$

where B = width of the base slab.

Once ΣM_R is known, the factor of safety can be calculated as

$$\text{FS}_{(\text{overturning})} = \frac{M_1 + M_2 + M_3 + M_4 + M_5 + M_6 + M_v}{P_a \cos \alpha (H'/3)} \tag{14.4}$$

The usual minimum desirable value of the factor of safety with respect to overturning is 2 to 3. Some designers prefer to determine the factor of safety against overturning with the formula

$$\text{FS}_{(\text{overturning})} = \frac{M_1 + M_2 + M_3 + M_4 + M_5 + M_6}{P_a \cos \alpha (H'/3) - M_v} \tag{14.5}$$

14.6 Check for Sliding along the Base

The factor of safety against sliding may be expressed by the equation

$$\text{FS}_{(\text{sliding})} = \frac{\Sigma F_{R'}}{\Sigma F_d} \tag{14.6}$$

where

$\Sigma F_{R'}$ = sum of the horizontal resisting forces
ΣF_d = sum of the horizontal driving forces

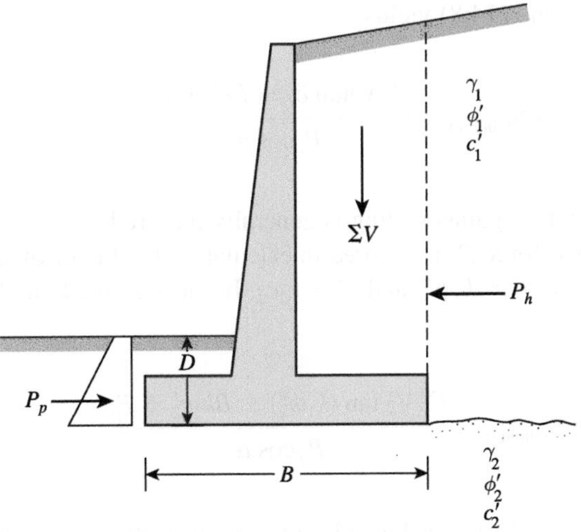

Figure 14.7 Check for sliding along the base

Figure 14.7 indicates that the shear strength of the soil immediately below the base slab may be represented as

$$s = \sigma' \tan \delta' + c_a'$$

where

δ' = angle of friction between the soil and the base slab
c_a' = adhesion between the soil and the base slab

Thus, the maximum resisting force that can be derived from the soil per unit length of the wall along the bottom of the base slab is

$$R' = s(\text{area of cross section}) = s(B \times 1) = B\sigma' \tan \delta' + Bc_a'$$

However,

$$B\sigma' = \text{sum of the vertical force} = \Sigma V(\text{see Table 14.1})$$

so

$$R' = (\Sigma V)\tan \delta' + Bc_a'$$

Figure 14.7 shows that the passive force P_p is also a horizontal resisting force. Hence,

$$\Sigma F_{R'} = (\Sigma V)\tan \delta' + Bc_a' + P_p \tag{14.7}$$

The only horizontal force that will tend to cause the wall to slide (a *driving force*) is the horizontal component of the active force P_a, so

$$\Sigma F_d = P_a \cos \alpha \tag{14.8}$$

Combining Eqs. (14.6), (14.7), and (14.8) yields

$$FS_{(sliding)} = \frac{(\Sigma V)\tan \delta' + Bc_a' + P_p}{P_a \cos \alpha} \tag{14.9}$$

A minimum factor of safety of 1.5 against sliding is generally required.

In many cases, the passive force P_p is ignored in calculating the factor of safety with respect to sliding. In general, we can write $\delta' = k_1\phi'_2$ and $c_a' = k_2c_2'$. In most cases, k_1 and k_2 are in the range from $\frac{1}{2}$ to $\frac{2}{3}$. Thus,

$$FS_{(sliding)} = \frac{(\Sigma V)\tan (k_1\phi_2') + Bk_2c_2' + P_p}{P_a \cos \alpha} \tag{14.10}$$

If the desired value of $FS_{(sliding)}$ is not achieved, several alternatives may be investigated (see Figure 14.8):

- Increase the width of the base slab (i.e., the heel of the footing).
- Use a key to the base slab. If a key is included, the passive force per unit length of the wall becomes

$$P_p = \frac{1}{2}\gamma_2 D_1^2 K_p + 2c_2' D_1 \sqrt{K_p}$$

where $K_p = \tan^2\left(45 + \frac{\phi_2'}{2}\right)$.

- Use a *deadman anchor* at the stem of the retaining wall.

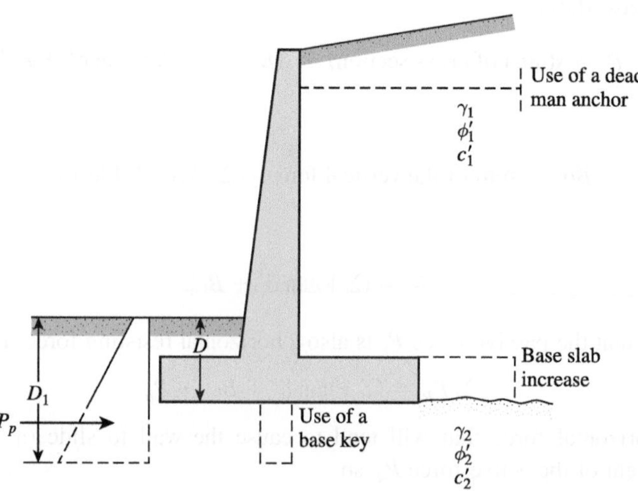

Figure 14.8 Alternatives for increasing the factor of safety with respect to sliding

14.7 Check for Bearing Capacity Failure

The vertical pressure transmitted to the soil by the base slab of the retaining wall should be checked against the ultimate bearing capacity of the soil. The nature of variation of the vertical pressure transmitted by the base slab into the soil is shown in Figure 14.9. Note that q_{toe} and q_{heel} are the *maximum* and the *minimum* pressures occurring at the ends of the toe and heel sections, respectively. The magnitudes of q_{toe} and q_{heel} can be given as

$$q_{max} = q_{toe} = \frac{\Sigma V}{B}\left(1 + \frac{6e}{B}\right) \tag{14.11}$$

and

$$q_{min} = q_{heel} = \frac{\Sigma V}{B}\left(1 - \frac{6e}{B}\right) \tag{14.12}$$

where e = eccentricity $= \dfrac{B}{2} - \dfrac{\Sigma M_R - \Sigma M_O}{\Sigma V}$.

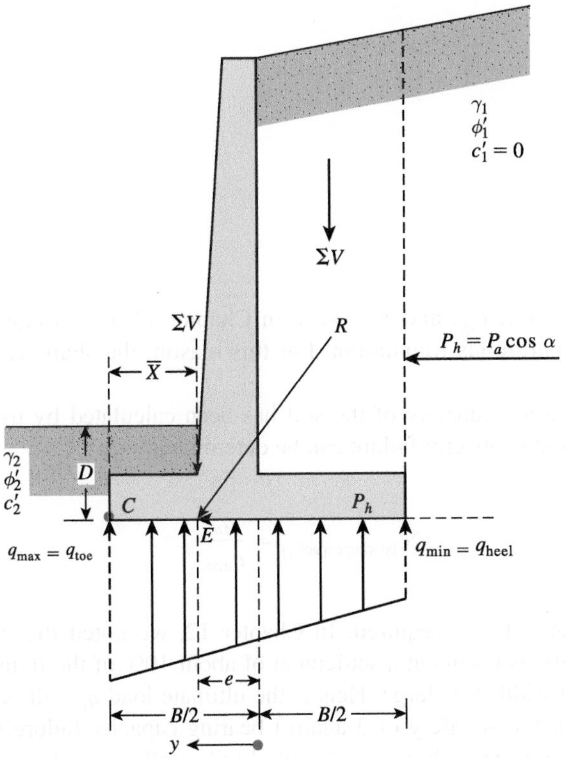

Figure 14.9 Check for bearing capacity failure

The magnitudes of ΣM_R and ΣM_O can be determined respectively from Table 14.1 and Eq. (14.2). Note that ΣV includes the weight of the soil, as shown in Table 14.1, and that when the value of the eccentricity e becomes greater than $B/6$, $q_{\min}$ [Eq. (14.12)] becomes negative. Thus, there will be some tensile stress at the end of the heel section. This stress is not desirable, because the tensile strength of soil is very small. If the analysis of a design shows that $e > B/6$, the design should be reproportioned and calculations redone.

The relationships pertaining to the ultimate bearing capacity of a shallow foundation were discussed in Chapter 12. Recall that [Eq. (12.20)]

$$q'_u = c'_2 N_c F_{cd} F_{ci} + q N_q F_{qd} F_{qi} + \tfrac{1}{2}\gamma_2 B' N_\gamma F_{\gamma d} F\gamma i \tag{14.13}$$

where

$$q = \gamma_2 D$$

$$B' = B - 2e$$

$$F_{cd} = 1 + 0.4\frac{D}{B}$$

$$F_{qd} = 1 + 2\tan\phi'_2(1 - \sin\phi'_2)^2\frac{D}{B}$$

$$F_{\gamma d} = 1$$

$$F_{ci} = F_{qi} = \left(1 - \frac{\psi^\circ}{90^\circ}\right)^2$$

$$F_{\gamma i} = \left(1 - \frac{\psi^\circ}{\phi'^\circ_2}\right)^2$$

$$\psi^\circ = \tan^{-1}\left(\frac{P_a\cos\alpha}{\Sigma V}\right)$$

Note that the shape factors F_{cs}, F_{qs}, and $F_{\gamma s}$ given in Chapter 12 are all equal to unity, because they can be treated as a continuous foundation. For this reason, the shape factors are not shown in Eq. (14.13).

Once the ultimate bearing capacity of the soil has been calculated by using Eq. (14.13), the factor of safety against bearing capacity failure can be determined:

$$\text{FS}_{(\text{bearing capacity})} = \frac{q'_u}{q_{\max}} \tag{14.14}$$

Generally, a factor of safety of 3 is required. In Chapter 12, we noted that the ultimate bearing capacity of shallow foundations occurs at a settlement of about 10% of the foundation width. In the case of retaining walls, the width B is large. Hence, the ultimate load q'_u will occur at a fairly large foundation settlement. A factor of safety of 3 against bearing capacity failure may not ensure that settlement of the structure will be within the tolerable limit in all cases. Thus, this situation needs further investigation.

Example 14.1

The cross section of a cantilever retaining wall is shown in Figure 14.10. Calculate the factors of safety with respect to overturning, sliding, and bearing capacity. Use 150 lb/ft^3 for the unit weight of concrete.

Solution

Refer to Figure 14.10. Note that

$$H' = H_1 + H_2 + H_3 = 6 \tan 10° + 18 + 2.75 = 21.81 \text{ ft}$$

$$P_a = \frac{1}{2}\gamma_1 H'^2 K_a$$

For $\phi'_1 = 34°$ and $\alpha = 10°$, the value of K_a is 0.294 (Table 11.2), so

$$P_a = \frac{\frac{1}{2}(117)(21.81)^2(0.294)}{1000} = 8.18 \text{ kip/ft}$$

$$P_v = P_a \sin 10° = 1.42 \text{ kip/ft}$$

$$P_h = P_a \cos 10° = 8.06 \text{ kip/ft}$$

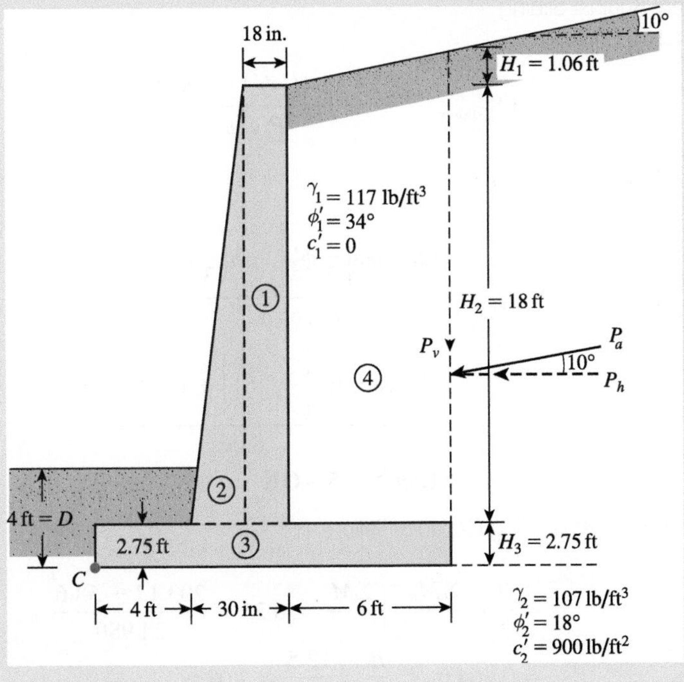

Figure 14.10

Factor of Safety Against Overturning

The following table can now be prepared for determination of the resisting moment.

Section	Weight (kip/ft)	Moment arm from *C* (ft)	Moment about *C* (kip.ft/ft)
1	$(1.5)(18)(0.15) = 4.05$	5.75	23.29
2	$\frac{1}{2}(1.0)(18)(0.15) = 1.35$	$4 + \frac{2}{3}(1) = 4.67$	6.3
3	$(12.5)(2.75)(0.15) = 5.156$	6.25	32.23
4	$\left(\frac{18 + 19.06}{2}\right)(6)(0.117) = 13.01$	$\approx 4 + 2.5 + \frac{6}{2} = 9.5$	123.6
	$P_v = 1.42$	12.5	17.75
	$\Sigma V = 24.986$		$\Sigma\, 203.17 = \Sigma M_R$

The overturning moment, M_O, is

$$M_O = P_h \frac{H'}{3} = (8.06)\left(\frac{21.81}{3}\right) = 58.6 \text{ kip.ft/ft}$$

So

$$\text{FS}_{(\text{overturning})} = \frac{\Sigma M_R}{M_O} = \frac{203.17}{58.6} = \mathbf{3.47 > 3 — OK}$$

Factor of Safety Against Sliding

From Eq. (14.10),

$$\text{FS}_{(\text{sliding})} = \frac{(\Sigma V)\tan(k_1 \phi_2') + Bk_2 c_2' + P_p}{P_a \cos\alpha}$$

Let $k_1 = k_2 = 2/3$ and $P_p = 0$. So

$$\text{FS}_{(\text{sliding})} = \frac{(\Sigma V)\tan\left[\frac{2}{3}\phi_2'\right] + B\left(\frac{2}{3}c_2'\right)}{P_a \cos\alpha}$$

$$= \frac{(24.986)\tan\left[\frac{2}{3}(18)\right] + (12.5)\frac{2}{3}(0.9)}{8.06}$$

$$= \mathbf{1.59 > 1.5—OK}$$

Factor of Safety Against Bearing Capacity Failure

$$e = \frac{B}{2} - \frac{\Sigma M_R - \Sigma M_o}{\Sigma V} = 6.25 - \frac{203.17 - 58.6}{24.986}$$

$$= 0.464 \text{ ft} < \frac{B}{6} = \frac{12.5}{6} = 2.08 \text{ ft}$$

Again, from Eqs. (14.11)

$$q_{\text{toe}} = \frac{\Sigma V}{B}\left(1 + \frac{6e}{B}\right) = \frac{24.986}{12.5}\left[1 + \frac{(6)(0.464)}{12.5}\right] = 2.44 \text{ kip/ft}^2$$

The ultimate bearing capacity of the soil can be determined from Eq. (14.13):

$$q'_u = c'_2 N_c F_{cd} F_{ci} + qN_q F_{qd} F_{qi} + \frac{1}{2}\gamma_2 B' N_\gamma F_{\gamma d} F_{\gamma i}$$

From Table 12.1 for $\phi'_2 = 18°$, $N_c = 13.1$, $N_q = 5.26$, and $N_\gamma = 4.07$,

$$q = \gamma_2 D = (4)(0.107) = 0.428 \text{ kip/ft}^2$$

$$B' = B - 2e = 12.5 - (2)(0.464) = 11.572 \text{ ft}$$

$$F_{cd} = 1 + 0.4\left(\frac{D}{B}\right) = 1 + (0.4)\left(\frac{4}{12.5}\right) = 1.128$$

$$F_{qd} = 1 + 2\tan\phi'_2(1 - \sin\phi'_2)^2\left(\frac{D}{B}\right) = 1 + (0.31)\left(\frac{4}{12.5}\right) = 1.099$$

$$F_{\gamma d} = 1$$

$$F_{ci} = F_{qi} = \left(1 - \frac{\psi°}{90°}\right)^2$$

$$\psi = \tan^{-1}\left(\frac{P_a \cos\alpha}{\Sigma V}\right) = \left(\frac{8.06}{24.986}\right) = 17.88°$$

So

$$F_{ci} = F_{qi} = \left(1 - \frac{17.88}{90}\right)^2 = 0.642$$

$$F_{\gamma i} = \left(1 + \frac{\psi}{\phi'_2}\right)^2 = \left(1 - \frac{17.88}{18}\right)^2 \approx 0$$

Hence,

$$\begin{aligned} q'_u &= (0.9)(13.1)(1.128)(0.642) + (0.428)(5.26)(1.099)(0.642) \\ &\quad + \frac{1}{2}(0.107)(11.572)(4.07)(1)(0) \\ &= 10.126 \text{ kip/ft}^2 \end{aligned}$$

So

$$\text{FS}_{(\text{bearing capacity})} = \frac{q'_u}{q_{\text{toe}}} = \frac{10.126}{2.44} = \mathbf{4.15 > 3\text{—OK}}$$

■

Example 14.2

Figure 14.11 shows a gravity retaining wall for a granular ($c' = 0$) backfill. The same soil is present at the bottom of the wall and on the left. The unit weight and the friction angle of the backfill are 18.5 kN/m^3 and 35°, respectively. The unit weight of the concrete is 24.0 kN/m^3. Determine the factors of safety with respect to overturning, sliding, and bearing capacity failure.

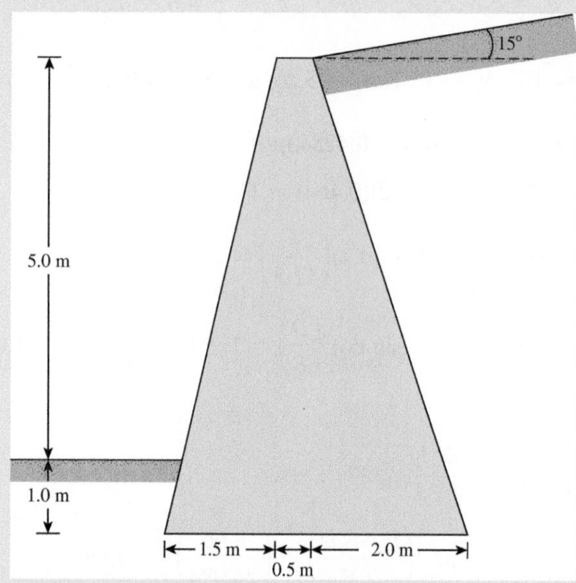

Figure 14.11

Solution

The retaining wall and the soil regions of interest are divided into the rectangles and triangles shown in Figure 14.12 for case of computation of the weights and moment arms.

For $\phi' = 35°$ and $\alpha = 15°$, we find from Table 11.2 that $K_a = 0.2968$. Thus,

$$P_a = \frac{1}{2}\gamma H'^2 K_a = \frac{1}{2}(18.5)(6.54)^2(0.2968) = 117.4 \text{ kN/m}$$

$$P_h = P_a \cos\alpha = 113.4 \text{ kN/m}$$

$$P_v = P_a \sin\alpha = 30.4 \text{ kN/m}$$

Section	Weight (kN/m)	Moment arm from C (m)	Moment about C (kN·m/m)
1	(0.5)(1.5)(6.0)(24.0) = 108.0	1.0	108.0
2	(0.5)(6.0)(24.0) = 72.0	1.75	126.0
3	(0.5)(2.0)(6.0)(24.0) = 144.0	2.67	384.5
4	(0.5)(2.0)(6.0)(18.5) = 111.0	3.33	369.6
5	(0.5)(2.0)(0.54)(18.5) = 10.0	3.33	33.3
	P_v = 30.4	4.0	121.6
	ΣV = 475.4		ΣM_R = 1143.0

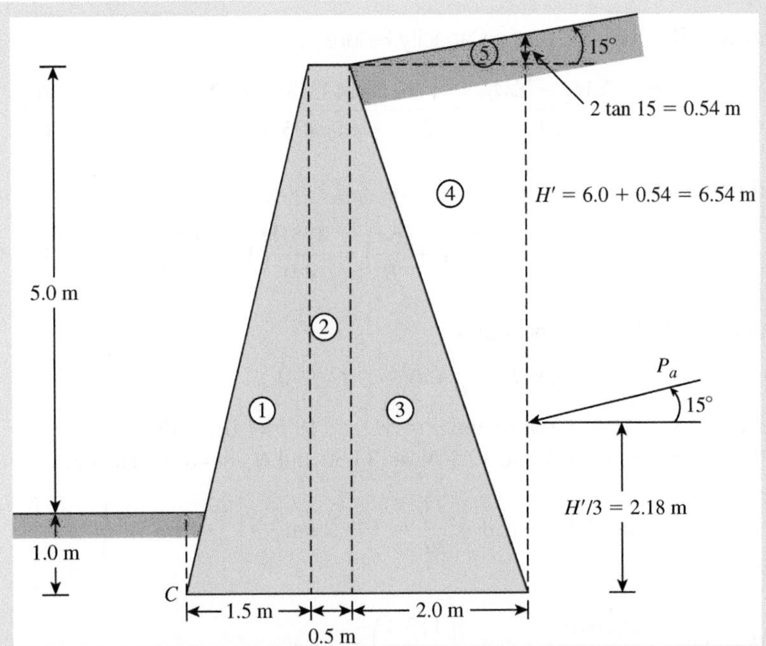

Figure 14.12

Stability with Respect to Overturning

The overturning moment is

$$M_O = P_h \frac{H'}{3} = 113.4 \times 2.18 = 247.2 \text{ kN/m}$$

The resisting moment is

$$\Sigma M_R = 1143.0 \text{ kN/m}$$

Therefore,

$$\text{FS}_{(\text{overturning})} = \frac{\Sigma M_g}{M_O} = \frac{1143.0}{247.2} = \mathbf{4.62}$$

Stability with Respect to Sliding

For $\phi' = 35°$, from Eq. 11.28,

$$K_p = \tan^2\left(45 + \frac{\phi'}{2}\right) = 3.690$$

$$P_p = \frac{1}{2}\gamma D^2 K_p = \frac{1}{2}(18.5)(1.0)^2(3.690) = 34.1 \text{ kN/m}$$

Take δ' as $\frac{2}{3}\phi'$. So $\delta' = 23.3°$, we have

$$\text{FS}_{(\text{sliding})} = \frac{(\Sigma V)\tan\delta' + P_p}{P_h} = \frac{475.4 \times \tan 23.3 + 34.1}{113.4} = \mathbf{2.11}$$

Stability with Respect to Bearing Capacity Failure

$$e = \frac{B}{2} - \frac{\Sigma M_R - \Sigma M_O}{\Sigma V} = \frac{4.0}{2} - \frac{1143.0 - 247.2}{475.4} = 0.116 \text{ m} < \frac{B}{6}\left(= \frac{4}{6} = 0.67 \text{ m}\right)$$

From Eq. (14.11),

$$q_{\max} = q_{\text{toe}} = \frac{\Sigma V}{B}\left(1 + \frac{6e}{B}\right) = \frac{475.4}{4.0}\left(1 + \frac{(6)(0.116)}{4.0}\right) = 139.5 \text{ kN/m}^2$$

In granular soil, Eq. 14.13 becomes

$$q'_u = qN_q F_{qd} F_{qi} + 0.5\gamma B' N_\gamma F_{\gamma d} F_{\gamma i}$$

where, $B' = B - 2e = 3.768$ m and $q = \gamma D = (18.5)(1.0) = 18.5 \text{ kN/m}^2$.

For $\phi' = 35°$, from Table 12.1 $N_q = 33.30$ and $N_\gamma = 48.03$. Hence,

$$F_{qd} = 1 + 2\tan\phi'(1 - \sin\phi')^2 \frac{D_f}{B} = 1 + 2\tan 35(1 - \sin 35)^2 \frac{1}{4} = 1.07$$

$$F_{\gamma d} = 1$$

$$\Psi = \tan^{-1}\left(\frac{P_a \cos\alpha}{\Sigma V}\right) = \tan^{-1}\left(\frac{113.4}{475.4}\right) = 13.4°$$

$$F_{qi} = \left(1 - \frac{\Psi}{90}\right)^2 = \left(1 - \frac{13.4}{90}\right) = 0.72$$

$$F_{\gamma i} = \left(1 - \frac{\Psi}{\phi'}\right)^2 = \left(1 - \frac{15}{35}\right)^2 = 0.33$$

$$q_u = (18.5)(33.30)(1.07)(0.72) + (0.5)(18.5)(3.768)(48.03)(1.0)(0.33) = 1027.0 \text{ kN/m}^2$$

$$\text{FS}_{\text{(bearing capacity)}} = \frac{q'_u}{q_{\max}} = \frac{1027.0}{139.5} = \mathbf{7.4}$$

■

Example 14.3

In Example 14.2, if there is a possibility that the soil in front of the wall will be removed sometime later, neglect the passive resistance and compute the factors of safety.

Solution

The passive resistance P_p was not considered in computing $\text{FS}_{\text{(overturning)}}$, which remains unchanged at **4.62**. However, neglecting P_p will reduce $\text{FS}_{\text{(sliding)}}$, which becomes

$$\text{FS}_{\text{(sliding)}} = \frac{(\Sigma V)\tan\delta' + P_p}{P_h} = \frac{475.4 \times \tan 23.3 + 0.0}{113.4} = \mathbf{1.81}$$

The $\text{FS}_{\text{(bearing capacity)}}$ remains the same at **7.4**.

■

Example 14.4

A concrete gravity retaininvg wall is shown in Figure 14.13. Determine:

a. The factor of safety against overturning
b. The factor of safety against sliding
c. The pressure on the soil at the toe and heel

(*Note*: Unit weight of concrete = γ_c = 150 lb/ft^3.)

Solution

$$H' = 15 + 25 = 17.5 \text{ ft}$$

$$K_a = \tan^2\left(45 - \frac{\phi'_1}{2}\right) = \tan^2\left(45 - \frac{30}{2}\right) = \frac{1}{3}$$

$$P_a = \frac{1}{2}\gamma(H')^2K_a = \frac{1}{2}(121)(17.5)^2\left(\frac{1}{3}\right) = 6176 \text{ lb/ft}$$

$$= 6.176 \text{ kip/ft}$$

Since $\alpha = 0$

$$P_h = P_a = 6.176 \text{ kip/ft}$$

$$P_v = 0$$

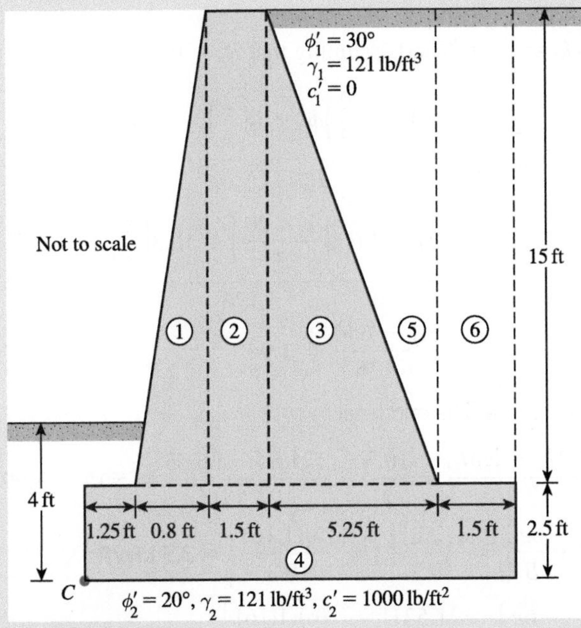

Figure 14.13

Part a: Factor of Safety Against Overturning

The following table can now be prepared to obtain ΣM_R.

Area (from Figure 14.13)	Weight (kip)	Moment arm from *C* (ft)	Moment about *C* (kip.ft/ft)
1	$\frac{1}{2}(0.8)(15)(\gamma_c) = 0.9$	$1.25 + \frac{2}{3}(0.8) = 1.783$	1.605
2	$(1.5)(15)(\gamma_c) = 3.375$	$1.25 + 0.8 + 0.75 = 2.8$	9.45
3	$\frac{1}{2}(5.25)(15)(\gamma_c) = 5.906$	$1.25 + 0.8 + 1.5 + \frac{5.25}{3} = 5.3$	31.30
4	$(10.3)(2.5)(\gamma_c) = 3.863$	$\frac{10.3}{2} = 5.15$	19.89
5	$\frac{1}{2}(5.25)(15)(0.121) = 4.764$	$1.25 + 0.8 + 1.5 + \frac{2}{3}(5.25) = 7.05$	33.59
6	$(1.5)(15)(0.121) = \frac{2.723}{21.531}$	$1.25 + 0.8 + 1.5 + 5.25 + 0.75 = 9.55$	$\frac{26.0}{121.84} = M_R$

The overturning moment

$$M_O = \frac{H'}{3}P_a = \left(\frac{17.5}{3}\right)(6.176) = 36.03 \text{ kip/ft}$$

$$\text{FS}_{(\text{overturning})} = \frac{121.84}{36.03} = \mathbf{3.38}$$

Part b: Factor of Safety Against Sliding

From Eq. (14.10), with $k_1 = k_2 = 2/3$ and assuming that $P_p = 0$,

$$\text{FS}_{(\text{sliding})} = \frac{\Sigma V \tan\left(\frac{2}{3}\right)\phi_2' + B\left(\frac{2}{3}\right)c_2'}{P_a}$$

$$= \frac{21.531 \tan\left(\frac{2 \times 20}{3}\right) + 10.3\left(\frac{2}{3}\right)(1.0)}{6.176}$$

$$= \frac{5.1 + 6.87}{6.176} = \mathbf{1.94}$$

Part c: Pressure on the Soil at the Toe and Heel

$$e = \frac{B}{2} - \frac{\Sigma M_R - \Sigma M_O}{\Sigma V} = \frac{10.3}{2} - \frac{121.84 - 36.03}{21.531} = 5.15 - 3.99 = 1.16 \text{ ft}$$

$$q_{\text{toe}} = \frac{\Sigma V}{B}\left[1 + \frac{6e}{B}\right] = \frac{21.531}{10.3}\left[1 + \frac{(6)(1.16)}{10.3}\right] = \mathbf{3.5 \text{ kip/ft}^2}$$

$$q_{\text{heel}} = \frac{\Sigma V}{B}\left[1 - \frac{6e}{B}\right] = \frac{21.531}{10.3}\left[1 - \frac{(6)(1.16)}{10.3}\right] = \mathbf{0.678 \text{ kip/ft}^2}$$

■

Example 14.5

Repeat Example 14.4 and use Coulomb's active pressure for calculation and $\delta' = 2\phi'/3$.

Solution

Refer to Figure 14.14 for the pressure calculation:

$$\delta' = \frac{2}{3}\phi' = \left(\frac{2}{3}\right)(30) = 20°$$

From Table 11.4, $K_a = 0.4794$ ($\alpha = 0°, \beta = 70°$), so

$$P_a = \frac{1}{2}(0.121)(17.5)^2(0.4794) = 8.882 \text{ kip/ft}$$
$$P_h = P_a \cos 40 = (8.882)(\cos 40) = 6.8 \text{ kip/ft}$$
$$P_v = P_a \sin 40 = 5.71 \text{ kip/ft}$$

Part a: Factor of Safety Against Overturning

Refer to Figures 14.15 and 14.13

Area (from Figures 14.13 and 14.15)	Weight (kip/ft)	Moment arm from *C* (ft)	Moment about *C* (kip.ft/ft)
1	0.9[a]	1.783[a]	1.605
2	3.375[a]	2.8[a]	9.46
3	5.906[a]	5.3[a]	31.30
4	3.863[a]	5.15[a]	19.89
	$P_v = 5.71$	1.25 + 0.8 + 1.5 + 5.25 − 1.21 = 7.59	43.34
	19.754		105.6

[a]Same as in Example 14.4

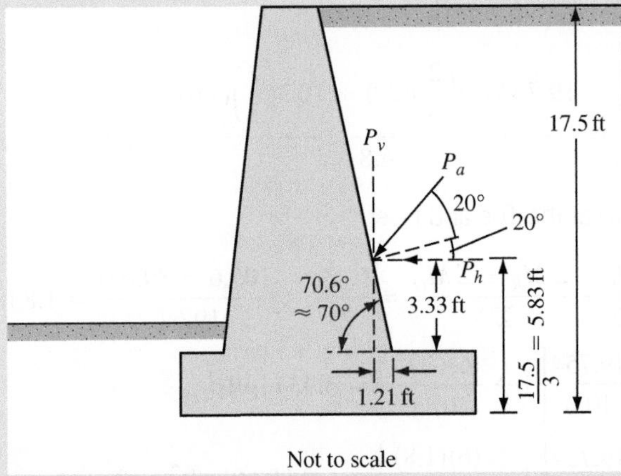

Figure 14.14

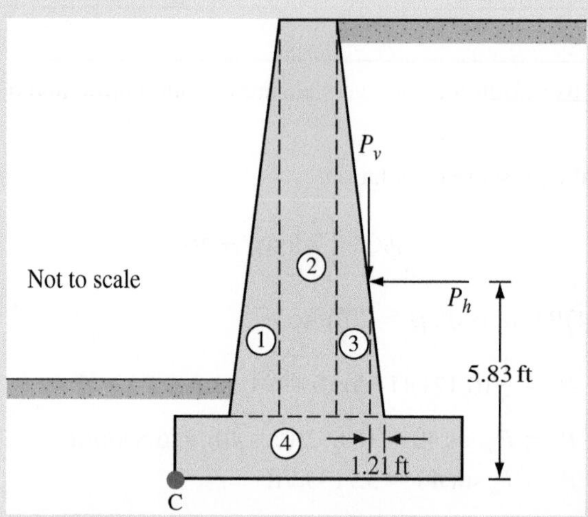

Figure 14.15

The overturning moment is

$$M_O = P_h \frac{H'}{3} = (6.8)\left(\frac{17.5}{3}\right) = 39.67 \text{ kip.ft/ft}$$

Hence,

$$\text{FS}_{(\text{overturning})} = \frac{105.6}{39.67} = \mathbf{2.66}$$

Part b: Factor of Safety Against Sliding

$$\text{FS}_{(\text{sliding})} = \frac{\Sigma V \tan\left(\frac{2}{3}\right)\phi_2' + B\left(\frac{2}{3}\right)c_2'}{P_h}$$

$$= \frac{19.754 \tan\left(\frac{2}{3}\right)(20) + 10.3\left(\frac{2}{3}\right)(1.0)}{6.8} = \mathbf{1.7}$$

Part c: Pressure on the Soil at the Toe and Heel

$$e = \frac{B}{2} - \frac{\Sigma M_R - \Sigma M_O}{\Sigma V} = \frac{10.3}{2} - \frac{(105.6 - 39.67)}{19.67} = 1.8 \text{ ft}$$

$$q_{\text{toe}} = \frac{19.754}{10.3}\left[1 + \frac{(6)(1.8)}{10.3}\right] = \mathbf{3.93\ kip/ft^2}$$

$$q_{\text{heel}} = \frac{19.754}{10.3}\left[1 - \frac{(6)(1.8)}{10.3}\right] = \mathbf{-0.093\ kip/ft^2} \approx \mathbf{0}$$

■

Example 14.6

Determine the safety factors for the cantilever retaining wall shown in figure 14.16 with respect to sliding and overturning, using Coulomb's active earth pressures. Assume δ' is 24° and the unit weight of concrete is 24 kN/m³. How would you improve the stability of this wall?

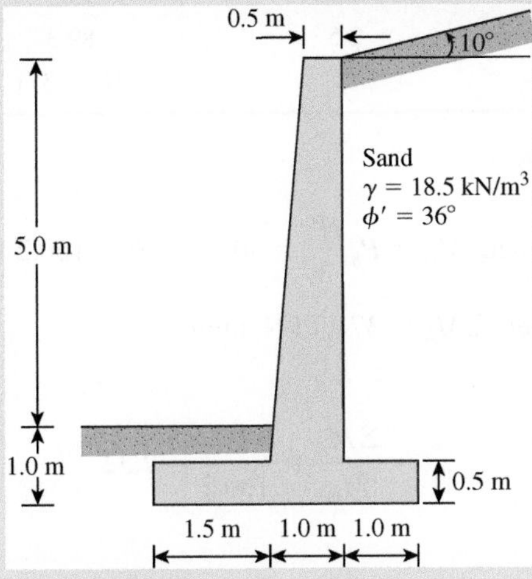

Figure 14.16

Solution

From Table 11.4 for $\alpha = 10°$, $\phi' = 36°$ and $\delta' = 24°$, Coulomb's active earth pressure coefficient is given by $K_a = 0.2633$

$$P_a = \frac{1}{2}\gamma H^2 K_a = \frac{1}{2}(18.5)(6.0)^2(0.2633) = 87.7 \text{ kN/m}$$

$$P_k = P_a \cos 24 = 80.1 \text{ kN/m}$$

$$P_v = P_a \sin 24 = 35.7 \text{ kN/m}$$

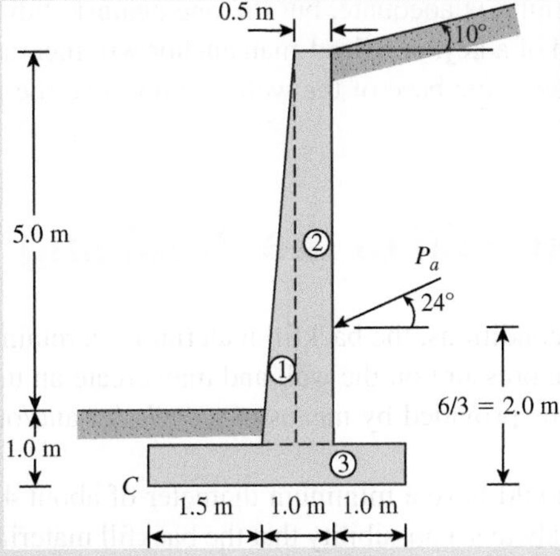

Figure 14.17

Section	Weight (kN/m)	Moment arm from C (m)	Moment about C (kN·m/m)
1	(0.5)(0.5)(5.5)(24.0) = 33.0	1.83	60.4
2	(0.5)(5.5)(24.0) = 66.0	2.25	148.5
3	(0.5)(3.5)(24.0) = 42.0	1.75	73.5
	$P_v = 35.7$	2.5	89.3
	$\Sigma V = 176.7$		$\Sigma M_R = 371.7$

Stability with Respect to Overturning

$$\text{Overturning moment } M_O = P_h \frac{H'}{3} = 80.1 \times 2.0 = 160.2 \text{ kN}\cdot\text{m/m}$$

$$\text{Resisting moment } \Sigma M_R = 371.7 \text{ kN}\cdot\text{m/m}$$

Therefore,

$$FS_{(\text{overturning})} = \frac{\Sigma M_R}{M_O} = \frac{371.7}{160.2} = \mathbf{2.32}$$

Stability with Respect Sliding

For $\phi' = 36°$ using Eq. 11.28, $K_p = \tan^2\left(45 + \frac{\phi'}{2}\right) = 3.852$

$$P_p = \frac{1}{2}\gamma D^2 K_p = \frac{1}{2}(18.5)(1.0)^2(3.852) = 35.6 \text{ kN/m}$$

Taking δ' as $\frac{2}{3}\phi'$, $\delta' = 24°$

$$FS_{(\text{sliding})} = \frac{(\Sigma V)\tan\delta' + P_p}{P_h} = \frac{176.7 \times \tan 24 + 35.6}{80.1} = \mathbf{1.43}$$

The factor of safety against overturning is adequate, but the one against sliding is not. Increasing the width of the base slab, provision of a key, or a dead-man anchor will increase the safety factors. In this case, provision of a small key at the base of the wall will improve the factor of safety with respect to sliding. ■

14.8 Drainage from the Backfill of the Retaining Wall

As the result of rainfall or other wet conditions, the backfill material for a retaining wall may become saturated. Saturation will increase the pressure on the wall and may create an unstable condition. For this reason, adequate drainage must be provided by means of *weepholes* and/or *perforated drainage pipes* (see Figure 14.18).

The *weepholes*, if provided, should have a minimum diameter of about 4 in. (100 mm) and be adequately spaced. Note that there is always a possibility that the backfill material may be washed into weepholes or drainage pipes and ultimately clog them. Thus a filter material needs to be placed behind the weepholes or around the drainage pipes, as the case may be; geotextiles now serve that purpose.

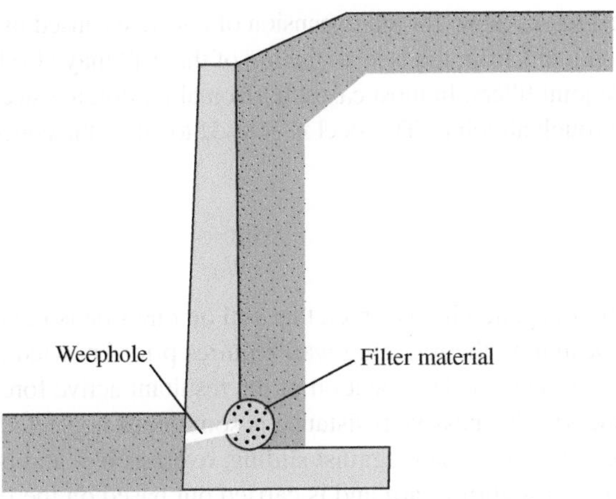

Figure 14.18 Drainage provisions for the backfill of a retaining wall

14.9 Provision of Joints in Retaining-Wall Construction

A retaining wall may be constructed with one or more of the following joints.

1. *Construction joints* (Figure 14.19a) are vertical and horizontal joints that are placed between two successive pours of concrete. To increase the shear at the joints, keys may be used. If keys are not used, the surface of the first pour is cleaned and roughened before the next pour of concrete.
2. *Contraction joints* (Figure 14.19b) are vertical joints (grooves) placed in the face of a wall (from the top of the base slab to the top of the wall) that allow the concrete to shrink without noticeable harm. The grooves may be about 0.25 to 0.3 in. (6 to 8 mm) wide and 0.5 to 0.6 in. (12 to 16 mm) deep.

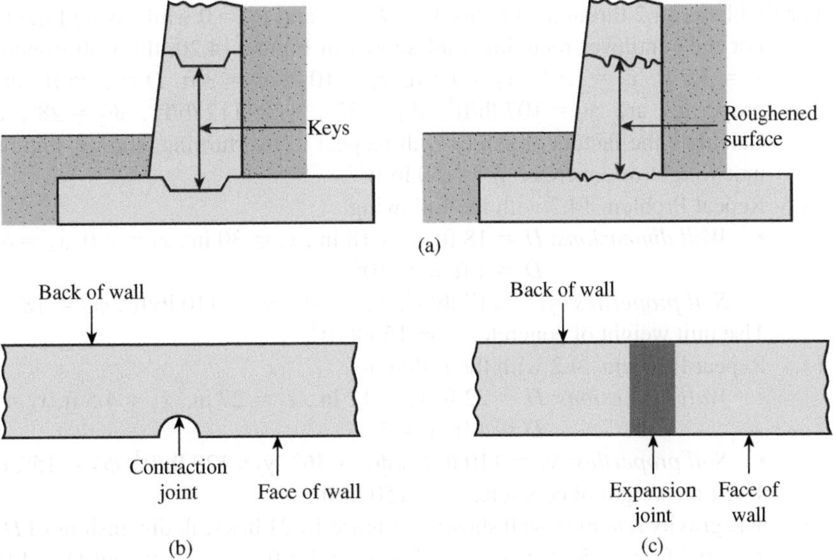

Figure 14.19 (a) Construction joints; (b) contraction joint; (c) expansion joint

3. *Expansion joints* (Figure 14.19c) allow for the expansion of concrete caused by temperature changes; vertical expansion joints from the base to the top of the wall may also be used. These joints may be filled with flexible joint fillers. In most cases, horizontal reinforcing steel bars running across the stem are continuous through all joints. The steel is greased to allow the concrete to expand.

14.10 Summary

In the design of retaining wall, it is generally assumed that soil on one side is in an active state and the other side in a passive state. Designing the retaining wall requires prior knowledge of the lateral earth pressures acting on both sides of the wall. The location of the resultant active force can be determined by Rankine's or Coulomb's theory. The passive resistance is sometimes neglected, conservatively.

The retaining wall is designed to be safe against sliding, overturning, and bearing capacity failure. The check for these is fairly straightforward and is carried out based on the free-body diagram of the retaining wall showing the forces and their locations. The nature of the loading is such that there is always some eccentricity and inclination for the bearing capacity check. It is suggested that the settlements also be checked.

Problems

14.1 State whether the following are true or false.

a. Cantilever retaining walls do not have steel reinforcements.

b. Cantilever retaining walls have smaller cross sections than gravity retaining walls.

c. Adhesion at the base of the retaining wall is generally less than the cohesion of the soil beneath the base.

d. Provision of a key at the base of the wall reduces the passive resistance.

e. The pressure distribution beneath the base of a retaining wall is never uniform.

For Problems 14.2 through 14.6, use $k_1 = k_2 = \frac{2}{3}$ and $P_p = 0$ while using Eq. (14.10).

14.2 For the cantilever retaining wall shown in Figure 14.20, the wall dimensions are $H = 24$ ft, $x_1 = 1.2$ ft, $x_2 = 1.8$ ft, $x_3 = 4.5$ ft, $x_4 = 10$ ft, $x_5 = 3$ ft, $D = 5.25$ ft, and $\alpha = 10°$. The soil properties are $\gamma_1 = 107$ lb/ft^3, $\phi'_1 = 32°$, $\gamma_2 = 112$ lb/ft^3, $\phi'_2 = 28°$, and $c'_2 = 630$ lb/ft^2. Calculate the factors of safety with respect to overturning, sliding, and bearing capacity. Use unit weight of concrete, $\gamma_c = 150$ lb/ft^3.

14.3 Repeat Problem 14.2 with the following:

- *Wall dimensions*: $H = 18$ ft, $x_1 = 18$ in., $x_2 = 30$ in., $x_3 = 4$ ft, $x_4 = 6$ ft, $x_5 = 2.75$ ft, $D = 4$ ft, $\alpha = 10°$
- *Soil properties*: $\gamma_1 = 117$ lb/ft^3, $\phi'_1 = 34°$, $\gamma_2 = 110$ lb/ft^3, $\phi'_2 = 18°$, $c'_2 = 800$ lb/ft^2

Use unit weight of concrete, $\gamma_c = 150$ lb/ft^3.

14.4 Repeat Problem 14.2 with the following:

- *Wall dimensions*: $H = 22$ ft, $x_1 = 12$ in., $x_2 = 27$ in., $x_3 = 4.5$ ft, $x_4 = 8$ ft, $x_5 = 2.75$ ft, $D = 4$ ft, $\alpha = 5°$
- *Soil properties*: $\gamma_1 = 110$ lb/ft^3, $\phi'_1 = 36°$, $\gamma_2 = 120$ lb/ft^3, $\phi'_2 = 15°$, $c'_2 = 1000$ lb/ft^2

Use unit weight of concrete, $\gamma_c = 150$ lb/ft^3.

14.5 The gravity retaining wall shown in Figure 14.21 has wall dimensions of $H = 15$ ft, $x_1 = 1.5$ ft, $x_2 = 0.8$ ft, $x_3 = 5.25$ ft, $x_4 = 1.25$ ft, $x_5 = 1.5$ ft, $x_6 = 2.5$ ft, and $D = 4$ ft, and soil properties of $\gamma_1 = 121$ lb/ft^3, $\phi'_1 = 30°$, $\gamma_2 = 121$ lb/ft^3, $\phi'_2 = 20°$, and $c'_2 = 1000$ lb/ft^2. Calculate the

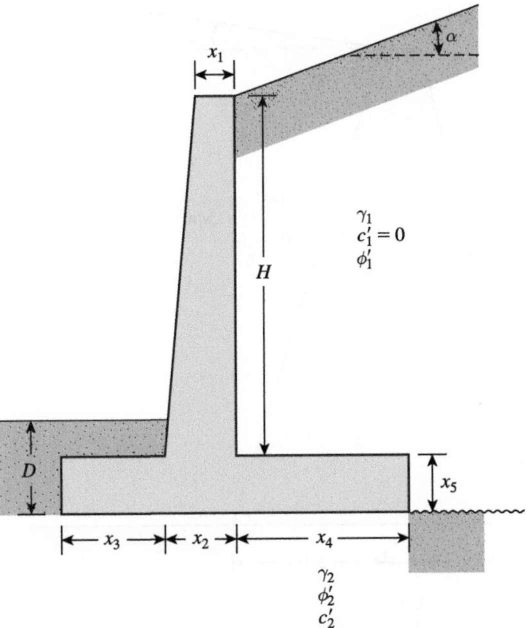

Figure 14.20

factor of safety with respect to overturning and sliding and bearing capacity. Use Rankine active pressure for the calculation.

14.6 Repeat Problem 14.5. Use Coulomb's active pressure calculation and $\delta' = \frac{2}{3}\phi'_1$.

14.7 Determine the factor of safety of the retaining wall shown in Figure 14.22 with respect to overturning sliding and bearing capacity failure. The unit weight of the concrete is 24 kN/m^3.

14.8 Redo Example 14.6 using Rankine's earth pressures.

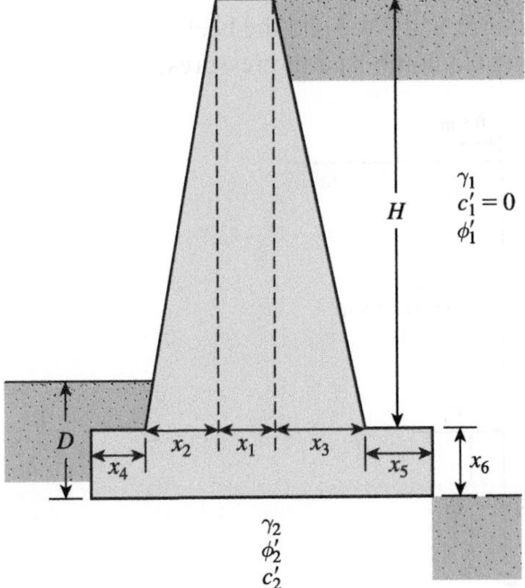

Figure 14.21

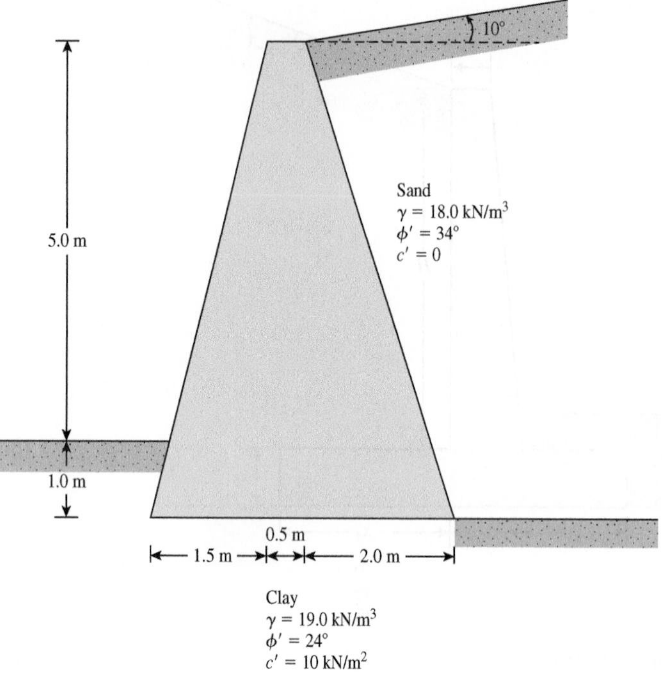

Figure 14.22

14.9 Using Rankine's earth pressures and neglecting the passive resistance, determine the safety factor with respect to sliding and overturning for the gravity retaining wall shown in Figure 14.23. The unit weight of the concrete 24 kN/m^3. Take δ' as $\frac{2}{3}\,\phi'$.

If the design requirement is that the eccentricity should be less than $B/6$, check whether this is satisfied.

Find the soil pressures at the toe and the heel.

14.10 Redo Problem 14.9 using Coulomb's earth pressures.

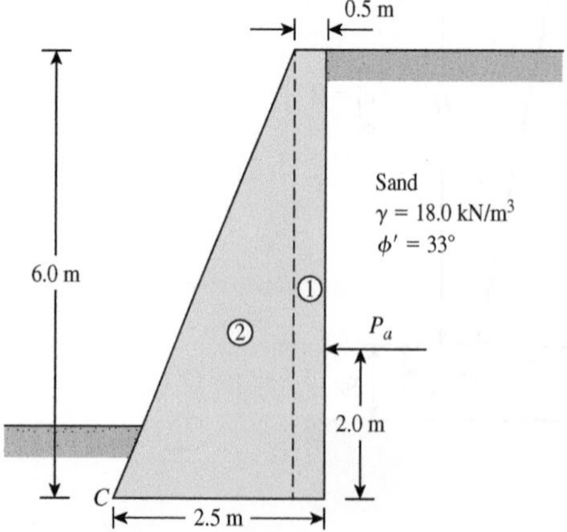

Figure 14.23

15 Slope Stability

15.1 Introduction

An exposed ground surface that stands at an angle with the horizontal is called an *unrestrained slope*. The slope can be natural or man-made. If the ground surface is not horizontal, a component of gravity will tend to move the soil downward as schematically shown in Figure 15.1. If the component of gravity is large enough, slope failure can occur—that is, the soil mass in zone *abcdea* can slide downward. The driving force overcomes the resistance from the shear strength of the soil along the rupture surface. Figure 15.2 shows a slope failure triggered by heavy rainfall, where the failure surface follows a circular arc.

Civil engineers often are expected to make calculations to check the safety of natural slopes, slopes of excavations, and compacted embankments. This check involves determining the shear stress developed along the most likely rupture surface and comparing it with the shear strength of the soil. This process is called *slope stability analysis*. The most likely rupture surface is the critical surface that has the minimum factor of safety.

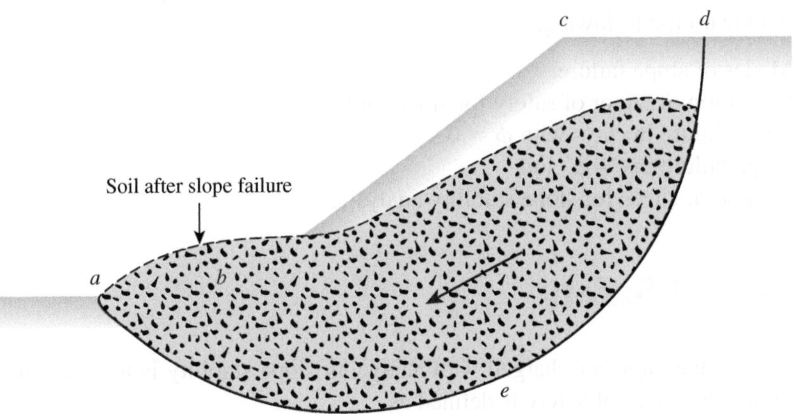

Figure 15.1 Schematic diagram of a slope failure

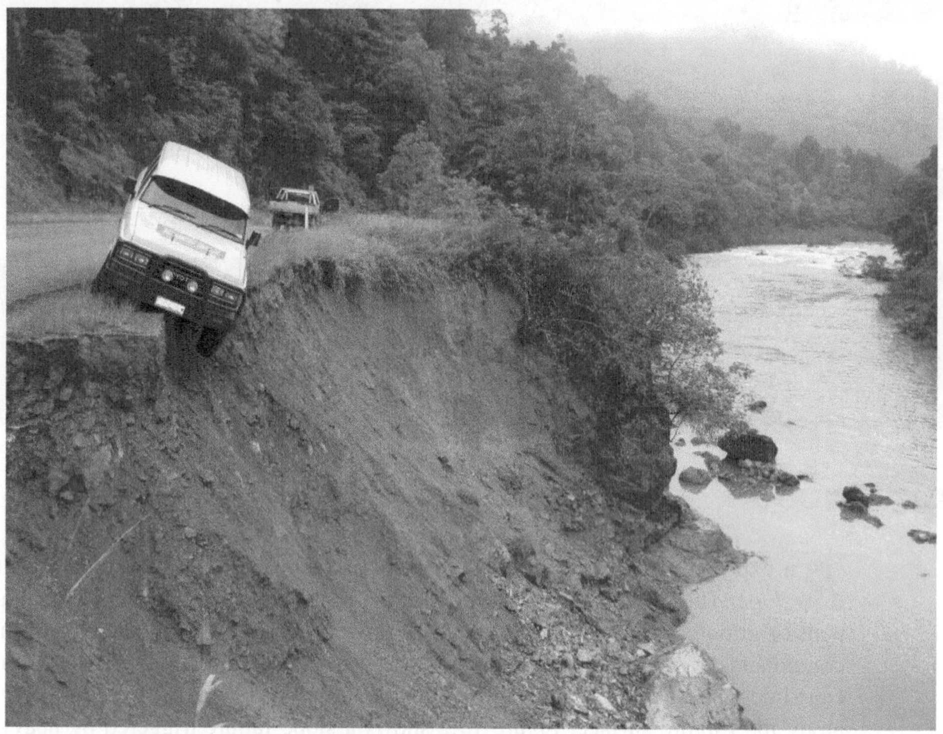

Figure 15.2 Photograph of a slope failure (*Courtesy of Kirralee Rankine, Golder Associates, Australia*)

The stability analysis of a slope is difficult to perform. Evaluation of variables such as the soil stratification and its in-place shear strength parameters may prove to be a formidable task. Seepage through the slope and the choice of a potential slip surface add to the complexity of the problem.

This chapter explains the basic principles involved in slope stability analysis. In this chapter, you will learn the following.

- Modes of slope failure.
- Definition of factor of safety for slope failure.
- Slope failure in clays with $\phi = 0$.
- Slope failure in $c'-\phi'$ soil.
- Method of slices for slope stability analysis.

15.2 Factor of Safety

The task of the engineer charged with analyzing slope stability is to determine the factor of safety. Generally, the factor of safety is defined as

$$F_s = \frac{\tau_f}{\tau_d} \tag{15.1}$$

where

F_s = factor of safety with respect to strength
τ_f = average shear strength of the soil
τ_d = average shear stress developed along the potential failure surface

The shear strength of a soil consists of two components, cohesion and friction, and may be written as

$$\tau_f = c' + \sigma' \tan \phi' \tag{15.2}$$

where

c' = cohesion
ϕ' = angle of friction
σ' = normal stress on the potential failure surface

In a similar manner, we can write

$$\tau_d = c'_d + \sigma' \tan \phi'_d \tag{15.3}$$

where c'_d and ϕ'_d are, respectively, the average cohesion and the angle of friction that develop along the potential failure surface. Substituting Eqs. (15.2) and (15.3) into Eq. (15.1), we get

$$F_s = \frac{c' + \sigma' \tan \phi'}{c'_d + \sigma' \tan \phi'_d} \tag{15.4}$$

Now we can introduce some other aspects of the factor of safety — that is, the factor of safety with respect to cohesion, $F_{c'}$, and the factor of safety with respect to friction, $F_{\phi'}$. They are defined as

$$F_{c'} = \frac{c'}{c'_d} \tag{15.5}$$

and

$$F_{\phi'} = \frac{\tan \phi'}{\tan \phi'_d} \tag{15.6}$$

When we compare Eqs. (15.4) through (15.6), we can see that when $F_{c'}$ becomes equal to $F_{\phi'}$, it gives the factor of safety with respect to strength. Or, if

$$\frac{c'}{c'_d} = \frac{\tan \phi'}{\tan \phi'_d}$$

then we can write

$$F_s = F_{c'} = F_{\phi'} \tag{15.7}$$

When F_s is equal to 1, the slope is in a state of impending failure. Generally, a value of 1.5 for the factor of safety with respect to strength is acceptable for the design of a stable slope.

15.3 Culmann's Method for Stability Analysis

Considerable evidence is available to suggest that slope failures usually occur on curved failure surfaces, as shown in Figure 15.1. Culmann (1875) approximated the surface of potential failure, as a plane as shown in Figure 15.3. In this figure a slope of height H makes an angle β with the horizontal. For the homogeneous soil constituting the slope, the unit weight is γ and the shear strength can be given as

$$\tau_f = c' + \sigma' \tan \phi'$$

Based on Culmann's analysis, the plane which has the *minimum factor of safety*, F_s (or critical plane), can be given by AC which makes an angle α_{cr} with the horizontal. Or

$$\alpha_{cr} = \frac{\beta + \phi'_d}{2} \tag{15.8}$$

The developed cohesion (c'_d) and angle of friction (ϕ'_d) can be given by the relation

$$c'_d = \frac{\gamma H}{4}\left[\frac{1 - \cos(\beta - \phi'_d)}{\sin\beta \cos\phi'_d}\right] \tag{15.9}$$

The maximum height of the slope, $H = H_{cr}$, for which critical equilibrium will exist (that is, $F_s = 1$) is obtained by substituting $c'_d = c'$ and $\phi'_d = \phi'$ in Eq. (15.9). Hence,

$$H_{cr} = \frac{4c'}{\gamma}\left[\frac{\sin\beta \cos\phi'}{1 - \cos(\beta - \phi')}\right] \tag{15.10}$$

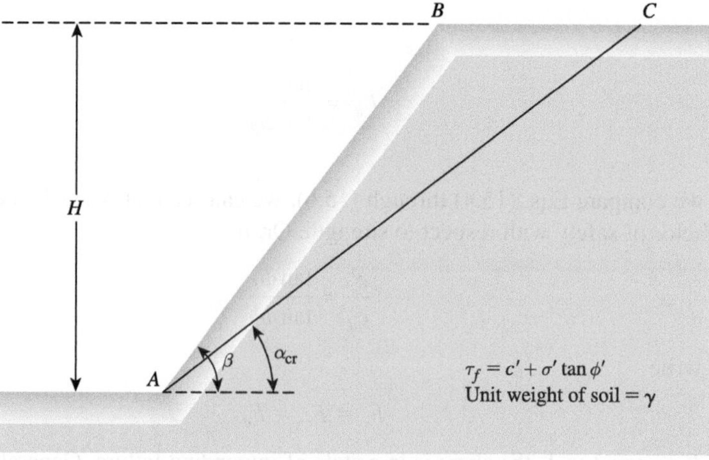

Figure 15.3 Finite slope analysis—Culmann's method

Example 15.1

Consider a slope with $\beta = 60°$. For the soil, given: $\gamma = 110\ \text{lb/ft}^3$, $\phi' = 28°$, and $c' = 400\ \text{lb/ft}^2$. What will be the critical height of the slope (that is, H_{cr})?

Solution

From Eq. (15.10),

$$H_{cr} = \frac{4c'}{\gamma}\left[\frac{\sin\beta\cos\phi'}{1-\cos(\beta-\phi')}\right] = \frac{(4)(400)}{110}\left[\frac{\sin 60 \cos 28}{1-\cos(60-28)}\right] = \mathbf{73.2\ ft}$$

■

Example 15.2

An excavation is made in a soil with a slope of 50°. The soil properties are $\gamma = 18.0\ \text{kN/m}^3$, $\phi' = 27°$, and $c' = 15.0\ \text{kN/m}^2$. How deep can the excavation continue before failure occurs? What is the inclination of the failure plane?

Solution

From Eq. (15.10),

$$H_{cr} = \frac{4c'}{\gamma}\left[\frac{\sin\beta\ \cos\phi'}{1-\cos(\beta-\phi')}\right] = \frac{(4)(15.0)}{(18.0)}\left[\frac{\sin 50 \cos 27}{1-\cos(50-27)}\right] = \mathbf{28.6\ m}$$

Inclination of the failure plane is [Eq. (15.8)]

$$\alpha_{cr} = \frac{\beta+\phi'_d}{2} = \frac{50+27}{2} = 38.5°$$

The critical plane where $F_s = 1$ is inclined at **38.5° to the horizontal**. ■

Example 15.3

A cut is to be made in a soil having $\gamma = 105\ \text{lb/ft}^3$, $c' = 600\ \text{lb/ft}^2$, and $\phi' = 15°$. The side of the cut slope will make an angle of 45° with the horizontal. What should be the depth of the cut slope that will have a factor of safety (F_s) of 3?

Solution

Given: $\phi' = 15°$; $c' = 600\ \text{lb/ft}^2$. If $F_s = 3$, then $F_{c'}$ and $F_{\phi'}$ should both be equal to 3.

$$F_{c'} = \frac{c'}{c'_d}$$

or

$$c'_d = \frac{c'}{F_{c'}} = \frac{c'}{F_s} = \frac{600}{3} = 200\ \text{lb/ft}^2$$

Similarly

$$F_{\phi'} = \frac{\tan \phi'}{\tan \phi'_d}$$

$$\tan \phi'_d = \frac{\tan \phi'}{F_{\phi'}} = \frac{\tan \phi'}{F_s} = \frac{\tan 15}{3}$$

or

$$\phi'_d = \tan^{-1}\left[\frac{\tan 15}{3}\right] = 5.1°$$

Substituting the preceding values of c'_d and ϕ'_d in Eq. (15.9)

$$H = \frac{4c'_d}{\gamma}\left[\frac{\sin\beta \cdot \cos\phi'_d}{1 - \cos(\beta - \phi'_d)}\right]$$

$$= \frac{4 \times 200}{105}\left[\frac{\sin 45 \cdot \cos 5.1}{1 - \cos(45 - 5.1\,)}\right]$$

$$= \mathbf{23.05\ ft}$$

■

Example 15.4

If it is required to ensure a factor of safety of 1.5 in the excavation made in Example 15.2, how deep can the excavation continue?

Solution

$$F_s = F_{c'} = F_{c'} = 1.5$$

$$c'_d = \frac{15.0}{1.5} = 10.0\ \text{kN/m}^2$$

$$\tan \phi'_d = \frac{\tan 27}{1.5}$$

Therefore, $\phi'_d = 18.8°$. Substituting the values of c'_d and ϕ'_d in Eq. (15.9),

$$c'_d = \frac{\gamma H}{4}\left[\frac{1 - \cos(\beta - \phi'_d)}{\sin\beta\cos\phi'_d}\right]$$

$$10 = \frac{(18.0)H}{4}\left[\frac{1 - \cos(50 - 18.8)}{\sin 50 \cos 18.8}\right]$$

$$H = \mathbf{11.14\ m}$$

■

15.4 Factor of Safety along a Plane

In some instances, it may be required to obtain the factor of safety against sliding (F_s) along a randomly selected plane AC as shown in Figure 15.4. Note that AC makes an angle α with the horizontal. Referring to the figure, the following procedure may be adopted.

1. Obtain the weight (W) of the wedge ABC per unit length of the slope as

$$W = \tfrac{1}{2}\gamma H^2(\cot\alpha - \cot\beta) \tag{15.11}$$

2. Calculate the normal and tangential components of W, or

$$N_a = \text{normal component} = W\cos\alpha$$
$$T_a = \text{tangential component} = W\sin\alpha$$

The force T_a is the force whose effect will be to destabilize the slope.

3. Calculate the resisting tangential force (T_r), or

$$T_r = \overline{AC}(c'_d + \sigma'\tan\phi'_d) = \overline{AC}\left(\frac{c'}{F_s} + \frac{N_a\tan\phi'}{\overline{AC}\ F_s}\right) = \frac{H}{\sin\alpha}\left[\frac{c'}{F_s} + \left(\frac{N_a\sin\alpha}{H}\right)\left(\frac{\tan\phi'}{F_s}\right)\right]$$

4. For equilibrium,

$$T_a = T_r$$

or

$$W\sin\alpha = \frac{H}{F_s\sin\alpha}\left[c' + \left(\frac{W\cos\alpha\,\sin\alpha}{H}\right)(\tan\phi')\right]$$

$$F_s = \frac{H}{W\sin^2\alpha}\left[c' + \frac{W\cos\alpha\,\sin\alpha}{H}\tan\phi'\right] \tag{15.12}$$

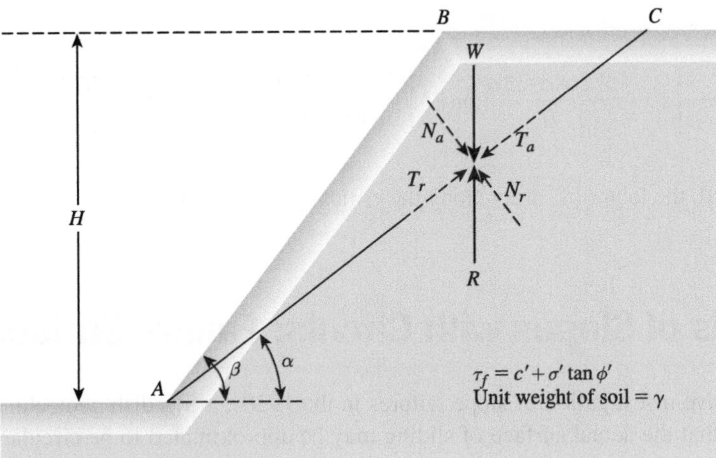

Figure 15.4 Factor of safety along a plane—Culmann's method

Example 15.5

Refer to Figure 15.4. Given: $H = 10$ ft, $\beta = 50°$, $\alpha = 30°$, $\gamma = 100$ lb/ft^3, $\phi' = 10°$, and $c' = 600$ lb/ft^2. Determine the factor of safety, F_s, against sliding along plane AC.

Solution

From Eq. (15.11)

$$W = \tfrac{1}{2}\gamma H^2(\cot\alpha - \cot\beta) = \left(\tfrac{1}{2}\right)(100)(10)^2(\cot 30° - \cot 50°) = 4465 \text{ lb}$$

Eq. (15.12)

$$F_s = \frac{H}{W\sin^2\alpha}\left[c' + \frac{W\cos\alpha\,\sin\alpha}{H}\tan\phi'\right]$$

$$= \frac{10}{(4465)(\sin^2 30)}\left[600 + \frac{(4465)(\cos 30)(\sin 30)}{10}\tan 10\right] = \mathbf{5.68}$$

■

Example 15.6

In Example 15.2, it was shown that the failure occurs along a plane inclined at 38.5° to horizontal when the soil is excavated to 28.6 m depth. Determine the safety factor along this plane using Eq. (15.12).

Solution

$$\gamma = 18.0 \text{ kN/m}^3,\ \phi' = 27°, \text{ and } c' = 15.0 \text{ kN/m}^2$$

Referring to Figure 15.4, $H = 28.6$ m, $\alpha = 38.5°$ and $\beta = 50°$. From Eq. (15.11),

$$W = \frac{1}{2}\gamma H^2(\cot\alpha - \cot\beta) = \frac{1}{2}(18)(28.6)^2(\cot 38.5 - \cot 50) = 3077.7 \text{ kN}$$

From Eq. (15.12),

$$F_s = \frac{H}{W\sin^2\alpha}\left[c' + \frac{W\cos\alpha\sin\alpha}{H}\tan\phi'\right] = \frac{28.6}{3077.7\sin^2 38.5}\left[15.0 + \frac{3077.7\cos 38.5\sin 38.5}{28.6}\tan 27\right]$$

$$= \mathbf{1.00}$$

As expected, the factor of safety along the critical plane is 1.00.

■

15.5 Analysis of Slopes with Circular Failure Surfaces—General

After extensive investigation of slope failures in the 1920s, a Swedish geotechnical commission recommended that the actual surface of sliding may be approximated to be circularly cylindrical. Since that time, most conventional stability analyses of slopes have been made by assuming that the curve of potential sliding is an arc of a circle. However, in many circumstances (for example, zoned dams and foundations on weak strata), stability analysis using plane failure of sliding is more appropriate and yields excellent results.

Modes of Failure

In general, slope failure occurs in one of the following modes (Figure 15.5).

1. When the failure occurs in such a way that the surface of sliding intersects the slope at or above its toe, it is called a *slope failure* (Figure 15.5a). The failure circle is referred to as a *toe circle* if it passes through the toe of the slope and as a *slope circle* if it passes above the toe of the slope. Under certain circumstances, a *shallow slope failure* can occur, as shown in Figure 15.5b.
2. When the failure occurs in such a way that the surface of sliding passes at some distance below the toe of the slope, it is called a *base failure* (Figure 15.5c). The failure circle in the case of base failure is called a *midpoint circle*. Here the center of the circle lies above the midpoint of the slope (see Figure 15.5c).

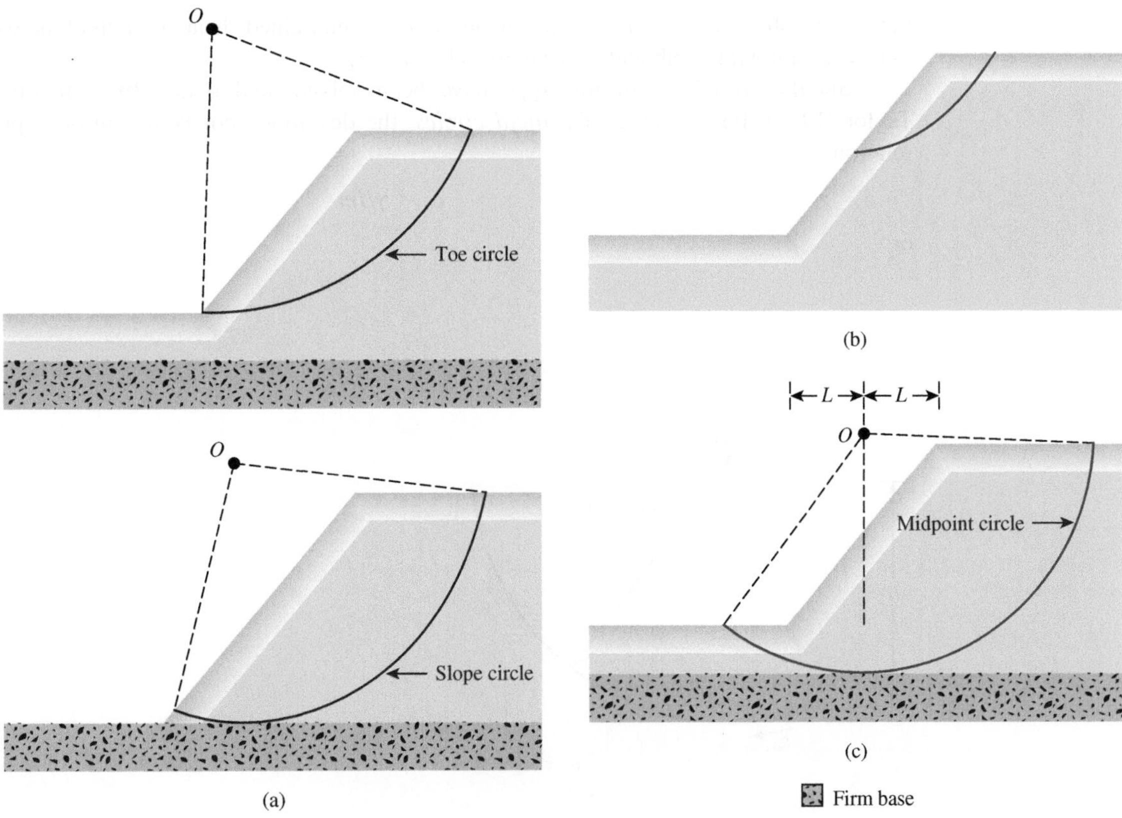

Figure 15.5 Modes of failure of finite slope: (a) slope failure; (b) shallow slope failure; (c) base failure

Types of Stability Analysis Procedures

Various procedures of stability analysis may, in general, be divided into two major classes:

1. *Mass procedure*: In this case, the mass of the soil above the surface of sliding is taken as a unit. This procedure is useful when the soil that forms the slope is assumed to be homogeneous, although this is not the case in most natural slopes.
2. *Method of slices*: In this procedure, the soil above the surface of sliding is divided into a number of vertical parallel slices. The stability of each slice is calculated separately. This is a versatile technique in which the nonhomogeneity of the soils and pore water pressure can be taken into consideration. It also accounts for the variation of the normal stress along the potential failure surface.

The fundamentals of the analysis of slope stability by mass procedure and method of slices are given in the following sections.

15.6 Mass Procedures—Slopes in Homogeneous Clay Soil with $\phi = 0$

Figure 15.6 shows a slope in a homogeneous soil. The undrained shear strength of the soil is assumed to be constant with depth and may be given by $\tau_f = c_u$.

Stability problems of this type have been solved analytically by Fellenius (1927) and Taylor (1937). For the case of *critical circles*, the developed cohesion can be expressed by the relationship

$$c_d = \gamma H m$$

or

$$\frac{c_d}{\gamma H} = m \qquad (15.13)$$

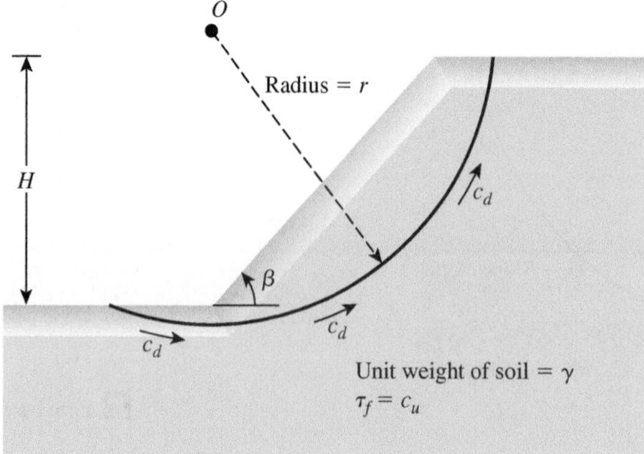

Figure 15.6 Stability analysis of slope in homogeneous saturated clay soil ($\phi = 0$)

Note that the term m on the right-hand side of the preceding equation is nondimensional and is referred to as the *stability number*. The critical height (i.e., $F_s = 1$) of the slope can be evaluated by substituting $H = H_{cr}$ and $c_d = c_u$ (full mobilization of the undrained shear strength) into the preceding equation. Thus,

$$H_{cr} = \frac{c_u}{\gamma m} \tag{15.14}$$

Values of the stability number, m, for various slope angles, β, are given in Figure 15.7 (Taylor, 1937).

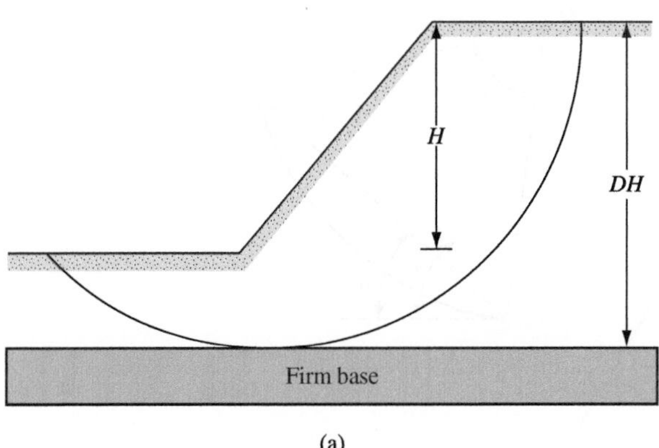

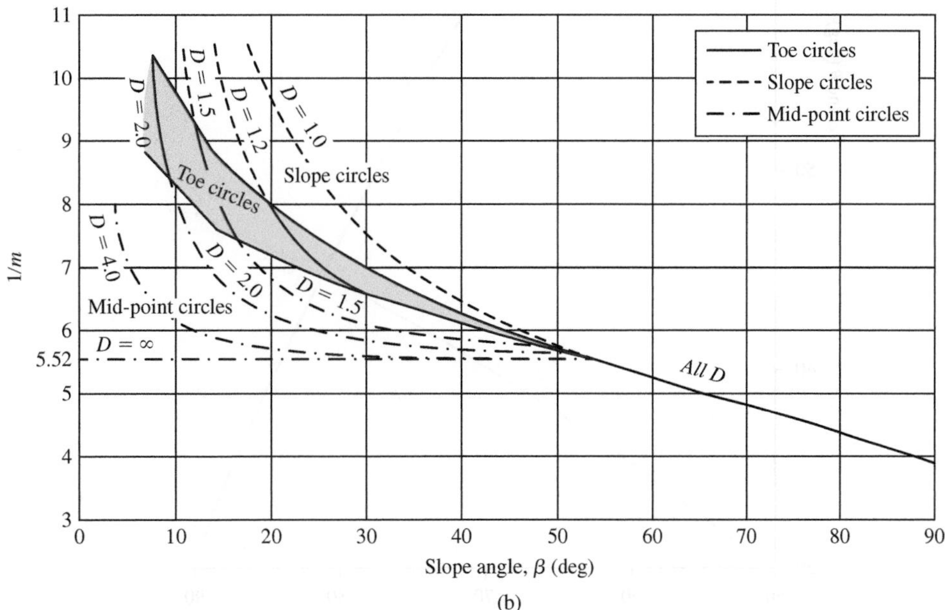

Figure 15.7 (a) Definition of parameters for midpoint circle-type failure, (b) chart for determining stability numbers. (*Based on Terzaghi and Peck, 1967*).

In reference to Figure 15.7, the following must be pointed out:

1. For a slope angle β greater than 53°, the critical circle is always a toe circle. The location of the center of the critical toe circle may be found with the aid of Figure 15.8.
2. For $\beta < 53°$, the critical circle may be a toe, slope, or midpoint circle, depending on the depth of the firm base under the slope. This is called the *depth function*, which is defined as

$$D = \frac{\text{vertical distance from top of slope to firm base}}{\text{height of slope}} \tag{15.15}$$

3. The maximum possible value of the stability number for failure as a midpoint circle is 0.181.

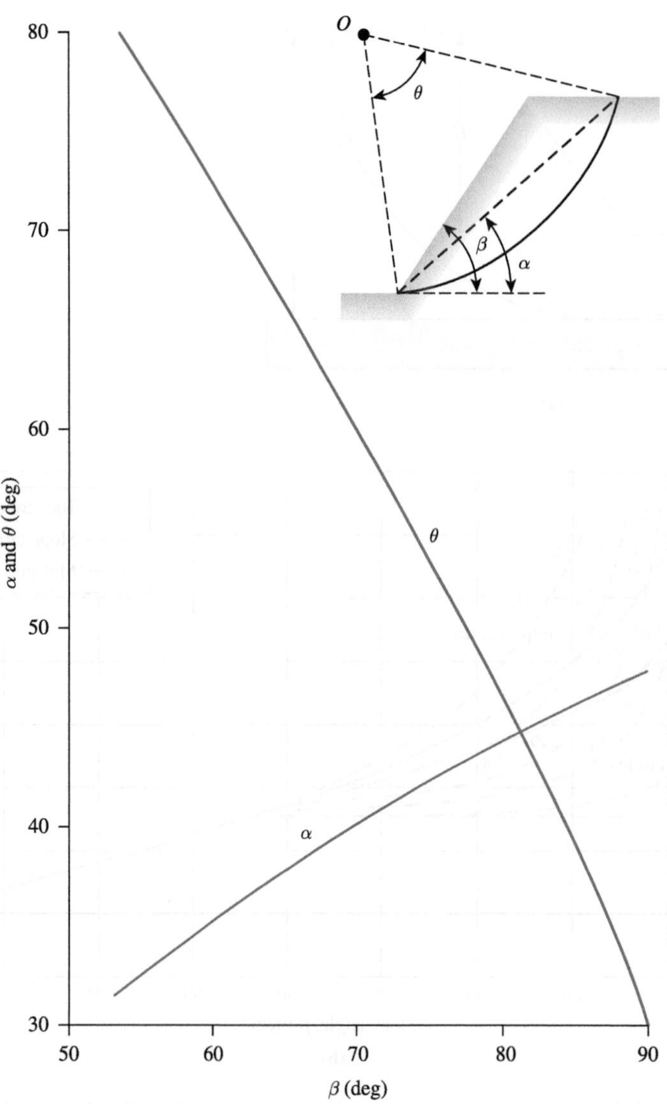

Figure 15.8 Location of the center of critical circles for $\beta > 53°$

Example 15.7

A 25-ft high cut slope has to be made in a saturated soft clay. Given: slope angle $\beta = 60°$, $c_u = 800\ \text{lb/ft}^2$, and $\gamma = 122\ \text{lb/ft}^3$. What will be the factor of safety (F_s) of the slope against sliding?

Solution

From Eq. (15.13)

$$m = \frac{c_d}{\gamma H}$$

From Figure 15.7, for $\beta = 60°$, the value of $\frac{1}{m}$ is 5.13. So $m = 0.195$. Thus,

$$0.195 = \frac{c_d}{(122)(25)};\ c_d = 594.75\ \text{lb/ft}^2$$

$$F_s = \frac{c_u}{c_d} = \frac{800}{594.75} \approx \mathbf{1.35}$$

■

Example 15.8

A cut slope was excavated in a saturated clay. The slope made an angle of 40° with the horizontal. Slope failure occurred when the cut reached a depth of 20 ft. Previous soil explorations showed that a rock layer was located at a depth of 30 ft below the ground surface.

Assuming an undrained condition and that $\gamma_{sat} = 110\ \text{lb/ft}^3$, Determine the undrained cohesion of the clay (use Figure 15.7).

Solution

We know that

$$D = \frac{30}{20} = 1.5$$

$$\gamma_{sat} = 110\ \text{lb/ft}^3$$

$$H_{cr} = \frac{c_u}{\gamma m}$$

From Figure 15.7, for $\beta = 40°$ and $D = 1.5$, $\frac{1}{m} \approx 5.71$. Hence, $m = 0.175$. Thus,

$$c_u = (H_{cr})(\gamma)(m) = (20)(110)(0.175) = \mathbf{385\ lb/ft^2}$$

■

Example 15.9

A site consists of a 12-m-thick clay deposit underlain by bedrock. The clay has $\gamma = 20.0\ \text{kN/m}^3$ and $c_u = 40.0\ \text{kN/m}^2$. If an excavation is to be made at a slope angle β of 25° to a depth of 10.0 m, what would be the factor of safety F_s? What type of failure circle would you expect?

Solution

$$D = 12/10 = 1.2$$

For $D = 1.2$ and $\beta = 25°$, from Figure 15.7, $1/m = 7.10$ and $m = 0.141$. From Eq. (15.13),

$$\frac{1}{m} = \frac{\gamma H}{c_d}$$

$$c_d = (20.0)(10.0)(0.141) = 28.2 \text{ kN/m}^2$$

$$F_s = \frac{c_u}{c_d} = \frac{40.0}{28.2} = \mathbf{1.42}$$

From Figure 15.7, it can be seen that the failure mode is a **toc circle.** ■

15.7 Mass Procedures—Slopes in Homogeneous $c'-\phi'$ Soil

For soils with ϕ' greater than about 3°, the critical circles of failure are toe circles. Figure 15.9a shows a slope of height H in homogeneous soil having a unit weight of γ. The shear strength of the soil is given by

$$\tau_f = c' + \sigma' \tan\phi'$$

The trial failure surface is $\widehat{AC}$ which is the arc of a circle having a radius r with the center at O. Considering a unit length of the slope perpendicular to the cross section shown, the forces to be considered for the stability of the soil wedge ABC are

- Weight of the wedge, W = (area of the soil wedge ABC)(γ)
- The resultant cohesive force developed along $\widehat{AC}$ (see Figure 15.9b) = $(c_d')(\widehat{AC}) = (c_d')(\overline{AC}) = C_d$
- The resultant of the normal and frictional forces along $AC = F$

It is important to point out that C_d acts in a direction parallel to the cord $\overline{AC}$ (see Figure 15.9b) and at a distance a from the center of the circle O such that

$$C_d(a) = c_d'(\widehat{AC})r$$

or

$$a = \frac{c_d'(\widehat{AC})r}{C_d} = \frac{\widehat{AC}}{\overline{AC}}r \tag{15.16}$$

Now, if we assume that full friction is mobilized ($\phi_d' = \phi'$ or $F_{\phi'} = 1$), the line of action of F will make an angle of ϕ' with a normal to the arc and thus will be a tangent to a circle with its center at O and having a radius of $r \sin\phi'$. This circle is called the *friction circle*. Actually, the radius of the friction circle is a little larger than $r \sin\phi'$.

Because the directions of W, C_d, and F are known and the magnitude of W is known, a force triangle, as shown in Figure 15.9c, can be drawn. The magnitude of C_d can be determined from the force polygon. So the cohesion per unit area developed can be found.

$$c_d' = \frac{C_d}{\overline{AC}}$$

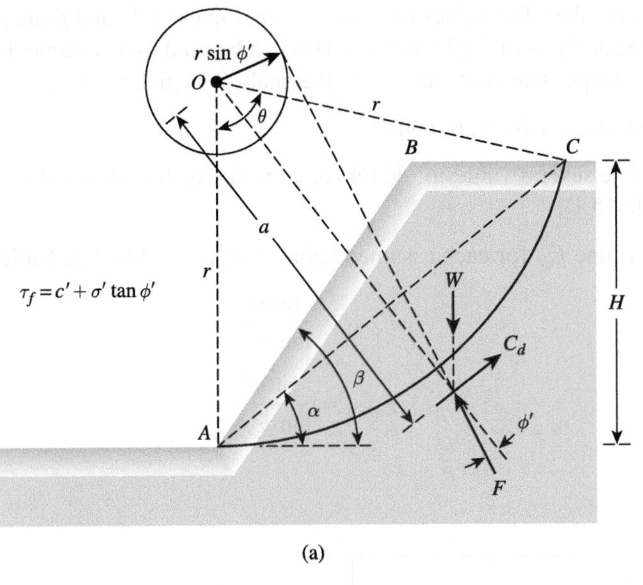

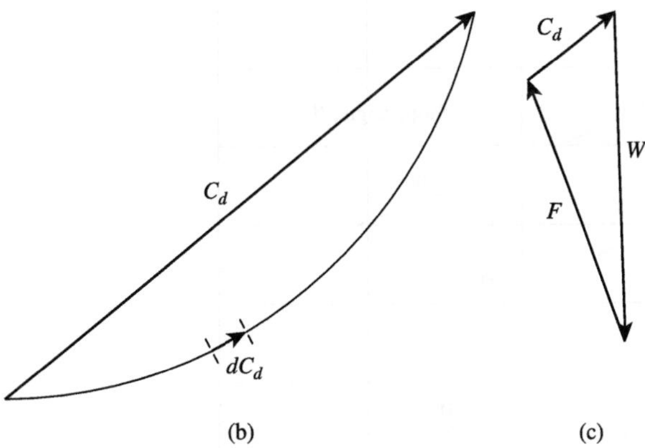

Figure 15.9 Stability analysis of slope in homogeneous c'–ϕ' soil

Determination of the magnitude of c'_d described previously is based on a trial surface of sliding. Several trials must be made to obtain the most critical sliding surface, along which the developed cohesion is a maximum. Thus, we can express the maximum cohesion developed along the critical surface as

$$c'_d = \gamma H[f(\alpha, \beta, \theta, \phi')] \tag{15.17}$$

For critical equilibrium—that is, $F_{c'} = F_{\phi'} = F_s = 1$—we can substitute $H = H_{cr}$ and $c'_d = c'$ into Eq. (15.17) and write

$$c' = \gamma H_{cr}[f(\alpha, \beta, \theta, \phi')]$$

or

$$\frac{c'}{\gamma H_{cr}} = f(\alpha, \beta, \theta, \phi') = m \tag{15.18}$$

where m = stability number. The values of m for various values of ϕ' and β are given in Figure 15.10 which is based on the analysis of Taylor (1937). This can be used to determine the factor of safety, F_s, of the homogeneous slope. The procedure to do the analysis is given below:

Step 1. Determine c', ϕ', γ, β, and H.

Step 2. Assume several values of ϕ'_d (Note: $\phi'_d \leq \phi'$, such as $\phi'_d(1)$, $\phi'_d(2)$ . . . . (Column 1 of Table 15.1).

Step 3. Determine $F_{\phi'}$ for each assumed value of ϕ'_d as (Column 2, Table 15.1)

$$F_{\phi'(1)} = \frac{\tan \phi'}{\tan \phi'_{d(1)}}$$

$$F_{\phi'(2)} = \frac{\tan \phi'}{\tan \phi'_{d(2)}}$$

$$\vdots$$

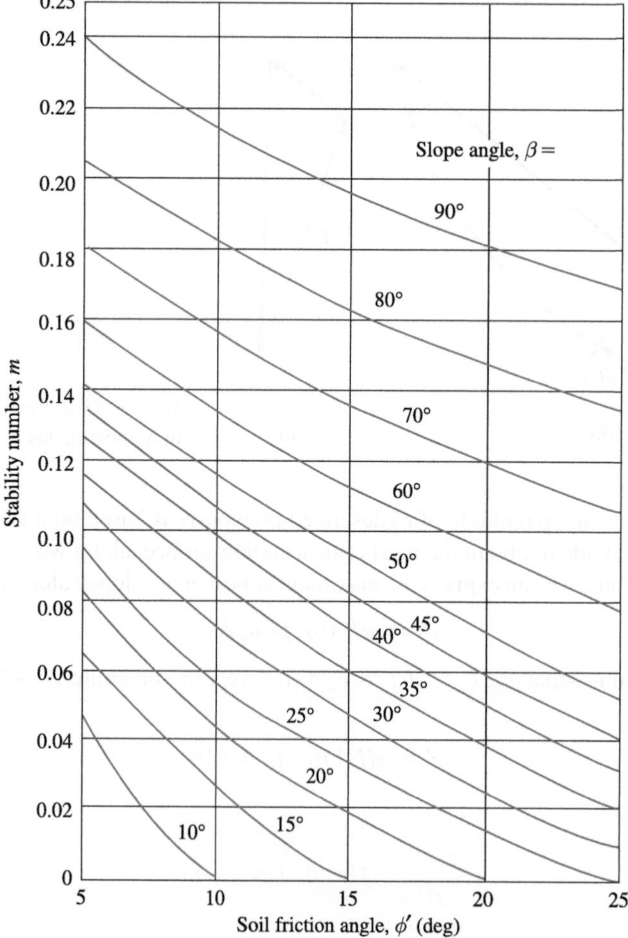

Figure 15.10 Taylor's stability number

Table 15.1 Determination of F_s by Friction Circle Method

ϕ'_d (1)	$F_{\phi'} = \dfrac{\tan\phi'}{\tan\phi'_d}$ (2)	m (3)	c'_d (4)	F'_c (5)
$\phi'_{d(1)}$	$\dfrac{\tan\phi'}{\tan\phi'_{d(1)}}$	m_1	$m_1\gamma H = c'_{d(1)}$	$\dfrac{c'}{c'_{d(1)}} = F_{c'(1)}$
$\phi'_{d(2)}$	$\dfrac{\tan\phi'}{\tan\phi'_{d(2)}}$	m_2	$m_2\gamma H = c'_{d(2)}$	$\dfrac{c'}{c'_{d(2)}} = F_{c'(2)}$
$\vdots$	$\vdots$	$\vdots$	$\vdots$	$\vdots$

Step 4. For each assumed value of ϕ'_d and β, determine m (that is, m_1, m_2, m_3, . . .) from Figure 15.10 (Column 3, Table 15.1).

Step 5. Determine the developed cohesion for each value of m as (Column 4, Table 15.1)

$$c'_{d(1)} = m_1\gamma H$$
$$c'_{d(2)} = m_2\gamma H$$
$$\vdots$$

Step 6. Calculate $F_{c'}$ for each value of c'_d (Column 5, Table 15.1), or

$$F_{c'(1)} = \frac{c'}{c'_{d(1)}}$$
$$F_{c(2)} = \frac{c'}{c'_{d(2)}}$$
$$\vdots$$

Step 7. Plot a graph of $F_{\phi'}$ versus the corresponding $F_{c'}$ (Figure 15.11) and determine $F_s = F_{\phi'} = F_{c'}$.

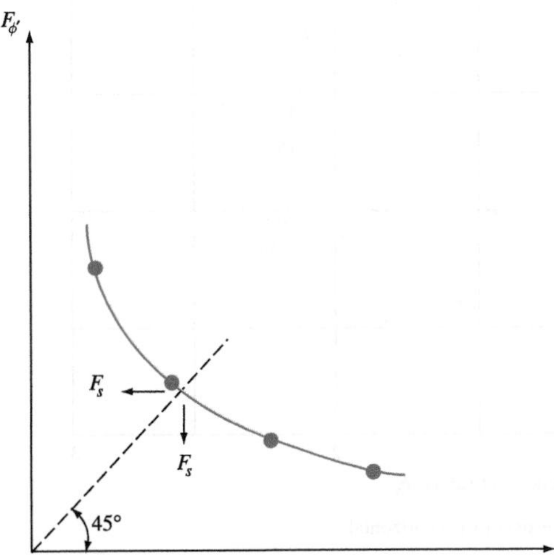

Figure 15.11 Plot of $F_{\phi'}$ versus $F_{c'}$ to determine F_s

An example of determining F_s using the procedure just described is given in Example 15.10.

Using Taylor's friction circle method of slope stability (as shown in Example 15.10), Singh (1970) provided graphs of equal factors of safety, F_s, for various slopes. Using the results of Singh (1970), the variations of $c'/\gamma H$ with factor of safety (F_s) for various friction angles (ϕ') are plotted in Figure 15.12.

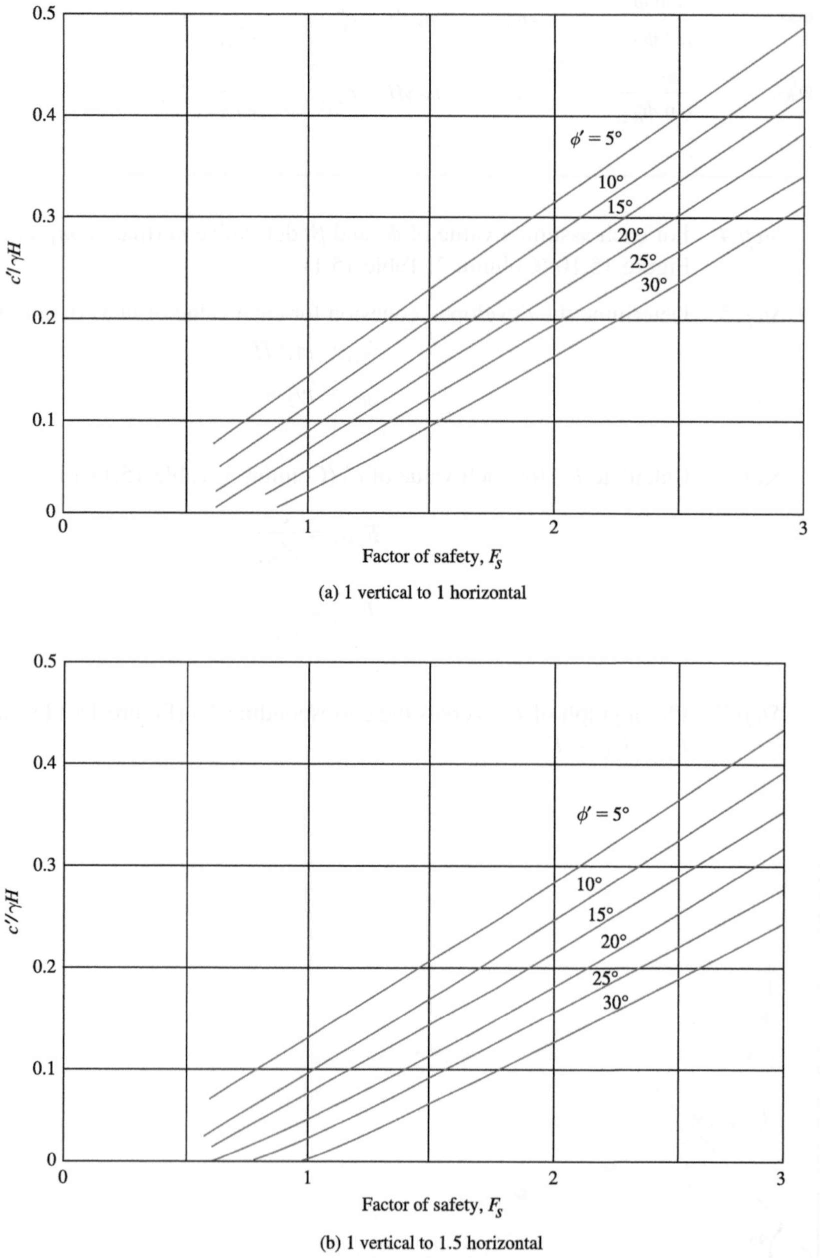

Figure 15.12 Plot of $c'/\gamma H$ against F_s for various slopes and ϕ' (Based on Singh, 1970)

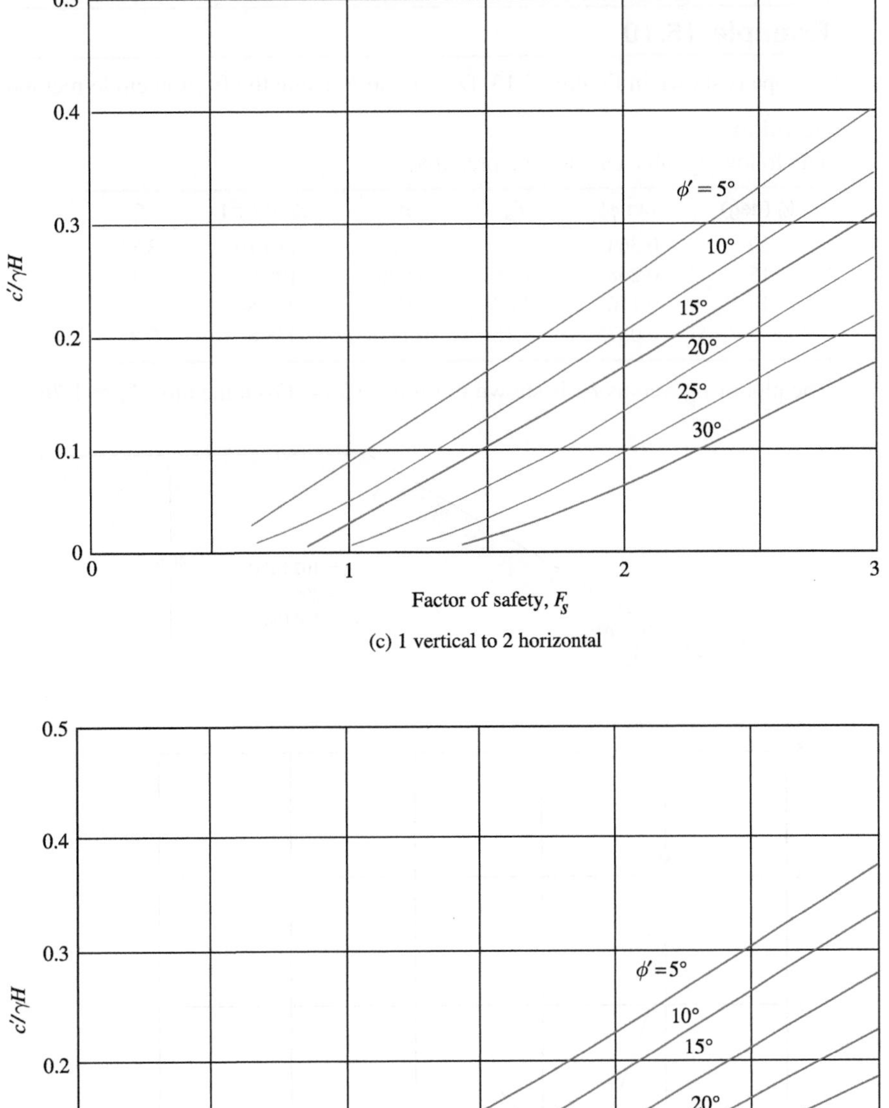

(c) 1 vertical to 2 horizontal

(d) 1 vertical to 2.5 horizontal

Figure 15.12 (*Continued*)

Example 15.10

A slope is shown in Figure 15.13. Determine F_s using the friction circle method.

Solution

The following table can now be prepared.

ϕ'_d (deg)	$\tan \phi'_d$	$F_{\phi'}$	m	c'_d (lb/ft²)	F'_c
20	0.364	1	0.027	110.16	3.81
15	0.268	1.36	0.048	195.84	2.14
10	0.176	2.07	0.073	297.84	1.41
5	0.0875	4.16	0.11	448.8	0.94

The plot of $F_{\phi'}$ versus $F_{c'}$ is shown in Figure 15.14. From the plot, $F_s \approx$ **1.76**.

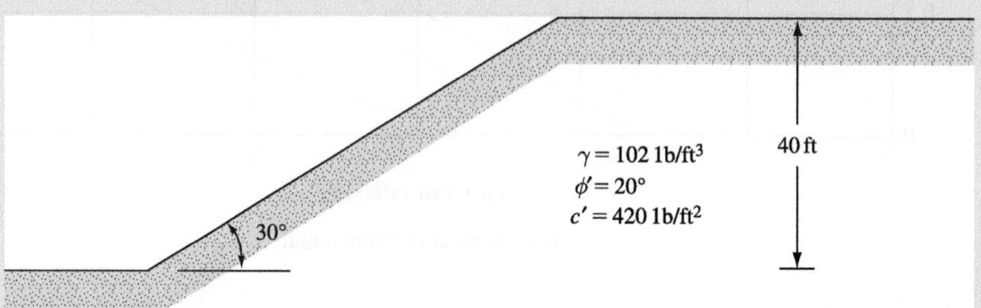

Figure 15.13

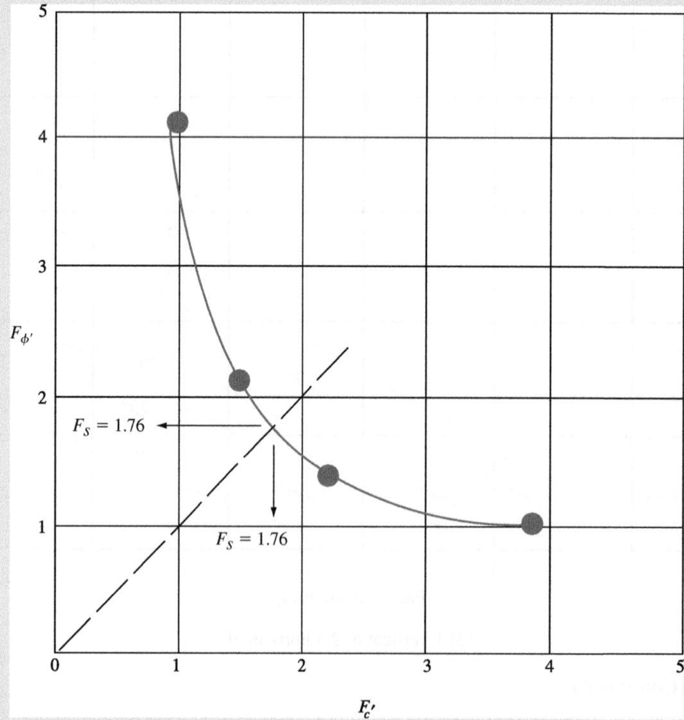

Figure 15.14 ■

Example 15.11

Find the factor of safety of a slope in a homogeneous soil where $c' = 15$ kN/m^2, $\phi' = 24°$, $\gamma =$ 20.0 kN/m^3, the depth of the cut is 10.0 m, and the slope is 35° to the horizontal.

Solution

Let's select some values for ϕ'_d and prepare the table shown.

ϕ'_d (deg)	tan ϕ'_d	$F_{\phi'}$	m	$c'_d = m\gamma H$ (kN/m^2)	$F_{c'} = \frac{c'}{c'_d}$
22	0.404	1.10	0.031	6.2	2.42
18	0.325	1.37	0.047	9.4	1.60
16	0.287	1.55	0.056	11.2	1.34
14	0.249	1.79	0.065	13.0	1.15

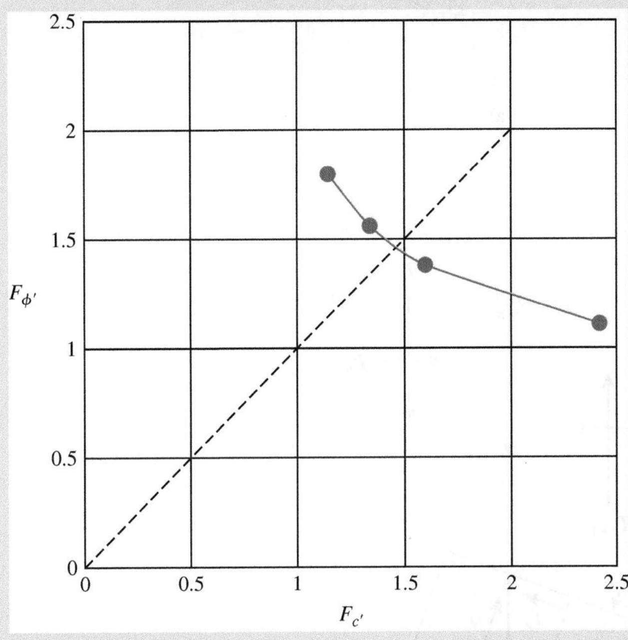

Figure 15.15

From Figure 15.15, F_s = **1.42** ■

15.8 Ordinary Method of Slices

Stability analysis by using the method of slices can be explained with the use of Figure 15.16a, in which AC is an arc of a circle representing the trial failure surface. The soil above the trial failure surface is divided into several vertical slices. The width of each slice need not be the same. Considering a unit length perpendicular to the cross section shown, the forces that act on a typical slice (nth slice)

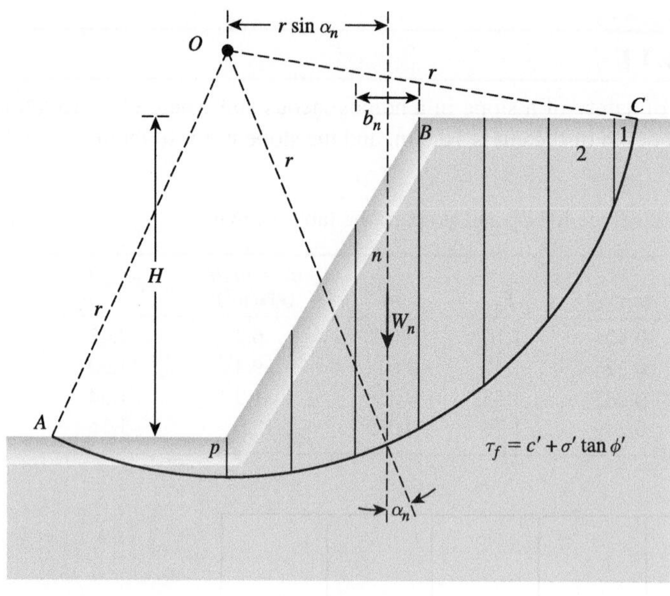

(a)

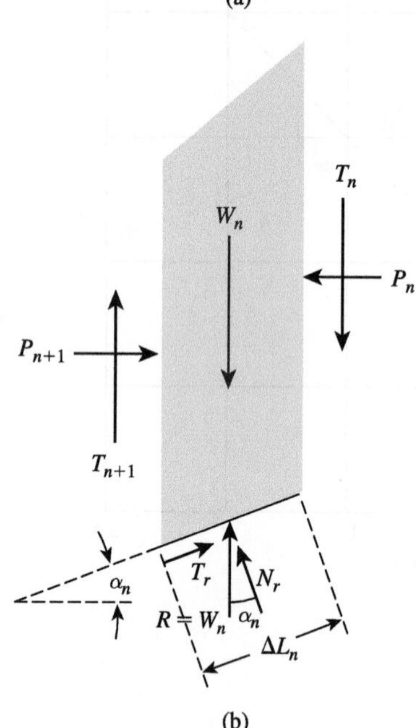

(b)

Figure 15.16 Stability analysis by ordinary method of slices: (a) trial failure surface; (b) forces acting on nth slice

are shown in Figure 15.16b. W_n is the weight of the slice. The forces N_r and T_r, respectively, are the normal and tangential components of the reaction R. P_n and P_{n+1} are the normal forces that act on the sides of the slice. Similarly, the shearing forces that act on the sides of the slice are T_n and T_{n+1}. For simplicity, the pore water pressure is assumed to be zero. The forces P_n, P_{n+1}, T_n, and T_{n+1} are difficult to determine. However, we can make an approximate assumption that the resultants of

P_n and T_n are equal in magnitude to the resultants of P_{n+1} and T_{n+1} and that their lines of action coincide.

For equilibrium consideration,

$$N_r = W_n \cos \alpha_n$$

The resisting shear force can be expressed as

$$T_r = \tau_d(\Delta L_n) = \frac{\tau_f(\Delta L_n)}{F_s} = \frac{1}{F_s}[c' + \sigma' \tan \phi']\Delta L_n \tag{15.19}$$

The normal stress, σ', in Eq. (15.19) is equal to

$$\frac{N_r}{\Delta L_n} = \frac{W_n \cos \alpha_n}{\Delta L_n}$$

For equilibrium of the trial wedge *ABC*, the moment of the driving force about *O* equals the moment of the resisting force about *O*, or

$$\sum_{n=1}^{n=p} W_n r \sin \alpha_n = \sum_{n=1}^{n=p} \frac{1}{F_s}\left(c' + \frac{W_n \cos \alpha_n}{\Delta L_n} \tan \phi'\right)(\Delta L_n)(r)$$

or

$$F_s = \frac{\sum_{n=1}^{n=p} (c' \Delta L_n + W_n \cos \alpha_n \tan \phi')}{\sum_{n=1}^{n=p} W_n \sin \alpha_n} \tag{15.20}$$

[*Note*: ΔL_n in Eq. (15.20) is approximately equal to $(b_n)/(\cos \alpha_n)$, where b_n = the width of the *n*th slice.]

Note that the value of α_n may be either positive or negative. The value of α_n is positive when the slope of the arc is in the same quadrant as the ground slope. To find the minimum factor of safety—that is, the factor of safety for the critical circle—one must make several trials by changing the center of the trial circle. This method generally is referred to as the *ordinary method of slices*.

For convenience, a slope in a homogeneous soil is shown in Figure 15.16. However, the method of slices can be extended to slopes with layered soil, as shown in Figure 15.17. The general procedure of stability analysis is the same. However, some minor points should be kept in mind. When Eq. (15.20) is used for the factor of safety calculation, the values of ϕ' and c' will not be the same for all slices. For example, for slice No. 3 (see Figure 15.17), we have to use a friction angle of $\phi' = \phi_3'$ and cohesion $c' = c_3'$; similarly, for slice No. 2, $\phi' = \phi_2$ and $c' = c_2'$.

There can be significant computational effort involved with the method of slices when the soil is nonhomogeneous and includes a water table. Nowadays, there are several software packages available for carrying out these computations.

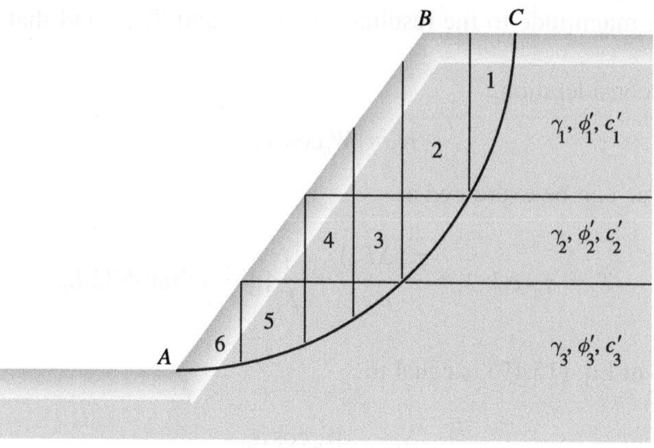

Figure 15.17 Stability analysis, by ordinary method of slices, for slope in layered soils

Example 15.12

A slope is shown in Figure 15.18. Using the ordinary method of slices, determine the factor of safety of the trial failure wedge *ABC*.

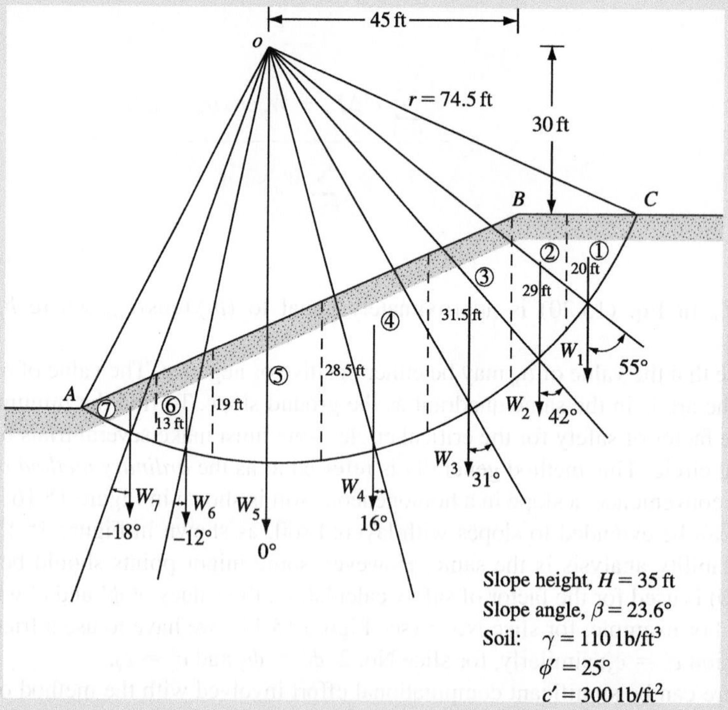

Figure 15.18

Solution

The wedge has been divided into seven slices. Now the following can be prepared (*Note:* $\gamma = 110\ \text{lb/ft}^3$.)

Slice No.	Weight of wedge, W (lb/ft)	Width of Wedge, b (ft)	α (deg)	$\Delta L = \dfrac{b}{\cos\alpha}$ (ft)	$W\cos\alpha$ (lb)	$W\sin\alpha$ (lb)
1	$\left(\dfrac{0+20}{2}\right)(12.5)(110) = 13{,}750$	12.5	55	21.79	7,886.7	11,263.3
2	$\left(\dfrac{20+29}{2}\right)(10)(110) = 26{,}950$	10	42	13.46	20,027.8	18,033.1
3	$\left(\dfrac{29+31.5}{2}\right)(15)(110) = 49{,}912.5$	15	31	17.5	42,783.4	25,706.8
4	$\left(\dfrac{31.5+28.5}{2}\right)(20)(110) = 66{,}000$	20	16	20.8	63,443.3	18,192.1
5	$\left(\dfrac{28.5+19}{2}\right)(20)(110) = 52{,}250$	20	0	20	52,250	0
6	$\left(\dfrac{19+13}{2}\right)(10)(110) = 17{,}600$	10	−12	10.22	17,215.3	−3659.2
7	$\left(\dfrac{0+13}{2}\right)(15)(110) = 10{,}725$	15	−18	15.77	10,200.1	−3314.2
				Σ119.54	Σ213,806.6	Σ66,221.9

$$F_s = \frac{\sum_{n=1}^{n=p}(c'\Delta L + W_n\cos\alpha_n\tan\phi')}{\sum_{n=1}^{n=p}W_n\sin\alpha_n} = \frac{(300)(119.54) + (213{,}806.6)(\tan 25)}{66{,}221.9}$$

$$= \frac{35{,}862 + 99{,}699.7}{66{,}221.9} = 2.047 \approx \mathbf{2.05}$$

■

15.9 Summary

A slope can be natural or man-made. Its stability is influenced by the inclination and height of the slope and the shear strength parameters of the soil. The slope failure generally is assumed to occur along a plane or a circular arc. The factor of safety is defined as the ratio of the shear strength to the shear stress developed along the failure surface.

In Section 15.2 and 15.3, determining the factor of safety assuming a failure plane was discussed. This was followed by Taylor's stability analysis for undrained clays ($\phi = 0$) and c'–ϕ' soils. This procedure is fairly straightforward and uses a stability number defined by Taylor as $\dfrac{c}{\gamma H}$. In each of these methods, it is assumed that the factor of safety (F_s) is the same as the one with respect to friction angle ($F_{\phi'}$) or cohesion (F_c).

This chapter also introduced the ordinary method of slices, where the soil enclosed within the failure surface is divided into slices and the factor of safety is determined by considering the equilibrium of the slices. This method also can be used in soils that are not homogeneous, where Culmann's or Taylor's methods will not work.

Problems

15.1 State whether the following are true or false.

a. Increasing the slope angle reduces the factor of safety.

b. With increasing factor of safety, there is increased development of shear resistance along the potential failure surface.

c. The factor of safety decrease when the depth of excavation is increased.

d. The failure mode in an undrained homogeneous clay ($\phi = 0$) with a slope of 60° to horizontal always will be a toe circle.

e. The midpoint circle passes through the toe.

15.2 A slope is shown in Figure 15.19. Assuming that the slope failure would occur along a plane (Culmann's assumption), find the height of the slope for critical equilibrium.

15.3 Referring to Figure 15.19, find the height of the slope H that will have a factor of safety of 2 against sliding. Assume that the critical surface for sliding is a plane.

15.4 Refer to Figure 15.19. If $H = 13$ ft, what will be the factor of safety with respect to sliding? Assume that the critical surface is a plane.

15.5 An embankment is to be built using a soil having $\gamma = 19.5\ \text{kN/m}^3$, $\phi' = 26°$, and $c' = 10.0\ \text{kN/m}^2$. What is the maximum height that can be achieved when maintaining a factor of safety of 2.0 and slope angle β of 50°? Assume that the failure is along a plane (Culmann's assumption).

15.6 A 10.0 m excavation at a 45° slope is proposed in a soil where $\gamma = 19.0\ \text{kN/m}^3$, $\phi' = 25°$, and $c' = 10.0\ \text{kN/m}^2$. Assuming a plane failure as suggested by Culmann, determine the factor of safety.

Also determine the inclination of the failure plane.

15.7 In problem 15.6, to what depth can the excavation be continued before the slope fails?

15.8 A 10.0 m excavation at a 45° slope is proposed in a soil where $\gamma = 19.0\ \text{kN/m}^3$, $\phi' = 25°$, and $c' = 10.0\ \text{kN/m}^2$. Assuming a plane failure as suggested by Culmann, it was shown in Problem 15.6 that the critical plane is inclined at 31.9°. Using Eq. (15.12), verify that the factor of safety along this plane is 1.37.

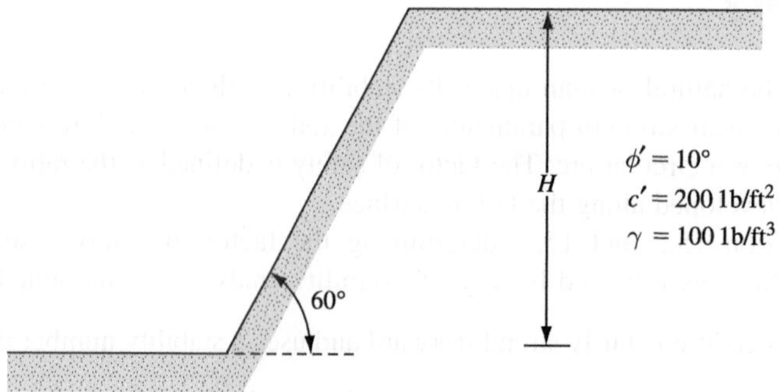

Figure 15.19

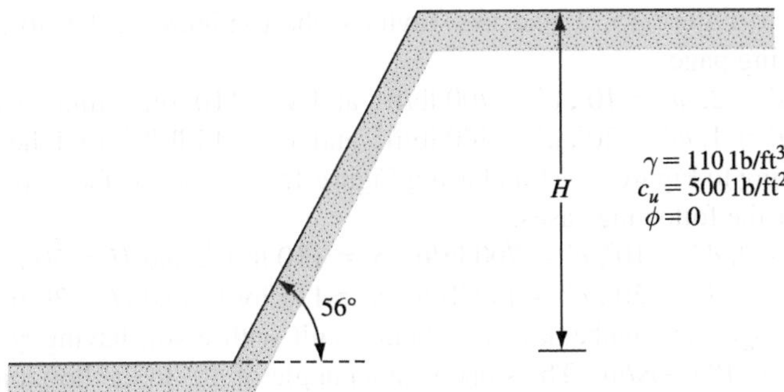

Figure 15.20

15.9 A cut slope is to be made in a clay (Figure 15.20). Given: c_u = 500 lb/ft² ($\phi = 0$ condition) and γ = 110 lb/ft³. The slope makes an angle of 56° with the horizontal. Determine the maximum depth up to which the cut could be made. Assume that the critical surface for sliding is circularly cylindrical. What will be the nature of the critical circle (that is, toe, slope, or midpoint)?

15.10 For the cut slope described in Problem 15.9, if we need a factor of safety of 2 against sliding, how deep should the cut be made?

15.11 Using the graph given in Figure 15.7, determine the height of a slope, 1 vertical to 1/2 horizontal, in saturated clay having an undrained shear strength of 680 lb/ft². The desired factor of safety against sliding is 2. Given: γ = 120 lb/ft³

15.12 Refer to Problem 15.11. What should be the critical height of the slope? What will be the nature of the critical circle? Also find the radius of the critical circle.

15.13 A cut slope was excavated in a saturated clay. The slope made an angle of 45° with the horizontal. Slope failure occurred when the cut reached a depth of 15 ft. Previous soil explorations showed that a rock layer was located at a depth of 35 ft below the ground surface. Assuming an undrained condition and γ_{sat} = 110 lb/ft³, find the following:

a. Determine the undrained cohesion of the clay (use Figure 15.7).

b. What was the nature of the critical circle?

15.14 If the cut slope described in Problem 15.13 is to be excavated in a manner such that H_{cr} = 25 ft, what angle should the slope make with the horizontal? [Use the results of Problem 15.13(a).]

15.15 A clay embankment is being built at a slope of 2 (horizontal): J (vertical) to a height of 10.0 m. The embankment is built on top of 5 m of the same clay, which is underlain bedrock. The clay has γ = 19.5 kN/m³ and c_u = 50 kN/m². What is the factor of safety against any potential slope failure?

15.16 An excavation is made in a homogeneous saturated clay (γ = 19.0 kN/m³, c_u = 45.0 kN/m², and $\phi = 0$) to a depth of 8.0 m at a slope angle β of 60°, Determine the factor of safety of the slope, and locate the critical circle.

15.17 15 m of homogeneous clay (γ = 19.0 kN/m³, c_u = 40.0 kN/m², and $\phi = 0$) deposit lies on top of bedrock. How deep can you excavate at a slope angle β of 35° before failure occurs?

15.18 An excavation is to be made in a homogeneous clay (γ = 20.0 kN/m³, c_u = 60.0 kN/m², and $\phi = 0$) where the bed rock is at a 18.0 m depth. What is the maximum possible slope angle for excavating to a depth of 12.0 m such that the factor of safety is above 1.5?

15.19 Refer to Figure 15.21 and use Taylor's chart (Figure 15.10) to solve the problems on the following page.

a. If $n' = 2$, $\phi' = 10°$, $c' = 700$ lb/ft^2, and $\gamma = 110$ lb/ft^3, find the critical height of the slope.

b. If $n' = 1$, $\phi' = 20°$, $c' = 400$ lb/ft^2, and $\gamma = 115$ lb/ft^3, find the critical height of the slope.

15.20 Referring to Figure 15.21 and using Figure 15.10, find the factor of safety with respect to sliding for the following cases:

a. $n' = 2$, $\phi' = 10°$, $c' = 700$ lb/ft^2, $\gamma = 110$ lb/ft^3, and $H = 50$ ft

b. $n' = 1$, $\phi' = 20°$, $c' = 400$ lb/ft^2, $\gamma = 115$ lb/ft^3, and $H = 30$ ft

15.21 A homogeneous embankment is being built with a soil having $\gamma = 19.5$ kN/m^3, $\phi' = 20°$, and $c' = 18.0$ kN/m^2. The slope is at an angle of 40° to the horizontal. What is the maximum possible height before failure occurs?

15.22 Determine the factor of safety of a 1.0 (vertical) to 1.5 (horizontal) slope where $H = 8.0$ m, $\gamma = 20.0$ kN/m^3, $\phi' = 20°$, and $c' = 15.0$ kN/m^2, using Figure 15.12.

15.23 Redo Problem 15.22 using Figure 15.10.

15.24 Referring to Figure 15.22 and using the ordinary method of slices, find the factor of safety against sliding for the following trial case:
$n' = 1$, $\phi' = 20°$, $c' = 400$ lb/ft^2, $\gamma = 115$ lb/ft^3, $H = 40$ ft, $\alpha = 30°$, and $\theta = 70°$

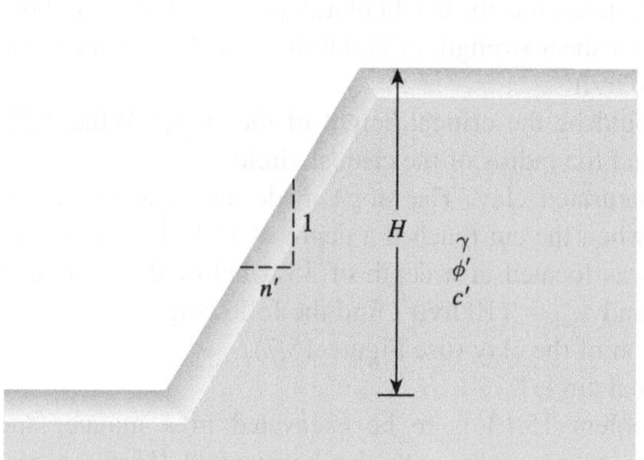

Figure 15.21

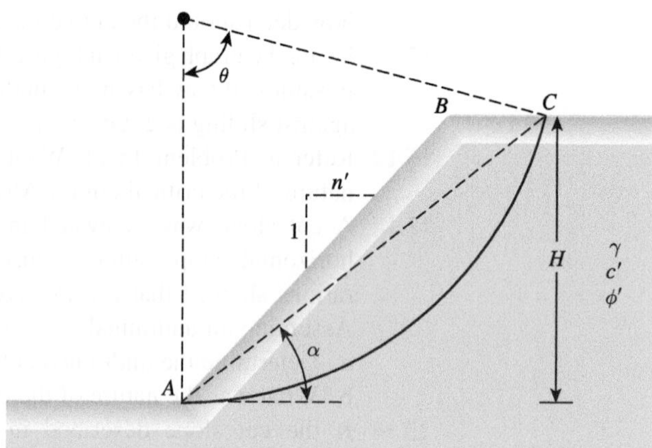

Figure 15.22

References

CULMANN, C. (1875). *Die Graphische Statik*, Meyer and Zeller, Zurich.

FELLENIUS, W. (1927). *Erdstatische Berechnungen*, Revised Edition, W. Ernst u. Sons, Berlin.

SINGH, A. (1970). "Shear Strength and Stability of Man-Made Slopes", *Journal of the Soil Mechanics and Foundations Division*, ASCE, Vol. 96, No. SM6, 1879–1892.

TAYLOR, D. W. (1937). "Stability of Earth Slopes", *Journal of the Boston Society of Civil Engineers*, Vol. 24, 197–246.

TERZAGHI, K. and PECK, R. B. (1967). *Soil Mechanics in Engineering Practice*, 2nd ed., Wiley, New York.

Answers to Selected Problems

Chapter 2

2.2 **a.**

U.S. sieve No.	Percent finer
4	100
10	95.2
20	84.2
40	61.4
60	41.6
100	20.4
200	7

b. $D_{60} = 0.41$ mm
$D_{30} = 0.185$ mm
$D_{10} = 0.09$ mm

c. 4.56

d. 0.928

2.4 **a.**

U.S. sieve No.	Percent finer
4	100
6	100
10	100
20	98.18
40	48.30
60	12.34
100	7.80
200	4.70

b. $D_{60} = 0.48$ mm
$D_{30} = 0.33$ mm
$D_{10} = 0.23$ mm

c. 2.09
d. 0.99

2.6 Gravel: 0% Silt: 31%
Sand: 46% Clay: 23%

2.8 Gravel: 0% Silt: 16%
Sand: 70% Clay: 14%

2.10 Gravel: 0% Silt: 20%
Sand: 66% Clay: 14%
AN-355-a

2.12 Gravel: 37%
Sand: 47%
Fines: 16%

Chapter 3

3.2 a. 122 lb/ft^3
b. 108.93 lb/ft^3
c. 0.56
d. 0.359
e. 58.3%
f. 1.31 lb

3.4 a. 111.2 lb/ft^3
b. 0.509
c. 0.337
d. 51.79%

3.6 $\gamma_{sat} = 113.6$ lb/ft^3
$\gamma_d = 81.1$ lb/ft^3

3.8 a. 104.7 lb/ft^3
b. 91.41 lb/ft^3
c. 15 lb

3.10 a. 112.61 lb/ft^3
b. 0.48
c. 0.324
d. 6.01 lb/ft^3

3.12 a. 103.31 lb/ft^3
b. 0.655
c. 0.396
d. 23.9%

3.14 a. 116.8 lb/ft^3
b. 88.6%

3.16 a. 117.7 lb/ft^3
b. 0.487
c. 2.32
d. 107.57 lb/ft^3

3.18 $\gamma = 115.56$ lb/ft^3
$\gamma_{d(\min)} = 85.24$ lb/ft^3
$\gamma_{d(\max)} = 124.33$ lb/ft^3

Chapter 4

4.2 **a.** 39.5
b. 20.8

4.4 **a.** 37
b. 15

4.6

Soil No.	Classification
1	ML
2	CH
3	CL
4	SC
5	ML
6	SC

4.8

Soil	Classification
A	SP
B	SW-SM
C	MH
D	CH
E	SC

4.10

Soil	Classification
A	SC
B	GM-GC
C	CH
D	ML
E	SM

4.12 **a.** GC—Clayey sandy gravel
b. MH—Gravelly sandy silt of high plasticity
c. SW-SM—Well graded clayey gravelly sand

Chapter 5

5.2 0.136 kg

5.4

w (%)	γ_{zav} (lb/ft^3)
5	143.6
10	128.8
15	116.7
20	106.7
25	98.3

5.6 Maximum dry unit weight
= 100.8 lb/ft^3
Optimum moisture content
= 15%

5.8 93.2%

5.10

Control Test No.	Specifications
1	No
2	Yes
3	No
4	Yes

Chapter 6

6.2 0.0138 in./s

6.4 h = 34 cm
v = 0.019 cm/s

6.6 **a.** 6.88×10^{-2} in./min
b. 17.3 in.

6.8 0.023 cm/s

6.10 17.6×10^{-3} $m^3/hr/m$

6.12 8.0×10^{-5} $m^3/s/m$

6.14 0.37 ft/min

6.16 0.075×10^{-6} cm/s

Chapter 7

7.2

	lb/ft^2		
Point	σ	u	σ'
A	0	0	0
B	230	0	230
C	702	249.6	452.4
D	1482	624	858

7.4 **a.**

	lb/ft^2		
Depth (ft)	σ	u	σ'
0	0	0	0
15	1683	0	1683
27	3121.8	748.8	2373

b. 627 lb/ft^2
c. 7.18 ft

7.6 $e = 0.675$
$\gamma_{sat} = 19.83 \text{ kN/m}^3$
$\sigma = 138.3 \text{ kN/m}^2$
$u = 88.3 \text{ kN/m}^2$
$\sigma' = 50.0 \text{ kN/m}^2$

7.8 15.1 ft

7.12

z (ft)	$\Delta\sigma_z$ (lb/ft²)
1.5	3450
3	3187.1
6	2262.8
9	1484
12	995.8

7.14 171 lb/ft²

Chapter 8

8.2 **b.** 1.18 ton/ft²
c. 0.451

8.4 **a.** 1.56
b. 75.7 mm

8.6 0.84 in.

8.8 36.2 mm

8.10 **a.** 782.2 days
b. 1600.35 days

8.12 **a.** 20 min
b. 53.8 min
c. From t_{50}, 3.178×10^{-5} in.²/s
From t_{90}, 5.086×10^{-5} in.²/s

8.14 33.6 days

Chapter 9

9.2 $\phi' = 32.2°$
Shear force = 50.4 lb

9.4 $c' = 65.5 \text{ kN/m}^2$
$\phi' = 23.2°$

9.6 20 lb/in.²

9.8 29°

9.10 76.73 lb/in.²

9.12 1 ton/ft²

9.14 $c = 7.25$ lb/in.²
$\phi = 9°$

9.16 1900 lb/ft²

9.18 60.4 kN/m²

Chapter 10

10.2 5.57%

10.4 1.83 mm

10.6

Depth (m)	$(N_1)_{60}$ Liao & Whitman	$(N_1)_{60}$ Skempton
3.5	15	15
8.8	6	6
12.4	10	9
18.5	10	9
23.6	10	8
26.9	11	9

10.8 36°

10.10

Depth (ft)	$(N_1)_{60}$
5	10
10	12
15	11
20	11
25	13
30	13

10.12 1473 lb/ft^2

10.14 **a.** 43°
b. 71%

10.16 D_r = **75%**
$\phi' = 40.4°$

10.18 54%

Chapter 11

11.2 **a.** $P_a = 1688.5$ lb/ft
$\bar{z} = 3.33$ ft
b. $P_a = 2547$ lb/ft
$\bar{z} = 4$ ft
c. $P_a = 4042.7$ lb/ft
$\bar{z} = 6$ ft

11.4 **a.** $P_p = 16{,}500$ lb/ft
$\sigma'_p = 3300$ lb/ft^2
b. $P_p = 45{,}276$ lb/ft
$\sigma'_p = 6468$ lb/ft^2

11.6 **a.** 42.5 kN/m
b. 44.2 kN/m
c. 51.3 kN/m

11.8 **a.** $P_p = 20{,}318.4$ lb/ft
$\bar{z} = 4.15$ ft
b. $P_p = 89{,}067.1$ lb/ft
$\bar{z} = 7.68$ ft
11.10 **a.** σ_a at top $= -800$ lb/ft^2
σ_a at bottom $= 943$ lb/ft^2
b. 6.426 ft
c. 1001 lb/ft
d. $P_a = 2571.3$ lb/ft
Resultant 2.53 ft from the bottom
11.12 **a.** σ_a at top $= -650$ lb/ft^2
σ_a at bottom $= 1093$ lb/ft^2
b. 5.22 ft
c. 3103.8 lb/ft
d. $P_a = 4798.3$ lb/ft
11.14 **a.** $P_a = 61.8$ kN/m
$\bar{z} = 1.67$ m
b. $P_a = 55.8$ kN/m
$\bar{z} = 1.67$ m (inclined at 20° to the horizontal)
11.16 58,504.8 lb/ft
11.18

δ' (deg)	$P_p = \frac{1}{2}K_p\gamma H^2$ (lb/ft)
0	54,106
10	76,638
20	117,320

Chapter 12

12.2 5723 lb/ft^2
12.4 15,715 lb/ft^2
12.6 966.2 kN/m
12.8 1092 kip
12.10 302.7 kN
12.12 $B = 6.75$ ft
12.14 **a.** $e = 113$ mm (vertical load)
b. $e = 30$ mm (load inclined at 1.9° to vertical)
12.16 0.424 in.
12.18 12.3 kip/ft^2
12.20 4.27 kip/ft^2
12.22 34.65 ft
12.24 D_f (fully compensated) $= 10$ m
$q_{u(\text{net})} = 333.9$ kN/m^2
FS $= 3.3$
12.26 4.9 in.

Chapter 13

13.2 **a.** 91 kip
b. 313.5 kip
13.4 257.2 kN
13.6 337 kN
13.8 96.2 kip
13.10 446 kN
13.12 229.3 kip
13.14 70.3%
13.16 90.2%
13.18 1092 kip
13.20 9.93 in.
13.22 567 kip
13.24 49.5 kip
13.26 12,590 kN
13.28 **a.** 5488 kip
b. 70.02 kip
c. 5560 kip
13.30 16,267 kN

Chapter 14

14.2 $FS_{(overturning)} = 3.21$
$FS_{(sliding)} = 1.5$
$FS_{(bearing\ capacity)} = 5.34$
14.4 $FS_{(overturning)} = 3.81$
$FS_{(sliding)} = 1.66$
$FS_{(bearing\ capacity)} = 3.73$
14.6 $FS_{(overturning)} = 2.66$
$FS_{(sliding)} = 1.7$
$FS_{(bearing\ capacity)} = 3.26$
14.8 $FS_{(overturning)} = 3.81$
$FS_{(sliding)} = 1.84$
14.10 $FS_{(overturning)} = 3.74$
$FS_{(sliding)} = 1.44$
$e < B/6$
$q_{max} = 101.1\ kN/m^2$
$q_{min} = 88.3\ kN/m^2$

Chapter 15

15.2 19.1 ft
15.4 1.35
15.6 $F_s = 1.37$
Inclination at 31.9° to the horizontal
15.8 $F_s = 1.37$
15.10 12.28 ft
15.12 $H_{cr} = 28.9$ ft
$r = 42.55$ ft
15.14 12°
15.16 $F_s = 1.55$
Radius of critical circle = 12.0 m
15.18 $\beta = 31.5°$
15.20 **a.** 1.5
b. 1.36
15.22 1.35
15.24 1.3

Index